Methods in Enzymology

Volume 382
QUINONES AND QUINONE ENZYMES
Part B

METHODS IN ENZYMOLOGY

EDITORS-IN-CHIEF

John N. Abelson   Melvin I. Simon

DIVISION OF BIOLOGY
CALIFORNIA INSTITUTE OF TECHNOLOGY
PASADENA, CALIFORNIA

FOUNDING EDITORS

Sidney P. Colowick and Nathan O. Kaplan

Methods in Enzymology

Volume 382

Quinones and Quinone Enzymes

Part B

EDITED BY

Helmut Sies

HEINRICH-HEINE-UNIVERSTAT
DUSSELDORF, GERMANY

Lester Packer

UNIVERSITY OF SOUTHERN CALIFORNIA
LOS ANGELES, CALIFORNIA

ELSEVIER
ACADEMIC
PRESS

AMSTERDAM • BOSTON • HEIDELBERG • LONDON
NEW YORK • OXFORD • PARIS • SAN DIEGO
SAN FRANCISCO • SINGAPORE • SYDNEY • TOKYO
Academic Press is an imprint of Elsevier

Elsevier Academic Press
525 B Street, Suite 1900, San Diego, California 92101-4495, USA
84 Theobald's Road, London WC1X 8RR, UK

This book is printed on acid-free paper. ∞

For all information on all Academic Press Publications
visit our Web site at www.academicpress.com

ISBN: 0-12-182786-0

PRINTED IN THE UNITED STATES OF AMERICA
04 05 06 07 08 9 8 7 6 5 4 3 2 1

Dedicated to the Pioneers of this field: Frederick L. Crane, Lars Ernster, Karl Folkers, and Andres O. M. Stoppani

Table of Contents

Section I. Mitochondrial Ubiquinone and Reductases

Section II. Anticancer Quinones and Quinone Oxidoreductases

Section III. Quinone Reductases: Chemoprevention and Nutrition

Section IV. Quinones and Age-Related Diseases

Contributors to Volume 382

Article numbers are in parentheses and following the names of contributors.
Affiliations listed are current.

L. MARIO AMZEL (9), *Interim Director, Department of Biophysics and Biophysical Chemistry, The Johns Hopkins University School of Medicine, Baltimore, Maryland 21205*

ŽILVINAS ANUSEVIČIUS (15), *Institute of Biochemistry, LT-2600 Vilnius, Lithuania*

GAD ASHER (16), *Department of Molecular Genetics, Weizmann Institute of Science, Rehovot 76100, Israel*

MICHEL AUBIER (4), *Laboratoire de Biophysique and IFR 02, Faculté X. Bichat, 75018 Paris, France*

ALESSANDRA BARACCA (1), *Dipartimento Di Biochimica, Universita Di Bologna, 40126 Bologna, Italy*

RAYMOND P. BAUMANN (12), *Department of Pharmacology and Developmental Therapeutics Program, Cancer Center, Yale University School of Medicine, New Haven, Connecticut 06520*

M. FLINT BEAL (26), *Department of Neurology and Neuroscience, Weill Medical College of Cornell University, New York, New York 10021*

ASHER BEGLEITER (18), *Manitoba Institute of Cell Biology, CancerCare Manitoba, Department of Internal Medicine; Department of Pharmacology and Therapeutics, University of Manitoba, Manitoba R3E 0V9, Canada*

ROSARIO I. BELLO (13), *Departamento de Biología Celular, Fisiología, e Immunología, Campus Rabanales, Universidad de Córdoba, 14014 Córdoba, Spain*

MARIO A. BIANCHET (9), *Department of Biophysics and Biophysical Chemistry, The Johns Hopkins University School of Medicine, Baltimore, Maryland 21205*

JORGE BOCZKOWSKI (4), *Laboratoire de Biophysique and IFR 02, Faculté X. Bichat, 75018 Paris, France*

MICAHEL K. BOWMAN (2), *William R. Wiley Environmental Molecular Sciences Laboratory, Pacific Northwest National Laboratory, Richland Washington 99352*

JOHN BUTLER (10), *Department of Biological Sciences, The University of Salford, Salford M5 4WT, United Kingdom*

JONATHAN CAPE (2), *Institute of Biological Chemistry, Washington State University, Pullman, Washington 99164-6340*

MARIA CECILIA CARRERAS (4), *Laboratory of Oxygen Metabolism, University Hospital, University of Buenos Aires, 1120 Buenos Aires, Argentina*

ERCOLE CAVALIERI (17), *Eppley Institute for Research in Cancer and Allied Diseases, University of Nebraska Medical Center, Omaha, Nebraska 68198-6805*

NARIMANTAS ČĖNAS (15), *Institute of Biochemistry, LT-2600 Vilnius, Lithuania*

DHRUBAJYOTI CHAKRAVARTI (17), *Eppley Institute for Research in Cancer and Allied Diseases, University of Nebraska Medical Center, Omaha, Nebraska 68198-6805*

TOM S. CHAN (6), *Department of Pharmaceutical Sciences, University of Toronto, Toronto, Ontario M5S 2S2, Canada*

SHIUAN CHEN (11), *Department of Surgery, Beckman Research Institute of the City of Hope, Duarte, California 91010-0269*

CHING KUANG CHOW (7), *Graduate Center for Nutritional Sciences and Kentucky Agricultural Experiment Station, University of Kentucky, Lexington, Kentucky 40506*

ANGELA DIFRANCESCO (10), *Division of Pediatric Oncology, Policlinico Gemelli, 00168 Rome, Italy*

ALBENA T. DINKOVA-KOSTOVA (14, 19, 23), *Department of Pharmacology and Molecular Science, Lewis B. and Dorothy Cullman Cancer Center Chemoprotection Center, Johns Hopkins University School of Medicine; Center for Human Nutrition, Johns Hopkins University Bloomberg School of Public Health, Baltimore, Maryland 21205-2185*

MANUCHAIR EBADI (27), *School of Medicine and Health Sciences, University of North Dakota, Grand Forks, North Dakota 58203*

JOSHUA EKEN (27), *School of Medicine and Health Sciences, University of North Dakota, Grand Forks, North Dakota 58203*

HESHAM EL RAFAEY (27), *School of Medicine and Health Sciences, University of North Dakota, Grand Forks, North Dakota 58203*

MARGARITA FAIG (9), *Department of Biophysics and Biophysical Chemistry, The Johns Hopkins University School of Medicine, Baltimore, Maryland 21205*

JED W. FAHEY (14, 23), *Department of Pharmacology and Molecular Science, Johns Hopkins University School of Medicine, Baltimore, Maryland 21205-2185*

ROMANA FATO (1), *Dipartimento Di Biochimica, Universita Di Bologna, 40126 Bologna, Italy*

JEANNE FOURIE (18), *Manitoba Institute of Cell Biology, CancerCare Manitoba, Department of Pharmacology and Therapeutics, University of Manitoba, Manitoba R3E 0V9, Canada*

MARIA LUISA GENOVA (1), *Dipartimento Di Biochimica, Universita Di Bologna, 40126 Bologna, Italy*

CONSUELO GÓMEZ-DÍAZ (13), *Departamento de Biología Celular, Fisiología, e Immunolgía, Campus Rabanales, Universidad de Córdoba, 14014 Córdoba, Spain*

CINDY HAGEN (27), *School of Medicine and Health Sciences, University of North Dakota, Grand Forks, North Dakota 58203*

ANDREW M. JAMES (3), *Medical Research Council, Dun Human Nutrition Unit, Cambridge CB2 2XY, United Kingdom*

ELIZABETH H. JEFFERY (25), *Department of Food Science and Human Nutrition, University of Illinois, Urbana, Illinois 61801*

YOUNG-HWA KANG (21), *College of Pharmacy, University of Illinois at Chicago, Chicago, Illinois 60612*

THOMAS W. KENSLER (22), *Department of Environmental Health Sciences, School of Public Health, Johns Hopkins University Bloomberg, Baltimore, Maryland 21205*

GEOFFREY F. KELSO (3), *Department of Chemistry, University of Otago, Dunedin, New Zealand*

MOHSEN KHERADPHEZHOU (27), *School of Medicine and Health Sciences, University of North Dakota, Grand Forks, North Dakota 58203*

RICHARD J. KNOX (11), *Enact Pharma PLC, Salisbury SP4 0JQ, United Kingdom*

DAVID M. KRAMER (2), *Institute of Biological Chemistry, Washington State University, Pullman, Washington 99164-6340*

MI-YOUNG KWAK (22), *Department of Environmental Health Sciences, School of Public Health, Johns Hopkins University Bloomberg, Baltimore, Maryland 21205*

GIORGIO LENAZ (1), *Dipartimento Di Biochimica, Universita Di Bologna, 40126 Bologna, Italy*

CONSTANZA L. LISDERO (4), *Laboratory of Oxygen Metabolism, University Hospital, University of Buenos Aires, 1120 Buenos Aires, Argentina*

JOSEPH LOTEM (16), *Department of Molecular Genetics, Weizmann Institute of Science, Rehovot 76100, Israel*

MARIANA MELANI (4), *Laboratory of Oxygen Metabolism, University Hospital, University of Buenos Aires, 1120 Buenos Aires, Argentina*

ALAIN MEULEMANS (4), *Laboratoire de Biophysique and IFR 02, Faculté X. Bichat, 75018 Paris, France*

LINA MISEVIČIENÈ (15), *Institute of Biochemistry, LT-2600 Vilnius, Lithuania*

ARNOLD MUNNICH (5), *Department of Genetics, Hôpital Necker-Enfants Malades, 75014 Paris, France*

FLORIAN MULLER (2), *Institute of Biological Chemistry, Washington State University, Pullman, Washington 99164-6340*

MICAHEL P. MURPHY (3), *Medical Research Council, Dun Human Nutrition Unit, Cambridge CB2 2XY, United Kingdom*

CHRISTINE M. MUNDAY (24), *AgResearch, Ruakura Agricultural Research Centre, Hamilton, New Zealand*

REX MUNDAY (20, 24), *AgResearch, Ruakura Agricultural Research Centre, Hamilton, New Zealand*

PLÁCIDO NAVAS (13), *Centro Andaluz de Biología del Desarrollo, Universidad Pablo de Olavide, 41013 Sevilla, Spain*

HENRIKAS NIVINSKAS (15), *Institute of Biochemistry, LT-2600 Vilnius, Lithuania*

PETER J. O'BRIEN (6), *Department of Pharmaceutical Sciences, University of Toronto, Toronto, Ontario M5S 2S2, Canada*

PHILIP J. PENKETH (12), *Department of Pharmacology and Developmental Therapeutics Program, Cancer Center, Yale University School of Medicine, New Haven, Connecticut 06520*

JOHN M. PEZZUTO (21), *Schools of Pharmacy, Nursing, and Health Sciences, Purdue University, West Lafayette, Indiana 47907-7880*

JUAN JOSÉ PODEROSO (4), *Laboratory of Oxygen Metabolism, University Hospital, University of Buenos Aires, 1120 Buenos Aires, Argentina*

MINERVA RAMOS-GOMEZ (22), *Department of Environmental Health Sciences, School of Public Health, Johns Hopkins University Bloomberg, Baltimore, Maryland 21205*

ARTHUR G. ROBERTS (2), *Institute of Biological Chemistry, Washington State University, Pullman, Washington 99164-6340*

ELEANOR ROGAN (17), *Eppley Institute for Research in Cancer and Allied Diseases, University of Nebraska Medical Center, Omaha, Nebraska 68198-6805*

DAVID ROSS (8), *Department of Pharmaceutical Sciences, School of Pharmacy, University of Colorado, Denver, Colorado 80262*

AGNÈS RÖTIG (5), *Department of Genetics, Hôpital Necker-Enfants Malades, 75014 Paris, France*

PIERRE RUSTIN (5), *Department of Genetics, Hôpital Necker-Enfants Malades, 75014 Paris, France*

LEO SACHS (16), *Department of Molecular Genetics, Weizmann Institute of Science, Rehovot 76100, Israel*

JONAS ŠARLAUSKAS (15), *Institute of Biochemistry, LT-2600 Vilnius, Lithuania*

ALAN C. SARTORELLI (12), *Department of Pharmacology and Developmental Therapeutics Program, Cancer Center, Yale University School of Medicine, New Haven, Connecticut 06520*

HELEN A. SEOW (12), *Department of Pharmacology and Developmental Therapeutics Program, Cancer Center, Yale University School of Medicine, New Haven, Connecticut 06520*

YOSEF SHAUL (16), *Department of Molecular Genetics, Weizmann Institute of Science, Rehovot 76100, Israel*

SUSHIL SHARMA (27), *School of Medicine and Health Sciences, University of North Dakota, Grand Forks, North Dakota 58203*

SHAIK SHAVALI (27), *School of Medicine and Health Sciences, University of North Dakota, Grand Forks, North Dakota 58203*

DAVID SIEGEL (8), *Department of Pharmaceutical Sciences, School of Pharmacy, University of Colorado, Denver, Colorado 80262*

ROBIN A.J. SMITH (3), *Department of Chemistry, University of Otago, Dunedin, New Zealand*

KATHERINE K. STEPHENSON (14), *Department of Pharmacology and Molecular Science, Lewis B. and Dorothy Cullman Cancer Center Chemoprotection Center, Johns Hopkins University School of Medicine, Baltimore, Maryland 21205-2185*

KRISTIN E. STEWART (25), *Division of Nutritional Sciences, University of Illinois, Urbana, Illinois 61801*

PAUL TALALAY (14, 19, 23), *Department of Pharmacology and Molecular Science, Lewis B. and Dorothy Cullman Cancer Center Chemoprotection Center, Johns Hopkins University School of Medicine; Center for Human Nutrition, Johns Hopkins University Bloomberg School of Public Health, Baltimore, Maryland 21205-2185*

JOSÉ M. VILLALBA (13), *Departamento de Biología Celular, Fisiología, e Immunología, Campus Rabanales, Universidad de Córdoba, 14014 Córdoba, Spain*

NOBUNAO WAKABAYASHI (22), *Department of Environmental Health Sciences, School of Public Health, Johns Hopkins University Bloomberg, Baltimore, Maryland 21205*

TIMOTHY H. WARD (10), *Drug Development Unit, Paterson Institute, Christie Hospital, Manchester M20 9BX, United Kingdom*

JOHN X. WILSON (6), *Department of Physiology, University of Western Ontario, London, Ontario N6A 5C1, Canada*

Preface

Developments in genomics and proteomics rapidly generated focus on new -*omics*, particularly metabolomics and phenomics. Quinones, hydroquinones, semiquinones and their metabolites are naturally occurring compounds that serve as wonderful examples for this new paradigm of interdigitating ,-*omics*. In addition to a role as substrates and products in metabolism, quinone compounds are intermediates in many pathways of gene regulation, enzyme protein induction, feedback control, and waste product elimination. Quinones play a pivotal role in energy metabolism (Peter Mitchell's proton-motive, *Q cycle'*), many other key processes, and even in chemotherapy where redox cycling drugs are utilized.

The present volume of *Methods in Enzymology* on quinones and quinone enzymes serves to bring together current methods and concepts on this topic. It focuses on the role in the so-called Phase II of drug metabolism (xenobiotics), but include aspects on Phase I (CYP, cytochromes P-450) and Phase III (transport systems) as well. This volume of *Methods in Enzymology, Part B* addresses mitochondrial ubiquinone and reductases, anticancer quinones, and the role of quinone reductases in chemoprevention and nutrition, as well as the role of quinones in age-related diseases, whereas (*Part A*) focused on quinones and quinone enzymes in terms of coenzyme Q (detection and quinone reductases), plasma membrane quinone reductases, and the role of quinones in cellular signaling and modulation of gene expression. *Phase II Enzymes, Part C*, will be focusing on glutathione, glutathione S-transferases, and other conjugation enzymes.

The enzyme, NAD(P)H:quinone oxidoreductase, is the subject of a major section in this volume. This enzyme, discovered in 1958 in Stockholm by Lars Ernster, and named DT-Diaphorase by him, has multiple roles, some of which were only recently discovered.

Human polymorphisms exist in these enzymes that relate to variations in cancer risk, and enzymes targeted by quinones are being investigated. Modern methods in assaying quinone reactions and, indeed, various quinones themselves, are also included in this volume.

Following its discovery in 1957, ubiquinone (coenzyme Q_{10}) as a major naturally occurring quinone became a highlight of scientific interest and an established role in mitochondrial electron transport by Frederick Crane. Fundamental contributions were made by Karl Folkers on its supplemental use for health benefits in disease prevention and by Andrés O.M. Stoppani, a pioneer of Argentinian biochemistry, in utilizing quinones for the treatment of Chagas disease.

We thank the Advisory Committee (Enrique Cadenas, Los Angeles; Gustav Dallner, Stockholm; Tom Kensler, Baltimore; Lars-Oliver Klotz, Düsseldorf; David Ross, Denver) for their valuable suggestions and wisdom in selecting the contributions for this volume.

HELMUT SIES AND LESTER PACKER

METHODS IN ENZYMOLOGY

VOLUME LIV. Biomembranes (Part E: Biological Oxidations)
Edited by SIDNEY FLEISCHER AND LESTER PACKER

VOLUME LV. Biomembranes (Part F: Bioenergetics)
Edited by SIDNEY FLEISCHER AND LESTER PACKER

VOLUME LVI. Biomembranes (Part G: Bioenergetics)
Edited by SIDNEY FLEISCHER AND LESTER PACKER

VOLUME LVII. Bioluminescence and Chemiluminescence
Edited by MARLENE A. DELUCA

VOLUME LVIII. Cell Culture
Edited by WILLIAM B. JAKOBY AND IRA PASTAN

VOLUME LIX. Nucleic Acids and Protein Synthesis (Part G)
Edited by KIVIE MOLDAVE AND LAWRENCE GROSSMAN

VOLUME LX. Nucleic Acids and Protein Synthesis (Part H)
Edited by KIVIE MOLDAVE AND LAWRENCE GROSSMAN

VOLUME 61. Enzyme Structure (Part H)
Edited by C. H. W. HIRS AND SERGE N. TIMASHEFF

VOLUME 62. Vitamins and Coenzymes (Part D)
Edited by DONALD B. MCCORMICK AND LEMUEL D. WRIGHT

VOLUME 63. Enzyme Kinetics and Mechanism (Part A: Initial Rate and
Inhibitor Methods)
Edited by DANIEL L. PURICH

VOLUME 64. Enzyme Kinetics and Mechanism (Part B: Isotopic Probes and
Complex Enzyme Systems)
Edited by DANIEL L. PURICH

VOLUME 65. Nucleic Acids (Part I)
Edited by LAWRENCE GROSSMAN AND KIVIE MOLDAVE

VOLUME 66. Vitamins and Coenzymes (Part E)
Edited by DONALD B. MCCORMICK AND LEMUEL D. WRIGHT

VOLUME 67. Vitamins and Coenzymes (Part F)
Edited by DONALD B. MCCORMICK AND LEMUEL D. WRIGHT

VOLUME 68. Recombinant DNA
Edited by RAY WU

VOLUME 69. Photosynthesis and Nitrogen Fixation (Part C)
Edited by ANTHONY SAN PIETRO

VOLUME 70. Immunochemical Techniques (Part A)
Edited by HELEN VAN VUNAKIS AND JOHN J. LANGONE

VOLUME 71. Lipids (Part C)
Edited by JOHN M. LOWENSTEIN

VOLUME 72. Lipids (Part D)
Edited by JOHN M. LOWENSTEIN

VOLUME 73. Immunochemical Techniques (Part B)
Edited by JOHN J. LANGONE AND HELEN VAN VUNAKIS

VOLUME 74. Immunochemical Techniques (Part C)
Edited by JOHN J. LANGONE AND HELEN VAN VUNAKIS

VOLUME 75. Cumulative Subject Index Volumes XXXI, XXXII, XXXIV–LX
Edited by EDWARD A. DENNIS AND MARTHA G. DENNIS

VOLUME 76. Hemoglobins
Edited by ERALDO ANTONINI, LUIGI ROSSI-BERNARDI, AND EMILIA CHIANCONE

VOLUME 77. Detoxication and Drug Metabolism
Edited by WILLIAM B. JAKOBY

VOLUME 78. Interferons (Part A)
Edited by SIDNEY PESTKA

VOLUME 79. Interferons (Part B)
Edited by SIDNEY PESTKA

VOLUME 80. Proteolytic Enzymes (Part C)
Edited by LASZLO LORAND

VOLUME 81. Biomembranes (Part H: Visual Pigments and Purple Membranes, I)
Edited by LESTER PACKER

VOLUME 82. Structural and Contractile Proteins (Part A: Extracellular Matrix)
Edited by LEON W. CUNNINGHAM AND DIXIE W. FREDERIKSEN

VOLUME 83. Complex Carbohydrates (Part D)
Edited by VICTOR GINSBURG

VOLUME 84. Immunochemical Techniques (Part D: Selected Immunoassays)
Edited by JOHN J. LANGONE AND HELEN VAN VUNAKIS

VOLUME 85. Structural and Contractile Proteins (Part B: The Contractile Apparatus and the Cytoskeleton)
Edited by DIXIE W. FREDERIKSEN AND LEON W. CUNNINGHAM

VOLUME 86. Prostaglandins and Arachidonate Metabolites
Edited by WILLIAM E. M. LANDS AND WILLIAM L. SMITH

VOLUME 87. Enzyme Kinetics and Mechanism (Part C: Intermediates, Stereo-chemistry, and Rate Studies)
Edited by DANIEL L. PURICH

VOLUME 88. Biomembranes (Part I: Visual Pigments and Purple Membranes, II)
Edited by LESTER PACKER

VOLUME 89. Carbohydrate Metabolism (Part D)
Edited by WILLIS A. WOOD

VOLUME 90. Carbohydrate Metabolism (Part E)
Edited by WILLIS A. WOOD

VOLUME 91. Enzyme Structure (Part I)
Edited by C. H. W. HIRS AND SERGE N. TIMASHEFF

VOLUME 92. Immunochemical Techniques (Part E: Monoclonal Antibodies and General Immunoassay Methods)
Edited by JOHN J. LANGONE AND HELEN VAN VUNAKIS

VOLUME 93. Immunochemical Techniques (Part F: Conventional Antibodies, Fc Receptors, and Cytotoxicity)
Edited by JOHN J. LANGONE AND HELEN VAN VUNAKIS

VOLUME 94. Polyamines
Edited by HERBERT TABOR AND CELIA WHITE TABOR

VOLUME 95. Cumulative Subject Index Volumes 61–74, 76–80
Edited by EDWARD A. DENNIS AND MARTHA G. DENNIS

VOLUME 96. Biomembranes [Part J: Membrane Biogenesis: Assembly and Targeting (General Methods; Eukaryotes)]
Edited by SIDNEY FLEISCHER AND BECCA FLEISCHER

VOLUME 97. Biomembranes [Part K: Membrane Biogenesis: Assembly and Targeting (Prokaryotes, Mitochondria, and Chloroplasts)]
Edited by SIDNEY FLEISCHER AND BECCA FLEISCHER

VOLUME 98. Biomembranes (Part L: Membrane Biogenesis: Processing and Recycling)
Edited by SIDNEY FLEISCHER AND BECCA FLEISCHER

VOLUME 99. Hormone Action (Part F: Protein Kinases)
Edited by JACKIE D. CORBIN AND JOEL G. HARDMAN

VOLUME 100. Recombinant DNA (Part B)
Edited by RAY WU, LAWRENCE GROSSMAN, AND KIVIE MOLDAVE

VOLUME 101. Recombinant DNA (Part C)
Edited by RAY WU, LAWRENCE GROSSMAN, AND KIVIE MOLDAVE

VOLUME 102. Hormone Action (Part G: Calmodulin and Calcium-Binding Proteins)
Edited by ANTHONY R. MEANS AND BERT W. O'MALLEY

VOLUME 103. Hormone Action (Part H: Neuroendocrine Peptides)
Edited by P. MICHAEL CONN

VOLUME 104. Enzyme Purification and Related Techniques (Part C)
Edited by WILLIAM B. JAKOBY

VOLUME 105. Oxygen Radicals in Biological Systems
Edited by LESTER PACKER

VOLUME 106. Posttranslational Modifications (Part A)
Edited by FINN WOLD AND KIVIE MOLDAVE

Section I

Mitochondrial Ubiquinone and Reductases

[1] Mitochondrial Quinone Reductases: Complex I

By GIORGIO LENAZ, ROMANA FATO,
ALESSANDRA BARACCA, and MARIA LUISA GENOVA

Introduction

The NADH:quinone (Coenzyme Q, CoQ, ubiquinone) oxidoreductase (Complex I) is the most complicated enzyme of the respiratory chain in mitochondria and aerobic bacteria.[1-3] Mitochondrial Complex I is a multisubunit enzyme that uses the energy associated to NADH oxidation by CoQ to pump hydrogen ions across the inner membrane, thus significantly contributing to the formation of an electrochemical proton gradient ($\Delta\mu_{H^+}$) and consequently to the efficiency of the oxidative phosphorylation process. The number of protons pumped by Complex I is estimated to be 2 to 5 (probably 4) $H^+/2e^-$.[4-8]

In spite of recent improvements of our knowledge on both structural and functional properties of Complex I,[9-11] the atomic structure and the detailed reaction mechanism of the enzyme are still unknown: it is the only enzyme of the membrane-bound respiratory chain to remain a "L shaped black box."[12] The reason for this lack of information is principally due to the complexity of this enzyme: in fact, the bovine enzyme consists of 46 different polypeptides[13,14] with different prosthetic groups: an FMN, 8 to 9 Fe-S clusters and one or more molecules of quinones. The need to clarify the molecular mechanism of the enzyme complex is strongly supported by different

[1] T. Friedrich, K. Stainmuller, and H. Weiss, *FEBS Lett.* **367,** 107 (1995).
[2] J. E. Walker, J. M. Skehel, and S. K. Buchanan, this series, Vol. 260, p. 14.
[3] K. L. Soole and R. I. Menz, *J. Bioenerget. Biomembr.* **27,** 397 (1995).
[4] F. Di Virgilio and G. F. Azzone, *J. Biol. Chem.* **257,** 4106 (1982).
[5] M. K. F. Wikstrom, *FEBS Lett.* **169,** 300 (1984).
[6] G. C. Brown and M. D. Brand, *Biochem. J.* **252,** 473 (1988).
[7] H. Weiss and T. Friedrich, *J. Bioenerg. Biomembr.* **23,** 743 (1991).
[8] T. Yano, *Mol. Aspects Med.* **23,** 345 (2002).
[9] N. Grigorieff, *Curr. Opin. Struct. Biol.* **9,** 476 (1999).
[10] A. D. Vinogradov, *Biochim. Biophys. Acta* **1364,** 169 (1998).
[11] M. Degli Esposti, A. Ngo, G. L. McMullen, A. Ghelli, F. Sparla, B. Benelli, M. Ratta, and A. W. Linnane, *Biochem. J.* **313,** 327 (1996).
[12] A. Matsuno-Yagi and T. Yagi, *J. Bioenerg. Biomembr.* **33,** 155 (2001).
[13] I. M. Fearnley, J. Carroll, R. J. Shannon, M. J. Runswick, J. E. Walker, and J. Hirst, *J. Biol. Chem.* **276,** 38345 (2001).
[14] J. Hirst, J. Cazzoll, I. M. Fearnley, R. J. Shannon, and J. E. Walker, *Biochim. Biophys. Acta* **1604,** 35 (2003).

research fields. Besides representing a major target of bioenergetics, since many neurodegenerative disorders as well as the aging process have been associated with Complex I deficiency,[15–17] knowledge of the enzyme structure-function relationship is essential for better understanding of these physiological-pathological dysfunctions.

Therefore the availability of reliable methods allowing the study of the Complex I activity is very important to investigate the mechanism of electron transfer and proton translocation by the enzyme and to assess cell bioenergetic damage occurring in mitochondrial diseases and aging.

Assay of Redox Activities of Complex I

NADH-CoQ Reductase

Investigation of electron transfer in the complex along its redox groups is very difficult to perform. Moreover, isolation of Complex I in an active form is not an easy task,[18] so the best way to assay its activity is to study the enzyme in situ, in mitochondrial membranes or in submitochondrial particles. From the functional point of view, Complex I activity can be isolated from the other respiratory complexes by the action of specific inhibitors such as antimycin A and mucidin for Complex III (acting respectively on center "i" and "o") and cyanide for Complex IV. Myxothiazol should be avoided, since it also inhibits Complex I.[19] Difficulties in assaying Complex I activity may arise from a limited permeability of its substrates; in particular, beef heart mitochondria, which are the choice material for Complex I studies, must be permabilized to NADH, and this may be achieved through freezing and thawing cycles (from 1 to 3 cycles). On the other hand, the complex uses Coenzyme Q_{10} as physiological electron acceptor; however, this molecule is too hydrophobic and cannot be used as exogenous substrate. For this reason Complex I activity is normally assayed by using short-chain analogs of CoQ_{10}: the most widely used are CoQ_1 (with only one isoprenoid unit in the side chain) and decylubiquinone (DB) with a 10-carbon-atom linear saturated side chain.[20,21] This lack of a suitable

[15] S. Rhaman, R. B. Blok, H. H. Dahl, D. M. Danks, D. M. Kirby, C. W. Chow, J. Christodoulou, and D. R. Thorburn, *Ann. Neurol.* **39,** 343 (1996).
[16] J. L. Loeffen, J. A. Smeitink, J. M. Trijbels, A. J. Janssen, R. H. Triepels, R. C. Sengers, and L. P. van den Heuvel, *Hum. Mutat.* **15,** 123 (2000).
[17] A. H. Schapira, *Biochim. Biophys. Acta* **1364,** 261 (1998).
[18] C. I. Ragan, *Curr. Top. Bioenerg.* **15,** 1 (1987).
[19] M. Degli Esposti, A. Ghelli, M. Crimi, E. Estornell, R. Fato, and G. Lenaz, *Biochem. Biophys. Res. Commun.* **190,** 1090 (1993).
[20] G. Lenaz, *Biochim. Biophys. Acta* **1364,** 207 (1998).
[21] E. Estornell, R. Fato, F. Pallotti, and G. Lenaz, *FEBS Lett.* **332,** 127 (1993).

assay method with endogenous substrates makes Complex I activity measurements considerably hampered. It has to be borne in mind that the activity of Complex I with these exogenous acceptors may be strongly underestimated.[20,22,23]

In Bovine Heart Mitochondria and Submitochondrial Particles

Beef heart mitochondria (BHM) are obtained by a large-scale procedure[24] and submitochondrial particles (SMP) by sonic irradiation of frozen and thawed mitochondria.[25] SMP obtained by this method are essentially broken membrane fragments[26]; alternatively, coupled closed particles (electron transfer particles or ETP_H) are prepared by the method of Hansen and Smith[27] or EDTA-particles by the method of Lee and Ernster.[28]

All preparations are kept frozen at $-80°$ at a stock concentration ranging between 40 and 60 mg/ml of protein detected by the biuret method by Gornall et al.[29] BHM are frozen and thawed two or three times before use, while SMP are used after thawing once: under these conditions the permeability barrier for NADH is completely lost, as demonstrated by the lack of further stimulation by detergents.

NADH-CoQ reductase is assayed essentially as described by Yagi[30] and modified by Degli Esposti et al.[19] and Estornell et al.[21]

REAGENTS.

Buffer, 50 mM KCl, 10 mM Tris-HCl, 1 mM EDTA, pH 7.4
NADH, 30 mM freshly prepared solution in water
Antimycin A, 2 mM in ethanol
KCN, 2 mM in buffer solution
All quinones used are kept as 20 to 30 mM ethanol stock solutions
Rotenone, 2 mM in ethanol

[22] G. Lenaz, R. Fato, M. L. Genova, G. Formiggini, G. Parenti Castelli, and C. Bovina, *FEBS Lett.* **366,** 119 (1995).

[23] M. L. Genova, C. Castelluccio, R. Fato, G. Parenti Castelli, M. Merlo Pich, G. Formiggini, C. Bovina, M. Marchetti, and G. Lenaz, *Biochem. J.* **311,** 105 (1995).

[24] A. L. Smith, *Methods Enzymol.* **10,** 81 (1967).

[25] R. E. Beyer, *Methods Enzymol.* **10,** 519 (1967).

[26] R. Fato, M. Cavazzoni, C. Castelluccio, G. Parenti Castelli, G. Palmer, and G. Lenaz, *Biochem. J.* **290,** 225 (1993).

[27] M. Hansen and A. L. Smith, *Biochim. Biophys. Acta* **81,** 214 (1964).

[28] C. P. Lee and L. Ernster, *Eur. J. Biochem.* **3,** 391 (1968).

[29] A. G. Gornall, C. J. Bardawill, and M. M. David, *J. Biol. Chem.* **177,** 752 (1949).

[30] T. Yagi, *Arch. Biochem. Biophys.* **281,** 305 (1990).

PROCEDURE. KCN and Antimycin A are added directly to the buffer to a final concentration of 2 mM and 2 μM, respectively; because of the alkaline hydrolysis of CN^-, it is necessary to adjust the pH to 7.4 to 7.5 with HCl. KCN and Antimycin A are added to avoid the electron flow through the cytochrome system (complexes III and IV).

The reaction is started by the addition of 10 to 20 μg/ml of mitochondrial protein (either BHM or SMP) and different amounts of the quinone analogs used as external electron acceptor and is followed spectrophotometrically by the decrease of absorbance at 340 minus 380 nm of a saturating amount of NADH (75 μM) in a double wavelength spectrophotometer (Jasco V-550 equipped with a double wavelength accessory and a rapid mixing apparatus) with an extinction coefficient of 3.5 mM^{-1} cm^{-1}. This extinction coefficient was directly calculated, taking in account that, in the double wavelength spectrophotometer used, the light transfer is due to an optical fiber device; the extinction coefficient may vary with other spectrophotometers.

We have used this method to study the kinetic parameters of different short chain quinone homologs such as CoQ_0, CoQ_1, CoQ_2, CoQ_3 and analogs having straight saturated chain such as 6-pentyl and 6-decylubiquinones (respectively PB and DB).[21,31] The results obtained are listed in Table I.

As shown in the table, different quinones elicited different NADH-CoQ reductase activity, and this is principally due to the capability of quinone analogs to interact with the physiological active site of Complex I. The widely used method to test this capability is to measure the effect of a specific Complex I inhibitor such as rotenone: NADH-CoQ_1, -CoQ_2, -CoQ_3, -DB, -PB are at least 90% sensitive to rotenone (2 μM), suggesting that all quinones are able to interact with the physiological site in Complex I. Nevertheless, only CoQ_1, PB and DB can elicit high activity; this observation could be explained taking in account that CoQ_2 is also a Complex I inhibitor,[31–33] whereas CoQ_3 and higher homologs are too hydrophobic to be used as external electron acceptors. On the other hand, CoQ_0 as well as the tetramethyl benzoquinone analog duroquinone (DQ) are too hydrophilic and can accept electrons also from a non-physiological site in the complex upstream with respect to the rotenone inhibition site.

[31] R. Fato, E. Estornell, S. Di Bernardo, F. Pallotti, G. Parenti Castelli, and G. Lenaz, *Biochemistry* **35,** 2705 (1996).
[32] G. Lenaz, P. Pasquali, E. Bertoli, and G. Parenti Castelli, *Arch. Biochem. Biophys.* **169,** 217 (1975).
[33] L. Landi, P. Pasquali, L. Cabrini, A. M. Sechi, and G. Lenaz, *J. Bioenerg. Biomembr.* **16,** 153 (1984).

TABLE I
KINETIC CONSTANTS FOR NADH-CoQ REDUCTASE IN BOVINE HEART SMP BY
USING VARIOUS ACCEPTORS[a]

Quinone homologs or analogs	V_{max}^b (μmol min^{-1} mg^{-1})	K_m^b (μM)
CoQ_0	0.18	65
CoQ_1	0.98 ± 0.40	20.1 ± 5.4
CoQ_2	0.29	1.3
CoQ_3	0.17	0.8
PB	0.93	21
DB	0.58 ± 0.15	1.8 ± 0.8

[a] R. Fato, E. Estornell, S. Di Bernardo, F. Pallotti, G. Parenti Castelli, and G. Lenaz, *Biochemistry* **35**, 2705 (1996).
[b] Data for CoQ_1 and DB are means $\pm$ SD of six different titrations. All other data are means of two titrations.

In Mitochondria From Different Rat Tissues

REAGENTS.

Reagents used are same as above.

PROCEDURE. Liver, heart and muscle mitochondria from rats are prepared by the procedure of Kun *et al.*,[34] and the enzyme activity is assayed at 30° after one cycle of freezing and thawing to remove the permeability barrier to NADH without damaging the enzyme. The assay medium is the same used for BHM or SMP. The only difference is a higher residual activity after rotenone inhibition in rat liver mitochondria (RLM) because of a NADH-dehydrogenase activity of the outer mitochondrial membrane. The rotenone-insensitive component is more evident when using CoQ_1 as substrate, so that DB seems to be a more situable acceptor. In Table II are reported some results when using a quasi-saturating concentration of DB (20 to 40 μM) and NADH (75 μM).

In Mitochondria From Human Platelets

REAGENTS.

Reagents used are same as above.

PROCEDURE. Platelets are obtained from venous blood samples and purified as described by Degli Esposti *et al.*[35] and mitochondrial membranes are prepared as described by Merlo Pich *et al.*[36]

---

[34] E. Kun, E. Kirsten, and W. N. Piper, *Methods Enzymol.* **55,** 115 (1979).

TABLE II

NADH-DB REDUCTASE ACTIVITY IN MITOCHONDRIA FROM
DIFFERENT RAT TISSUES

Mitochondria source	NADH-DB (nmol min^{-1} mg^{-1})
Liver	154 ± 21
Heart	292 ± 81
Muscle	52 ± 20
Brain cortex[a]	227 ± 66

[a] Mitochondria are prepared according to M. Battino, A. Gorini, R. F. Villa, M. L. Genova, C. Bovina, S. Sassi, G. P. Littarru, and G. Lenaz, *Mech. Ageing Dev.* **78**, 173 (1995), except that protease inhibitors are omitted as in M. L. Genova, C. Bovina, M. Marchetti, F. Pallotti, C. Tietz, G. Biagini, A. Pugnaloni, C. Viticchi, A. Gorini, R. F. Villa, and G. Lenaz, *FEBS Lett.* **410**, 467 (1997).

The activity is measured at 32° by utilizing a quasi saturating concentration of DB (about 20 to 40 μM) as external electron acceptor and NADH (75 μM) with an activity value of about 4.6 ± 1.6 nmol min^{-1} mg^{-1}.

In Cultured Cells

The major problem in this case is the preparation of the mitochondrial fraction from cells: we have prepared mitochondria from the human osteosarcoma 143B cell line by using the method described by Trounce *et al.*,[37] starting from about 1 to 2×10^8 cells (6 to 12 semiconfluent 150-mm dishes). The mitochondrial fraction is kept frozen at $-80°$. Complex I activity is assayed after only one thawing process in a buffer containing 250 mM sucrose, 1 mM EDTA, 50 mM Tris-HCl (pH 7.4), 75 μM NADH, and 2 mM KCN. The reaction is started with 50 μM DB and followed spectrophotometrically at 340 minus 380 nm for 3 minutes at 30° in a Jasco V-550, a double-wavelength spectrophotometer, with the extinction coefficient of 3.5 mM^{-1} cm^{-1}; in another sample, the activity is measured after the addition of 2 μM rotenone; the activity in the presence of rotenone is subtracted from the previous one, giving the specific Complex I activity (40 to 50 nmol min^{-1} mg^{-1}).

[35] M. Degli Esposti, V. Carelli, A. Ghelli, M. Ratta, M. Crimi, S. Sangiorgi, P. Montagna, G. Lenaz, E. Lugaresi, and P. Cortelli, *FEBS Lett.* **352**, 375 (1994).

[36] M. Merlo Pich, C. Bovina, G. Formiggini, G. G. Cometti, A. Ghelli, G. Parenti Castelli, M. L. Genova, M. Marchetti, S. Semeraro, and G. Lenaz, *FEBS Lett.* **380**, 176 (1996).

[37] I. A. Trounce, Y. L. Kim, A. S. Jun, and D. C. Wallace, *Methods Enzymol.* **264**, 484 (1996).

Determination of NADH Dehydrogenase by Using Water-Soluble Electron Acceptors

Complex I may transfer electrons to water soluble acceptors such as ferricyanide and 2,6-dichlorophenolindophenol (DCIP), but the reduction of these compounds is not coupled to energy transduction.[38] Among these acceptors, ferricyanide accepts electrons before the physiological reduction site, as shown by the lack of inhibition by 2 μM rotenone of NADH-ferricyanide reductase activity[39] and by retention of ferricyanide reduction activity by the solubilized type I NADH dehydrogenases[40] that lack the hydrophobic sector of the enzyme.

NADH-Ferricyanide Reductase Activity

REAGENTS.

Buffer, 50 mM KCl, 10 mM Tris-HCl, 1 mM EDTA, pH 7.4
NADH, 30 mM freshly prepared solution in water
Antimycin A, 2 mM in ethanol
KCN, 2 mM in buffer solution
Rotenone, 2 mM in ethanol
Variable amounts of potassium ferricyanide

PROCEDURE. The reaction is started by adding 150 μM NADH to the reaction mixture containing 10 to 20 μg/ml of mitochondrial protein (either BHM or SMP), 2 μM of antimycin A and 2 mM of KCN and following spectrophotometrically the decrease of ferricyanide absorbance at 420 minus 500 nm in a Jasco V-550 spectrophotometer equipped with a double wavelength apparatus with an extinction coefficient of 1 mM^{-1} cm^{-1} at 30° and under continuous mixing.

The rate of ferricyanide reduction is several times that of Coenzyme Q$_1$, the most effective quinone acceptor either in BHM or in SMP, with values ranging between 20 to 25 μmol min^{-1} mg^{-1} in BHM and 40 to 50 μmol min^{-1} mg^{-1} in SMP. The method has been successfully used also in mitochondrial membranes from human platelets.[41]

The specific activity of ferricyanide reductase with 2 mM potassium ferricyanide was used to estimate the content of active Complex I in the membrane by considering half of maximum turnover (8 × 10^5 min^{-1} as

[38] A. B. Kotlyar and M. Gutman, *Biochim. Biophys. Acta* **1140,** 169 (1992).
[39] Y. Hatefi, A. G. Haavik, and D. E. Griffiths, *J. Biol. Chem.* **237,** 1676 (1962).
[40] R. L. Ringler, S. Minakami, and T. P. Singer, *J. Biol. Chem.* **238,** 352 (1963).
[41] V. Carelli, A. Ghelli, L. Bucchi, P. Montagna, A. De Negri, V. Leuzzi, C. Carducci, G. Lenaz, E. Lugaresi, and M. Degli Esposti, *Ann. Neurol.* **45,** 320 (1999).

in Cremona and Kearney[42]), since this concentration of ferricyanide is approximately equal to the K_m.[43,44]

Nevertheless, the amount of Complex I determined by this method may be overestimated so that it is better to estimate the FMN content by the high-performance liquid chromatography (HPLC) method of Light et al.[45] after water extraction at 80° according to Yagi,[46] with a Nova-Pack C18 column, 3.9 × 150 mm (Waters) and 20% methanol in water containing 5 mM ammonium acetate (pH 6) as mobile phase, at a flow rate of 0.8 ml/min, in a Waters Millennium 2010 Chromatography Manager equipped with a Waters 996 Photodiode Array Detector.

NADH-2,6-Dichlorophenolindophenol (DCIP) Reductase Activity

REAGENTS.

Buffer, 50 mM KCl, 10 mM Tris-HCl, 1 mM EDTA, pH 7.4
NADH, 30 mM freshly prepared solution in water
Antimycin A, 1 mg/ml in ethanol
KCN, 2 mM in buffer solution
Rotenone, 2 mM in ethanol
DCIP 20 mM in water

PROCEDURE. The sample is prepared by mixing 2 ml of buffer added with 2 mM KCN, 2 μM antimycin A and 10 to 20 μg/ml of mitochondrial protein (either BHM or SMP); the reaction is started with 75 μM NADH addition and the decrease in absorbance of DCIP (100 to 150 μM) is followed spectrophotometrically in a Jasco V-550 equipped with a double wavelength apparatus at 600 nm minus 700 nm by using an extinction coefficient of 21 mM^{-1} cm^{-1} at 30° under continuous mixing.

Metabolic Flux Control

It is known that Complex I may undergo functional alterations affecting mitochondrial bioenergetics, particularly in aging and in mitochondrial diseases.[47–49] The decrease in an individual enzyme activity in a metabolic

[42] T. Cremona and E. B. Kearney, *J. Biol. Chem.* **239,** 2328 (1964).
[43] S. Smith, I. R. Cottingham, and C. I. Ragan, *FEBS Lett.* **110,** 279 (1980).
[44] M. Degli Esposti, A. Ghelli, M. Ratta, D. Cortes, and E. Estornell, *Biochem. J.* **301,** 161 (1994).
[45] D. R. Light, C. Walsh, and A. Merletta, *Anal. Biochem.* **109,** 87 (1980).
[46] K. Yagi, *Methods Enzymol.* **18B,** 290 (1971).
[47] G. Lenaz, C. Bovina, C. Castelluccio, R. Fato, G. Formiggini, M. L. Genova, M. Marchetti, M. Merlo Pich, F. Pallotti, G. Parenti Castelli, and G. Biagini, *Mol. Cell. Biochem.* **174,** 329 (1997).
[48] G. Lenaz, M. D'Aurelio, M. Merlo Pich, M. L. Genova, B. Ventura, C. Bovina, G. Formiggini, and G. Parenti Castelli, *Biochim. Biophys. Acta* **1459,** 397 (2000).

pathway is meaningful only if it is able to affect the rate of the whole pathway, and this will depend on the degree of flux control exerted by the individual step itself[50]; in other words, it depends on how much the step is rate controlling on the whole pathway. Analysis of the literature shows that mitochondrial Complex I activity is among the major rate-controlling steps in electron transfer chain,[51] but the control of respiration is shared between different biochemical steps and its distribution changes according to the metabolic rate performed by mitochondria.[52,53]

Quantitative Analysis of Metabolic Flux Control

The metabolic control exerted by a single enzyme in a metabolic pathway can be expressed by a coefficient (Cv_i calculated from the experimentally measured percent changes in the enzyme activities [global activity and individual step]) on addition of small concentrations of a specific inhibitor. Thus,

$$Cv_i = (dJ/dI)_{I=0}/(dv_i/dI)_{I=0} \qquad (1)$$

that is, the ratio of the initial slope of the inhibition curve of the global activity (J) to the initial slope of the inhibition curve of the individual step (v_i). The titration curves can be obtained by a nonlinear regression-fitting procedure, performed on the experimental data with a commercial application program.[54] The initial slope of each curve is calculated as the limit of the derivative of the function for inhibitor concentration tending to zero.

The profile, particularly the initial slope, of the global flux curve is indicative of the control exerted by the examined enzyme: the steeper the curve, the greater the similarity to the hyperbolic titration curve of the single enzyme step and the higher (closer to 1) the flux control coefficient. If a metabolic pathway is composed of distinct enzymes, the extent to which each enzyme is rate-controlling may be different and the

[49] G. Lenaz, A. Baracca, C. Bovina, M. Cavazzoni, M. D'Aurelio, S. Di Bernardo, R. Fato, G. Formiggini, M. L. Genova, A. Ghelli, M. Merlo Pich, F. Pallotti, G. Parenti Castelli, and B. Ventura, *in* "Recent Research Developments in Bioenergetics" (S. G. Pandalai, ed.), Vol. 1, p. 63. Transworld Research Network, Trivandrum (India), 2000.

[50] A. Kacser and J. A. Burns, *Biochem. Soc. Trans.* **7,** 1149 (1979).

[51] B. Ventura, M. L. Genova, C. Bovina, G. Formiggini, and G. Lenaz, *Biochim. Biophys. Acta* **1553,** 249 (2002).

[52] T. Letellier, M. Malgat, and J-P. Mazat, *Biochim. Biophys. Acta* **1141,** 58 (1993).

[53] G. P. Davey and J. B. Clark, *J. Neurochem.* **66,** 1617 (1996).

[54] Jandel SigmaPlot® Scientific Graphing Software, Jandel Corporation.

sum of all the flux control coefficients for the different enzymes should be equal to unity.[55]

Metabolic Control of Complex I Over the Respiratory Chain Activity

This requires the preliminary determination of experimental inhibition curves of the NADH oxidase (global activity) and the NADH-ubiquinone oxidoreductase activity (individual step). The two activities can be determined in succession[51] by using a dual-wavelength spectrophotometer equipped with a temperature-controlled compartment and a rapid mixing device.

REAGENTS.

Buffer, 50 mM KCl, 10 mM Tris-HCl, 1 mM EDTA, pH 7.4
NADH, 30 mM freshly prepared solution in water
Antimycin A, 2 mM in ethanol
KCN, 0.2 M freshly prepared in buffer solution
DB, 30 mM solution in ethanol
Rotenone, 12 μM, solution in ethanol

PROCEDURE. Frozen and thawed mitochondria (2 mg protein/ml) are pulse sonicated five times at 10-sec periods (150 W) with 50-sec intervals in an ice water bath under nitrogen gas. An aliquot of the mitochondrial suspension (i.e., 80 μg protein/ml, final concentration for rat liver mitochondria) is diluted in buffer solution. The reaction is started with 75 μM NADH, and the oxidation of the substrate at 30° is followed at the wavelength couple of 340 minus 380 nm ($\varepsilon = 3.5$ mM^{-1} cm^{-1}), first in the absence and then in the presence of antimycin A (2 μM) plus K-cyanide (1 mM) as downstream inhibitors and decylubiquinone (DB, 60 μM) as electron acceptor: the former assay condition indicates the NADH oxidase activity of the whole mitochondrial respiratory chain, the latter represents the specific activity of Complex I. Inhibition curves for both activities are determined experimentally by employing rotenone-titrated mitochondrial suspensions, preincubated for 5 min with increasing amounts (0 to 100 pmoles/mg protein) of a concentrated solution of the specific inhibitor for Complex I. The activity rates are expressed as percentage of the control that has no inhibitor present and plotted against the rotenone concentration.

[55] A. K. Groen, R. J. A. Wanders, H. V. Westerhoff, R. van der Meer, and J. M. Tager, *J. Biol. Chem.* **257,** 2754 (1982).

Metabolic Control of Complex I Over Coupled Mitochondrial Respiration

To determine flux control exerted by Complex I over coupled respiration in intact mitochondria, experiments are performed of stepwise inhibition of oxygen consumption in the presence of NAD-linked substrates (global activity), in comparison with stepwise inhibition of NADH-ubiquinone oxidoreductase activity (individual step). The latter is necessarily determined in mitochondrial membrane fragments in the presence of Complex I substrates (NADH and DB) and downstream inhibitors (antimycin A and KCN), essentially as described in the previous section, while the former is assayed in freshly prepared mitochondria by using a thermostatically controlled oxygraph apparatus equipped with a Clark's electrode and a rapid mixing device.[51]

REAGENTS.

> Buffer, 100 mM KCl, 75 mM mannitol, 25 mM sucrose, 10 mM KH$_2$PO$_4$, 2 mM MgCl 10 mM Tris-HCl and 50 μM EDTA, pH 7.4
> Rotenone, 40 μM, solution in ethanol
> Na-Glutamate, 0.8 M, Na-Malate 0.4 M solution in water.
> ADP, 0.5 M

PROCEDURE. In a typical experiment, mitochondria (1 to 2 mg protein) are incubated for 5 min at 30° in a small volume (about 800 μl) of respiration buffer in the absence or in the presence of increasing amounts of rotenone (0 to 100 pmoles/mg protein), and then diluted 1:1 with the same buffer solution. State 4 respiration is started by addition of 10 mM glutamate plus 5 mM malate and monitored for approximately 2 min, then state 3 is induced by the addition of 0.5 mM ADP.[56] Activity rates are expressed as percentage of the control where no inhibitor is present and plotted against the rotenone concentration for determination of the flux control coefficient.

Other Activities of Complex I

Proton Translocation

The proton pumping activity of Complex I can be studied by a spectrofluorometric assay in submitochondrial particles (EDTA particles) prepared from mitochondria of different sources as beef heart and

---

[56] B. Chance and G. R. Williams, *Adv. Enzymol.* **17**, 65 (1956).

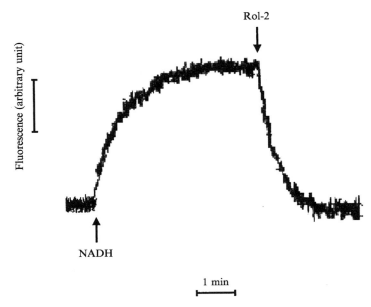

FIG. 1. Quenching of ACMA fluorescence induced by the protonophoric activity of Complex I on addition of 100 μM NADH to platelet-coupled submitochondrial particles. Experimental details are described in the text. The addition of 1 μM rolliniastatin-2 completely restores the initial ACMA fluorescence.

horse platelets.[57,58] The enzyme activity is analyzed by following the fluorescence quenching of the monoamine 9-amino-6-chloro-2-metoxyacridine (ACMA), a probe highly sensitive to the transmembrane ΔpH[59,60] (Fig. 1). When using the pH-sensitive fluorescent dye, both the extent and the initial rate of Complex I proton pumping can be monitored, allowing one to get information on the dependence of the enzyme proton pumping capacity on different experimental parameters, such as, for example, the CoQ analog used as electron acceptor.[57] According to the following relationship,

[57] L. Helfenbaum, A. Ngo, A. Ghelli, A. W. Linnane, and M. Degli Esposti, *J. Bioenerget. Biomembr.* **29,** 71 (1997).
[58] A. Baracca, L. Bucchi, A. Ghelli, and G. Lenaz, *Biochem. Biophys. Res. Commun.* **235,** 469 (1997).
[59] R. Casadio, *Eur. Biophys. J.* **19,** 189 (1991).
[60] A. Baracca, S. Barogi, V. Carelli, G. Lenaz, and G. Solaini, *J. Biol. Chem.* **275,** 4177 (2000).

$$\Delta pH = A\,Q/(B{-}Q)\exp(Q/[B - Q] + C\,Q) \qquad (2)$$

where A, B and C are three empirical fitting parameters, the ACMA fluorescence quenching measurement is referred to be strictly related to transmembrane ΔpH; then to accurately quantify the enzyme proton transport activity, the ACMA signal has to be calibrated by fluorescence quenching measurements at ΔpHs of known extent artificially induced in the membrane system.[59,61]

REAGENTS.

Buffer, 250 mM sucrose, 2.5 mM Mg Cl$_2$, 50 mM KCl, 50 mM Tricine, pH 8
Valinomycin 2 mM, in 100% ethanol
Carboxin 10 mM, in 100% ethanol
Antimycin A 2 mM, in 100% ethanol
Methoxyacrylate (MOA)-stilbene 2 mM, in 100% ethanol
Oligomycin 2 mM, in 100% ethanol
Rolliniastatin-2 2 mM, in 100% ethanol
ACMA 1 mM, in 100% ethanol
NADH 20 mM (freshly prepared)
CoQ analog (undecyl-ubiquinone, UBQ) 10 mM, in 100% ethanol

PROCEDURE. Complex I proton transport activity is measured at room temperature following the ACMA fluorescence changes induced on addition of 100 μM NADH under continuous stirring. The reaction mixture consists of 1 ml buffer containing 1 μM valinomycin, 20 μM carboxin (Complex II inhibitor),[62] 1 μM antimycin A and 1 μM methoxyacrylate (MOA)-stilbene to inhibit electron transfer through Complex III,[19] 1 μM oligomycin to inhibit ATP synthase proton transport during ATP synthesis,[63] 75 μg of protein (EDTA-particles), 0.5 μM ACMA and 20 to 30 μM CoQ analog (UBQ).

The dye fluorescence quenching induced by H$^+$-transport activity of Complex I is completely abolished by the addition of 1 μM rolliniastatin-2, a potent and specific inhibitor of the enzyme proton translocation.[64]

The assay has to be carried out by using a spectrofluorometer equipped with a thermostating system, and the excitation and emission

[61] V. Fregni and R. Casadio, *Biochim. Biophys. Acta* **1143**, 215 (1993).
[62] P. C. Mowery, B. A. C. Ackrell, and T. P. Singer, *Biochem. Biophys. Res. Commun.* **71**, 354 (1976).
[63] E. C. Slater, *Methods Enzymol.* **10**, 48 (1967).
[64] M. Degli Esposti, *Biochim. Biophys. Acta* **1346**, 222 (1998).

wavelengths used to measure the ACMA fluorescence are 412 and 510 nm, respectively.

Production of Superoxide Radical

Superoxide radical ($O_2^{\bullet-}$) and hydrogen peroxide (H_2O_2) are constantly produced in aerobic cells. Among the various intracellular organelles, the production of both species of molecules has been most studied and well documented in mitochondria, accounting for 1% ($O_2^{\bullet-}$) and 0.5% (H_2O_2), respectively, of organ oxygen uptake in rat liver and heart under physiological pO_2 conditions. It has long been understood that mitochondria do not release $O_2^{\bullet-}$ to the reaction medium or to the cytosolic space.[65] The release of $O_2^{\bullet-}$ toward the intermembrane space has been difficult to assess; however, it has been recently claimed that rat liver mitoplasts produce superoxide in the reaction medium.[66] Conversely, the vectorial release of $O_2^{\bullet-}$ into the matrix space of mitochondria is a well-established notion supported by experimental observations in submitochondrial particles (SMP), in which the inner mitochondrial membrane is inverted by sonication of mitochondria, or in fragments of mitochondrial membranes, usually obtained by freezing and thawing mitochondrial suspensions. NADH-ubiquinone oxidoreductase (Complex I) and ubiquinol-cytochrome c oxidoreductase (Complex III) are the major sources of superoxide anion in the mitochondrial respiratory chain: Complex I–supported production of $O_2^{\bullet-}$ in mitochondrial membranes is about 50% of the superoxide generation at the Complex III site.[67]

Quantitation of $O_2^{\bullet-}$ Production in SMP

SMP are obtained by pulse sonication of mitochondria, essentially as described in previous contributions to this series.[68,69] Two or three washings in isotonic buffer solution yield acceptable preparations of open (non-phosphorylating) particles devoid of endogenous Mn-superoxide dismutase (Mn-SOD). The rates of $O_2^{\bullet-}$ generation are best assessed by spectrophotometric determination with a highly sensitive spectrophotometer, such as a dual-wavelength model (i.e., V-550 extended model [Jasco Europe, Cremella-LC, Italy]) equipped with a temperature-controlled compartment and a rapid mixing device. Basically, the spectrophotometric

[65] B. Chance, H. Sies, and A. Boveris, *Physiol. Rev.* **59,** 527 (1979).

[66] D. Han, F. Antunes, F. Daneri, and E. Cadenas, *Methods Enzymol.* **349,** 271 (2002).

[67] J. F. Turrens and A. Boveris, *Biochem. J.* **191,** 421 (1980).

[68] A. Boveris, *Methods Enzymol.* **105,** 429 (1984).

[69] C. Gregg, *Methods Enzymol.* **10,** 181 (1967).

indicator is adrenochrome, produced by the $O_2^{\bullet-}$ -dependent oxidation of epinephrine, or ferricytochrome c (cytochrome c^{2+}) formed on reaction of superoxide with ferrocytochrome c (cytochrome c^{3+}).

To be assayed for superoxide generation, SMP are suspended in reaction buffer solution (KCl 50 mM, Tris 10 mM, EDTA 1 mM, pH 7.8 at 25° is a common choice) and supplemented with appropriate substrates and inhibitors in order to obtain maximal rates of $O_2^{\bullet-}$ production at the NADH dehydrogenase region.

Adrenochrome Formation

The rate of adrenochrome formation is monitored spectrophotometrically at $485 - 575$ nm ($\varepsilon = 2.96$ mM^{-1} cm^{-1}). Although the mechanism for the reaction of $O_2^{\bullet-}$ with epinephrine is still debated, it has been determined that the assay can detect one adrenochrome formed per one $O_2^{\bullet-}$ -produced[70]; the addition of Cu,Zn-superoxide dismutase provides the rates of SOD-sensitive adrenochrome formation, which is taken stoichiometrically as the rate of $O_2^{\bullet-}$ production. Catalase is an advisable addition to prevent H_2O_2-dependent side chain reactions that lead to overestimation of superoxide generation rates. Mucidin (strobilurin A),[71] a respiratory inhibitor acting at the level of Complex III, is added to functionally isolate Complex I from the downstream segments of the respiratory chain contained in the mitochondrial membranes. Use of antimycin A should be avoided,[72] since it is known that center i inhibitors enhance ROS production at the level of Complex III when added to mitochondria respiring on NAD-linked substrates.

Reagents.

Buffer, KCl 50 mM, Tris 10 mM, EDTA 1 mM, pH 7.8
Catalase, c8531 by SIGMA
Mucidin, 1.8 mM solution in ethanol (not commercially available)
Epinephrine sodium bitartrate, 0.1 M in 0.1 N HCl, freshly prepared
    solution to be kept in ice and to be used within 1 hour
NADH, 30 mM freshly prepared solution in water
Cu,Zn-SOD, s7008 by SIGMA

[70] E. Cadenas, A. Boveris, C. I. Ragan, and A. O. M. Stoppani, *Arch. Biochem. Biophys.* **180**, 248 (1977).

[71] G. Von Jagow, G. W. Gribble, and B. L. Trumpower, *Biochemistry* **25**, 775 (1986).

[72] M. L. Genova, B. Ventura, G. Giuliano, C. Bovina, G. Formiggini, G. Parenti Castelli, and G. Lenaz, *FEBS Lett.* **505**, 364 (2001).

PROCEDURE. The reaction medium (2.5 ml final volume) consists of buffer solution, 150 U/ml catalase (EC 1.11.1.6), 1.8 μM mucidin and beef heart submitochondrial particles (0.25 mg protein/ml). An aliquot of epinephrine is diluted in the reaction medium immediately before use (1 mM, final concentration). The reaction is started by addition of 125 μM NADH as Complex I substrate, at room temperature. Eventual addition of Complex I acceptors (60 μM Coenzyme Q_{10} analogs and homologs) or inhibitors (i.e, rotenone 0.2 nmol/mg protein) should be performed before supplementation with the substrate. The SOD-sensitive activity can be calculated by subtracting the residual activity in presence of 40 to 50 U/ml Cu,Zn-SOD (EC 1.15.1.1).

We have found that when NADH is present in the reaction mixture, the rate of adrenochrome formation increases as the reaction proceeds during the approximately 45 min monitoring time (Fig. 2).

This behavior cannot be ascribed to a lag time that is due to limited access of NADH to the enzyme, since there are no substrate diffusion barriers in SMP and, furthermore, a second NADH addition yields a straight line.

The hypothesis of an autocatalytic process, accounting for a progressively greater superoxide generation, deserves some comments. In our case, it can be also excluded that the above mentioned process is due to

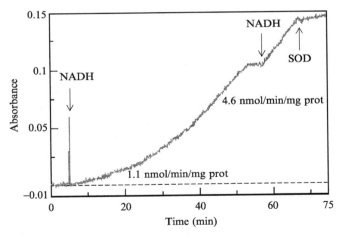

FIG. 2. Time course of adrenochrome formation during superoxide-dependent oxidation of epinephrine after addition of 125 μM NADH to Complex I in bovine heart SMP. Numbers near the trace indicate the initial and final rate of adrenochrome production (see text for details). Superoxide dismutase (SOD) is used to give assay specificity.

TABLE III
Effect of Quinones and Enzyme Inhibitors on Superoxide Production by
Complex I in Bovine Heart SMP[a]

Quinone acceptor	Inhibitor	Superoxide production ($nmol\ min^{-1}\ mg\ prot^{-1}$)
None	None	1.10 ± 0.21
None	Rotenone	2.76 ± 0.19
DB	None	2.36 ± 1.71
DB	Rotenone	5.71 ± 1.46

[a] The activity was assayed in mucidin-inhibited SMP by monitoring the oxidation of ephinephrine to adrenochrome by the superoxide radical produced after the addition of NADH (125 μM), decylubiquinone (DB, 60 μM) and rotenone (500 pmol/mg protein) as indicated.

the spontaneous oxidation of epinephrine by molecular oxygen: control of the baseline during the assay is achieved by working at pH 7.8, since it has been reported that epinephrine autoxidation is rather slow at physiological pH whereas it occurs much more rapidly as the pH is raised from 7.8 toward alkaline values.[73] On the contrary, one should not neglect the ability of adrenochrome, which is the product accumulated during the SOD-sensitive oxidation of epinephrine, in undergoing a redox cycle with further production of superoxide.[74] A similar cycle can arise enzymatically in the presence of the mitochondrial respiratory chain, supplemented with NADH, when adrenochrome is reduced to the corresponding semiquinone by Complex I (NADH-adrenochrome oxidoreductase activity by Complex I in mucidin-inhibited beef heart SMP: Vmax = 0.278 μmoles/min/mg prot, Km = 67 μM[75]). The o-semiquinone species was shown to rapidly react with oxygen, forming the superoxide anion. Consequently, when the present method is employed for assaying superoxide production by Complex I, only initial rates (at about 5 to 10 minutes) are considered to represent true rates of the direct $O_2^{\bullet-}$ generation by the respiratory complex (Table III).

Cytochrome c Reduction

Native cytochrome c is an excellent quantitative trap for $O_2^{\bullet-}$ but when added to submitochondrial particles, it is susceptible to reduction by Complex III more effectively than by superoxide. Besides, the ferrocytochrome

[73] H. P. Misra and I. Fridovich, *J. Biol. Chem.* **247**, 3171 (1972).

[74] A. Bindoli, D. J. Deeble, M. P. Rigobello, and L. Galzigna, *Biochim. Biophys. Acta* **1016,** 349 (1990).

[75] M. L. Genova, A. Bernacchia, and G. Lenaz, unpublished data, (2003).

c produced is oxidized by cytochrome oxidase. It is therefore advisable to use acetylated cytochrome c that, although not as rapidly reduced by $O_2^{\bullet-}$ as native cytochrome c, is much more slowly reduced by the reductase activity of the inner mitochondrial membrane and is not efficiently oxidized by cytochrome oxidase.[76] Nevertheless, since a residual enzymatic activity is present, the addition of respiratory inhibitors acting at the level of Complex III (mucidin) and Complex IV (potassium cyanide) is suggested to functionally isolate Complex I from the downstream segments of the respiratory chain. One should, however, be aware that KCN can influence the Cu,Zn-SOD, therefore the Mn-enzyme should be preferred when cyanide is present in the reaction medium.[67] Reduction of acetylated cytochrome c is followed spectrophotometrically with a dual wavelength spectrophotometer at 550 minus 540 nm ($\varepsilon = 19.1$ mM^{-1} cm^{-1}).

REAGENTS.

Buffer, KCl 50 mM, Tris 10 mM, EDTA 1 mM, KCN 1 mM, pH 7.8
Acetylated cytochrome c (from equine heart, c4186 by SIGMA), 7 mM solution in water
Mucidin, 1.8 mM solution in ethanol (not commercially available)
NADH, 30 mM freshly prepared solution in water
Cu,Zn-SOD, s7008 by SIGMA

PROCEDURE. An aliquot of partially acetylated cytochrome c is diluted to 180 μM in freshly prepared buffer solution in the presence of 1.8 μM mucidin and beef heart submitochondrial particles (3 μg protein/ml). The reaction is initiated by the addition of 125 μM NADH, at room temperature. The addition of 40 to 50 U/ml Cu,Zn-superoxide dismutase gives the SOD-sensitive rate of acetylated cytochrome c reduction, which gives the stoichiometric rate of $O_2^{\bullet-}$ generation by Complex I.

---

[76] A. Azzi, C. Montecucco, and C. Richter, *Biochem. Biophys. Res. Commun.* **65**, 597 (1975).

[2] Q-Cycle Bypass Reactions at the Q_o Site of the Cytochrome bc_1 (and Related) Complexes

By David M. Kramer, Arthur G. Roberts, Florian Muller, Jonathan Cape, and Michael K. Bowman

The Q-Cycle and Its Bypass Reactions

The cytochrome (cyt) bc_1 complex plays a central role in chemiosmotic energy conversion in mitochondria and many bacteria, oxidizing quinol (QH_2)—ubihydroquinone (UQH_2) in the case of mitochondria—and reducing a soluble electron carrier (cyt c in mitochondria), while acting as a proton "shuttle" or translocator to store energy in an electrochemical proton gradient, or proton motive force *(pmf)*.[1,2] The *pmf* in turn drives the synthesis of ATP at the F_0-F_1-ATP synthase. The structurally analogous cyt $b_6 f$ complex of chloroplasts plays a similar role—but oxidizes plastohydroquinone (PQH_2)—during oxygenic photosynthesis. The cyt $bc_1/b_6 f$ enzymes are dimeric integral membrane complexes composed of as few as three subunits in some prokaryotes and up to eleven subunits in mitochondria. The proteins house four essential redox-active, metal centers in two distinct chains. The "low potential chain" consists of two b-type hemes, cyt b_H and cyt b_L with relatively higher and lower redox potentials, housed in a single cyt b protein. The "high potential chain" consists of a "Rieske" iron-sulfur complex (2Fe2S) in the Rieske iron-sulfur protein (ISP) and another relatively high potential carrier, cyt c_1 in the cyt bc_1 complex, cyt f in the cyt $b_6 f$ complex, and an assortment of other carriers in complexes from distantly related bacteria.[3,4] The cyt bc_1 and $b_6 f$ complexes also possess two quinone/quinol binding sites. One of these, the Q_o site is located on the p-side of the membrane (the intermembrane space in mitochondria), at the interface between cyt b_L and the ISP and acts during normal turnover to oxidize QH_2 to quinone (Q). The other site, termed the Q_i site, is located close to cyt b_H, towards the n-side of the membrane, and acts to reduce Q to QH_2.

Biochemical studies and X-ray crystal structures of mitochondrial cyt bc_1 complexes show that the Q_o site possesses proximal and distal Q_o site

[1] Z. Zhang, L. Huang, V. Shulmeister, Y. Chi, K. Kim, L. Hung, A. Crofts, E. Berry, and S. Kim, *Nature* **392,** 677–684 (1998).
[2] B. L. Trumpower and R. B. Gennis, *Annu. Rev. Biochem.* **63,** 675–716 (1994).
[3] A. Niebisch and M. Bott, *J. Biol. Chem.* **278,** 4339–4346 (2003).
[4] J. Xiong, K. Inoue, and C. Bauer, *Proc. Natl. Acad. Sci. USA* **95,** 14851–14856 (1998).

METHODS IN ENZYMOLOGY, VOL. 382

niches.[5,6] The proximal Q_o niche is located close to the cyt b_L heme, while the distal niche is located close to the ISP.[6] The two niches bind different classes of inhibitors.[5-7] "Distal niche" inhibitors such as stigmatellin, 5-n-undecyl-6-hydroxy-4,7-dioxobenzothiazole (UHDBT)[8,9] strongly interact with the Rieske ISP and bind at the distal niche.[6] "Proximal niche inhibitors" such as MOA-stilbene,[10] mucidin (strobilurin A),[11,12] famoxadone[13] and myxothiazol[9] bind closer to cyt b_L[5,7] and do not interact strongly with the ISP.

Electron transfer through the cyt bc_1 and b_6f complexes likely occurs through the so-called "Q-cycle," first proposed by Mitchell[14,15] and later modified by several groups.[16-18] The central step in the Q-cycle is the "bifurcated" oxidation of quinol at the Q_o site (Fig. 1A), resulting in one QH_2 electron being transferred to the 2Fe2S, leaving an unstable semiquinone (SQ) species at the Q_o site, termed SQ_o, which is oxidized by cyt b_L, forming Q. The two protons from the quinol oxidized at the Q_o site are released to the p-side of the membrane. After two turnovers of the Q_o site, the two electrons sent down the low potential chain reduce Q at Q_i site to QH_2, with an uptake of two protons from the n-side of the membrane (i.e., the matrix in mitochondria). Overall, for each electron transferred to the 2Fe2S cluster, two protons are shuttled to the p-side of the membrane.

One of the central questions in understanding the cyt bc_1 and b_6f complexes is: why is the yield of the bifurcated reaction (and thus proton translocation) so high? In fact, there are at least four short-circuiting "bypass"

[5] T. A. Link, U. Haase, U. Brandt, and G. von Jagow, *J. Bioenerg. Biomembr.* **25**, 221–232 (1993).

[6] A. Crofts, S. Hong, N. Ugulava, B. Barquera, R. Gennis, M. Guergova-Kuras, and E. Berry, *Proc. Natl. Acad. Sci. USA* **96**, 10021–10026 (1999).

[7] G. von Jagow and T. A. Link, *Methods Enzymol.* **126**, 253–271 (1986).

[8] G. von Jagow and T. Ohnishi, *FEBS Lett.* **185**, 311–315 (1985).

[9] G. Thierbach, B. Kunze, H. Reichenbach, and G. Hofle, *Biochim. Biophys. Acta* **765**, 227–235 (1984).

[10] U. Brandt, H. Schagger, and G. von Jagow, *Eur. J. Biochem.* **173**, 499–506 (1988).

[11] J. Subik, M. Behun, P. Smigan, and V. Musilek, *Biochim. Biophys. Acta* **343**, 363–370 (1974).

[12] P. Sedmera, V. Musilek, F. Nerud, and M. Vondracek, *J. Antibiot. (Tokyo)* **34**, 1069 (1981).

[13] D. B. Jordan, K. T. Kranis, M. A. Picollelli, R. S. Schwartz, J. A. Sternberg, and K. M. Sun, *Biochem. Soc. Trans.* **27**, 577–580 (1999).

[14] P. Mitchell, *J. Theor. Biol.* **62**, 327–367 (1976).

[15] P. Mitchell, *FEBS Lett.* **59**, 137–139 (1975).

[16] A. R. Crofts and Z. Wang, *Photosynth. Res.* **22**, 69–87 (1989).

[17] B. L. Trumpower, *J. Biol. Chem.* **265**, 11409–11412 (1990).

[18] H. Ding, C. C. Moser, D. E. Robertson, M. K. Tokito, F. Daldal, and P. L. Dutton, *Biochemistry* **34**, 15979–15996 (1995).

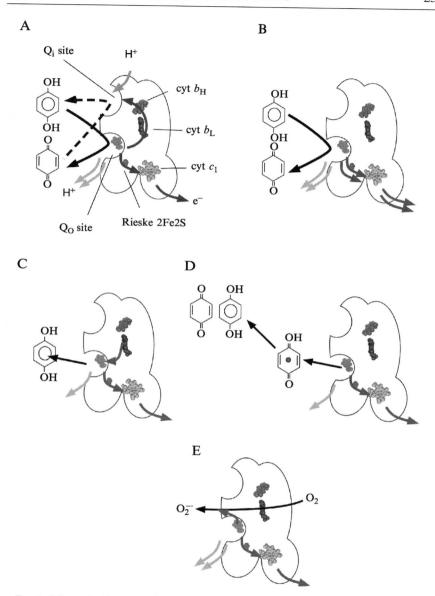

FIG. 1. Schematic diagrams of the Q-cycle and its bypass reactions. Figure 1A illustrates the movements of substrates, products, electrons and protons in the Q-cycle. Figure 1B–E show the same for Q-cycle bypass reactions 1 through 4, respectively. Movements of electrons are indicated by dark gray arrows, those of protons are indicated by light gray arrows. Details are provided in the text.

reactions that are clearly favored by thermodynamics.[19] Any of these reactions would decrease the efficiency of proton pumping, robbing the cell of ATP, while at least one these produces the highly reactive oxygen species, superoxide.

In the Q-cycle, cyt b_L oxidizes the reactive SQ_o. Each of the four Q-cycle bypass reactions discussed here is characterized by an alternate fate for the Q_o site SQ.[19]

Bypass Reaction 1

SQ_o could be oxidized by a second electron transfer reaction to the 2Fe2S cluster—that is, resulting in both quinol electrons being directly transferred to the 2Fe2S cluster (Fig. 1B). This would result in a single proton released on the *p*-side per electron transferred to the high potential chain.

Bypass Reaction 2

SQ_o is both a good reductant and a good oxidant, and thus should oxidize pre-reduced low potential chain components, via cyt b_L, reforming QH_2 (Fig. 1C). This second turnover of the Q_o site would result in only one proton released on the *p*-side of the membrane per electron transferred to the high potential chain.

Bypass Reaction 3

SQ_o could escape from the site and disproportionate (i.e., $2SQ \rightarrow Q + QH_2$) or otherwise react in the membrane (Fig. 1D).

Bypass Reaction 4

SQ_o can also directly reduce O_2, forming superoxide (Fig. 1E). Q-cycle bypass reactions are typically assayed in the presence of elicitors—that is, inhibitors that increase the rates of bypass reactions. The most commonly used of these is antimycin A, which blocks the Q_i site. Steady-state turnover in the presence of antimycin A results in reduction of the low potential chain components and subsequent inhibition of the bifurcated reaction. Such conditions are expected to increase the rates of all four bypass reactions.[19,20] Although the addition of such inhibitors is somewhat artificial, they are thought to mimic the effects of backpressure caused by high *pmf in vivo*, which is known to also induce bypass reactions.[20–23]

[19] F. Muller, A. R. Crofts, and D. M. Kramer, *Biochemistry* **41**, 7866–7874 (2002).
[20] F. Muller, *Age: Journal of the American Aging Society* **23**, 227–256 (2000).

The availability of high-resolution crystal structures, a large array of mutant strains in a number of species and new spectroscopic approaches should greatly facilitate elucidation of the mechanisms by which the various bypass reactions are prevented. To aid this approach, this chapter describes the application and limitations of assays for detection of the various bypass reactions. Also described is a method for probing conformational changes in the ISP head domain, which are thought to be critical for preventing some of the bypass reactions.

Estimating the Concentrations of cyt bc_1 and b_6f Complexes

Estimates of cyt bc_1 and b_6f Concentrations

Most estimates of bypass kinetics are normalized to the concentration of cyt bc_1 or b_6f complex in the assay suspension. The best estimates of bc_1/b_6f concentrations are obtained spectrophotometrically via the reduced minus oxidized spectra of cyt c_1/f and the two cyt b hemes.[24–26] A suspension of bc_1/b_6f complex, typically in the concentration range of a few micromolar in a well-buffered suspension medium at pH from 6 to 8, is placed in a standard glass, quartz or plastic cuvette in a spectrophotometer. Approximately 100 μM potassium or sodium ferricyanide is added to oxidize all the electron carriers in the complex. After incubation for a few minutes to ensure full oxidation, a baseline absorbance spectrum is taken from about 530 to 575 nm, covering the region of the cyt α-absorbance bands. Potassium ascorbate (from a freshly prepared or frozen 1 M stock solution made in suspension medium and brought to the pH of the complex suspension with KOH) is added to a final concentration of 10 mM with mixing to chemically reduce the high potential chain carriers. Absorbance spectra are repeated, about every minute, until stable cyt c_1/f α-band absorbance extents are obtained. Next, approximately 1 mg of sodium hydrosulfite (dithionite) powder is added to fully reduce the complex. After the dithionite is rapidly dissolved, the sample is allowed to equilibrate without

[21] V. D. Longo, L. L. Liou, J. S. Valentine, and E. B. Gralla, *Arch. Biochem. Biophys.* **365**, 131–142 (1999).

[22] D. M. Guidot, J. M. McCord, R. M. Wright, and J. E. Repine, *J. Biol. Chem.* **268**, 26699–26703 (1993).

[23] R. M. Lebovitz, H. Zhang, H. Vogel, J. Cartwright, L. Dionne, N. Lu, S. Huang, and M. M. Matzuk, *Proc. Natl. Acad. Sci. USA* **93**, 9782–9787 (1996).

[24] P. O. Ljungdahl, J. D. Pennoyer, D. E. Robertson, and B. L. Trumpower, *Biochim. Biophys. Acta* **891**, 227–241 (1987).

[25] D. S. Bendall, *Biochim. Biophys. Acta* **683**, 119–151 (1982).

[26] S. U. Metzger, W. A. Cramer, and J. Whitmarsh, *Biochim. Biophys. Acta* **1319**, 233–241 (1997).

further mixing. The dithionite will reduce O_2, followed by any oxidized carriers in the complex. Again, spectra are measured at approximately 1-min intervals until a stable redox state is achieved. The ascorbate-ferricyanide and the dithionite-ascorbate difference traces will give the absorbance changes for cyt c_1/f and the two cyt b hemes respectively. Ljungdahl and coworkers[27] present formulae for calculating the cyt concentrations of isolated cyt bc_1 complex; [cyt c_1] = $5.365 \cdot 10^{-2}$ $(\Delta A_{553-540}) - 9.564 \cdot 10^{-3}$ $(\Delta A_{562-577})$ (mM); [cyt b] = $3.539 \cdot 10^{-2}$ $(\Delta A_{562-577}) - 1.713 \cdot 10^{-3}$ $(\Delta A_{553-540})$ (mM).

Measurements of the Bifurcated Oxidation of QH_2

One approach to measuring Q-cycle bypass reactions is to probe the extent of reduction of the high and low potential chains following QH_2 oxidation. A good example of this approach was presented by Kramer and Crofts[28] for the cyt $b_6 f$ complex in intact thylakoid membranes. The oxidation of cyt f is blocked by removal of Cu from plastocyanin by incubation in suspension buffer (330 mM sorbitol, 10 mM KCl, 1 mM EDTA, 5 mM MgCl$_2$, 0.4% [w/v] bovine serum albumin and 50 mM HEPES) with 50 mM KCN and 0.3 mM Na ferricyanide at pH 7.8 for one hour on ice in darkness, followed by centrifugation for 2 min at 2000g and resuspension in suspension buffer at pH 7.5. This treatment also pre-oxidizes the electron carriers of the complex. Oxidation of the cyt b chain is prevented by addition of 80 μM MOA-stilbene, which blocks the Q_i site in the cyt $b_6 f$ complex (but blocks the Q_o site in cyt bc_1 complex). Plastoquinone is reduced to PQH_2 *in situ*, by sequential flash activations of the photosystem II reaction center, and the extents of reduction of cyt f and cyt b_H + cyt b_L are monitored by their characteristic (kinetically resolved) absorbance changes at 545, 554, 563 and 573 nm. The absorbance changes can be analyzed to yield estimates of absorbance change associated with the reduction of cyt f ($\Delta A_{cyt\,f}$) and cyt b ($\Delta A_{cyt\,b}$) by using the formulae: $\Delta A_{cyt\,f} = \Delta A_{554} - \Delta A_{545} - 0.32(\Delta A_{573} - \Delta A_{545})$; $\Delta A_{cyt\,f} = \Delta A_{563} - \Delta A_{545} - 0.63(\Delta A_{573} - \Delta A_{545})$.

The relative extinction coefficients of cyt b and f when using the above formulae are nearly equal and thus the bifurcated oxidation of PQH_2 will result in approximately equal absorbance changes at 554 and 563 nm. In contrast, the bypass reactions will result in preferential reduction of the high potential chain (i.e., cyt f).

[27] P. O. Ljungdahl, J. D. Pennoyer, D. E. Robertson, and B. L. Trumpower, *Biochim. Biophys. Acta* **891**, 227–241 (1987).
[28] D. M. Kramer and A. R. Crofts, *Biochim. Biophys. Acta* **1183**, 72–84 (1993).

The best discrimination between bifurcated oxidation and bypass reactions will be observed at lower pH, where the overall driving force for the bifurcated reaction is low.[28] For example, at pH 6, the equilibrium constant for the bifurcated oxidation of quinol, with cyt f and cyt b_H as the final electron acceptors, is about 1, whereas the equilibrium constant for sending both electrons to cyt f is greater than 10^5. With partial reduction of the PQ pool, the bifurcated reaction will result in only fractional reduction of both cytochromes, whereas the bypass reactions will result in essentially full reduction of cyt f. The appearance of excess reductant in the high potential chain was observed to be rather slow (on the seconds timescale) and may have reflected the release of MOA-stilbene from the Q_i pocket rather than a true bypass reaction.[28] Nevertheless, this type of measurement sets an upper limit for the sum of all possible bypass reaction rates in the cyt $b_6 f$ complex of about 1 s^{-1}.

A suitable kinetic spectrophotometer with millisecond time resolution and facilities for flash activation of photosystem II is required for assays on thylakoid membranes. Description of this type of instrumentation is beyond the scope of this chapter, but the reader is referred to references below.[29,30] However, similar results can be obtained with conventional spectrophotometers, on isolated cyt $b_6 f$ or bc_1 complexes. In this case, the complexes can be pre-oxidized by incubation with 1 mM potassium ferricyanide followed by dialysis against suspension buffer to remove excess reductant. In the case of the cyt bc_1 complex, oxidation of cyt b_H can be inhibited by 10 μM antimycin A. The reaction is started by rapid addition of QH$_2$, preferably in a stopped flow apparatus, while following the absorbance changes in the 540–570 nm region, as above.

Measurements of H^+/e^- Stoichiometries

The Q-cycle will result in the translocation of two protons per electron passed from quinol to the mobile electron acceptor. Bypass reactions will translocate less than one proton per electron. Thus measurements of the stoichiometry of protons translocated per electron transferred (H^+/e^-) should reflect the extent of bypass reactions. This type of assay is probably restricted to photosynthetic material where the electron transfer reactions can be rapidly initiated and stopped by switching on and off a light source.

[29] J. Amesz, in "Biophysical Techniques in Photosynthesis" (J. Amesz and A. J. Hoff, eds.), Vol. 4, pp. 3–10. Kluwer Academic Publishers, The Netherlands, 1996.
[30] D. M. Kramer and A. R. Crofts, in "Photosynthesis and the Environment. Advances in Photosynthesis" (N. Baker, ed.), Vol. 5, pp. 25–66. Kluwer Academic Publishers, Dordrecht, The Netherlands, 1996.

Thus it will not be covered here, but see recent reviews in the references below.[31,32]

Inhibitor-Insensitive cyt c Reduction

One of the most useful and straightforward assays for bypass reactions is to monitor inhibitor-insensitive cyt c reduction. It was noted some time ago that even at saturating concentrations, antimycin A does not completely inhibit cyt c reduction.[19,33] In the presence of inhibitors like antimycin A, which block the Q_i site, two turnovers of the Q_o site should lead to essentially complete reduction of the cyt b hemes. Further turnover of the Q-cycle should be prohibited, and any measured cyt c reduction should occur via a bypass reaction. The maximal total rate of bypass reactions in the presence of antimycin A in isolated yeast cyt bc_1 complex is on the order of a few turnovers per second (about 2% of maximal normal turnover).

It has recently been shown that myxothiazol, as well as a number of other proximal Q_o inhibitors (including E-β-methoxyacrylate-stilbene, mucidin and famoxadone) also induce superoxide production in isolated yeast cyt bc_1 complex, at about 50% the V_{max} observed in the presence of antimycin A.[19,34] It has been proposed that proximal Q_o site inhibitors induce superoxide production because they allow formation, but not oxidation of SQ_o. This is very similar to the mode of action of antimycin A except that the cyt b hemes will remain oxidized in the presence of the proximal niche Q_o site inhibitors during steady-state turnover.

The inhibitor-insensitive cyt c reduction assay not only probes the sum total of all bypass reactions but can also allow one to partially identify which bypass mechanism predominates. For all bypass reactions, one of the two QH_2 electrons is transferred, as in the normal Q-cycle, down the high potential chain through the Rieske 2Fe2S cluster and cyt c_1 then onto soluble cyt c. The fate of the second (SQ_o/Q) electron depends on which bypass reaction operates. Under aerobic conditions, the Q_o site SQ can reduce O_2 to superoxide, which in turn, readily reduces cyt c.[35]

Bypass via superoxide can be distinguished from other bypass reactions by addition of saturating levels of superoxide dismutase (SOD), which will

[31] S. Berry and B. Rumberg, *Biochim. Biophys. Acta* **1410**, 248–261 (1999).
[32] C. A. Sacksteder, A. Kanazawa, M. E. Jacoby, and D. M. Kramer, *Proc. Natl. Acad. Sci. USA* **97**, 14283–14288 (2000).
[33] A. Boveris and E. Cadenas, *FEBS Lett.* **54**, 311–314 (1975).
[34] F. Muller, A. G. Roberts, M. K. Bowman, and D. M. Kramer, *Biochemistry* Submitted (2003).
[35] B. H. Bielski, *Philos. Trans. R. Soc. Lond. B. Biol. Sci.* **311**, 473–482 (1985).

divert the superoxide electron from cyt c to H_2O_2.[36] If all bypass reactions occurred via superoxide reduction, the rate of cyt c reduction should be decreased by 50% on addition of SOD. This is precisely what is observed in the presence of proximal Q_o site inhibitors, such as myxothiazol.[19,34] In contrast, in the presence of antimycin A, only about 70% of the total cyt c reduction rate could be ascribed to the superoxide bypass on the basis of the SOD effect.[19] The most likely candidate for the remaining 30% was bypass reaction 3, since the major steady-state difference between the cyt bc_1 complex in the presence of antimycin and proximal niche Q_o site inhibitors is the redox state of cyt b.

The cyt c reduction assay is essentially a variation of that normally used to measure cyt bc_1 turnover, except that inhibitors are added, and that the concentration of cyt bc_1 is raised (typically to 50 nM final concentration) to give good steady-state rates of cyt c reduction. We found that beef heart cyt c purified with trichloroacetic acid (obtained from Sigma) had the lowest background reduction rate (i.e., in the presence of quinol but the absence of cyt bc_1 complex),[19] and thus we suggest that it be used exclusively. Assays are conducted in suspension buffer [typically 50 mM MOPS (3-[N-Morpholino]propane-sulfonic acid), 100 mM KCl, pH 6 to 8] in the presence of 50 μM beef heart cyt c, 50 μM decyl-ubiquinol (other water-soluble quinol analogs—for example, UQH_2-2—work as well, although decyl-ubiquinol gives a low background cyt c reduction rate), 10 μM antimycin A. Potassium cyanide (1 mM) is added to inhibit trace amounts of cyt c oxidase, which co-purifies with the cyt bc_1 complex. As a control, 10 μM stigmatellin is added to completely abolish the bypass reactions.

The progress of the reaction is monitored at 550 nm—that is, the α-band of cyt c with an extinction coefficient of 17.5 mM^{-1} cm^{-1}, for 1 to 2 minutes to obtain the background rate (non-enzymatic reduction of cyt c by decyl-ubiquinol), after which point 50 nM cyt bc_1 (previously equilibrated with 10 μM antimycin A) is added to the cuvette. This experiment is repeated in the absence and presence of Mn-SOD. It is important to use Mn-SOD rather than the cheaper Cu Zn SOD because the later is sensitive to KCN. The difference in rates in the presence and absence of SOD is taken as the rate of superoxide formation, with a stoichiometry of 1 mol superoxide:1 mol cyt c reduced. It is noteworthy that the antimycin A bypass reaction is sensitive to pH, increasing especially at pH > 8.[37,38]

An interesting finding is that complete removal of oxygen under Ar(g) does not slow the rate of cyt c reduction, although it abolishes superoxide

[36] I. Fridovich, *Ann. N. Y. Acad. Sci.* **893**, 13–18 (1999).
[37] P. K. Jensen, *Biochim. Biophys. Acta* **122**, 167–174 (1966).
[38] E. Cadenas, A. Boveris, and B. Chance, *Biochem. J.* **186**, 659–667 (1980).

production and the effect of SOD on the rate. This likely indicates that O_2 oxidation of the SQ_o is not the rate-limiting step in either superoxide formation or Q-cycle bypass.[19] In the absence of O_2, the bypass reaction still occurs, but via a secondary, slower pathway, possible via release of the SQ from the Q_o pocket followed by direct reduction of cyt c or disproportionation. Alternatively, the O_2 and inhibitor-insensitive pathway may occur by reoxidation of the Rieske 2Fe2S cluster, allowing it to oxidize the SQ.[19]

Measuring the Q Cycle Bypass Reactions via Superoxide or H_2O_2 Formation

Good estimates of the rates of superoxide production under a range of conditions are important for assessing the physiological consequences of Q-cycle bypass reactions. Although superoxide can, in principle, be detected directly by using EPR, its steady state concentrations are usually too low for this to be feasible. Instead, more stable and readily-measured products of superoxide reactions are usually measured.

EPR Spin Traps

Spin traps (e.g., 5,5-dimethyl-1-pyroline-N-oxide) are frequently used to detect superoxide production. A reagent is added to the suspension, which readily reacts with superoxide, producing an adduct which is a more stable, EPR-detectable radical. Unfortunately, most of the detectable species have very short half-lives,[39] making them of marginal use for measuring slow rates of superoxide production, such as those encountered with the Q-cycle bypass reactions. Also, many of the older spin traps do not allow one to differentiate between superoxide and hydroxyl radical as the source of the spin adduct. The short half-life problem has been addressed in two ways. First, a new spin trap called DEPMPO (5-diethoxyphosphoryl-5-methyl-1-pyrroline-n-oxide), which has a significantly longer half life ($t_{1/2} = \sim 10$ minutes at pH 8.2).[39,40] One drawback of DEPMPO is that it rather slowly reacts with superoxide, thus requiring high concentrations (relative to cyt c) to compete with cyt c reduction. Roubaud and co-workers[41] found that about 10 mM was required to trap 60% of superoxide formed by the xanthine/xanthine oxidase system. A second approach has

[39] C. Frejaville, H. Karoui, B. Tuccio, F. Le Moigne, M. Culcasi, S. Pietri, R. Lauricella, and P. Tordo, *J. Med. Chem.* **38**, 258–265 (1995).

[40] B. Tuccio, R. Lauricella, C. Fréjaville, J.-C. Bouteiller, and P. Tordo, *J. Chem. Soc., Perkin Trans.* **2**, 295–298 (1995).

[41] V. Roubaud, S. Sankarapandi, P. Kuppusamy, P. Tordo, and J. L. Zweier, *Anal. Biochem.* **247**, 404–411 (1997).

been to rapidly freeze samples in liquid N_2, preventing the decay of the EPR-detectable radical.[42] After freezing has occurred, the EPR-detectable signal is stable nearly indefinitely, but the signal itself is broadened. Nevertheless, reasonable sensitivity was obtained for the production of superoxide in the presence of 0.08 U/ml xanthine oxidase, or approximately 1.33 μmol superoxide L^{-1} s^{-1}.[42] This is about the same as the maximal superoxide production rate in the presence of 1 μM cyt bc_1 complex inhibited by antimycin A. Our initial attempts to use DEPMPO were only partly successful, perhaps because the radical adduct resulting from reaction with superoxide reacted with other species in the cyt bc_1 complex (F. Muller, A. Roberts and D. M. Kramer, unpublished results).

Epinephrine/Adrenochrome Assay

Superoxide will oxidize catechols with rates up to 10^4 M^{-1} s^{-1},[35,43] and the SOD-inhibitable oxidation of epinephrine to adrenochrome has been employed extensively as simple spectrophotometric assay to measuring rates of superoxide production in mitochondria.[38] The assay measures the appearance of adrenochrome at 485 to 575 nm, which has an extinction coefficient of 2.96 M^{-1} cm^{-1}.[43] Superoxide dismutase inhibits this reaction by around 90%. The buffer for this assay contains 100 mM KCl, 50 mM MOPS, 1 mM EDTA, 2 mM epinephrine (Sigma) and 0.5 μM catalase (Sigma).

We note that to employ this assay the oxidation of epinephrine by superoxide has to compete with reduction of cyt c, requiring the use of relatively high concentrations of epinephrine. In addition, epinephrine also slowly reduces cyt c directly, which can complicate interpretation of results from slower assays. It is thus recommended that a series of cyt bc_1 concentrations be used to determine if significant interference occurs. Despite this drawback, we have obtained data with this assay comparable to other techniques. This assay is best used as a control to validate more-sensitive assays and is particularly useful for assays with the uninhibited cyt bc_1 complex, which interferes with some other assays, notably the Amplex red assay (see following text).

Measurements of Superoxide Production via Detection of H_2O_2

A number of assays detecting H_2O_2 have been developed that employ a fluorogenic substrate coupled to horseradish peroxidase (HRP), such as HRP/homovanillic acid (or HRP/scopoletin).[43–45] More recently, the

[42] M. Dambrova, L. Baumane, I. Kalvinsh, and J. E. Wikberg, *Biochem. Biophys. Res. Commun.* **275**, 895–898 (2000).

[43] G. Loschen, A. Azzi, C. Richter, and L. Flohe, *FEBS Lett.* **42**, 68–72 (1974).

[44] G. Barja, *Free Radic. Biol. Med.* **33**, 1167–1172 (2002).

[45] H. H. Ku, U. T. Brunk, and R. S. Sohal, *Free Radic. Biol. Med.* **15**, 621–627 (1993).

sensitivity of this approach has been improved by using the non-fluorescent substrate Amplex[TM] red,[46,47] which on HRP/H_2O_2 oxidation yields the highly stable and brightly fluorescent resorufin red. An advantage of this approach is the high stability of the product of Amplex red, even in the presence of HRP. We have employed this technique to measure superoxide (as H_2O_2) produced from yeast mitochondrial membranes and isolated cyt bc_1.[19]

The Amplex red–horseradish peroxidase kit can be obtained from Molecular Probes (product #A-12212). Hydrogen peroxide oxidizes Amplex[TM] red to stable, fluorescent resorufin in the presence of HRP. Increase in resorufin red fluorescence is followed at excitation of 530 nm and emission at 590 nm by using any simple fluorimeter (the signal is quite strong). The rates of H_2O_2 are determined from the initial slopes of fluorescence increase and the H_2O_2 standards provided in the kit. The data is then corrected for the concentration of cyt bc_1 complex (expressed as mol H_2O_2 mol^{-1} cyt bc_1 s^{-1}), which is measured spectrophotometrically via the reduced minus oxidized spectra of cyt b and cyt c_1.[27] To convert H_2O_2 measurements to $O_2^{\bullet-}$ production, the latter have to be multiplied by 2, since there is 1 mol of H_2O_2 formed for every 2 mol of $O_2^{\bullet-}$ that is dismutated by SOD.[35,43]

The H_2O_2 assay system and buffer is essentially the same as for the cyt c assay except that 300 to 15,000 units Mn-SOD/ml, 400 μM Amplex red and 10 u/ml horseradish peroxidase are added. Typically about 50 nM cyt bc_1 is used per reaction. Although KCN does not change the relative rates of H_2O_2 formation, it did decrease the absolute fluorescence yield, likely by interaction with the fluorophore, and is thus best omitted. The presence of Mn-SOD is critical to convert all superoxide into H_2O_2. Even in small amounts, superoxide can interfere with H_2O_2 detection, since superoxide can react (rapidly) with HRP and HRP Compound I.[35] As a control, the concentration of SOD is varied (see previous text) to determine if conversion of superoxide to H_2O_2 is rate limiting. When used, antimycin and stigmatellin were at about 10 to 30 μM.

Despite its convenience, there are several potential problems with Amplex red assay, mostly derived from the redox chemistry of Amplex red or resorufin. The protocol provided by Molecular Probes suggests avoiding thiol concentrations above 10 μM because of redox interactions. We have found that, at concentration over 20 μM, myxothiazol will react appreciably with Amplex red (in the absence of cyt bc_1 and cyt c), increasing the background rate and making H_2O_2 measurements problematic. We

[46] M. Zhou and N. Panchuk-Voloshina, *Anal. Biochem.* **253**, 169–174 (1997).
[47] M. Zhou, Z. Diwu, N. Panchuk-Voloshina, and R. P. Haugland, *Anal. Biochem.* **253**, 162–168 (1997).

also found that selected batches of stigmatellin directly catalyzed the transfer of electrons from decyl-UQH_2 to Amplex red in the complete absence of the cyt bc_1. No other inhibitor tested (MOA-stilbene, famoxadone, mucidin, antimycin A) exhibited this type of behavior, and we speculate that selected batches of stigmatellin contain a redox-active impurity.

Last, great care should be taken when employing the Amplex red assay with uninhibited cyt bc_1 complex, since we have found that resorufin itself can interact with the cyt bc_1 complex (J. Cape and D. M. Kramer, unpublished observations). When antimycin A was excluded from the assay mixture, we detected a small but significant initial mono-exponential increase in fluorescence, followed by a mono-phasic decay of the fluorescence signal. These signals were completely abolished by addition of 20 μM stigmatellin, showing that this behavior was coupled to the oxidation of QH_2 at the Q_o site. Surprisingly, similar results were found when 13 μM resorufin (in the absence of Amplex red) was included in the superoxide assay mixture. In this case, the fluorescence signals were inhibited by the addition of antimycin A, suggesting that electrons from the Q_i site were responsible for the observed interference.

Probing the Involvement of ISP Domain Movements in Restricting Bypass Reactions

Transferring both electrons down the high-potential chain of cyt bc_1 complexes is one of the primary bypass reactions and is heavily favored by thermodynamics.[19,28,48] To force electron flow down both the high- and low-potential chains, it has been proposed that a "catalytic switch" gates electron transfer at the Q_o site.[49,50]

Recent structure and function studies have revealed an interesting mechanism for the proposed "catalytic switch." X-ray crystal structures of the cyt bc_1 complex from different crystal forms and crystals made in the absence and presence of Q_o site inhibitors showed that the hydrophilic "head" domain of the ISP occupies several different positions.[51–53] One position, termed ISP_B, places the 2Fe2S cluster close to the Q_o site and

[48] U. Brandt, *Biochim. Biophys. Acta* **1365,** 261–268 (1998).

[49] U. Brandt, U. Haase, H. Schägger, and G. V. Jagow, *J. Biol. Chem.* **266,** 19958–19964 (1991).

[50] U. Brandt, *Biochim. Biophys. Acta* **1275,** 41–46 (1996).

[51] E. A. Berry, M. Guergova-Kuras, L. S. Huang, and A. R. Crofts, *Annu. Rev. Biochem.* **69,** 1005–1075 (2000).

[52] S. Iwata, J. W. Lee, K. Okada, J. K. Lee, M. Iwata, B. Rasmussen, T. A. Link, S. Ramaswamy, and B. K. Jap, *Science* **281,** 64–71 (1998).

[53] C. Hunte, J. Koepke, C. Lange, T. Rossmanith, and H. Michel, *Structure with Folding and Design* **8,** 669–684 (2000).

cyt b_L; another, termed ISP_C, places the ISP close to cyt c_1, while intermediate positions have also been found. Similar conformational changes have been shown to occur in the chloroplast cyt b_6f complex and bc-type complexes.[54,55]

The crystal structures of the cyt bc_1 complexes show that in the ISP_B position, the 2Fe2S cluster is too distant from the cyt c_1 heme to allow for efficient electron transfer.[56] Similarly, when in the ISP_C position, the 2Fe2S cluster is too distant from the Q_o site to effectively interact with QH_2. Motion of the water-soluble ISP head domain would allow the 2Fe2S cluster to interact with both ubiquinol at the Q_o site and cyt c_1.[57] It is thought that the lack of mutual simultaneous interactions of 2Fe2S and its two interacting sites (cyt c_1/f and the Q_o pocket) only allows one electron at a time to be sent to the high potential chain. If the ISP head is constrained to the ISP_B position during QH_2 oxidation, SQ_o would be prevented from reducing the high potential chain components, effectively preventing Bypass Reaction 1. Recent independent evidence also supports the involvement of ISP pivoting both in cyt bc_1 and cyt b_6f catalysis.[54,58]

Mylar Orientation of Cytochrome bc_1/b_6f Complexes

One particularly useful technique for probing changes in orientation of the ISP head domain is EPR of oriented samples. Such studies have been performed with a range of materials, including the cyt b_6f complex,[55,59] cyt bc-type complexes from bacteria.[60,61] The complexes are partially ordered onto plastic sheets, and information on the orientation of anisotropic electron paramagnetic resonance (EPR) signals is determined by rotating the sheets in the magnetic field of an EPR instrument.

Partially ordered samples can be obtained by a simple procedure.[62–66] Many complexes become orientated in this procedure because, at low

[54] A. Roberts, M. K. Bowman, and D. M. Kramer, *Biochemistry* In Press (2001).

[55] B. Schoepp, M. Brugna, A. Riedel, W. Nitschke, and D. Kramer, *FEBS Lett.* **450**, 245–250 (1999).

[56] Z. Zhang, L. Huang, V. M. Shulmeister, Y. Chi, K. K. Kim, L. Hung, A. R. Crofts, E. A. Berry, and S. Kim, *Nature* **392**, 677–684 (1998).

[57] A. Crofts, E. Berry, R. Kuras, M. Guergova-Kuras, S. Hong, and N. Ugulava, *in* "Photosynthesis: Mechanisms and Effects" (G. Garab, ed.), Vol. III, pp. 1481–1486. Kluwer Academic Publ., Dordrecht, 1999.

[58] E. Darrouzet, M. Valkova-Valchanova, and F. Daldal, *Biochemistry* **39**, 15475–15483 (2000).

[59] A. G. Roberts, M. K. Bowman, and D. M. Kramer, *Biochemistry* **41**, 4070–4079 (2001).

[60] M. Brugna, S. Rogers, A. Schricker, G. Montoya, M. Kazmeier, W. Nitschke, and I. Sinning, *Proc. Natl. Acad. Sci. USA* **97**, 2069–2074 (2000).

[61] M. Brugna, W. Nitschke, M. Asso, B. Guigliarelli, D. Lemesle-Meunier, and C. Schmidt, *J. Biol. Chem.* **274**, 16766–16772 (1999).

detergent concentrations, membrane proteins tend to spontaneously form two-dimensional arrays, or "sheets" because of association of the hydrophobic, transmembrane portions of the proteins. On dehydration, the sheets become aligned like "leaves" that have fallen from a tree. Each "leaf" lies flat on the ground but the underside of the "leaf" surfaces face both up and down, while the stems of the leaves point in all directions of the compass. Only the angle between the membrane perpendicular and the applied magnetic field is needed to specify the orientation of the sample—that is, they have uniaxial ordering. Freeze-fracture electron micrographs of membranes, such as thylakoids, sometimes show highly ordered 2D crystals of proteins in the membranes. However, these crystalline patches are microscopic in size but can be used to determine low-resolution structures—for example, the cyt $b_6 f$ complex.[67] Even if 2D crystals formed, an EPR sample of roughly 2 mm by 1 cm would be a patchwork of crystalline fragments, each oriented differently from its neighboring patches and giving the same EPR spectrum as a sample with only uniaxial ordering.

To make partially ordered complexes, a stock suspension of >30 μM of the cyt $b_6 f$ or cyt bc_1 complex is dissolved in an appropriate buffer with or without inhibitors (e.g., 30 mM HEPES, pH 7.6), to a final concentration of approximately 300 nM complex. This is centrifuged for 14 hours at 141,000g and 4°. The pellet is then suspended in 25 ml of distilled, deionized water but maintaining original inhibitor concentrations and is centrifuged for 30 min at 141,000g and 4°. The resulting pellet is resuspended with 0 to 200 μl of de-ionized distilled water. This is applied to strips of Mylar (type A, Dupont, Co., Wilmington, DE) with a small paintbrush and allowed to dry for 3 days at 4° under argon above 80% w/v ZnCl$_2$ to maintain humidity. We recommend performing this step in a glass dessicator containing a wide beaker with 400 ml of the ZnCl$_2$ solution, and over this, the Mylar strips. The Mylar strips are clipped to a plastic frame to prevent them from falling. Better orientation is achieved with repeated thin applications over a period of hours or days rather than a single thick

[62] A. Riedel, A. W. Rutherford, G. Hauska, A. Müller, and W. Nitschke, *J. Biol. Chem.* **266,** 17838–17844 (1991).

[63] A. W. Rutherford and P. Setif, *Biochim. Biophys. Acta* **1019,** 128–132 (1990).

[64] J. K. Blasie, M. Erecinska, S. Samuels, and J. S. Leigh, *Biochim. Biophys. Acta* **501,** 33–52 (1978).

[65] B. Guigliarelli, J. Guillaussier, C. More, P. Setif, H. Bottin, and P. Bertrand, *J. Biol. Chem.* **268,** 900–908 (1993).

[66] C. More, V. Belle, M. Asso, A. Fournel, G. Roger, and B. Guigliarelli, *Biospectroscopy* **5,** S3–S18 (1999).

[67] C. Breyton, *J. Biol. Chem.* **275,** 13195–13201 (2000).

application. The resulting partially oriented sample is quite stable and can be stored for months at $-80°$.

Similar procedures can be performed by using intact, isolated membranes.[68,69] The procedure is simpler than with the isolated complexes, requiring no protein purification, but with two drawbacks. First, the concentration of complexes is usually lower, decreasing signal-to-noise ratios. Second, bioenergetic membranes usually contain other species with EPR signals overlapping at least some 2Fe2S transitions, possibly complicating interpretation of results. The procedure is essentially the same as for isolated complexes, except that dilution of detergent is unnecessary. We have found that washing intact membranes with 10 mM EDTA improves resolution, eliminating EPR signals from loosely associated metals.

To chemically reduce the 2Fe2S cluster and cyt c_1/f, while maintaining the cyt b hemes oxidized, the strips are dipped momentarily in a 10 mM solution of sodium ascorbate buffered with 30 mM HEPES (pH 7.6). To obtain oxidized samples, where the cyt b and cyt c_1/f are oxidized, strips are briefly dipped in 5 mM sodium ferricyanide buffered with 30 mM HEPES (pH 7.6). After dipping, strips are allowed to dry under Ar(g) or N_2(g) for several minutes. Several strips are then stacked and placed in an EPR tube and rapidly frozen at $-80°$ for storage.

The orientation of the ISP 2Fe2S complex with respect to the membrane plane can be estimated by taking EPR spectra with the sample rotated to different orientations with respect to the magnetic field. Rotation can be achieved with a commercial goniometer, or simply by turning the sample by hand. In the later case, the orientation is roughly estimated by marking the sample tube with an indicator line, and comparing the position of the line with a fixed protractor disc. Typically, spectra are recorded at 5° to 10° intervals. It is necessary to take measurements over at least 180° rotation to determine the overall orientation of the plastic sheets with respect to the sample tube (see following text).

Interpreting Oriented EPR Spectra

The entire EPR spectral shape varies as the membrane-coated sheets are rotated in the magnetic field of the spectrometer. Most authors estimate changes in orientation qualitatively, and we describe later a procedure for doing this. We describe in another publication[59] an approach for quantitative estimation of orientation, based on rapid simulation of such

[68] U. Liebl, S. Pezennec, A. Riedel, A. Kellner, and W. Nitschke, *J. Biol. Chem.* **1992,** 14068–14072 (1992).
[69] J. Bergström, *FEBS Lett.* **183,** 87–90 (1985).

spectra. We present here a semi-quantitative approach that should allow most authors to estimate both orientation and order parameters.

When the EPR spectrum is dominated by anisotropy of the g-factor or a large hyperfine coupling, a complete analysis can be done by examining only the EPR signal intensity at the three principal g-values rather than the entire spectrum as a function of membrane orientation. The usual practice is to make polar plots of the signal intensity at the three principal values, termed g_x, g_y and g_z. It is important to note that in fully anisotropic materials, such as the Rieske 2Fe2S cluster, each of these signals will be largest when the magnetic field is along some direction relative to the transition. The orientation that maximizes one signal is orthogonal to those of the other two. In the case of the Rieske 2Fe2S cluster, the g_x, g_y and g_z transitions are observed at about $g = 2.01$, 1.89 and 1.78, though these vary depending on conditions.

The particular shape of a polar plot for a specific axis is a result of several factors, including the flatness of the membranes; the extent of protein alignment within the membrane; conformational heterogeneity; 2D ordering of proteins within the membrane; and sample geometry in the spectrometer. To completely analyze the polar plots, each of these considerations would have to be taken into account.

Effects of Orientation of the EPR Transitions

The orientation of the transition with respect to the membrane plane will predominantly determine the angular dependence of the signal. The starting model is that in every complex there is some particular direction pointing directly out of the perfectly flat and perfectly ordered membrane. If the membrane-containing sheets were oriented perpendicular to the magnetic field, the axis of each protein would be parallel or anti-parallel to the magnetic field and a single-crystal-like EPR spectrum would be seen. In fact, neither membranes nor the protein will be perfectly oriented, so that some broadening of the single-crystal-like EPR spectrum occurs. Because the membranes are oriented both "up" and "down," two maxima will appear in the polar plots, one at 90° and another at 270°, where the membranes are orthogonal to the magnetic field.

A transition oriented within the membrane plane will exhibit its largest EPR signal when the membrane is parallel to the magnetic field. But because the sample is rotationally disordered around the normal to the membrane plane (see previous text), all the transitions can never be parallel to the field at the same time and the maximal amplitude of the signal will be smaller than for a transition oriented perpendicular to the membrane. Again, two signal maxima will be observed, one at 0° and another at

180°, the two angles where the membranes are parallel to the magnetic field. In general, transitions closer to perpendicular to the membrane will give larger maximal signals because more of these transitions can be simultaneously aligned parallel to the magnetic field.

The situation is more complex when a transition has an intermediate orientation (that is, between parallel and normal to the membrane plane), say at angle α. In these cases, rotational disorder will allow the transition to be rotated about an axis perpendicular to the membrane plane, but always keeping the same angle, α, with respect to the plane. Essentially, the transition will appear in the EPR like a set of two mirror-image cones, one on each side of the membrane. As this type of transition is rotated in the field, a set of four maxima will appear. If mosaic spread is minimal (but see the following), the maxima will appear at $+/- \alpha$ and $\alpha +/-180°$. (This is a general solution for transitions with low mosaic spread, since for those found at 0° and 90°, signals will be found at 0° and 180°, and 90° and 270°, respectively.)

A further complicating factor when interpreting oriented EPR data, but one that may yield important information, is that the transitions are not perfectly ordered. The "disorder" may have several sources.

Membrane Disorder

The membranes as they lie on the sheets may not be perfectly flat and smooth. This will result in "mountains" and "valleys" disordering the sample.

The hydrophobic and hydrophilic regions on the surface of membrane proteins interact with the solvent and with the hydrophobic lipids of the membrane to embed the protein in the membrane or to anchor it to the surface. The interplay of hydrophobic and hydrophilic interactions may uniquely orient each protein in the sample relative to the membrane surface. More likely, some amount of wobble will occur that may be influenced by the lipid composition of the membranes or interactions between stacked membranes. In the case of proteins with a soluble domain tethered to a hydrophobic tail inserted into the membrane, there may be significant flexibility between these two domains that enables a wide range of orientations of the soluble domain relative to the membrane surface.

Conformational Heterogeneity

Some proteins have more than one conformation, and the EPR-active center may be in more than one orientation relative to the membrane, a notable example being the Rieske head domain in the cyt bc_1 complex.

When sample preparation conditions can be found so that only a single conformation is present in the sample, EPR of oriented membranes can be a powerful tool to characterize the conformational change at the EPR-active site.

Quantitative interpretation of oriented membrane EPR becomes quite difficult if there are two conformations present that have similar EPR parameters. Calculated polar plots based on a single conformation will often show systematic deviations from the experimental data, a wide range of orientation giving equally good fits or require unreasonably large amounts of mosaic spread or disorder. Including two conformations in the calculations more than doubles the number of variable parameters and although better fits are obtained, strong correlations between parameters make it difficult to achieve a unique fit. Even when one of the conformations has been independently characterized, a unique fit for the orientation of the other conformation can be problematic because the orientation is often highly correlated with the relative amounts of the two conformations.

Sample Geometry

The physical sample itself is the final consideration that must be considered in the distribution function. The exercise of some care in assembly of the sample and in the measurement can largely eliminate this factor. It is important that the sheets supporting the membranes are stacked parallel to each other in the sample; that the sheets are flat and not bent or wrinkled; and that the sheets are rotated about an axis that is both parallel to the sheets and perpendicular to the magnetic field of the spectrometer.

Mosaic Spread

The apparent disorder of the system—that is, a function of all the previously-mentioned effects—is often modeled by a single, simplified "mosaic spread" term, which is simply defined as a Gaussian or normal distribution of orientations with a standard deviation of θ around the average orientation. The standard deviation θ will have two effects. First it will obviously broaden the angular distribution of the signal. Second, since transitions oriented more perpendicular to the membrane plane are accentuated, large mosaic spread will tend to skew the maximal observed signal to orientations nearer to the membrane normal.

As a note of caution, the simple mosaic spread parameter will clearly not account for all situations. For instance, an inhomogeneous system containing a well-ordered and a disordered fraction will appear as a weighted sum of anisotropic and isotropic signals.

A Simple Method for Estimating Changes in the Orientation and Ordering of Membrane-Bound Anisotropic EPR Signals

In this section, we describe a simple approach for analyzing EPR of partially ordered samples for information about the orientation of a redox center. We use the Rieske 2Fe2S center as an example but note that the technique should be generally applicable to anisotropic signals. Our approach here is to present calibration data in a form that can be used by non-experts without resorting to complex simulations.

Figures 2A and 2B show simulated polar plots of EPR intensity for transitions at a range of α, as a function of the angle between the Mylar plane and the magnetic field. The mosaic spread (θ) was set to a moderate $18°$, but even so, there are pronounced effects on the breadth of the orientation-dependence and on the deviation between α and orientation of the maximal signal. Figures 3A through 3C show simulated polar plots of EPR intensity as a function of angle at various mosaic spreads. With a narrow mosaic spread (e.g., $18°$), the polar plots appear very anisotropic. By broadening mosaic spread from $\theta = 18°$ to $360°$, the lobes of the polar plots become progressively more isotropic (gray lines).

Step-by-Step Procedure for Estimating θ and α

In our procedure, α and θ are estimated by a graphical procedure (see Fig. 4). A line segment (AB) is drawn from a maximal signal value to the origin of the polar plot. A second line (CD) is drawn perpendicular to and bisecting AB. A third line (AD) is drawn from the origin (i.e., A) to the intersection between CD and the intensity profile occurring closest to the $0°$ axis. The angle (θ') between AD and AB is taken and compared to the data in Fig. 5 to estimate mosaic spread—that is, θ. As can be seen in Fig. 5, the relationship between θ and θ' depends only marginally on α, and thus for these purposes the average relationship should be sufficiently accurate.

Next the angle between AB and the $0°$ axis (α') is measured and a correlation function with the appropriate value of θ is used to estimate α. A series of relationship relating α to α' at a range of θ values is shown in Fig. 6. The large deviations as α approaches 0 and $90°$ are due to an inability to distinguish between changes in θ and α when the EPR transition is nearly parallel or perpendicular to the membrane plane. In these cases, one could deduce the orientation from data on the other two transitions in the g tensor, keeping in mind that the three transitions in an anisotropic signal are mutually orthogonal. Alternatively, a more complete simulation of the data[59] may allow resolution of orientation.

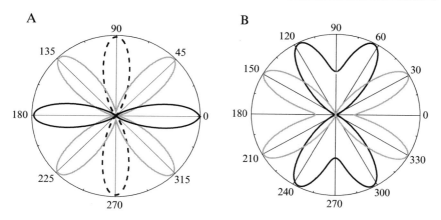

Fɪɢ. 2. Simulated angular dependence of an EPR signal in partially ordered samples. Simulation of the normalized EPR intensity as a function of angle (α) at (A) $0°$ (black, solid), $90°$ (black, dashed) and $45°$ (gray) and (B) $60°$ (black) and $30°$ (gray). The mosaic spread of these simulations was $18°$.

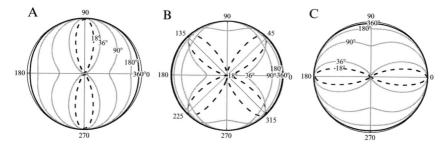

Fɪɢ. 3. Simulation of the effects of mosaic spread of the normalized EPR intensity as a function of angle. Figure 3A, 3B, and 3C are for EPR transitions with average orientation at $90°$, $45°$, $0°$ with respect to the membrane plane. The extents of mosaic spread are indicated shown on each trace.

Baseline Subtraction

Great care must be taken with the baselines in the original spectra when obtaining data for polar plots. It is very dangerous to assume that any orientation has no contribution from a particular transition, as this may artifactually decrease the apparent mosaic spread.

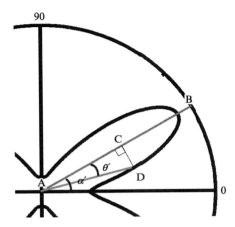

FIG. 4. Illustration of technique for estimating the orientation and mosaic spread of an EPR transition in a partially ordered sample. As described in the text, the angles α' and θ' are estimated from a polar plot of EPR intensity as a function of the angle of the membrane plane with respect to the magnetic field.

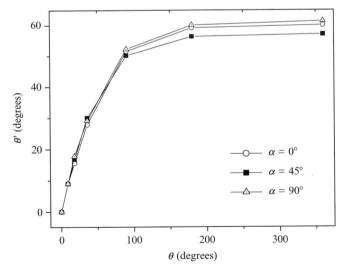

FIG. 5. Simulated relationships between estimated angle θ' and the mosaic spread (θ). See text for description.

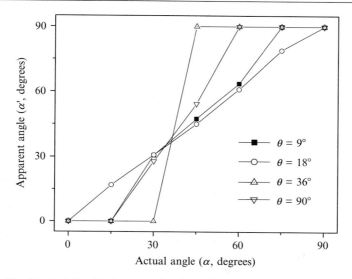

FIG. 6. Simulated relationship between α', estimated from polar plots of EPR intensity, and the angular orientation of the EPR transition (α) at several mosaic spread values. Slope at 9° is 1.03; slope at 18° is 1.42; slope at 36° is 1.97; slope at 90° is 6.

The Interpretation of θ and α

In general, a larger value of θ implies a lower ordering of the transition. As stated above, increases in disorder may have several sources. A major research focus for the cyt bc_1 and b_6f complexes is the observed multiple conformations of the ISP. In similar biochemical systems, multiple but distinct conformations have been clearly resolved using oriented EPR.[70] However, more commonly, the conformations will not be easily separated and will appear simply as a large θ. In this case, contributions from conformational changes do not have to affect each EPR transition equally. If, for instance, a conformation occurs predominantly about one principle transitional axis, the θ values for the other two axes should be affected to a larger extent. In these cases, without detailed simulations or supporting data or analysis, the origins of θ will be ambiguous.

Orientation of the g-Factor Axis With the Molecular Axis

The EPR measurements determine how the g-factor axes of the paramagnetic center are oriented with respect to the membrane, but one usually wants to know how the protein is oriented in the membrane. That

[70] B. Schoepp-Cothenet, M. Schutz, F. Baymann, M. Brugna, W. Nitschke, H. Myllykallio, and C. Schmidt, *FEBS Lett.* **487,** 372–376 (2001).

requires knowledge of how the g-factor axes are oriented relative to the protein or to the paramagnetic center. For some paramagnetic centers that occur in proteins or that are used as probes or labels of proteins, there is some information about the orientation of one of the g-factor axes. In almost all heme proteins, one g-factor axis is close to perpendicular to the heme plane. Likewise, the orientation of one g-factor and hyperfine axis is well established for protein probes such as VO(II) or nitroxide radicals. It is through that the lack of mutual simultaneous interactions of 2Fe2S at its two interacting sites (cyt c1/f and the Qo pocket) only allows one electron at a time to be sent to the high potential chain. In proteins, the paramagnetic centers observed by EPR rarely lie on rigorous symmetry axes, and thus symmetry constraints on the orientations of g-factors axes that are so useful in simple inorganic systems are absent. Consequently the g-factor axes often deviate strongly from the apparent molecular axes of the paramagnetic center and must be determined experimentally by single-crystal EPR measurements of that protein or a fragment. We have determined the g-factor axes of the Rieske 2Fe2S cluster, Fig. 7, in the

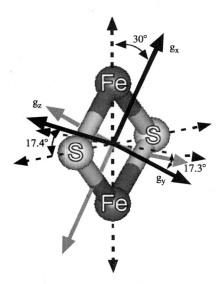

FIG. 7. Orientation of the g_x, g_y, and g_z axes with respect to the 2Fe2S molecular axis. Dotted arrows indicate the molecular axes—that is, they are drawn through the S-S and Fe-Fe axes, and through the plane of the cluster. The solid gray and black lines show the EPR g-tensor axes with respect to the molecular axes. The g_x transition is shifted $30°$ from the axis formed by the irons. The g_y transition axis is shifted $17.3°$ away from the through plane axis. The g_z transition is shifted $17.4°$ from the axis formed by the sulfurs.

bovine cyt bc_1 complex and found that the g-factor axes are skewed relative to the obvious Fe-Fe and S-S directions in the cluster (M. K. Bowman, A. G. Roberts, and D. M. Kramer, Biochemistry, in press.). Although the protein environment around the cluster has a small influence on the precise orientation of the g-factor axes, we were able to show that the motion of the Rieske subunit in the membrane is consistent with conformations observed in diffraction structures. Thus, to a first approximation, the relationship between g-tensor and molecular axes in Fig. 7 can be used to estimate the orientation of the ISP with respect to the membrane, keeping in mind the limitations imposed by the uniaxial ordering of the system (see previous text).

Acknowledgments

Research in the authors' groups was supported by U.S. Department of Energy Grant DE-FG03-98ER20299 (D.M.K.) and by NIGMS Grant GM61904 (M.K.B.). The authors wish to thank Drs. W. Nitsckke, J. Cooley, E. Berry, I. Forquer, and F. Daldal for important discussions and data.

[3]   Targeting Coenzyme Q Derivatives to Mitochondria

By Robin A. J. Smith, Geoffrey F. Kelso,
Andrew M. James, and Michael P. Murphy

Introduction

Coenzyme Q is an essential electron carrier and an important antioxidant in the mitochondrial inner membrane. Consequently, manipulating coenzyme Q content within mitochondria may help in understanding its functions and exploring its therapeutic potential. One way to increase the mitochondrial concentration of coenzyme Q is to administer it to isolated mitochondria, cells, or organisms. However, the hydrophobicity of native coenzyme Q and its consequent low solubility and bioavailability limit this approach.[1,2] The hydrophobicity of coenzyme Q is due to the long isoprenoid chain at ring position 6 (in humans this usually comprises 10 isoprenoid units) (Fig. 1, Table I). Less hydrophobic but still active derivatives

[1] M. Bentinger, G. Dallner, T. Chojnacki, and E. Swiezewska, *Free Rad. Biol. Med.* **34,** 563 (2003).
[2] M. P. Murphy, *Expert Opinion in Biological Therapy* **1,** 753 (2001).

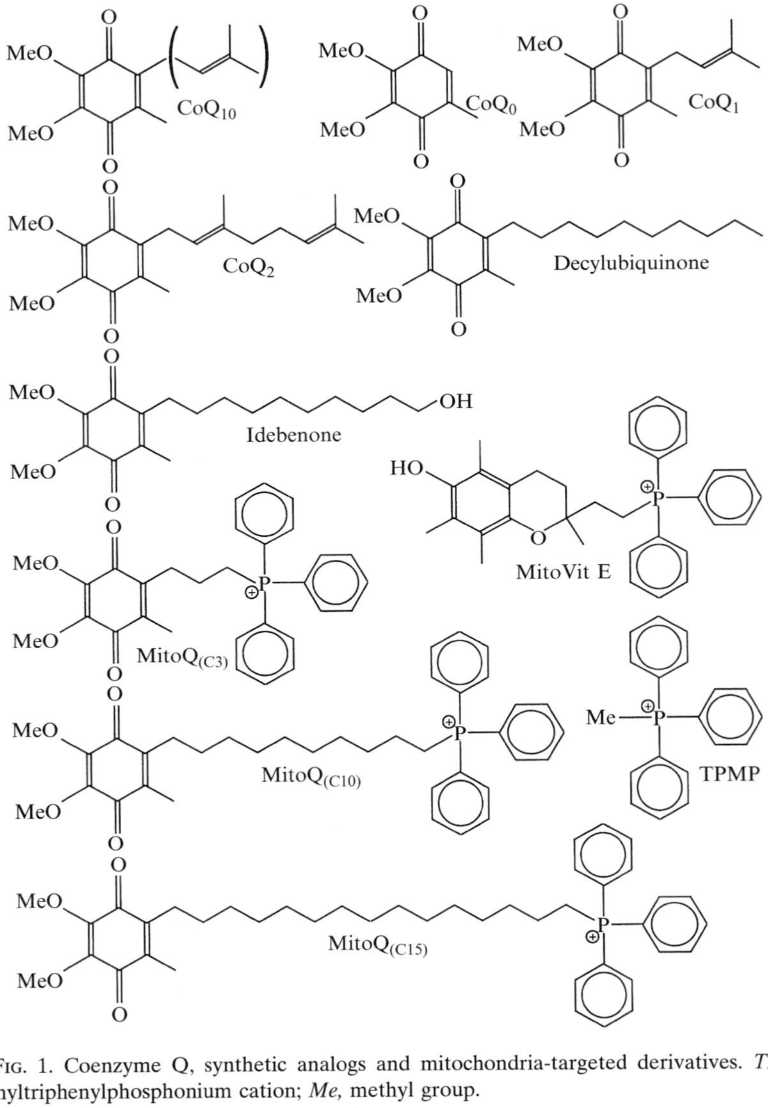

FIG. 1. Coenzyme Q, synthetic analogs and mitochondria-targeted derivatives. *TPMP*, methyltriphenylphosphonium cation; *Me*, methyl group.

of coenzyme Q can be made by replacing this carbon chain with more water soluble moieties, and these are widely used as artificial electron donors and acceptors in enzyme assays and for supplementing endogenous coenzyme Q levels in mitochondria, cells and organisms. Typically these

TABLE I

PARTITION COEFFICIENTS OF COENZYME Q, TARGETED DERIVATIVES AND RELATED COMPOUNDS

Compound	Partition Coefficient
Methyltriphenylphosphonium (TPMP)	[a]0.35 ± 0.02
MitoVit E	[b]7.4 ± 1.6
4-Bromobutyltriphenylphosphonium	[b]3.83 ± 0.22
4-Iodobutyltriphenylphosphonium	[c]4.0 ± 0.4
MitoQ$_{(C10)}$	[a]160 ± 9
MitoQ$_{(C5)}$	[c]2.8 ± 0.3
α-Tocopherol	[b]27.4 ± 1.9
Bromodecylubiquinone	[d]310 ± 60
Idebenone	[d]3.1×10^3
Decylubiquinone	[d]3.1×10^5
Coenzyme Q$_0$	[d]1.33
Coenzyme Q$_1$	[d]409
Coenzyme Q$_2$	[d]4.44×10^4
Ubiquinone (Coenzyme Q$_{10}$)	[d]1.82×10^{20}
Ubiquinol	[d]4.53×10^{20}
Decylubiquinol	[d]7.91×10^5
Idebenol	[d]7.82×10^3

Data[a–c] are octan-1-ol/phosphate buffered saline partition coefficients determined at 25° or 37°, or octanol/water partition coefficients.

[a] G. F. Kelso, C. M. Porteous, C. V. Coulter, G. Hughes, W. K. Porteous, E. C. Ledgerwood, R. A. J. Smith, and M. P. Murphy, *J. Biol. Chem.* **276**, 4588 (2001).

[b] R. A. J. Smith, C. M. Porteous, C. V. Coulter, and M. P. Murphy, *Eur. J. Biochem.* **263**, 709 (1999).

[c] R. A. J. Smith, C. M. Porteous, A. M. Gane, and M. P. Murphy, *Proc. Natl. Acad. Sci. USA* **100**, 5407 (2003).

[d] calculated by using Advanced Chemistry Development (ACD) Software Solaris V4.67 as described in M. L. Jauslin, T. Wirth, T. Meier, and F. Schoumacher, *Hum. Mol. Genet.* **11**, 3055 (2002).

derivatives have short chains of one or two isoprenoid units at position 6 (e.g., coenzyme Q$_1$ or coenzyme Q$_2$), a simple aliphatic side chain (e.g., decylubiquinone), or a modified alkyl side chain such as 10-hydroxydecyl in idebenone (Fig. 1). Even so, these compounds are still hydrophobic (Table I) and when administered to biological systems will partition non-specifically into all phospholipid bilayers, with only a small proportion ending up in mitochondria.

To minimize the cytoplasmic interactions of administered coenzyme Q and focus on its potential mitochondrial functions, we developed a strategy to direct coenzyme Q derivatives to mitochondria within cells and organisms.[3-5] For this, coenzyme Q analogs were covalently coupled to an alkyltriphenylphosphonium cation, yielding a class of molecules called MitoQ (Fig. 1).[6,7] This approach takes advantage of the large membrane potential across the mitochondrial inner membrane (-150 to -170 mV, negative inside), which drives the accumulation of lipophilic cations into the mitochondrial matrix (Fig. 2). Lipophilic cations such as alkyltriphenyl-phosphonium pass easily through phospholipid bilayers by a non–carrier-mediated process.[3] This is because the phosphonium in these lipophilic cations is surrounded by three hydrophobic phenyl groups and therefore has a large effective ionic radius (~4 Å) that lowers their activation energy barrier for movement through lipid bilayers.[5] The Nernst equation indicates that these cations will accumulate about 10-fold for every 61.5 mV of membrane potential, thus accumulating several-hundred fold within isolated mitochondria (Fig. 2). Cells will also accumulate lipophilic cations into the cytoplasm about 5 to 10-fold driven by the plasma membrane potential (-30 to -60 mV, negative inside), from where they will be further concentrated several-hundred fold inside mitochondria (Fig. 2). This approach to targeting enables mitochondrial antioxidant content to be manipulated independently of that in the rest of the cell, thus allowing mitochondrial oxidative damage to be blocked selectively. Here we outline how to synthesize and handle mitochondria-targeted coenzyme Q derivatives, and discuss experimental procedures for using them to study mitochondrial function.

Synthesis and Handling of Mitochondria-Targeted Coenzyme Q Derivatives

Synthetic Strategies

Mitochondria-targeted coenzyme Q derivatives with different length alkyl chains are referred to generically as MitoQ, with the number of aliphatic carbons in the linker chain indicated in a subscript (Fig. 1). The redox state of MitoQ is undefined; mitoquinol or mitoquinone are used when discussing a particular redox form. A straightforward procedure to

[3] M. P. Murphy, *Trends in Biotechnology* **15**, 326 (1997).
[4] M. P. Murphy and R. A. J. Smith, *Adv. Drug Deliv. Rev.* **41**, 235 (2000).
[5] R. F. Flewelling and W. L. Hubbell, *Biophys. J.* **49**, 531 (1986).
[6] G. F. Kelso, C. M. Porteous, C. V. Coulter, G. Hughes, W. K. Porteous, E. C. Ledgerwood, R. A. J. Smith, and M. P. Murphy, *J. Biol. Chem.* **276**, 4588 (2001).
[7] G. F. Kelso, C. M. Porteous, G. Hughes, E. C. Ledgerwood, A. M. Gane, R. A. Smith, and M. P. Murphy, *Ann. N.Y. Acad. Sci.* **959**, 263 (2002).

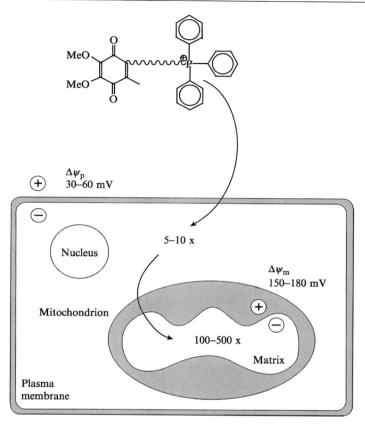

FIG. 2. Uptake of mitochondria-targeted Coenzyme Q derivatives by mitochondria within cells. This schematic shows the uptake of MitoQ of an unspecified alkyl linker length into cells driven by the plasma membrane potential ($\Delta\psi_p$) before being further accumulated into mitochondria driven by the mitochondrial membrane potential ($\Delta\psi_m$).

synthesize MitoQ$_{(C10)}$ is to start with commercially available idebenone and add on a triphenylphosphine (Ph$_3$P) by a displacement reaction to form the alkyltriphenylphosphonium salt (Fig. 3). It is preferable to form these salts from Ph$_3$P or triphenylphosphonium hydrogen bromide (Ph$_3$P.HBr) by using melts at 70 to 100° without additional solvent. Triphenylphosphonium cations containing hydrophobic alkyl groups are troublesome to purify because of their hygroscopicity, tendency to form viscous oils on exposure to air, and reluctance to crystallize. A simple work-up that utilizes the ready solubility of alkyltriphenylphosphonium cations in dichloromethane and their insolubility in diethyl ether that is useful for

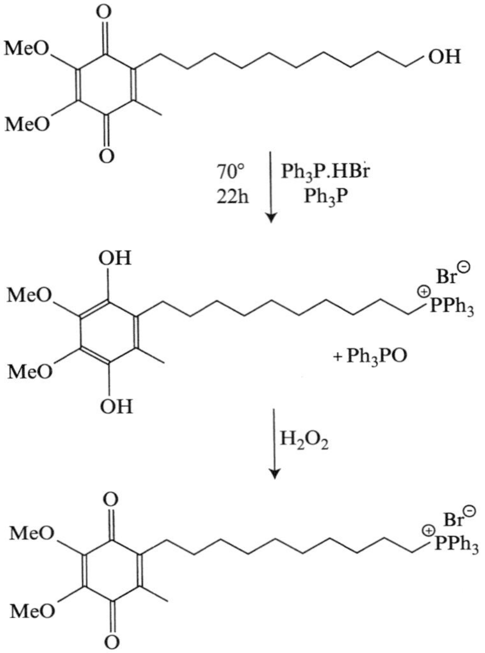

FIG. 3. Synthesis of MitoQ$_{(C10)}$ from Idebenone. Idebenone (0.678g, 2 mmol), Ph$_3$P (0.524g, 2 mmol) and Ph$_3$P.HBr (0.686g, 2 mmol) (prepared from Ph$_3$P and aqueous HBr as described[32]) were placed in a KIMAX tube flushed with nitrogen, then sealed and heated at 70° with magnetic stirring. After 22 hours the mixture was cooled to give a black, glass-like solid, which was dissolved in dichloromethane (4 ml) then transferred to a round-bottomed flask (25 ml) and the solvent evaporated *in vacuo* to give a dark red oil (2.446g) containing mitoquinol and triphenylphosphine oxide (Ph$_3$PO) along with lesser amounts of TPMP and Ph$_3$P. This was mixed with ethyl acetate (20 ml) and held at 70° for 5 minutes, then cooled, and the solvents decanted. This process was repeated twice more to remove Ph$_3$PO. The solid (~1.7g) was then dried under vacuum, dissolved in methanol (7 ml) and stirred with 30% aqueous hydrogen peroxide (7 ml) and pyridine (0.13 ml) at room temperature for 21 hours. The organic solvent was then evaporated *in vacuo* and the crude mixture was dissolved in dichloromethane (16 ml) and extracted with 2% aqueous HBr (4 × 10 ml). The organic layer was dried over anhydrous magnesium sulfate and applied directly to a silica gel column (12g silica gel). The column was eluted with dichloromethane (10 ml), then 5% rectified spirits in dichloromethane (100 ml) and 10% rectified spirits in dichloromethane (300 ml). Evaporation of the 5% rectified spirits in dichloromethane gave pure mitoquinone (1.06g) as an orange foam. *Ph*, phenyl. *Ph$_3$P*, triphenylphosphine. *Ph$_3$PO*, triphenylphosphine oxide.

the purification of MitoQ and related compounds is described in the legend to Fig. 3. Remaining contaminating organic material can be removed by column chromatography on silica gel by using an organic solvent containing 95% ethanol/water (rectified spirit). For final crystallization we have found it best to selectively dissolve the material in dichloromethane or chloroform then precipitate the product with ether or hexanes.

Aliphatic chains other than 10 carbons require more complicated synthetic procedures. For these it is convenient to start from Coenzyme Q_0, and some of the strategies that we have used to synthesize MitoQ with short (3) and long (15) carbon aliphatic linkers are shown in Fig. 4.

MitoQ is usually obtained in the oxidized, mitoquinone form as the bromide salt and is stable indefinitely when stored at $-20°$ in the dark under an inert gas. Further experiments have shown that pure $MitoQ_{(C10)}$ can survive in air for up to 20 days at room temperature and that it slowly decomposes over 12 to 24 hours in air above $50°$. Therefore all manipulations should be carried out close to room temperature with minimum exposure to air and light. Any water that is taken up because of the compound's hygroscopicity can be removed by lyophilization with protection from light. Stock solutions are generally prepared in ethanol (10 to 100%) or DMSO and are stored at $-20°$ in the dark under an inert gas. Dilute ($\sim$500 μM) aqueous solutions are stable for up to 14 days at room temperature in the dark.

Analytical Procedures

^{31}P NMR gives useful diagnostic shifts for analyzing the progression of MitoQ syntheses and purifications (Fig. 5A). The resonance peaks for Ph₃P, methyltriphenylphosphonium (TPMP), MitoQ and Ph₃PO are characteristic and unambiguous (Fig. 5A). Another method to assess the production of an alkyltriphenylphosphonium salt from Ph₃P is the UV absorbance change from a broad absorption peak around 260 nm to a distinctive spectrum with an absorption peak at 268 nm ($\varepsilon \sim 3,000$ M^{-1} cm^{-1}) and two local maxima at 263 nm and 275 nm (Fig. 5B). This change also occurs during MitoQ syntheses but is complicated by the absorbance in this region of ubiquinone/ubiquinol and by the absorbance shift that occurs on redox changes (see Fig. 6). TLC is also useful, particularly in following syntheses and analyzing the purity of radiolabeled compounds. While alkyltriphenylphosphonium cations are difficult to separate cleanly by TLC, the use of 95% ethanol/water in conjunction with dichloromethane on silica gel is reasonably effective (Fig. 5C). For definitive product identification ^{1}H-NMR is of course essential, but the results obtained by elemental

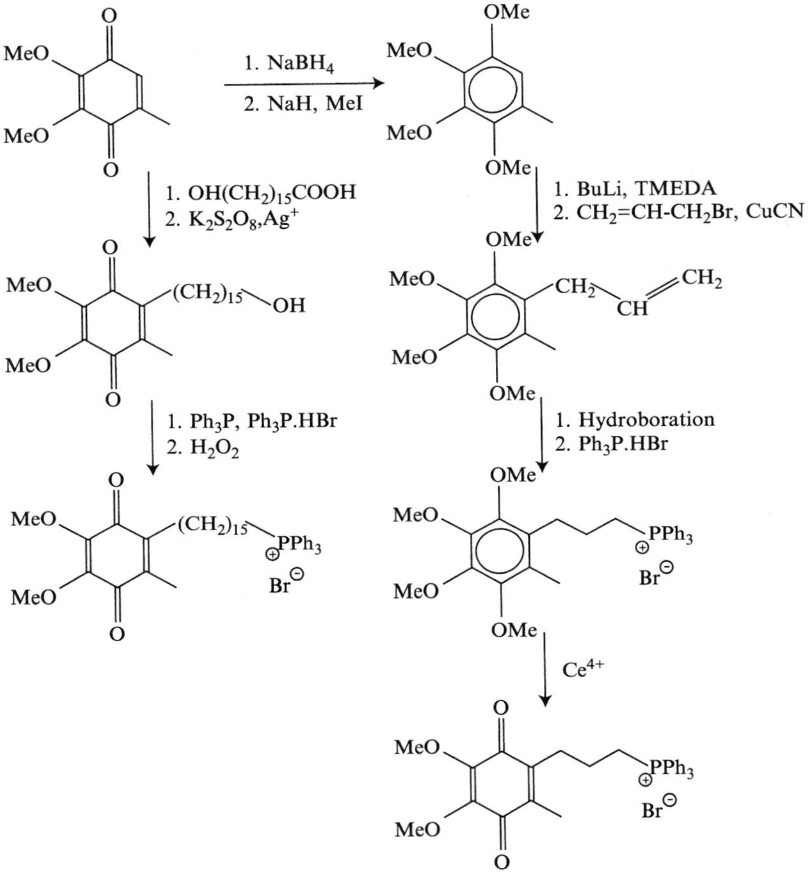

1. NaBH$_4$
2. NaH, MeI

1. OH(CH$_2$)$_{15}$COOH
2. K$_2$S$_2$O$_8$,Ag$^+$

1. BuLi, TMEDA
2. CH$_2$=CH-CH$_2$Br, CuCN

1. Ph$_3$P, Ph$_3$P.HBr
2. H$_2$O$_2$

1. Hydroboration
2. Ph$_3$P.HBr

Ce^{4+}

FIG. 4. Syntheses of MitoQ$_{(C3)}$ and MitoQ$_{(C15)}$. Outline strategies to synthesise MitoQ$_{(C3)}$ or MitoQ$_{(C15)}$ from Coenzyme Q$_0$ are shown. *TMEDA*, tetramethylethylenediamine. *BuLi*, butyllithium.

analysis tend to be affected by solvent incorporation. Therefore we routinely obtain high-resolution electrospray mass spectra to confirm product assignment. Electrospray mass spectrometry is also a sensitive procedure for detecting alkyltriphenylphosphonium salts in biological distribution studies, and when coupled with reverse-phase liquid chromatography with acetonitrile/1% aqueous formic acid as eluant, sensitivities in the order of 10 pg/ml can be obtained.

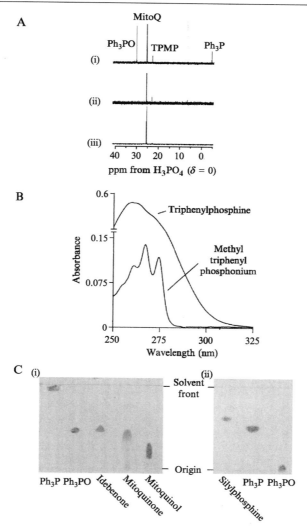

FIG. 5. Analytical procedures for monitoring the synthesis of alkyltriphenylphosphonium cations. (A) ^{31}P NMR (CDCl$_3$) spectra of: (i) crude reaction product from MitoQ$_{(C10)}$ synthesis; (ii) after washing with ethyl acetate; (iii) after oxidation and chromatography. The resonance peaks for Ph$_3$P, TPMP, MitoQ and Ph$_3$PO are characteristic and unambiguous at -5, 23, 25 and 30 ppm, respectively. (B) UV spectra of triphenylphosphine and methyltriphenylphosphonium (50 μM of each in ethanol). Care must be taken as Ph$_3$P oxidizes to Ph$_3$PO in aerobic ethanolic solutions giving a three-peaked absorption spectrum qualitatively similar to those of alkyltriphenylphosphonium salts. (C) Thin-layer chromatography of MitoQ$_{(C10)}$ and related compounds on silica gel. The solvent in (i) is 95% dichloromethane/5% aqueous ethanol (95%) and in (ii) is 75% hexane/25% dichloromethane. Visualization was under UV (254 nm) light.

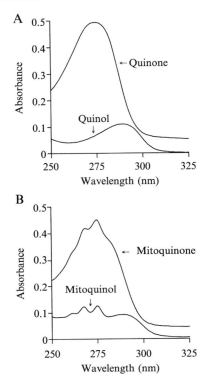

FIG. 6. Spectra of the different redox forms of Coenzyme Q_2 and of MitoQ$_{(C10)}$. (A) The UV absorption spectrum of reduced (ubiquinol) and oxidized (ubiquinone) forms of Coenzyme Q_2 (~40 μM of each in ethanol). (B) The UV absorption spectrum of reduced (mitoquinol) and oxidized (mitoquinone) forms of MitoQ$_{(C10)}$ (~40 μM of each in ethanol).

Incorporation of Radioactive and Stable Isotopes into Mitochondria-Targeted Coenzyme Q Derivatives

To measure the uptake by mitochondria of MitoQ and its distribution within cells or organisms, [^{3}H]-radiolabeled compounds are required.[6,8] The incorporation of stable isotopes, typically [^{2}H], into MitoQ is also essential for internal standards to spike tissue homogenates and thus enable quantitation of the distribution of MitoQ within the body by liquid chromatography-mass spectrometry (LC-MS). It is best to introduce the isotope at the last step of the synthesis to save costs, prevent isotope losses and

[8] R. A. J. Smith, C. M. Porteous, A. M. Gane, and M. P. Murphy, *Proc. Natl. Acad. Sci. USA* **100,** 5407 (2003).

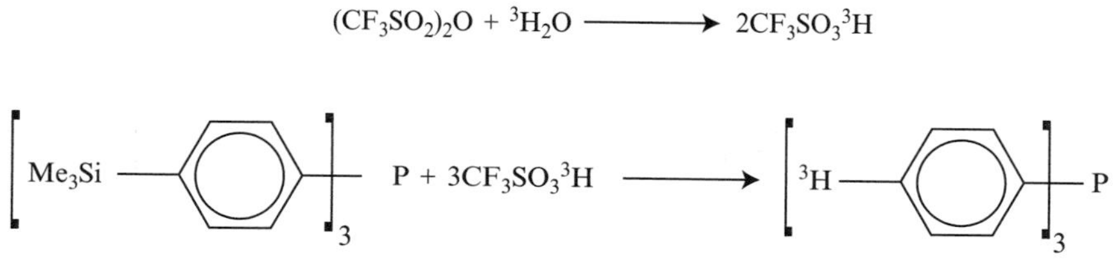

$$(CF_3SO_2)_2O + {}^3H_2O \longrightarrow 2CF_3SO_3{}^3H$$

$$\left[Me_3Si \rule{1cm}{0pt} \bigcirc \rule{1cm}{0pt} \right]_3 P + 3CF_3SO_3{}^3H \longrightarrow \left[{}^3H \rule{1cm}{0pt} \bigcirc \rule{1cm}{0pt} \right]_3 P$$

FIG. 7. Incorporation of hydrogen isotopes into triphenylphosphine. Preparation of [³H]triphenylphosphine ([³H]Ph₃P) is described. Trifluoromethanesulfonic anhydride (1.3 ml, 7.6 mmol) was added to a pre-dried KIMAX tube containing ${}^3H_2O$ (200 μL, 11.1 mmol, 1 Ci) and the mixture was purged with nitrogen, then sealed and stirred for 2 hours at 70 to 80°. The resultant [³H] trifluoromethanesulfonic acid was transferred via Pasteur pipette to a pre-dried KIMAX tube containing tris(p-trimethylsilylphenyl)phosphine (1 g, 2.09 mmol) (prepared as described[33]), purged with nitrogen (caution: fumes), sealed and stirred for 2 hours at 80°. The resultant mixture was basified with saturated sodium bicarbonate and then extracted with dichloromethane (2 × 10 ml). The organic extracts were combined, dried over anhydrous sodium sulfate and filtered. The organic solvent was removed under a stream of nitrogen, leaving a pale yellow solid (0.358 g), which was chromatographed on silica gel (2 g) and eluted with dichloromethane (10 × 4 ml). The fractions were analyzed by TLC (silica gel, 3:1 hexane/dichloromethane, UV visualization) and those containing PPh₃ were combined. The solvent was removed under a stream of nitrogen, giving [³H]PPh₃ as a white solid (0.301 mg, 55% yield). Radiopurity was estimated by TLC (silica gel, 3:1 hexane/dichloromethane, UV visualisation) separation followed by liquid scintillation counting. Most of the radiation (91%) was found in the PPh₃ region ($R_f = 0.21 - 0.61$) of the plate. The remainder of the radiation was found in the triphenylphosphine oxide (PPh₃O) region ($R_f = 0 - 0.20$). The specific activity of the sample (21 Ci mol^{-1}) was determined by assuming 90% of the total activity was due to PPh₃ and the remainder was from PPh₃O. Repetition of this reaction by using ${}^2H_2O$ gave [²H]-triphenylphosphine in ∼90% yield and isotopic purity of 75% [²H]₃ and 20% [²H]₂ as determined by electrospray mass spectroscopy.

minimize radioactive contamination. It is also useful if the same isotopically enriched compound can be used to make a range of products. With mitochondria-targeted compounds both of these are possible by using [³H]- or [²H]-Ph₃P in the final step of the synthesis (Figs. 3 and 4). Isotopically enriched Ph₃P can be prepared easily from commercially available ³H₂O or ²H₂O by exploiting a protiodesilylation reaction (Fig. 7). This procedure yields pure compounds that have relatively high levels of incorporation of [²H] or [³H] with minimal manipulation of labeled compounds and is easily adapted to other mitochondria-targeted compounds.

Modifying and Measuring Coenzyme Q Redox State

Generally, MitoQ is synthesized in the oxidized, mitoquinone form, although stocks can be a variable mixture of the two redox states. Ubiquinone has an intense absorption peak around 275 nm, while the reduced form, ubiquinol, has a weaker absorption peak around 290 nm (Fig. 6A). The spectra of mitoquinone and mitoquinol have similar features, but the three local absorption maxima of the alkyltriphenylphosphonium cation are superimposed (Fig. 6B).[6] To prepare a mitoquinol stock solution, a small amount (~0.5 ml) of a MitoQ solution (~10 mM) is prepared in ethanol. To this is added a few grains (~100 μg) of solid sodium borohydride followed by 0.5 ml H_2O, instantaneously reducing the yellow mitoquinone to the almost colorless mitoquinol. This is rapidly extracted by vortexing with diethyl ether:dichloromethane (2:1 v/v, 1.5 ml), followed by centrifugation. The upper organic phase is collected, vortexed with 2 M NaBr (0.5 ml), separated by centrifugation and evaporated to dryness under nitrogen. The whole process should be carried out quickly to avoid air oxidation, and the mitoquinol residue is stored under inert gas at $-20°$ in the dark. Solid mitoquinol, or stock solutions in 1 ml acidified ethanol, can be stored under argon at $-20°$ and remain colorless and appear relatively stable for several months, as autoxidation requires an initial deprotonation. Large-scale preparations of mitoquinol are best carried out by agitating the solutions under argon gas flow.

While MitoQ is usually present in the mitoquinone form, traces of mitoquinol may occur. Completely oxidized solutions can be prepared by adding 0.5 ml 50 mM KP$_i$ buffer (pH 8.5 to 9; KOH) to 0.5 ml of 10 mM MitoQ in ethanol and incubating at $25°$ until the UV spectrum shows complete oxidation (~10 minutes). The resulting mitoquinone solution can be stored under argon at $-20°$. Note, however, that MitoQ solutions are unstable during air exposure under alkaline conditions, forming a colored product. Therefore, great care must be taken to minimize the exposure of these compounds to high pH.

Changes in the redox state of MitoQ on interaction with mitochondria can be analyzed by measuring the rates of reduction or oxidation of the compounds spectrophotometrically, typically by following the 275 nm ubiquinone peak, in conjunction with appropriate respiratory substrates and inhibitors.[6] Such experiments indicated that MitoQ$_{(C10)}$ could both take up electrons from the vicinity of Complexes I and II and pass them to Complex III. However, it was unclear whether the compound was picking up electrons from the endogenous Q pool or directly from respiratory complexes. To investigate this, measurements of electron transport were carried out by using mitochondrial membrane fragments from which the endogenous Coenzyme Q pool had been removed by repeated pentane extractions.[9] However, as pentane extraction may damage membranes and

---

[9] F. L. Crane and R. Barr, *Methods Enzymol.* **18C**, 137 (1971).

because it is difficult to eliminate the possibility of residual Coenzyme Q, an alternative approach was also used in which mitochondria were prepared from yeast lacking Coenzyme Q through disruption of its biosynthetic pathway.[10] In both systems there was electron transfer to $MitoQ_{(C10)}$ in the absence of endogenous Coenzyme Q. These Coenzyme Q-deficient yeasts cannot carry out oxidative phosphorylation and so do not grow on non–fermentable carbon sources, but growth can be restored by addition of Coenzyme Q analogs. $MitoQ_{(C10)}$ did not restore growth, suggesting that while MitoQ can pick up and donate electrons directly to the respiratory chain, the rate is insufficient to restore oxidative phosphorylation.[6] One possibility is that the 10 carbon aliphatic chain of $MitoQ_{(C10)}$ does not penetrate far enough into the phospholipid bilayer, slowing its access to the active sites of respiratory complexes I, II and III. Whether longer alkyl chains have easier access to the respiratory chain active sites and are thus able to restore respiration is being investigated.

Experiments with Mitochondria-Targeted Coenzyme Q Derivatives

General Considerations

A wide range of experiments with MitoQ have been carried out on cells and isolated mitochondria.[6,7,11–13] In designing these experiments, it is important that any effects measured are due to the specific targeting of the Coenzyme Q moiety to mitochondria and not to non-specific interactions. Excessive uptake of alkyltriphenylphosphonium cations into mitochondria disrupts mitochondrial function by non-specific damage to the mitochondrial inner membrane, respiratory chain and ATP synthesis and export machinery. Such non-specific effects can be easily assessed by measuring the effect of compounds on mitochondrial respiration in an oxygen electrode. Concentrations of alkyltriphenylphosphonium cations that disrupt the inner membrane causing proton leak will increase State 4 respiration.[14] Disruption to the respiratory chain and/or the ATP synthesis system will decrease State 3 respiration.[14] Inhibition of uncoupled respiration indicates non-specific disruption to the respiratory chain, but this will only occur at

[10] C. M. Grant, F. H. MacIver, and I. W. Dawes, *FEBS Lett.* **410,** 219 (1997).
[11] P. M. Hwang, F. Bunz, J. Yu, C. Rago, T. A. Chan, M. P. Murphy, G. F. Kelso, R. A. J. Smith, K. W. Kinzler, and B. Vogelstein, *Nature Med.* **7,** 1111 (2001).
[12] G. Saretzki, M. P. Murphy, and T. von Zglinicki, *Aging Cell* **2,** 141 (2003).
[13] K. S. Echtay, M. P. Murphy, R. A. Smith, D. A. Talbot, and M. D. Brand, *J. Biol. Chem.* **277,** 47129 (2002).
[14] M. P. Murphy, *Biochim. Biophys. Acta* **1504,** 1 (2001).

very high concentrations as there is no selective uptake of the compounds into mitochondria on uncoupling. Mitochondria-targeted compounds usually start to disrupt mitochondrial respiration at around 10 to 20 μM, depending on the cation under consideration, corresponding to millimolar concentrations within the mitochondrial matrix. Thus MitoQ incubations with isolated mitochondria should not exceed low micromolar concentrations. Even if these MitoQ concentrations have no effect on respiration, it is still important to control for non-specific effects of lipophilic cations by carrying out control incubations with TPMP or other appropriate alkyltriphenylphosphonium cations, such as ethyl- or decyltriphenylphosphonium.

With cell incubations, MitoQ will accumulate 5 to 10 fold into the cytoplasm relative to the extracellular medium driven by the plasma membrane potential, and then further accumulate into the mitochondria (Fig. 2). Consequently, concentrations 5 to 10-fold lower should be used in cell incubations rather than with isolated mitochondria. Serum albumin binds MitoQ and thus affects its free concentration and may lead to differences between cell incubations in the presence or absence of serum. Incubations with isolated mitochondria usually last a few minutes, while cell incubations can persist for several days. Therefore it is first vital to establish MitoQ concentrations that do not have any toxicity on long-term incubation by using standard assays such as the release of cytosolic enzymes, the uptake of membrane-impermeant nuclear stains, or the induction of apoptosis. It is equally important to confirm that the compounds do not affect the rate or extent of cell division by assessing cell growth and clonogenicity. Even when a safe concentration range has been determined, control experiments should still be carried out with alkyltriphenylphosphonium cations such as TPMP. Another factor is that cell lines have a range of densities, volumes, mitochondrial contents, growth rates, and degrees of reliance on oxidative phosphorylation. Consequently, we find that different cell lines have distinct thresholds for MitoQ toxicity. So while experiments to date show that protective effects against oxidative stress are evident at MitoQ$_{(C10)}$ concentrations of 100 nM to 1 μM, the appropriate concentration range will vary and must be established for any new cell line under investigation.

As well as controlling for the non-specific interactions of alkyltriphenylphosphonium cations, it is also important to consider any effects of the ubiquinone moiety distinct from its mitochondrial localization. This can be done by comparing the MitoQ with an appropriate untargeted coenzyme Q analog, for example, by using decylubiquinone or idebenone as a control for MitoQ$_{(C10)}$. As the untargeted compound will be distributed throughout the cell, the same concentration of MitoQ should be more effective in protecting against mitochondrial oxidative damage. In contrast, if the effects of MitoQ were non-mitochondrial, then the untargeted compound should show similar potency. Addition of an uncoupler prevents the selective uptake of MitoQ into mitochondria and will decrease the potency of MitoQ several-hundred fold, if it is acting on mitochondria. While all of these are essential controls, it should be noted that the location of MitoQ on the surface of the phospholipid bilayer differs from that of other coenzyme derivatives that will partition into the center of the phospholipid bilayer (see discussion later).

Coenzyme Q is thought to block oxidative damage through its reduced, ubiquinol form, which prevents lipid peroxidation by acting as a chain-breaking antioxidant by donating $H^\bullet$. The resulting ubisemiquinone radicals are disproportionate to a ubiquinone and a ubiquinol[6]. The ubiquinone can then be recycled back to its active ubiquinol form by the respiratory chain. The antioxidant effects of MitoQ can be analyzed in isolated mitochondria exposed to an exogenous stressor such as ferrous iron/H_2O_2 by measuring oxidative damage markers or mitochondrial function.[6] In all cases it is important to carry out controls for the uptake of MitoQ to ensure that the added compound is in, or is converted to, the effective ubiquinol form, and to carry out controls with simple alkyltriphenylphosphonium cations.[6] Finally, as with all redox active compounds, under some conditions partially or fully reduced coenzyme Q or its derivatives can act as pro-oxidants, and care must be taken to eliminate this as a possible explanation of unexpected results.[13]

Analysis of Mitochondrial Uptake of MitoQ

The membrane potential-dependent uptake of MitoQ by mitochondria and cells is a vital property of these compounds. This uptake can be measured by a number of techniques to demonstrate that the novel compound is accumulated by mitochondria, or to confirm uptake when mitochondrial polarization or stability is uncertain. One straightforward procedure to measure uptake is by incubating energized mitochondria with radiolabeled MitoQ. For this, the procedures used to measure the uptake of [³H]TPMP by mitochondria to quantitate the membrane potential are easily adapted.[15] In brief, isolated mitochondria (e.g., rat liver mitochondria at 1 to 2 mg protein/ml) are incubated with 1 to 5 μM [³H]Mito Q in Eppendorf tubes containing 1 ml of an appropriate incubation medium supplemented with respiratory substrates. After incubation for 1 to 5 minutes, the mitochondria are pelleted by centrifugation (30 sec at 10,000 × g), and a sample of the supernatant is removed for scintillation counting.

[15] M. D. Brand, *in* "Bioenergetics – a practical approach" (G. C. Brown and C. E. Cooper, eds.), p. 39. IRL, Oxford, 1995.

Residual liquid is carefully removed from the mitochondrial pellet by using a rolled-up piece of tissue, and the pellet is resuspended by vortexing with 200 μl 20% (v/v) Triton. The end of the Eppendorf tube is clipped off into a scintillation vial by using a commercially available microcentrifuge tube cutter, and the radioactive content is quantitated by scintillation counting. The ratio of the uptake into mitochondria and the amount in the supernatant can be used to calculate an accumulation ratio by using a literature value for the mitochondrial volume (e.g., 0.5 to 0.9 μl/mg protein for rat liver mitochondria[16,17]) or by measuring the mitochondrial volume under the experimental conditions used.[15] The specific activity of the initial incubation medium is used to quantitate uptake. The effect of an uncoupler such as carbonylcyanide-p-trifluoromethoxy-phenylhydrazone (FCCP) on accumulation can be used to confirm that uptake is membrane potential dependent and to estimate non-specific binding to mitochondrial membranes. Comparison of the uptake of MitoQ with TPMP or ^{86}Rb can also be used to determine how the relative uptake is affected by the hydrophobicity of the MitoQ compound and to see whether uptake depends on the membrane potential in a Nernstian fashion.[6,18] For these experiments dual channel scintillation counting is required to quantitate ^{86}Rb and ^{3}H or ^{14}C and ^{3}H uptake simultaneously.[19]

The advantages of measuring the uptake of radiolabeled compounds are that they can be easily quantitated, that non-specific binding can be corrected for and that the uptake of various cations can be compared. However, for routine measurements with isolated mitochondria it is often more convenient to measure uptake continuously. For this, the ion-selective TPMP electrodes that are used to measure mitochondrial membrane potential are easily adapted.[20,21] Electrodes are constructed with an oxygen electrode in the base of the incubation chamber and an ion-selective and reference electrode inserted through a sealed Perspex lid that also contains an injection port.[20,21] This enables the oxygen consumption and cation concentration to be measured simultaneously.[21] When these electrodes are used with the relatively water-soluble cation TPMP they give the expected Nernstian response, and on addition of mitochondria there is negligible mitochondrial-binding of TPMP (Fig. 8A). Once the respiratory substrate

[16] G. C. Brown and M. D. Brand, *Biochem. J.* **225**, 399 (1985).

[17] R. P. Hafner and M. D. Brand, *Biochem. J.* **250**, 477 (1988).

[18] R. A. J. Smith, C. M. Porteous, C. V. Coulter, and M. P. Murphy, *Eur. J. Biochem.* **263**, 709 (1999).

[19] R. J. Burns, R. A. J. Smith, and M. P. Murphy, *Arch. Biochem. Biophys.* **322**, 60 (1995).

[20] N. Kamo, M. Muratsugu, R. Hongoh, and Y. Kobatake, *J. Membr. Biol.* **49**, 105 (1979).

[21] G. P. Davey, K. F. Tipton, and M. P. Murphy, *Biochem. J.* **288**, 439 (1992).

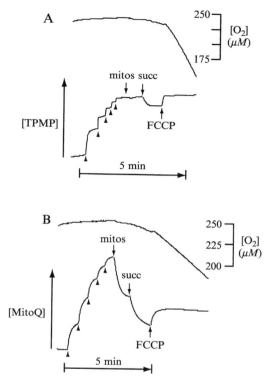

FIG. 8. Uptake of TPMP and MitoQ by energized mitochondria measured by using an ion-selective electrode. An electrode selective for lipophilic cations was constructed as described[21] and inserted though a Perspex lid into a stirred, thermostatted 3 ml chamber with a Clark oxygen electrode (Rank Brothers, Bottisham, Cambridgeshire, UK) built into the base. This enables the simultaneous measurement of oxygen consumption and concentration of the lipophilic cation. Additions to the chamber can be made through an injection port. An incubation medium of 120 mM KCl, 10 mM HEPES, 1 mM EGTA (pH 7.2) supplemented with rotenone (5 μg/ml) was stirred at 30° in the chamber. Then 5 $\times$ 1 μM additions of lipophilic cation were added from a 1 mM stock solution in ethanol to calibrate the electrode response. Rat liver mitochondria were then added to the chamber, energized by addition of succinate (5 mM) and uncoupled by addition of 333 nM FCCP. (A) The lipophilic cation was TPMP, and 1 mg mitochondrial protein/ml was added. (B) The lipophilic cation was MitoQ$_{(C10)}$, and 0.5 mg mitochondrial protein/ml was added.

succinate is added to establish the slow respiration rate and high membrane potential characteristic of State 4, there is a dramatic uptake of TPMP that comes to equilibrium with the membrane potential within a few seconds. When the uncoupler FCCP is added to abolish the membrane potential, the respiration rate increases, and all the accumulated TPMP is rapidly lost from the mitochondria, returning the ion-selective electrode trace to its initial base line. A similar approach can be used to measure the uptake of MitoQ by mitochondria. However, for more hydrophobic molecules such as MitoQ$_{(C10)}$, the electrode response is not simply described by the Nernst equation, presumably because of the hydrophobicity of the compound, and there is also considerable non-specific binding of MitoQ to mitochondrial membranes (Fig. 8B). Although this makes absolute quantitation of uptake problematic, these measurements show clearly that MitoQ is taken up by mitochondria driven by the membrane potential (Fig. 8B). Thus to fully characterize the uptake of MitoQ, particularly the more hydrophobic versions, both electrode and radioisotope measurements are required.

To measure MitoQ uptake by mitochondria within cells, it is best to use radiolabeled compounds.[6] While MitoQ uptake by cells can be measured by using an ion-selective electrode, the drift in electrode response over the 10 to 30 minutes required for accumulation by cells makes this approach unsuitable for more than qualitative demonstrations of uptake. Cells can be incubated with [^{3}H]MitoQ and pelleted by centrifugation after a certain time, then their radioactive content can be quantitated by scintillation counting as for isolated mitochondria.[6] To correct for non-specific binding, it is important to inhibit both the plasma membrane potential by addition of ionophores (valinomycin or gramicidin) and the Na-K ATPase inhibitor ouabain and the mitochondrial membrane potential with mitochondrial inhibitors (myxothiazol and oligomycin) and uncoupler (FCCP).[22] These measurements give total cell uptake, and the intracellular location of the compound is uncertain. However, decreased uptake on addition of uncoupler suggests that the compounds are mainly within the mitochondria. A related approach is to compare cells that have normal mtDNA (ρ^+) and consequently have a substantial mitochondrial membrane potential with cells that lack mtDNA (ρ^0) and therefore have a far lower mitochondrial membrane potential. This was done using ρ^+ and ρ^0 lines derived from a 143B osteosarcoma cell line, which had similar plasma membrane potentials, cell and mitochondrial volumes, and only differed in their mitochondrial membrane potential.[18] The substantially greater uptake of the mitochondria-targeted antioxidant by the ρ^+ over the ρ^0 cell line was solely due to the increased mitochondrial accumulation.[23]

It is also possible to confirm directly that the MitoQ within the cell is predominantly in the mitochondria by separating the cell into cytosolic and mitochondrial fractions. This is done by rapidly exposing cells to media

[22] A. J. James, P. W. Sheard, Y.-H. Wei, and M. P. Murphy, *Eur. J. Biochem.* **259,** 462 (1999).
[23] R. D. Appleby, W. K. Porteous, G. Hughes, A. M. James, D. Shannon, Y.-H. Wei, and M. P. Murphy, *Eur. J. Biochem.* **262,** 108 (1999).

containing digitonin, which ruptures the cholesterol-rich plasma membrane but not the mitochondrial inner membrane.[24,25] The mitochondria are then immediately separated from the rest of the cell by centrifugation through an oil layer into buffer. This enables the mitochondrial fraction to be separated from the cytoplasm within a few seconds, and the two fractions can then be assessed for MitoQ content. Cross contamination and the yields of the mitochondrial and cytoplasmic fractions can be inferred using enzyme assays specific for the two fractions, for example by using lactate dehydrogenase for the cytosol and citrate synthase for the mitochondria.[18]

Location of Targeted Coenzyme Q Derivatives Within Mitochondria

It is important to consider how what is known about the uptake through phospholipid bilayers and the membrane binding of alkyltriphenyl-phosphonium cations applies to MitoQ. The mitochondrial membrane potential drives the uptake of alkyltriphenylphosphonium cations, with the Nernst equation describing their distribution between the matrix and ex-tramitochondrial space (Fig. 9A). However, within mitochondria, alkyltri-phenylphosphonium cations are in a dynamic equilibrium between those bound on the matrix surface of the inner membrane and those free in solution, with about 60% of TPMP and 85% of tetraphenylphosphonium (TPP) within mitochondria being membrane adsorbed.[15,16,26] Therefore, TPMP uptake at a given membrane potential is greater than predicted by the Nernst equation, and a binding correction is required to calculate the membrane potential. This is determined by comparing TPMP uptake with that of $^{86}Rb^+$, which is not membrane bound.[15,26]

The movement of lipophilic cations such as TPP or TPMP across phospholipid bilayers is described by a three-step model of adsorption, translocation and desorption[27,28] (Fig. 9A). In the absence of a membrane potential the cation binds to the outer surface of the membrane, then permeates the hydrophobic potential energy barrier of the lipid bilayer, coming to bind to the inner surface of the membrane before finally desorbing to the mitochondrial matrix.[27] The essential features of uptake are the same in the presence of a membrane potential although now the internal potential and the energy barrier are altered[5] (Fig. 9B, top). Cations

[24] R. J. Burns and M. P. Murphy, *Arch. Biochem. Biophys.* **339,** 33 (1997).
[25] M. N. Berry, A. M. Edwards, G. J. Barrit, M. B. Grivell, H. J. Halls, B. J. Gannon, and D. S. Friend, *in* "Laboratory techniques in biochemistry and molecular biology" (R. H. Burdon and van P. H. Knippenberg, eds.), Vol. 21, p. 386. Elsevier, Amsterdam, 1991.
[26] C. Shen, C. C. Boens, and S. Ogawa, *Biochem. Biophys. Res. Commun.* **93,** 243 (1980).
[27] B. Ketterer, B. Neumcke, and P. Laeuger, *J. Membr. Biol.* **5,** 225 (1971).
[28] D. S. Cafiso and W. L. Hubbell, *Annu. Rev. Biophys. Bioeng.* **10,** 217 (1981).

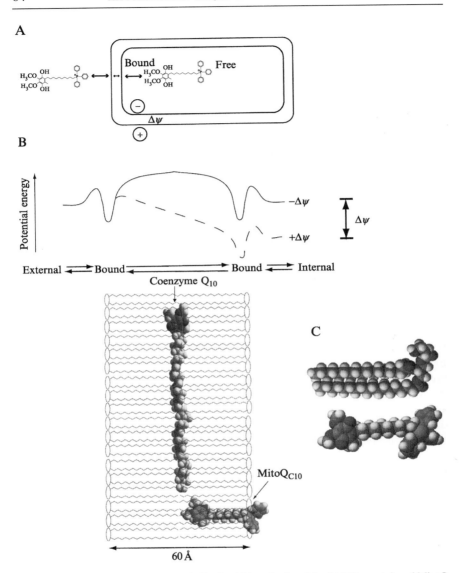

FIG. 9. Uptake and distribution of MitoQ within mitochondria. (A) The uptake of MitoQ by mitochondria in response to a membrane potential ($\Delta\psi$). The molecule is initially bound to the outside of the membrane, then translocates through the phospholipid interior to bind to the matrix facing surface of the membrane. The internally bound MitoQ is in equilibrium with that free in the matrix. The top section of (B) shows the energy profile in the absence of a $\Delta\psi$ (continuous line) for the movement of an alkyltriphenylphosphonium cation through the phospholipid bilayer. The molecules are bound in the potential energy wells close to the

such as TPP bind to both sides of the membrane as a monolayer without affecting the subsequent binding of other cations, at least at low concentrations (i.e., binding is described by a Langmuir adsorption isotherm). This binding to the phospholipid membrane occurs at potential energy wells close to the membrane surfaces (Fig. 9B, top). These potential wells are located at the level of the carbonyls of the membrane phospholipids, are at the hydrophobic side of the aqueous/lipid interface and have a dielectric constant of about 5 to 30.[5] The binding site is at the same electrostatic potential as the bulk phase. Equilibrium dialysis measurements of TPP binding to phosphatidylcholine vesicles showed that adsorption was linear with concentration up to about 600 μM TPP, above which there were deviations from linearity caused by repulsive interactions between adsorbed cations.[5] This onset of saturation of binding occurred when about 2×10^{12} TPP molecules/cm^2 phospholipid were bound, corresponding to about 1 TPP molecule bound per 100 phospholipid molecules and occurred at similar concentrations for other alkyltriphenylphosphonium cations.[5]

Triphenylphosphonium cations with hydrophobic alkyl chains interact with phospholipid bilayers in a similar way to TPP, with the triphenylphosphonium cation binding to the same location in the membrane.[5] The more-hydrophobic compounds adsorb to the membrane surface to a greater extent because of the entropic driving force of incorporation of the alkyl side chain into the membrane.[5] Increasing the length and hydrophobicity of the alkyl chain increased binding to soybean phospholipid bilayers,[29] and intravesicular membrane binding in *Halobacterium halobium* membrane vesicles increased from 72% for TPMP to 95% for a five-carbon alkyl chain.[30] That this model applies to mitochondria-targeted antioxidants has been confirmed by parallel measurements of the uptake of Mito-Vit E, and TPMP by mitochondria. These suggested that about 84% of accumulated MitoVit E was bound to the matrix surface of the inner membrane, in contrast to about 60% for TPMP,[18] and this is consistent

---

[29] A. Ono, S. Miyauchi, M. Demura, T. Asakura, and N. Kamo, *Biochemistry* **33,** 4312 (1994).
[30] M. Demura, N. Kamo, and Y. Kobatake, *Biochim. Biophys. Acta* **820,** 207 (1985).

---

membrane surfaces. When a $\Delta\psi$ is present the profile is altered (dashed line) and the potential energy of the MitoQ within mitochondria is lowered. The middle section shows the various equilibria between free and bound MitoQ. The bottom section is a schematic cross-section of a phospholipid bilayer showing the location of MitoQ on the matrix-facing surface of the inner membrane, where the phosphonium cation is bound at the level of the fatty acid carbonyls and the alkyl tail is inserted into the membrane. In contrast, Coenzyme Q$_{10}$ is located entirely within the hydrophobic core of the phospholipid bilayer. (C) Comparison of space filling models of phosphatidylethanolamine with two unsaturated C$_{18}$ fatty acids (i.e., distearate) with MitoQ$_{(C10)}$.

with their relative partition coefficients (Table I). Therefore, the triphenyl-phosphonium cation of all MitoQ derivatives binds at a potential well on the membrane surface and is in equilibrium with compound free in solution. The amount bound increases for the more hydrophobic molecules, which insert their hydrophobic alkyl group into the membrane. Experiments with $MitoQ_{(C10)}$ showed that its uptake by mitochondria increased with the membrane potential at low values as predicted by the Nernst equation. However, at higher membrane potentials, $MitoQ_{(C10)}$ binding started to saturate because of its greater hydrophobicity, causing a deviation from the Nernst equation.[6] While this occurs for hydrophobic molecules such as $MitoQ_{(C10)}$, it may not occur for shorter chain analogs such as $MitoQ_{(C3)}$.

The location of $MitoQ_{(C10)}$ associated with the mitochondrial inner membrane is likely to be as shown in Fig. 9B (bottom), with the triphenyl-phosphonium cation binding at the level of the fatty acid carbonyls while the alkyl chain is inserted into the lipid bilayer. This is further illustrated in Fig. 9C, where the fatty acid carbonyl groups of a typical phospholipid are aligned with the triphenylphosphonium of $MitoQ_{(C10)}$. This orientation enables the antioxidant moiety to access the lipid bilayer and thus prevent lipid peroxidation. In addition, most of the MitoQ is present on the matrix-facing leaflet of the inner membrane where most of the mitochondrial production of reactive oxygen species occurs.[31] In contrast, coenzyme Q and hydrophobic coenzyme Q derivatives will be in the hydrophobic core of the membrane (Fig. 9B, bottom). Consequently we expect that the longer the hydrophobic tail on the MitoQ, the greater the average steady state concentration will be at the lipid-water interface and the greater the insertion of the ubiquinone moiety into the membrane. However, the ubiquinone head group is somewhat polar and it is possible that the active moiety of MitoQ may loop out of the phospholipid bilayer and be present at the water-lipid interface. In summary, these compounds are positioned mostly on the matrix face of the mitochondrial inner membrane. Because this is where most of the active components of the oxidative phosphorylation machinery reside, the location of MitoQ may be fortuitous in offering protection against oxidative damage.

Conclusion

The use of targeted coenzyme Q derivatives as probes to investigate the biology of mitochondria is just starting. While work so far has shown that this method of targeting is of use in blocking some forms of mitochondrial

---

[31] J. St-Pierre, J. A. Buckingham, S. J. Roebuck, and M. D. Brand, *J. Biol. Chem.* **277**, 44784 (2002).

oxidative damage, much still needs to be explored about the mitochondrial production of reactive oxygen species and their contribution to signaling, aging and apoptosis. It is also hoped that MitoQ and related compounds can be used to explore whether these compounds have potential therapeutic effects. For example, the development of procedures to test these compounds in mouse models of human diseases is ongoing[8] but is beyond the scope of this chapter. The use of targeted coenzyme Q derivatives, in conjunction with increased understanding of their effects on mitochondrial function and redox behavior, may lead to new insights into the biology of mitochondria.

Acknowledgments

We thank Meredith F. Ross, Jordi Asin-Cayuela and Karim S. Echtay for helpful comments and discussions. This work was supported by grants from the Health Research Council of New Zealand, the Marsden Fund, administered by the Royal Society of New Zealand, and the Medical Research Council (UK).

[32] A. Hercouet and M. Le Corre, *Synthesis* 157 (1988).
[33] B. Richter, E. de Wolf, G. van Koten, and B. J. Deelman, *J. Org. Chem.* **65,** 3885 (2000).

[4] The Mitochondrial Interplay of Ubiquinol and Nitric Oxide in Endotoxemia

By Constanza L. Lisdero, Maria Cecilia Carreras,
Alain Meulemans, Mariana Melani, Michel Aubier,
Jorge Boczkowski, and Juan José Poderoso

Introduction

Sepsis is a common cause of morbidity and mortality, particularly in the elderly, immune-compromised, and critically ill patients. Almost 25 years ago, we showed that clinical sepsis and septic shock were associated with acquired mitochondrial dysfunction; a critical inhibition of mitochondrial complex I was observed in skeletal muscle of patients with sepsis.[1] Recently, mitochondrial dysfunction was ascribed to mitochondrial

[1] J. J. Poderoso, A. Boveris, M. A. Jorge, C. R. Gherardi, A. W. Caprile, J. Turrens, and A. O. Stoppani, *Medicina.* **38,** 371 (1978).

overproduction of reactive oxygen species and nitric oxide (NO), which may play a pivotal role in the pathophysiology of organ failure.[1-3]

Ubiquinone (UQ, coenzyme Q) is endogenously synthesized in every organ in a specific pathway branched from cholesterol biosynthesis[4]; organs have different ubiquinol content depending on the activity of this biosynthetic pathway.[5] In addition to its role as electron carrier in the mitochondrial respiratory chain,[6] ubiquinone exhibits prooxidant[7] and antioxidant[8] properties. Ubisemiquinone ($UQ^{\bullet-}$) participates in oxygen radical formation upon electron-transfer to oxygen,[9] whereas ubiquinol may be ascribed antioxidant properties upon its reaction with either peroxyl radicals or upon recovery of vitamin E radical.

In addition to the reversible inhibition of cytochrome oxidase, nitric oxide (NO) contributes to mitochondrial ubiquinol oxidation, within a reaction that yields $UQ^{\bullet-}$ and nitroxyl anion (NO^-) as well as oxygen- and nitrogen reactive species derived from $UQ^{\bullet-}$ autoxidation[10] (reactions [1–3]).

$$NO + UQH^- \rightarrow NO^- + UQ^{\bullet-} + H^+ \tag{1}$$

$$UQ^{\bullet-} + O_2 \rightarrow UQ + O_2^{\bullet-} \tag{2}$$

$$O_2^{\bullet-} + NO \rightarrow ONOO^- \tag{3}$$

The second-order rate constant for reactions [1–3] were calculated as $10^3–10^4 \ M^{-1} \ s^{-1}$,[10,11] $8 \times 10^3 \ M^{-1} \ s^{-1}$, and $1.9 \times 10^{10} \ M^{-1} \ s^{-1}$, respectively. Reactions [1] and [3] are the most efficient routes for NO utilization in

[2] D. Brealey, M. Brand, I. Hargreaves, S. Heales, J. Land, R. Smolenski, N. A. Davies, C. E. Cooper, and M. Singer, *Lancet*. **360,** 219 (2002).

[3] J. Boczkowski, C. L. Lisdero, S. Lanone, A. Samb, M. C. Carreras, A. Boveris, M. Aubier, and J. J. Poderoso, *FASEB J*. **13,** 1637 (1999).

[4] F. Gibson and I. G. Young, *Methods Enzymol*. **53,** 600 (1978).

[5] R. Alleva, M. Tomasetti, S. Bompadre, and G. P. Littarru, *Mol. Aspects Med*. **18,** S105 (1997).

[6] G. Lenaz, R. Fato, C. Castelluccio, M. L. Genova, C. Bovina, E. Estornell, V. Valls, F. Pallotti, and G. Parenti Castelli, *J. Clin. Invest*. **71,** S66 (1993).

[7] H. Nohls, L. Gille, and A. V. Kozlov, in "Subcellular Biochemistry" (Quinn and Kagan, eds.), Vol. 30, p. 509. Fat-soluble vitamins, Plenum Press, New York, 1998.

[8] J. J. Maguire, V. E. Kagan, E. A. Serbinova, B. A. Ackrell, and L. Packer, *Arch. Biochem. Biophys*. **292,** 47 (1992).

[9] A. Boveris, E. Cadenas, and A. O. M. Stoppani, *Biochem. J*. **156,** 435 (1976).

[10] J. J. Poderoso, M. C. Carreras, C. Lisdero, F. Schöpfer, C. Giulivi, A. Boveris, A. Boveris, and E. Cadenas, *Free Rad. Biol. Med*. **26,** 925 (1999).

[11] J. J. Poderoso, C. L. Lisdero, F. Schopfer, N. Riobo, M. C. Carreras, E. Cadenas, and A. Boveris, *J. Biol. Chem*. **274,** 37709 (1999).

mitochondria and lead to the accumulation of peroxynitrite ($ONOO^-$), a species that modulates redox signaling and contributes to cell damage.[12] Accumulation of $ONOO^-$ has been associated with different pathological conditions, such as neurodegeneration[13] and sepsis.[3,14] Deleterious effects of $ONOO^-$ are likely a function of its strong oxidative capacity leading to protein oxidation and nitration.[12] In mitochondria, $ONOO^-$ may be scavenged by glutathione peroxidase or by direct interaction with electron donors, such as glutathione, NADH, or ubiquinol, present in either the matrix or the inner membrane[15] (reaction [4]). The reaction of $ONOO^-$ (as ONOOH) with UQH^- is first-order in $ONOO^-$ and zero-order in UQH^- indicating that it involves a *cage* rearrangement to give OH^- and NO_2^- (reaction [4])[15]:

$$ONOOH[OH^\bullet \ldots NO_2^\bullet] + UQH^- \rightarrow NO_2^\bullet + UQ^{\bullet-} + H^+ + OH^- \qquad (4^*)$$

The generation of $UQ^{\bullet-}$ (reaction [4]) suggests that $ONOO^-$ may facilitate free radical propagation, supported by auto-oxidation of $UQ^{\bullet-}$ (reaction [2] following reaction [4]). Accordingly, we have shown that $ONOO^-$ increases O_2^- production in mitochondria.[15] However, increasing mitochondrial ubiquinol content clearly protects mitochondria from $ONOO^-$ effects.[15] This suggests the occurrence and significance of radical-radical termination reactions among the terms of reaction [4] and involving the NO_2^-/NO_2^-, $UQ^{\bullet-}/UQ$, and OH^-/OH^- redox pairs. These termination reactions intercept effects of aggressive radicals on mitochondrial components. For instance, tyrosine nitration by $ONOO^-$ proceeds in two steps: (a) H-abstraction by $NO_2^\bullet$ or $OH^\bullet$ to form tyrosyl radical, and (b) tyrosyl radical reaction with $NO_2^\bullet$, to form nitrotyrosine. Therefore, although mitochondrial production of reactive oxygen species and $ONOO^-$ are related to ubiquinol concentration (*pro-oxidant* effects),[8,9] the features of reaction [4] support an *antioxidant* activity of ubiquinol. Reactions [1–4] in mitochondria have important implications for the regulation of the levels of NO in the mitochondrial matrix and production of oxyradicals, processes intimately associated with mitochondrial integrity and function.

At physiological matrix NO levels (20 to 40 nM) with about 2 to 3 nmol ubiquinone/mg mitochondrial protein,[11] this sub-system is likewise capable of removing all NO in excess. In this setting, almost all NO is consumed to form O_2^-, and its Mn-SOD-mediated dismutation product

[12] R. Radi, A. Cassina, and R. Hodara, *Biol. Chem.* **383,** 401 (2002).

[13] H. Ischiropoulos and J. S. Beckman, *J. Clin. Invest.* **111,** 163 (2003).

[14] M. L. Johnson and T. R. Billiar, *World J. Surg.* **22,** 187 (1998).

[15] F. Schöpfer, N. A. Riobó, M. C. Carreras, B. Alvarez, R. Radi, A. Boveris, E. Cadenas, and J. J. Poderoso, *Biochem. J.* **349,** 35 (2000).
*Alternatively, ONOOH $[OH^\bullet \ldots NO_2^\bullet] + UQH^- \rightarrow NO_2^- + UQ^{\bullet-} + H^+ + OH^\bullet$

H_2O_2.[11,16] However, at high and sustained matrix NO concentration (≈ 500 nM), these reactions could not afford the cost of NO detoxification. In the context of mitochondrial derangement in septic shock, an important point to consider is the source of mitochondrial NO. It is noteworthy that iNOS is induced by inflammatory mediators in different organs and in macrophages. We have recently reported an increase in mitochondrial protein nitration during iNOS induction in *E. coli* endotoxemia and described the functional correlation between the loss of diaphragmatic force and mitochondrial respiratory impairment.[3,17] Moreover, recent data showed that NO can be synthesized within mitochondria by a mtNOS[18–20] that could be activated during sepsis; accordingly, Boveris *et al.* and Escámes *et al.* reported an increase of mtNOS activity in rat endotoxemia.[21,22]

This chapter provides the methodological bases that support *(a)* the relation between UQ content, mitochondrial NO steady-state concentration and peroxynitrite-mediated mitochondrial damage in representative organs and *(b)* the contribution of mtNOS to peroxynitrite formation in mitochondria.

Experimental Model: Endotoxemic Animals and Sample Preparation

Endotoxemic Animals

Sprague-Dawley male albino rats (weight: ~ 350 g) from Charles River France Inc. were divided into 2 groups, which received either sterile 0.9%, NaCl (control animals) or 10 mg/kg *E. coli* endotoxin suspension in NaCl (serotype 0.26 B6; DIFCO, Detroit, MI), intraperitoneally (septic animals). Animals were sacrificed 6 hours after LPS inoculation and liver, lung, diaphragm and heart were excised. Liver of normal C57BL/6J mice and transgenic mice deficient in iNOS gene C57BL/6-*Nos*2[tm1Lau] from Jackson Laboratories (Bar Harbor, ME) were also utilized. Animals were maintained on a 12-hr light/dark cycle with free access to food and water and were fasted one night before the experiment with water *ad libitum*.

[16] J. J. Poderoso, M. C. Carreras, C. Lisdero, N. Riobó, F. Schopfer, and A. Boveris, *Arch. Biochem. Biophys.* **328,** 85 (1996).

[17] J. Boczkowski, S. Lanone, D. Ungureanu-Longrois, T. Fournier, and M. Aubier, *J. Clin. Invest.* **98,** 1550 (1996).

[18] A. J. Kanai, L. L. Pearce, P. R. Clemens, L. A. Birder, M. M. VanBibber, S. Y. Choi, W. C. de Groat, and J. Peterson, *Proc. Natl. Acad. Sci.* **98,** 14126 (2001).

[19] C. Giulivi, J. J. Poderoso, and A. Boveris, *J. Biol. Chem.* **273,** 11038 (1998).

[20] M. C. Carreras, J. G. Peralta, D. P. Converso, P. V. Finocchietto, I. Rebagliati, A. A. Zaninovich, and J. J. Poderoso, *Am. J. Physiol. Heart Circ. Physiol.* **281,** H2282 (2001).

[21] A. Boveris, S. Alvarez, and A. Navarro, *Free Rad. Biol. Med.* **33,** 1186 (2002).

[22] G. Escámes, J. León, M. Macías, H. Khaldy, and D. Acuña-Castroviejo, *FASEB J.* **17,** 932 (2003).

Homogenate and Mitochondrial Preparations

Frozen tissue samples were homogenized with an Ultraturrax T25 (Janke and Kunkel, IKA Works, Cincinnati, OH) in lysis buffer (50 mM Tris HCl [pH 7.4], 0.1 mM EDTA, 1 μM leupeptin, 1 μM PMSF, 1 μM aprotinin). The crude homogenates were centrifuged at 4000g for 20 min at 4°, and stored at −70° until use. Mitochondria were isolated as previously described[20]; excised organs were placed in an ice-cold homogenization medium (1 g/10 ml) consisting of 100 mM KCl, 50 mM Tris-HCl, 1 mM EDTA, 5 mM MgCl$_2$, 1 mM ATP (pH 7.2) at 4° for isolation of diaphragm mitochondria and 0.23 M mannitol, 70 mM sucrose, 10 mM Tris-HCl, 1 mM EDTA (medium A) for isolation of liver, heart, and lung mitochondria. Mitochondria were purified by Percoll gradient centrifugation and finally resuspended in 0.25 M sucrose at a 20 mg/ml protein concentration.

Sources of NO in Endotoxemia: iNOS Expression and Activity

Measurement of NOS Expression and Activity

Proteins of tissue homogenates (100 μg/lane) and mitochondria (50 μg/lane) were separated by electrophoresis on precast 7.5% SDS-polyacrylamide gels (Bio-Rad, Richmond, CA) and transferred to a PVDF membrane (Bio-Rad). The membranes were incubated with a rabbit anti-mouse iNOS polyclonal antibody (1:1000) (Transduction Laboratories, Lexington, KY), and blotted with a goat anti-rabbit IgG (1:3000) conjugated to alkaline phosphatase (Bio-Rad), followed by detection of immunoreactive proteins by a chemiluminescence method (Bio-Rad). *E. coli* LPS-stimulated rat alveolar macrophages obtained by bronchoalveolar lavage were used as positive controls.

NOS activity was determined by the conversion of ^{3}H-L-arginine to ^{3}H-L-citrulline as follows: mitochondrial and homogenate activities were measured in 50 mM potassium phosphate buffer, pH 7.5 in the presence of either 100 μM (mitochondria) or 20 μM (homogenates) L-arginine, 0.1 μM ^{3}H-L-arginine, 0.1 mM NADPH, 0.3 mM CaCl$_2$, 0.1 μM calmodulin, 10 μM BH$_4$, 1 μM FAD, 1 μM FMN, 50 mM L-valine and 0.1 mg protein. Specific activity is calculated by subtracting the remaining activity in the presence of 50-fold excess concentration of the NOS inhibitor L-NMMA. Calcium-independent NOS activity is measured in the presence of 2 mM EGTA.

Tissue homogenates from LPS-treated animals expressed iNOS protein, detected in the Western blot analysis as an homogeneous band

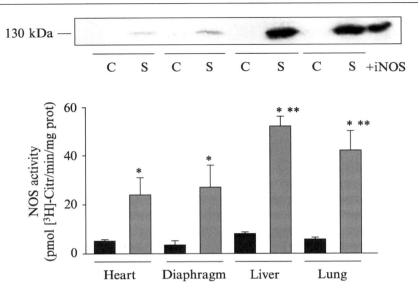

FIG. 1. iNOS expression and Ca^{2+}-independent activity of rat tissue homogenates in endotoxemia. Proteins were separated by electrophoresis in 7.5% SDS-PAGE and iNOS protein was detected with a rabbit anti-mouse iNOS polyclonal antibody (1:1000); *denotes $p < 0.05$; C, control; S, septic. *$p < 0.001$ vs controls by Student t test; **$p < 0.05$ vs endotoxemic heart by ANOVA and Dunnett test.

with molecular identity respect to control iNOS expressed by *in vitro* LPS-stimulated rat alveolar macrophages (Fig. 1). The level of expressed iNOS was ten-fold higher in liver and lung than in heart and diaphragm homogenates. iNOS specific activity resulted about two-fold higher in liver and lung than in diaphragm or cardiac muscle (Fig. 1). These data agree with early reports of Szabó *et al.*[23] in endotoxic shock.[24]

iNOS is Translocated to Mitochondria

Considering the recent reports on the presence of Ca^{2+}-dependent nNOS-α in liver mitochondria (mtNOS[25]), we analyzed a putative contribution of this enzyme to matrix NO concentration in liver tissue during endotoxemia. Immunoblotting analysis of mitochondrial proteins against

[23] C. Szabó, *New Horizons* **3**, 2 (1995).
[24] S. L. Elfering, T. M. Sarkela, and C. Giulivi, *J. Biol. Chem.* **277**, 38079 (2002).
[25] S. Ikenoya, M. Takada, T. Yuzuriha, K. Abe, and K. Katayama, *Chem. Pharm. Bull.* **29**, 158 (1981).

anti-nNOS antibodies confirmed an mtNOS band at 130 kDa, as previously described.[25] In addition, septic samples exhibited a lower expression of this constitutive mtNOS isoform and a higher expression of iNOS respect to the control samples (Fig. 2A, $p < 0.02$). Reciprocally, calcium-dependent mtNOS activity was 50% lower in mitochondria from septic animals than in the control ones, while calcium-independent activity, which was absent in the control samples, increased markedly during endotoxemia (more than 100% of total NOS activity) (Fig. 2B). The selective modulation of each specific NOS isoform in septic samples avoids the possibility of a contamination with cytosolic proteins and also excludes the contribution of the cross-reactivity of the utilized antibodies. Accordingly, immunoprecipitation of mitochondrial proteins with anti-iNOS antibodies showed that iNOS protein translocates to mitochondria in endotoxemia (Fig. 2C). Moreover, after endotoxemia, iNOS expression increased in cytosol and mitochondria of wild type but not in mice with iNOS gene disruption (Fig. 2D).

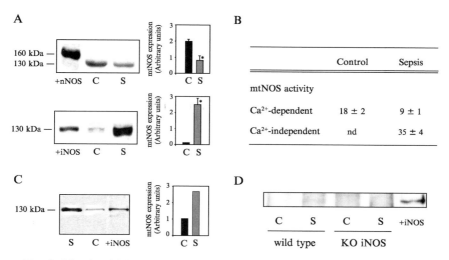

FIG. 2. Mitochondrial expression and activity of liver NOS isoforms in endotoxemia. (A) Proteins were separated as in Fig. 1 and revealed with monoclonal anti-nNOS antibodies and polyclonal anti-iNOS antibodies. The respective densitometries are placed on (A), right, and NOS activities in (B). In (C), it is shown as a representative Western blot of immunopurified mitochondrial proteins. In (D), mitochondria from controls and iNOS gene deficient mice. Abbreviations as in Fig. 1. *$p < 0.02$ vs controls and **$p < 0.01$ vs Ca^{2+}-dependent by Student t test.

These results indicate that, (a) matrix NO concentration is differentially contributed by both constitutively expressed mtNOS and increased cytosolic iNOS translocated to mitochondria and (b) targeted iNOS protein could be the main source of matrix NO in endotoxemia.

The Role of Ubiquinol in Endotoxemia

Ubiquinone Content and NO-Induced Hydrogen Peroxide Production by Mitochondria

This was performed as previously described[25]: 5 to 10 mg of mitochondrial protein was homogenized at 4° with 4 volumes (v/w) of water in an Ultraturrax for 20 sec. One ml of the homogenate was poured into a test tube containing 7 ml of a mixture of ethanol-n-hexane (2:5), and the tube was rapidly shaken for 10 minutes to extract UQ. This extraction was repeated three times, and the combined n-hexane layer was evaporated to dryness under a stream of nitrogen. The resulting residue was dissolved in 0.2 ml of ethanol and subjected to HPLC. Total ubiquinone determination was performed with reverse-phase chromatography as carried out in a Hypersil C-18 column (15 cm × 4.6 mm I.D., 5 μm) (Hypersil, England). The mobile phase was prepared by dissolving 6.1 g of $NaClO_4 \cdot H_2O$ in 1000 ml of ethanol-methanol-70% $HClO_4$ (700:300:1). The flow rate was 1.2 ml/min. The HPLC measurements were performed at 30°. The HPLC system consisted of a 510 pump (Waters, France) with an automatic injector (717 Autosamples, Waters). The UV detector for measuring oxidized ubiquinone was a photodiode array detector (996, Waters). For chromatography, 10 μl of the extracts or standards were injected into the apparatus; retention times were 4.3 min for UQ-9 and 5.3 min for UQ-10, determined at 275 nm with UV detector.

HPLC analysis of ubiquinone extracted from mitochondria (UQ-9 + UQ-10) showed that heart organelles have the highest content (3.7 ± 0.5 nmol/mg mitochondrial protein), liver and diaphragm contain intermediate amounts of ubiquinone (1.7 ± 0.3 and 1 ± 0.2, respectively) and lung samples had the lowest content (0.14 ± 0.03) (Fig. 3A). The mitochondrial ubiquinone content of tissues from septic animals was similar to that of organelles from the control ones (not shown).

The mitochondrial H_2O_2 production rate was continuously monitored by the horse-radish (HRP)/p-hydroxyphenyl acetic acid (p-HPA) assay[16] with a Hitachi F-2000 fluorescence spectrophotometer (Hitachi Ltd.; Tokyo, Japan) supplemented with 12 U/ml HRP, 250 μM p-HPA, at 0.1 to 0.5 mg protein/ml, and with excitation and emission wavelengths of

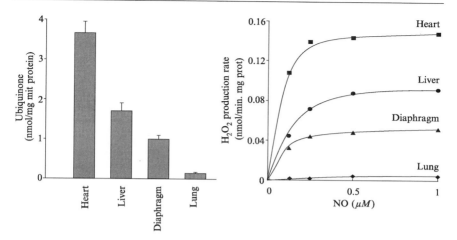

FIG. 3. NO-dependent mitochondrial H_2O_2 production rate and ubiquinol content. (A) Mitochondrial total ubiquinone of rat tissues, as measured by HPLC. (B) Hydrogen peroxide production rate of mitochondria in the presence of crescent NO concentration. Data are different from each other by two-factor ANOVA.

315 and 425 nm, respectively. The utilized medium consisted of 0.23 M mannitol, 70 mM sucrose, 30 mM Tris-HCl, 5 mM Na_2HPO_4/KH_2PO_4, and 1 mM EDTA (pH 7.4) with 8 mM succinate as substrate; to assess maximal mitochondrial H_2O_2 production at complex III, antimycin was added at 2 nmol/mg protein in selective experiments.

The mitochondrial H_2O_2 production rate was achieved at 0.1 to 1 μM NO; nitric oxide solutions (1.2 to 1.8 mM) were obtained by bubbling NO gas of 99.9% purity (AGA GAS Inc., Maumee, OH) in water degassed with He for 30 min at room temperature and stored for a week at 4°. Maximal H_2O_2 yield was achieved at 0.25 μM NO; H_2O_2 production rate was highest in heart organelles followed by liver, diaphragm and lung (Fig. 3B). According to reactions [1] and [2] and after Mn-SOD catalyzed dismutation, differences in NO-dependent H_2O_2 production rates of mitochondria from the different tissues were strictly related to their respective ubiquinone content (Fig. 4). A similar relationship was obtained when heart submitochondrial particles depleted of ubiquinone were subsequently reconstituted with known concentrations to achieve a similar ubiquinone content as the membranes from the analyzed rat tissues.[26] Submitochondrial particles were resuspended in 0.15 M KCl at a concentration of 20 mg protein/ml and lyophilized for 9 h to completely

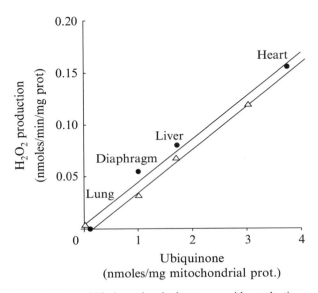

FIG. 4. Correlation between NO-dependent hydrogen peroxide production and ubiquinone content. (●), H_2O_2 production of isolated mitochondria from the different tissues in the presence of 1 μM NO (△): H_2O_2 production of heart submitochondrial particles depleted of endogenous ubiquinone and reconstituted with UQ-*10* at same NO concentration (both, $r^2 = 0.98$, $p = 0.011$ and 0.009, respectively).

dehydrate the samples. Mitochondrial ubiquinone was removed by suspending the lyophilized particles in n-pentane by gentle homogenization, and the suspension was shaken in a glass-stoppered tube for 5 min at 0°, 5 times. Extracted ubiquinone was 4 to 5 nmol UQ/mg protein. To incorporate ubiquinone at different concentrations, the depleted particles were gently homogenized in a small volume of n-pentane (1 to 2 ml) containing UQ-*10* at a concentration of 50 to 100 nmol/mg protein, and the suspension was shaken in an iced bath for 30 min. The particles were centrifuged, dried by evaporation for one hour and stocked. Depleted particles showed a markedly decreased oxygen uptake and NADH cytochrome *c* reductase activity. These parameters were normalized after reconstitution of mitochondrial UQ-*10*. The UQ-*10* content of ubiquinone-reconstituted particles (1 to 5 nmol/mg mitochondrial protein)

[26] F. Aberg, E-L. Appelkvist, G. Dallner, and E. Lars, *Arch. Biochem. Biophys.* **295**, 230 (1992).

was determined by HLPC with electro-chemical detection with an amperometric detector.

In the presence of 1 μM NO, mitochondrial H_2O_2 production rates of organelles with spontaneous or exogenously manipulated ubiquinone content were closely similar and thus, both, resulted well correlated with ubiquinol content ($r = 0.98$, $p < 0.05$) (Fig. 4).

The Endotoxemic Mitochondrial Damage

Mitochondrial Activities

Rats treated with LPS showed a significant decrease of the respiratory control ratio (state 3 O_2 uptake/state 4 O_2 uptake rates) of mitochondria from liver (-28%) and diaphragm (-35%) ($p < 0.05$), while in heart and lung mitochondria, this index resulted slightly modified (Table I). A lowering of respiratory control ratio is consistent with a partial uncoupling of oxidative phosphorylation to electron transfer rates that is currently associated to mitochondrial impairment. Moreover, these results reveal tissue-specific endotoxemic effects on the different mitochondrial populations.

Mitochondrial Hydrogen Peroxide Production

Mitochondria from endotoxemic animals supplemented with antimycin showed an increased mitochondrial H_2O_2 production rate reflecting a higher intramitochondrial concentration of superoxide radical, the product of the univalent reduction of oxygen.[9] After endotoxemia, the mitochondrial production rate of H_2O_2 was significantly increased in diaphragm ($+ 100\%$) and liver mitochondria ($+34\%$), with respect to their

TABLE I
MITOCHONDRIAL RESPIRATORY CONTROL RATIO IN ENDOTOXEMIA

	Heart	Diaphragm	Liver	Lung
Control	4.7 ± 0.3	4.5 ± 0.4	3.6 ± 0.3	2.8 ± 0.9
LPS	3.8 ± 0.3	$2.9 \pm 0.2^*$	$2.6 \pm 0.2^*$	2.9 ± 0.5

Data are expressed as mean $\pm$ S.E.M of samples from 5–7 animals by duplicate; *denotes significantly different from respective control group by Student's t test ($p < 0.05$).

TABLE II
MITOCHONDRIAL H_2O_2 PRODUCTION RATE IN ENDOTOXEMIA

	Heart	Diaphragm	Liver	Lung
Control	0.31 ± 0.02	0.13 ± 0.02	0.15 ± 0.01	n.d.
LPS	0.32 ± 0.01	0.24 ± 0.03[*]	0.18 ± 0.01[*]	n.d.

Maximal H_2O_2 production rate was determined fluorometrically in the presence of 6 mM succinate as substrate and 2 μM antimycin as complex III inhibitor. Data are expressed in nmol/min/mg protein as the mean ± S.E.M of samples from 7 animals by duplicate;

[*] denotes significantly different from respective control group by Student t test ($p < 0.05$).
n.d.: not detected.

respective controls ($p < 0.05$, Table II). No differences were assessed in heart mitochondria; H_2O_2 could not be detected in lung organelles. The increased rate of H_2O_2 production of antimycin-supplemented mitochondria is consistent with selective functional impairment of organelles in the compromised tissues and indicates the previous inhibition of electron transfer rate at mitochondrial complexes II–III.

Detection of $ONOO^-$ in Mitochondria

Presence of $ONOO^-$ was accomplished by detection of 3-nitrotyrosine residues in mitochondrial proteins by Western blot with a monoclonal anti-nitrotyrosine antibody (Upstate Biotechnology Incorporated, Lake Placid, MA) developed by Beckman and co-workers.[27] In brief, 60 μg of proteins of the mitochondrial suspension were separated by electrophoresis on precast 7.5% SDS-polyacrylamide gel (Bio-Rad, Richmond, CA), transferred to a nitrocellulose membrane (Bio-Rad, Richmond, CA) and revealed by using the anti-nitrotyrosine antibody. Bovine serum albumin (BSA) nitrated after 30 minutes of incubation with 1 mM SIN-1, was used as a positive control. Incubation of the antibody with 10 mM nitrotyrosine, prior to the membrane incubation, was used to ensure the specificity of the antibody.

Western blot analysis of mitochondrial homogenates from endotoxemic animals exhibited a reproducible pattern of protein nitration for several

[27] J. S. Beckman, Y. Z. Ye, P. Anderson, J. Chen, M. A. Accavetti, M. M. Tarpey, and C. R. White, *Biol. Chem. Hoppe-Seyler* **375**, 81 (1994).

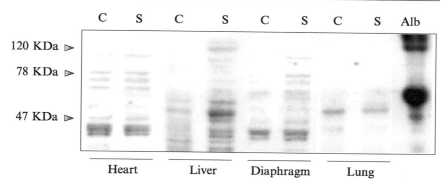

Fig. 5. Mitochondrial protein nitration. Nitration of mitochondrial proteins after 6 h of endotoxemia. *C*, control, *S*, septic. *Alb*, bovine serum albumin nitrated after 30 min exposure to 1 m*M* SIN-1.

bands (Fig. 5). Interestingly, immunoblot intensity was highest in liver and diaphragm mitochondria and lowest in heart samples, which showed only weak bands similar to control samples. No protein nitration was observed in lung mitochondria. The preincubation of the anti-nitrotyrosine antibody with free nitrotyrosine significantly attenuated the blotting signal, thus ensuring the specificity of the antibody (not shown).

Considering that increases in tissue and mitochondrial [NO]ss depend on iNOS activity and mitochondrial translocation, it is expected that in rat endotoxemia liver and lung would have twice the [NO]ss of heart and diaphragm. In diaphragm in similar conditions, [NO]ss was reported to be 0.473 μM.[3] Assuming a similar [NO]ss in heart and 0.9 μM in liver and that $d[ONOO^-]/dt/d[H_2O_2]/dt$ ratio increases at higher [NO]ss, being about 6 at 0.5 μM NO,[11] and considering that measured mitochondrial $d[O_2^-]/dt$ and $d[H_2O_2]/dt$ depends on [NO]ss and ubiquinol concentration (reactions [1] and [2]; Figs. 3 and 4), $d[ONOO^-]/dt$ should be higher in heart than in liver or diaphragm. Thus, very low nitration in heart organelles at high $d[ONOO^-]/dt$, may indicate the significance of high ubiquinol in heart protection as surmized from reaction [4]. In contrast, low ubiquinol in lung mitochondria decreases $d[O_2^-]/dt$ and $ONOO^-$ formation (reaction [1–3]), thus limiting mitochondrial protein nitration (Fig. 5).

The results show that, *(a)* in endotoxemia, oxidative stress, and the production of reactive oxygen species depend on increased NO production as released by iNOS; *(b)* relative amounts of NO production and ubiquinol concentrations determine distinct H_2O_2 and peroxynitrite production rates; and *(c)* intermediate concentrations set UQ-centered reactions in rates that favor subsequent impairment of mitochondrial respiratory functions by nitration of mitochondrial proteins. It is worth noticing that the highest cytosolic iNOS levels are followed by a proportional increase of mitochondrial iNOS, which should enhance the vectorial release of NO to the matrix.

These concepts are in agreement with the improvement in liver biochemical parameters of endotoxemic rats previously treated with UQ-*10*.[28] Administration of UQ-*10* decreased the overall mortality of endotoxemic dogs.[29] On the basis of iNOS activity/UQ content ratios, it is estimated that 3 to 10 fold increased UQH^- would be required in diaphragm and liver to limit critical $ONOO^-$ effects. It is noteworthy that the rates of the sequential reactions [1] and [2] depend on the levels of reduced UQ rather than total UQ in membranes. It is accepted that normal rat heart has a low degree of reduction of ubiquinone (30 to 40%).[24] However, UQH^-/UQ ratio is markedly increased in heart mitochondria in endotoxemia because of NO inhibition of cytochrome oxidase which augments the level of reduced intermediaries on the substrate-side of the electron transfer chain.

In this model of endotoxemia, the occurrence and pattern of nitration in different tissues is consistent with the clinical features and the evolution of septic multiorganic failure. Liver, skeletal muscle and diaphragm are compromised early in the evolution of sepsis, while heart compromise occurs at a later stage[23] or is less related to NO increase.[30] Endotoxemic lung distress is probably not related to primary mitochondrial damage. This fact suggests that lung damage, a common finding in clinical and experimental sepsis, is a consequence of neutrophil and macrophage recruitment and activation[31] rather than a result of early mitochondrial impairment. Although in those situations associated with excessive NO production a therapeutic utilization of UQ-*10* could be suggested, the magnitude and efficiency of the potential protection will depend on tissue bioavailability, incorporation to the mitochondrial membranes, and redox condition of ubiquinol.

[28] K. Sugino, K. Dohi, K. Yamada, and T. Kawasaki, *Surgery* **101,** 746 (1987).

[29] K. Yasumoto and Y. Inada, *Crit. Care. Med.* **14,** 570 (1986).

[30] M. Iqbal, R. I. Cohen, K. Marzouk, and S. F. Liu, *Crit. Care. Med.* **30,** 1291 (2002).

[31] B. J. Czermak, M. Breckwoldt, Z. B. Ravage, M. Huger-Lang, H. Schmal, N. M. Bless, H. P. Friedl, and P. A. Ward, *Am. J. Pathol.* **154,** 1057 (1999).

Acknowledgments

This work was supported by research grants from SECyT-ECOS-Sud A01S01, University of Buenos Aires (UBACyT M026), the Secretary for Science, Technology and Productive Innovation, Préstamo BID 1201 OC-AR PICT 08468, CONICET PIP 58 and the Fundación Perez Companc, Buenos Aires, Argentina.

[5] Mitochondrial Respiratory Chain Dysfunction Caused by Coenzyme Q Deficiency

By Pierre Rustin, Arnold Munnich, and Agnès Rötig

Introduction

Coenzyme Q_{10} (CoQ_{10}; ubiquinone 50) is an extremely hydrophobic molecule that plays a critical role in both antioxidant defenses and electron transfer activity of cell membranes, especially in the respiratory chain located in the mitochondrial inner membrane.[1,2] CoQ_{10} content varies strongly between tissues. Human heart, kidney, liver (114, 66.5, 55 $\mu g/g$ tissue respectively) and intestine, colon, testis or lung (11.5, 10.7, 10.5, 7.9 $\mu g/g$ tissue respectively) CoQ_{10} contents differ by more than one order of magnitude.[1] Skeletal muscle, pancreas, thyroid, spleen and brain (40, 33, 24.7, 24.6, 13.4 $\mu g/g$ tissue respectively) show median values. In addition, decreased ubiquinone content with age has been reported, although the decrease is variable depending on the tissue studied.[1] On the other hand, the redox status of CoQ_{10}, rather than its absolute amount, may be the crucial parameter to be examined under most physiological conditions.[3] On one hand, decreased CoQ_{10} content and/or increased oxidation status might denote a slowdown of oxidative metabolism activity in tissues, therefore a decreased need for electron transfer and antioxidant capacities, especially with age. On the other hand, decreased CoQ_{10} content might be one of the causative events in a cascade leading to the decrease of oxidative and antioxidant capacity with age. In other words, CoQ_{10} decrease over time could be seen either as a positive adjustment to a decreased metabolic demand, or as a harmful progressive deficiency with age.

[1] L. Ernster and G. Dallner, *Biochim. Biophys. Acta* **1271**, 195–204 (1995).
[2] F. L. Crane, *J. Am. Coll. Nutr.* **20**, 591–598 (2001).
[3] A. Kontush, S. Schippling, T. Spranger, and U. Beisiegel, *Biofactors* **9**, 225–229 (1999).

Respiratory Chain and Ubiquinone

CoQ_{10} plays a pivotal role in the mitochondrial respiratory chain, distributing the electrons between the various dehydrogenases and the cytochrome segments of the respiratory chain (Fig. 1).[4,5] Being in large excess compared to any other component of the respiratory chain (RC), it forms a kinetically compartmentalized pool which elevated redox status tightly decreases the activity of most of the RC dehydrogenases, aside from the succinate dehydrogenase.[6] According to the pioneer description by P. Mitchell, a CoQ_{10} cycle at RC complex III allows protons to be extruded from the mitochondrial matrix to the intermembrane space along with electron flow through the complex.[7]

Depending on its redox and protonation status, CoQ_{10} can react with molecular oxygen to produce superoxides (Fig. 1).[1] In particular, the full reduction of b-type cytochromes (favored by the oxidation of cytochrome c_1) and the increased formation of unstable ubisemiquinone ensures maximal superoxide production.[8] Accordingly, both ATPase inhibition (oligomycin) and genetic defects (NARP, Neuropathy, Ataxia with Retinitis Pigmentosa—mutation in the mitochondrial ATPase 6 gene) have been shown to favor high membrane potential with increased quinone reduction and subsequent superoxide overproduction.[9] Besides the pro-oxidant effect of CoQ_{10}, the fully reduced species (ubiquinol) can reduce a number of radical species, including superoxides, therefore acting as an antioxidant (Fig. 1). Noticeably, because CoQ_{10} is highly hydrophobic, hydrophobic radicals are presumably its natural radical targets—for example, lipoperoxy or tocopheryl radicals. CoQ_{10} can thus act as both a pro- and an antioxidant molecule.[10]

The partition of CoQ_{10} in the various compartments of a cell and in the membranes themselves (membrane core, membrane surfaces, and soluble phase) largely determines its efficiency as antioxidant *in vivo*. By manipulating the hydrophobicity of CoQ_{10}, a number of analogs have been

[4] A. Tzagoloff, "Mitochondria." Plenum Press, New York, 1982.

[5] D. Tyler, "The Mitochondria in Health and Disease." VCH Publishers, Inc., New York, 1992.

[6] M. Gutman, *in* "Bioenergetics of Membrane" (L. Packer, G. C. Pappageorgiou, and A. Trebs, eds.), pp. 165–175. Elsevier, Amsterdam, 1977.

[7] P. Mitchell, *FEBS Lett.* **59**, 137–139 (1975).

[8] J. H. Forman and A. Boveris, *in* "Free Radicals in Biology" (W. A. Pryor, ed.), pp. 65–90. Academic Press, New York, 1982.

[9] V. Geromel *et al.*, *Hum. Mol. Genet.* **10**, 1221–1228 (2001).

[10] H. Nohl, L. Gille, and A. V. Kozlov, *Free Radic. Biol. Med.* **25**, 666–675 (1998).

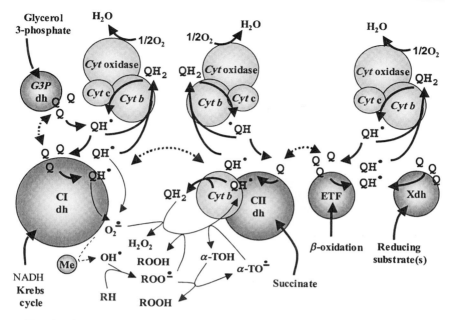

Fig. 1. The central role of CoQ_{10} (Q) in the distribution of electrons between dehydrogenases (dh) and cytochrome segments of the respiratory chain. Thick lines symbolize electron flow through the respiratory chain. Dotted lines correspond to electron exchanges between kinetically compartmentalized CoQ_{10} pools. Thin lines represent the antioxidant activity of CoQ_{10}. Cyt, Cytochrome; G3P, glycerol 3-phosphate; CI-II, complexes I and II of the respiratory chain, respectively; RH, ROOH, lipids and their peroxidized form; α-TO, α-tocopherol.

generated that target various intracellular locations. This has provided a choice of quinone analogs that can be selected as a function of the exact location of radicals to be scavenged.[11,12]

Finally, CoQ_{10} has been claimed to interfere with several additional mitochondrial functions, including intermediate metabolism steps, biosynthesis pathways, and apoptosis, either through the regulation of dehydrogenase activities, the generation of superoxides, or through direct interaction with other mitochondrial membrane components.[13,14]

[11] G. F. Kelso et al., J. Biol. Chem. 276, 4588–4596 (2001).
[12] G. F. Kelso et al., Ann. N. Y. Acad. Sci. 959, 263–274 (2002).
[13] E. Fontaine and P. Bernardi, J. Bioenerg. Biomembr. 31, 335–345 (1999).
[14] T. Yamamura et al., Antioxid. Redox. Signal. 3, 103–112 (2001).

Cellular Consequences of CoQ_{10} Depletion

Since the quinone pool constitutes a turntable in the respiratory chain and CoQ_{10} is a major antioxidant compound, a severe quinone depletion can result in an increase of reducing equivalents in both mitochondria and cytosol (e.g., elevated NADH/NAD ratio), and/or a decrease in mitochondrial ATP formation, and/or an increase of monovalent reduction of oxygen resulting in superoxide formation, and/or the functional impairment of numerous metabolic pathways requiring RC function (e.g., the tricarboxylic acid cycle and the β-oxidation). In addition, as quinone-dependent oxidative phosphorylation supplies most organs and tissues with energy, from the prenatal period onward, quinone depletion should theoretically give rise to various clinical manifestations, the severity of which would be dependent on the residual CoQ_{10} content.[5] The consequences of constitutive and profound ubiquinone depletion have been studied *in vitro* on cultured skin fibroblasts from a patient presenting with encephalomyopathy and widespread CoQ_{10} depletion.[15] Surprisingly enough, only partial decrease of cell respiration, mitochondrial oxidation of various substrates, and cell growth, was observed in these ubiquinone-deficient fibroblasts. Furthermore, these cells did not appear to overproduce either superoxide anions or lipoperoxides. Finally, apoptotic features did not increase as compared to control cells, even after serum deprivation. On the whole, these observations strongly suggest that ubiquinone does not play a major role in the antioxidant defenses of these cells. They may also denote that CoQ_{10}'s role in controlling oxidative stress and apoptosis varies greatly between cell types.[16] This assertion is supported by the tissue-specific involvement observed in the patient despite widespread ubiquinone depletion (e.g., no liver or heart involvement).[17]

Clinical Presentation of Coenzyme Q_{10} Depletion

Primary coenzyme Q_{10} (CoQ_{10}) deficiency is a rare, possibly treatable, autosomal recessive disorder with a clinical spectrum that encompasses three major phenotypes: (a) a myopathic form characterized by exercise intolerance, mitochondrial myopathy, myoglobinuria, epilepsy and ataxia; (b) a generalized infantile variant with severe encephalopathy and renal disease; and (c) an ataxic form, dominated by ataxia, seizures and either cerebral atrophy or anomalies of the basal ganglia.

[15] V. Geromel, A. Rotig, A. Munnich, and P. Rustin, *Free Radic. Res.* **36**, 375–379 (2002).

[16] S. Di Giovanni *et al.*, *Neurology* **57**, 515–518 (2001).

[17] V. Geromel *et al.*, *Mol. Genet. Metab.* **77**, 21–30 (2002).

The myopathic form associates the triad of central nervous system involvement, recurrent myoglobinuria and ragged-red fibers.[18-20] Lately, Di Giovanni et al. studied two brothers presenting at age 12 and 15 years with severe exercise intolerance, myoglobinuria, progressive weakness of trunk and proximal limb muscles, generalized seizures, elevated CK and lactic acidosis.[16] Muscle biopsies, performed respectively at age 15 and 17, showed ragged-red fibers (RRFs), lipid storage and mitochondrial respiratory chain complex I+III and II+III deficiency. Muscle CoQ_{10} concentration was reduced to 39% of the normal in patient 1 and 35% in patient 2. CoQ_{10} levels were nonetheless normal in fibroblasts and serum from both patients.

A more-generalized and infantile variant with severe encephalopathy and renal disease was recently described by Rötig et al.[21] Respiratory-chain function in two siblings with severe encephalomyopathy and renal failure suggesting a deficiency of CoQ_{10}-dependent respiratory-chain activities was assessed in muscle biopsy, circulating lymphocytes, and cultured skin fibroblasts. High-performance liquid chromatography (HPLC) analyses, combined with radiolabeling elements, allowed the quantification of cellular CoQ_{10} content. Undetectable CoQ_{10} and results of radiolabeling experiments in cultured fibroblasts supported the diagnosis of primary CoQ_{10} deficiency. The same analyses also disclosed a low content in deca-prenyl pyrophosphate, an intermediate compound in the synthesis of the lateral chain of CoQ_{10}. However, sequencing of the trans-prenyltrans-ferase cDNA (encoding the enzyme that elongates the prenyl side-chain of CoQ_{10}) failed to detect a disease-causing mutation in the coding sequence.[21] A somewhat related, neonatal and systemic presentation has also been reported in a Pakistani family.[22] Muscle CoQ_{10} depletion was established by both enzyme analysis and HPLC quantification. This systemic and severe presentation associating neurological symptoms, renal and cardiac features was found to poorly respond to CoQ_{10} oral supplementation.[22]

Finally, Musumeci et al. described six patients with cerebellar ataxia and cerebellar atrophy associated with low CoQ_{10} concentrations in skeletal muscle (<10 μg/g tissue, normal value: 25 ± 3).[23] As all known genetic causes of spinocerebellar atrophy (SCA) had been excluded, they attributed this apparently autosomal recessive syndrome to a primary defect of

[18] S. Ogasahara, A. G. Engel, D. Frens, and D. Mack, *Proc. Natl. Acad. Sci. USA* **86**, 2379–2382 (1989).

[19] C. Sobreira et al., *Neurology* **48**, 1238–1243 (1997).

[20] E. Boitier et al., *J. Neurol. Sci.* **156**, 41–46 (1998).

[21] A. Rotig et al., *Lancet* **356**, 391–395 (2000).

[22] S. Rahman, I. Hargreaves, P. Clayton, and S. Heales, *J. Pediatr.* **139**, 456–458 (2001).

[23] O. Musumeci et al., *Neurology* **56**, 849–855 (2001).

CoQ_{10}. Onset was in childhood in all but one patient, all patients had weakness, cerebellar ataxia, and cerebellar atrophy. Additional clinical features included seizure (3/6) and mental retardation (2/6). In 2000, the authors studied muscle biopsy samples from 110 patients with unexplained cerebellar atrophy and found CoQ_{10} concentrations <15 μg/g in 19/110. Eleven of these 19 patients had features similar to the original six patients: all had onset in childhood and showed weakness, cerebellar atrophy, and cerebellar ataxia. Seizures were present in three patients, mental retardation in four, pyramidal signs in two, and myoclonus in one. All patients seem to respond to CoQ_{10} administration. These observations confirm the existence of a CoQ_{10} deficiency syndrome dominated by cerebellar ataxia and atrophy, and raise the question of why the cerebellum would be especially vulnerable to CoQ_{10} deficiency. Lastly, a variant of the ataxic form has been reported in two siblings aged 29 and 32 years respectively, showing no cerebellar atrophy but involvement of the basal ganglia, and presenting the typical neuroradiological features of the Leigh syndrome associated with both pyramidal and extra-pyramidal signs.[24]

Detecting CoQ_{10} Deficiency

If severe, CoQ_{10} depletion can be detected indirectly through the measurement of CoQ_{10}-dependent enzyme activities.[21] On the other hand, it can be readily determined directly through the quantification of CoQ_{10} content in biological samples after extraction and HPLC analysis.[25,26] In the context of routine procedures for detecting RC deficiencies, a severe depletion of CoQ_{10} can be detected by reduced rates of both substrate-dependent oxygen consumption and RC CoQ_{10}-dependent enzyme activities. Adding a catalytic amount of oxidized exogenous quinone analog, such as decylubiquinone, should result in the *in vitro* normalization of deficient enzyme activities. Noticeably, glycerol-3 phosphate- and dihydroorotate cytochrome *c* reductase activities have been found especially sensitive to CoQ_{10} depletion.[21] The location of these dehydrogenases on the outer surface of the inner mitochondrial membrane and/or a lower affinity to CoQ_{10} may account for a higher dependency towards the CoQ_{10}-pool function, as compared to other RC dehydrogenases. All functional tests must be performed preferably on fresh material as to avoid membrane lipid

---

[24] L. Van Maldergem *et al.*, *Ann. Neurol.* **52,** 750–754 (2002).
[25] S. Imabayashi *et al.*, *Anal. Chem.* **51,** 534–536 (1979).
[26] K. Katayama, M. Takada, T. Yuzuriha, K. Abe, and S. Ikenoya, *Biochem. Biophys. Res. Commun.* **95,** 971–977 (1980).

disorganization that would result in artefactual quinone-pool dysfunction. In keeping with this, apparent CoQ_{10} depletion is frequently observed when studying freeze/thaw samples.

Supplementation Therapy

Several authors have reported a positive effect of CoQ_{10} therapy in patients with the myopathic variant and in patients with an atypical ataxic form. Lately, supplementation therapy initiated at a maximum daily dose of 300 mg for an adult was reported to improve most symptoms.[16] All symptoms and muscle weakness resolved, CK and lactic acid value returned to normal after two months of therapy and remained normal during a three-year follow-up. A second muscle biopsy was performed after 8 months of therapy. CoQ_{10} concentration returned to normal and activities of respiratory chain enzymes increased 2 to 3 times above pretreatment values. The authors concluded that muscle CoQ_{10} depletion could be corrected, which appears to stimulate mitochondrial proliferation and prevent apoptosis.[16]

The same group reported a good response to therapy in the recently described ataxic variant. They studied a 33-year old woman who was diagnosed with cerebral palsy in the early months of life. At 18 months, severe spastic tetraparesis and cognitive impairment were certified. Neurological symptoms remained stable and at the age of 31 years she developed psychiatric disturbances, well controlled by therapy. When examined, she was wheelchair-bound and presented trunk and limb ataxia, nystagmus, dysarthria, Babinski sign and mild mental retardation. EMG was myopathic. A sural nerve biopsy demonstrated axonal neuropathy. Brain MRI showed cerebellar atrophy and hyperintensity of white matter. Muscle morphology and respiratory chain enzymes were normal except for severe decrease of muscle CoQ_{10} (9.2 $\mu g/g$, normal value = 25 $\pm$ 3). Supplementation therapy was started at a daily dose of 900 mg/day with improvement in trunk ataxia. Compared with previously reported cases this patient showed distinctive clinical features as predominant spastic tetraparesis, extremely slow progression, axonal neuropathy and brain with matter abnormalities. Patients with CoQ_{10} deficiency syndrome dominated by cerebellar ataxia and atrophy, or presenting a pseudo-Leigh syndrome in adulthood also responded to CoQ_{10} supplementation (300 to 800 mg/day).[24]

Stimulation of respiration and fibroblast enzyme activities by exogenous quinones *in vitro* also prompted treatment of the two siblings with severe encephalomyopathy and renal failure with oral CoQ_{10} (ubidecarenone

5 mg/kg daily).[21] Clinical follow-up and detailed biochemical investigations of respiratory chain activity during the 5 years of oral quinone administration revealed a substantial improvement of their condition.[21]

Conclusion

Most frequently, CoQ_{10} depletion causes a mitochondrial encephalo-myopathy dominated by either myopathic or cerebellar involvement. A generalized infantile variant with severe encephalopathy and renal disease has also been reported. Noticeably, except for this latter case where the blockade in the biosynthesis of CoQ_{10} has been located, no information on the mechanism leading to CoQ_{10} depletion has been obtained. In view of the numerous potential causes of CoQ_{10} depletion putatively associated with a large number of human diseases,[27,28,29] the primary nature of the reported CoQ_{10} depletion remains to be established. Indeed, only the identification of disease-causing genes would allow establishment of the primary nature of CoQ_{10} depletion. Unfortunately, very few of the numerous genes involved in CoQ_{10} biosynthesis have been identified in human beings so far, and linkage analysis allowing for disease-gene identification is largely hampered by the absence of large families with CoQ_{10} depletion.

Among the large spectrum of recognized mitochondrial OXPHOS diseases resulting from deficiencies of various electron carriers of the mitochondrial respiratory chain, CoQ_{10} depletion (widespread or not) is distinguished by the frequently spectacular effect of oral CoQ_{10} supplementation. Further investigation should help decide the respective roles of antioxidant effects and electron-transfer restoration in the clinical improvement of the patients given quinone as a therapy. Whatever the mechanism, it is imperative that quinone-dependent multiple respiratory-chain enzyme deficiency be recognized, because in most cases this form of mitochondrial dysfunction is unique in responding well to oral quinone administration.

[27] K. Folkers et al., Int. Z. Vitaminforsch. **40**, 380–390 (1970).
[28] K. Folkers et al., Res. Commun. Chem. Pathol. Pharmacol. **28**, 145–152 (1980).
[29] G. Balercia et al., Andrologia **34**, 107–111 (2002).

[6] Coenzyme Q Cytoprotective Mechanisms

By Tom S. Chan, John X. Wilson, and Peter J. O'Brien

Coenzyme Q (CoQ, ubiquinone) cytoprotective mechanisms can be divided into bioenergetic function and antioxidant activity. To discern to what extent these processes are occurring in cells during cell stress it is necessary to measure the ability of the CoQ hydroquinone to both restore bioenergetic status (such as ATP generation and lactate/pyruvate ratios) and prevent membrane peroxidation and reactive oxygen species formation. This report reviews the current state of knowledge on the mechanism of CoQ-mediated cytoprotection. Also presented in this report are findings that suggest that CoQ_1 (ubiquinone-5) may be the optimal analog capable of protecting cells against cytotoxicity ensuing from Complex I inhibition. The mechanisms by which this could occur are discussed.

Introduction

All cell functions can be traced back to redox reactions taking place between cellular macromolecules. This is especially true for the mitochondrial mediated synthesis of ATP from electron transport that is exemplified by Peter Mitchell's Q cycle.[1] The redox active, lipophilic chemical, coenzyme Q (ubiquinone:CoQ) is present at high concentrations in the mitochondrial matrix and is necessary for the aerobic generation of ATP through the transport of electrons from Complex I and II to Complex III in the mitochondrial electron transport chain (ETC).

CoQ is synthesized through the melavonate pathway (shared by the cholesterol and dolichol synthesis) from 4-hydroxybenzoate, pyrophosphate and other cofactors.[2] The clinical importance of this pathway may underlie some of the toxicological effects observed with HMG-CoA reductase inhibitor (statin) therapy that not only inhibits cholesterol synthesis but may inhibit CoQ synthesis as well.[3] Clinical depletion of CoQ has been linked to many degenerative diseases such as myopathies[4] and neuropathies.[5]

---

[1] P. Mitchell, *FEBS Lett.* **59**, 137 (1975).

[2] G. Dallner and P. J. Sindelar, *Free Radic. Biol. Med.* **29**, 285 (2000).

[3] G. De Pinieux, P. Chariot, M. Ammi-Said, F. Louarn, J. L. Lejonc, A. Astier, B. Jacotot, and R. Gherardi, *Br. J. Clin. Pharmacol.* **42**, 333 (1996).

[4] T. Matsuoka, H. Maeda, Y. Goto, and I. Nonaka, *Neuromuscul. Disord.* **1**, 443 (1991).

[5] M. Ebadi, P. Govitrapong, S. Sharma, D. Muralikrishnan, S. Shavali, L. Pellett, R. Schafer, C. Albano, and J. Eken, *Biol. Signals Recept.* **10**, 224 (2001).

Recent attention has focused on other roles of CoQ in the cell. CoQ has been found to exist within virtually all phospholipid membranes, where it acts as a potent antioxidant by scavenging membrane radicals and regenerating other antioxidants.[6,7] Therefore, the cytoprotective activity of CoQ has become a multifaceted subject consisting of both antioxidant and bioenergetic processes.

The distribution of endogenous CoQ is limited by the lipophilic membrane content of the cell. With a calculated $logD > 20$, endogenous CoQ can exist only within the lipophilic core of the phospholipid bilayer.[8] Exogenously administered CoQ_{10} to rats did not significantly accumulate in muscle or the central nervous system but rather remained in the blood.[9] It may be that for this reason, CoQ therapy for Parkinson's disease required extremely high and frequent dosing before clinical effects could be observed (1200 mg/d).[10] Also, studies involving the endogenous activity of CoQ in cells have been difficult to perform, because its extraction (requiring the use of hydrophobic solvent) is a multistep process.[11] Fortunately, shorter chain CoQ analogs have been either found or synthesized which can act as surrogates for endogenous CoQ in the electron transport chain. A few studies using isolated mitochondria, submitochondrial particles (SMP) or electron transporting particles (ETP) (from here on collectively called mitochondrial subcellular fractions [MSF]) have shown that these analogs possess Complex I dependent NADH oxidase and cytochrome C reducing potential, suggesting that they are indeed able to participate in the ETC.[12] However, the mechanisms by which these CoQ analogues carry out this role are still under debate (Table I).

Cytoprotective Activities of CoQ: Antioxidant Activity

The cytoprotective activity of endogenous CoQ is mediated by its hydroquinone, which can be formed from the activity of mitochondrial Complex I (NADH dependent ubiquinone oxidoreductase [NQO$_2$]),

[6] A. Arroyo, F. Navarro, C. Gomez-Diaz, F. L. Crane, F. J. Alcain, P. Navas, and J. M. Villalba, *J. Bioenerg. Biomembr.* **32**, 199 (2000).

[7] P. Forsmark, F. Aberg, B. Norling, K. Nordenbrand, G. Dallner, and L. Ernster, *FEBS Lett.* **285**, 39 (1991).

[8] M. Jemiola-Rzeminska, J. Kruk, M. Skowronek, and K. Strzalka, *Chem. Phys. Lipids* **79**, 55 (1996).

[9] Y. Zhang, F. Aberg, E. L. Appelkvist, G. Dallner, and L. Ernster, *J. Nutr.* **125**, 446 (1995).

[10] R. N. Rosenberg, *Arch. Neurol.* **59**, 1523 (2002).

[11] G. Rousseau and F. Varin, *J. Chromatogr. Sci.* **36**, 247 (1998).

[12] Y. P. Wan, R. H. Williams, K. Folkers, K. H. Leung, and E. Racker, *Biochem. Biophys. Res. Commun.* **63**, 11 (1975).

TABLE I
STRUCTURE OF SHORT CHAIN CoQ ANALOGS

CoQ Analog	Log P (cyclohexane/water)	Structure
CoQ_0	0.39	
CoQ_1	2.65	
CoQ_2	5.1	
CoQ_{10}	20.9	

CoQ$_{10}$ and several shorter chain analogs. Calculated Log P of CoQ_{10} is from ALOGPS 2.1 (http://146.107.217.178/lab/alogps/).

Complex II (succinate dependent FADH:ubiquinone oxidoreductase), NAD(P)H dependent quinone oxidoreductase (NQO_1)[13] and NADPH dependent ubiquinone reductase (UQR).[14,15] Hydrophilic quinones prefer to be reduced by NQO_1,[16] whereas hydrophobic and endogenous CoQ

[13] R. E. Beyer, J. Segura-Aguilar, S. di Bernardo, M. Cavazzoni, R. Fato, D. Fiorentini, M. C. Galli, M. Setti, L. Landi, and G. Lenaz, *Mol. Aspects Med.* **18** (Suppl. S15) (1997).
[14] T. Kishi, T. Takahashi, S. Mizobuchi, K. Mori, and T. Okamoto, *Free Radic. Res.* **36**, 413 (2002).

analogs are preferentially reduced by complex I or UQR.[14] CoQ hydro-quinone has been shown to have potent antioxidant activity, preventing lipid peroxidation initiated by a variety of insults. For example, mitochondria depleted of CoQ by extraction with pentane have been shown to have increased susceptibility to Fe-ADP or adriamycin-mediated loss in respiration. Reconstitution of the mitochondria with CoQ was able to restore mitochondrial function.[7] The antioxidant activity of endogenous CoQ however, seems to be limited to preventing lipid peroxidation-mediated protein damage as CoQ hydroquinone supplementation was unable to protect against damage to the integral membrane protein, nicotinamide nucleotide transhydrogenase from peroxynitrite damage.[17]

Exogenously added CoQ_{10} to cellular systems does not seem to exhibit a significant level of protection against oxidative stress (at least not as high as that observed with the addition of vitamin E).[18] Instead, the cytoprotective role of CoQ_{10} tended to supplement that of other antioxidants. This could be explained by its ability to mediate the regeneration of spent antioxidant radicals. CoQ has been shown to be involved with the regeneration of both α-tocopherol (vitamin E) and ascorbate (vitamin C).[6,19]

Unlike ascorbate, α-tocopherol cannot be regenerated by GSH or NAD(P)H. However, it was later discovered that because CoQ was lipophilic and could be reduced in mitochondria, CoQ hydroquinone could mediate the regeneration of α-tocopherol from its chromanoxyl radical. This was substantiated by a study demonstrating that α-tocopherol was a more effective inhibitor of lipid peroxidation in mitochondrial membranes in the presence of succinate which acts to convert CoQ to CoQ hydroquinone, implying that a continuous supply of CoQ hydroquinone helps to restore mitochondrial α-tocopherol, presumably by recycling α-tocopherol.[6,20] On a clinical level, antioxidant synergy could be observed in healthy volunteers as a decrease in the number of apoptotic CD8 lymphocytes was observed when administered α-tocopherol and CoQ.[21]

However, the CoQ semiquinone radical is able to autoxidize to form superoxide if allowed access to H_2O. This paradoxical pro-oxidant activity was implicated in the lower antioxidant effect of a given concentration of

[15] T. Kishi, T. Takahashi, A. Usui, and T. Okamoto, *Biofactors* **10,** 131 (1999).

[16] L. Ernster, *Methods Enzymol.* **10,** 309 (1967).

[17] P. Forsmark-Andree, B. Persson, R. Radi, G. Dallner, and L. Ernster, *Arch. Biochem. Biophys.* **336,** 113 (1996).

[18] H. Shi, N. Noguchi, and E. Niki, *Free Radic. Biol. Med.* **27,** 334 (1999).

[19] E. Niki, *Mol. Aspects Med.* **18** (Suppl. S63) (1997).

[20] A. Lass and R. S. Sohal, *Arch. Biochem. Biophys.* **352,** 229 (1998).

[21] L. Mosca, S. Marcellini, M. Perluigi, P. Mastroiacovo, S. Moretti, G. Famularo, I. Peluso, G. Santini, and C. De Simone, *Biochem. Pharmacol.* **63,** 1305 (2002).

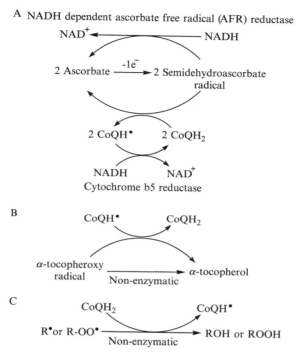

FIG. 1. The antioxidant activities of CoQH$_2$. (A) NADH dependent regeneration of ascorbate from the ascorbyl radical. (B) NAD(P)H dependent regeneration of α-tocopherol from its chromanoxyl radical. (C) Scavenging of lipid peroxyl and carbon-centered radicals in a direct lipid peroxidation chain breaking role.

CoQ in comparison to the same concentration of vitamin E in peroxidizing liposomes.[6,22] However, the pro-oxidant activity of the semi-quinone radical of CoQ may only be significant when cellular antioxidant mechanisms are severely compromised (e.g., low catalase, GSH peroxidase or GSH present). The presence of other reducing systems in the membranes of other cellular organelles may also play a part in fortifying the antioxidant role of vitamin E through the generation of CoQ hydroquinone.

Ascorbate on the other hand, is regenerated across the plasma membrane with the aid of cytochrome b$_5$ reductase, a heme containing enzyme responsible for mediating the transfer of electrons from intracellular NADH to extracellular electron acceptors such as the ascorbyl radical and ferricyanide (Fig. 1).[23]

[22] H. Nohl, L. Gille, and A. V. Kozlov, *Free Radic. Biol. Med.* **25,** 666 (1998).

Diversity of Antioxidant/Pro-Oxidant Capacity Between Different CoQ Analogs

Short chain analogs of CoQ found in microorganisms have been used for the study of mitochondrial function because of their increased solubility in aqueous solution and their ability to participate in the ETC. However, only a few studies have pointed out that the hydroquinones of shorter chain CoQ analogs possess greater antioxidant potential against lipid peroxidation.[24] This antioxidant activity, however, may be offset by the increased propensity of these analogs to react with unbound transition metals and oxygen. For example, the hydrophilicity and autoxidizability of reduced CoQ_0 (which lacks a position 6 side chain) could result in the oxidation and depletion of ascorbate in the cell.[25]

Furthermore, hydrophilic CoQ analogs may be more amenable to biotransformation in the cell. For example, we have recently found that the hydroquinone of CoQ_1 underwent metabolic deactivation by sulfotransferases to form the CoQ_1 sulfate conjugate, which decreased its cytoprotective and antioxidant activity.[26]

Cytoprotective Activities of CoQ: Protection Against Reductive Stress Caused by Complex I Inhibition in Isolated Rat Hepatocytes

Complex I inhibition has been associated with several neurological disorders including Parkinson's disease, Lebers hereditary optic neuropathy and Leigh's syndrome.[27,28] Mitochondrial toxicity with Complex I inhibition is associated with a marked increase in the cellular reduction potential of the cell caused by the accumulation of NADH that could be ameliorated by the addition of micromolar levels of menadione[29] through NQO_1 catalyzed NADH oxidation. Since shorter-chain analogs of CoQ were found to be able to participate in Complex I and III proton translocation, we have concentrated on studying the effects of CoQ_1 analogs on the cytotoxic effects of Complex I inhibition in isolated rat hepatocytes.

[23] J. M. Villalba, F. Navarro, C. Gomez-Diaz, A. Arroyo, R. I. Bello, and P. Navas, *Mol. Aspects Med.* **18** (Suppl. S7) (1997).

[24] V. E. Kagan, E. A. Serbinova, G. M. Koynova, S. A. Kitanova, V. A. Tyurin, T. S. Stoytchev, P. J. Quinn, and L. Packer, *Free Radic. Biol. Med.* **9,** 117 (1990).

[25] V. A. Roginsky, T. K. Barsukova, G. Bruchelt, and H. B. Stegmann, *Free Radic. Res.* **29,** 115 (1998).

[26] T. S. Chan and P. J. O'Brien, *Biofactors* (in press) (2003).

[27] A. Majander, K. Huoponen, M. L. Savontaus, E. Nikoskelainen, and M. Wikstrom, *FEBS Lett.* **292,** 289 (1991).

[28] M. Y. Huang, Y. J. Jong, J. L. Tsai, G. C. Liu, C. H. Chiang, C. Y. Pang, and Y. H. Wei, *J. Formos. Med. Assoc.* **95,** 325 (1996).

[29] F. A. Wijburg, N. Feller, C. J. de Groot, and R. J. Wanders, *Biochem. Int.* **22,** 303 (1990).

CoQ$_1$ was able to normalize the lactate:pyruvate ratio in Complex I inhibited cells, but not in NQO$_1$ inhibited cells (using the inhibitor dicumarol). The restoration of NADH/NAD$^+$ ratio in the cell probably had a 2-fold effect; inhibiting the Complex I–mediated NADH-dependent generation of reactive oxygen species that prevented the release of iron and further ROS (reactive oxygen species) generation as well as abolishing the feedback inhibition that increased NADH levels would have on the glycolytic ATP supply. Restoration of mitochondrial membrane potential also indicated that CoQ$_1$H$_2$ was acting at the level of the mitochondria, probably by acting as an electron bypass and supplying electrons to Complex III (Fig. 2).

Cytoprotective Activities of CoQ: Re-establishing Mitochondrial Function in Complex I Inhibited Isolated Rat Hepatocytes

A few studies in the past have acknowledged the specificity of shorter chain CoQ analogs that participate in the electron transport chain through Complex I, II and III[30–32] in subcellular mitochondrial fractions. In particular, the mitochondrial proton pumping efficiency of shorter chain CoQ analogs is poor compared with those with longer chains, presumably because of their inability to access the physiological sites of Complex I and III that are likely buried deep within the hydrophobic portion of the membrane.[25,33,34] Furthermore, in support of the H$_2$O-dependent pro-oxidant role of CoQ, shorter chain (more hydrophilic) analogs generally have less-stable semiquinone radicals that can autoxidize more readily. Thus the antioxidant activity of these shorter-chain analogs would depend on the maintenance of the levels of their corresponding hydroquinones and minimizing their oxidation via one electron process.

Because shorter chain analogs of CoQ are more rapidly reduced by NQO$_1$, their intracellular hydroquinone forms may be entirely maintained by this system. NQO$_1$ has a k$_{cat}$ for CoQ$_1$ approximately 100 times greater than for CoQ$_{10}$.[13] This is evidenced by the marked reduction in cytoprotection observed with Complex I inhibited hepatocytes in the presence of dicumarol, an NQO$_1$ inhibitor.[35]

[30] E. M. Degli, A. Ngo, G. L. McMullen, A. Ghelli, F. Sparla, B. Benelli, M. Ratta, and A. W. Linnane, *Biochem. J.* **313**(Pt 1), 327 (1996).
[31] K. Sakamoto, H. Miyoshi, M. Ohshima, K. Kuwabara, K. Kano, T. Akagi, T. Mogi, and H. Iwamura, *Biochemistry* **37**, 15106 (1998).
[32] Y. P. Wan, R. H. Williams, K. Folkers, K. H. Leung, and E. Racker, *Biochem. Biophys. Res. Commun.* **63**, 11 (1975).
[33] D. Xia, C. A. Yu, H. Kim, J. Z. Xia, A. M. Kachurin, L. Zhang, L. Yu, and J. Deisenhofer, *Science* **277**, 60 (1997).
[34] A. Dupuis, I. Prieur, and J. Lunardi, *J. Bioenerg. Biomembr.* **33**, 159 (2001).

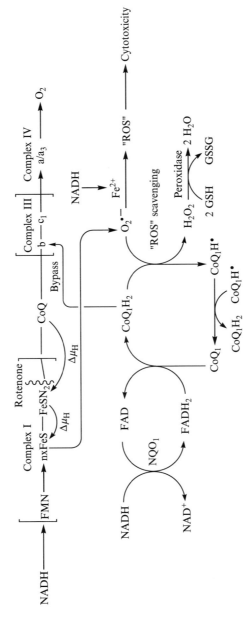

Fig. 2. NQO₁-mediated reduction of CoQ_1 to CoQ_1H_2 re-establishes cellular redox potential, provides antioxidant activity and restores mitochondrial function in Complex I inhibited hepatocytes. (T. S. Chan, S. Teng, J. X. Wilson, G. Galati, S. Khan, P. J. O'Brien, Coenzyme Q cytoprotective mechanisms for mitochondrial Complex I cytopathies involves NAD(P)H: quinone oxidoreductase 1 (NQO₁), *Free Rad. Res.* **36**(4), 421–427.

Diversity of Mitochondrial Function Between Different CoQ Analogs

Although endogenous CoQ_{10} has been shown to have the highest mitochondrial NADH-dependent proton motive activity, it is poorly suited for both *in vitro* and *in vivo* supplementation because its cellular incorporation requires the use of liposomal formulations. Furthermore, exogenously added CoQ_{10} may not even reach its intended site of action (the mitochondria) but instead prefer to stay within the plasma membrane. For this reason, shorter chain CoQ analogs have been considered to be a reasonable alternative for the study of mitochondrial function in isolated mitochondria and submitochondrial fractions. However, the use of CoQ analogs in cells may not necessarily support the data from subcellular fractions, especially since factors such as NQO_1 activity and glucose metabolism play vital roles. In an effort to address this obscurity, the mitochondrial activity of short chain CoQ analogs such as CoQ_0, CoQ_1 and CoQ_2, which differ only in the length of their position 6 isoprenyl chain, have been investigated for cytoprotective activity against rotenone-induced cell death and ATP depletion.

Materials and Methods

CoQ analogs (CoQ_0, CoQ_1 and CoQ_2) were purchased from Sigma Chemical Co.

Animals

Male Sprague Dawley rats were kept at a regular light and dark cycle and fed ad libitum with standard rat chow. Their livers were excised and placed immediately on ice-cold KCl (1.15%) and rinsed in ice-cold phosphate (pH 7.4, 0.05 M) buffered KCl. The liver was then cut into small pieces in ice-cold buffered KCl and homogenized with three strokes in a Heidolph Teflon homogenizer. The resulting homogenate was centrifuged at 8000 rpm. This step was repeated twice. The resulting supernatant was centrifuged at 106,000 × g for 60 minutes, and the cytosolic supernatant was separated from the microsomal pellet. The supernatant was washed in a second centrifugation at 106,000 × g for 60 minutes. The subcellular fractions were stored at −70° until used.[36]

---

[35] T. S. Chan, S. Teng, J. X. Wilson, G. Galati, S. Khan, and P. J. O'Brien, *Free Radic. Res.* **36,** 421 (2002).

[36] E. Reid and R. Williamson, *Methods Enzymol.* **31,** 713 (1974).

Spectrophotometric Determination of Cellular NQO_1 Activity

CoQ analogs were prepared in methanol at a concentration of 2 mM stock. Ten μl of cytosol (1.54 mg/ml) and 10 μl of 10 mM NADH were added to 970 μl of Tris-HCl with 1 mM DETAPAC (pH 7.4). The mixtures was monitored for 5 minutes at 340 nm on a Pharmacia ultrospec 1000 with SWIFT software. Afterwards, 10 μl of CoQ analog was added. The cuvette was inverted to allow for complete mixture and placed back into the spectrophotometer. The reaction was monitored for an additional 10 minutes. The rate of NADH oxidation was calculated from the slope of the line (Fig. 3).

Measuring the Cytoprotective Effect of CoQ Analogs on Complex I Inhibited Hepatocytes

CoQ_0, CoQ_1 and CoQ_2 were added to isolated rat hepatocytes that were pretreated for 30 minutes with rotenone (35 μM) (Fig. 4). Cytotoxicity was measured 0.5, 1, 2, 3, 4, 5, 6 and 7 hours after the addition of CoQ analogs by measuring the percentage of cells taking up trypan blue 0.2%.

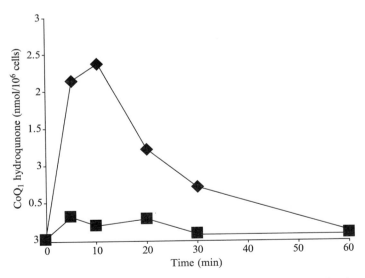

FIG. 3. Reduction of CoQ_1 in isolated hepatocytes. Hepatocytes were incubated at a density of 0.5×10^6 cells at $37°$ under an atmosphere of 95% O_2 and 5% CO_2. CoQ hydroquinone was determined as described in the Materials and Methods section.

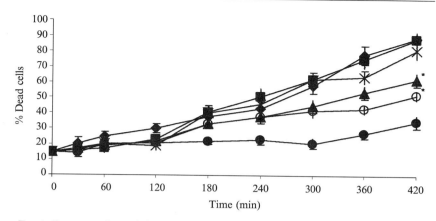

FIG. 4. Cytoprotective activity of CoQ_1 in Complex I–inhibited hepatocytes. Hepatocytes were incubated at a density of 0.5×10^6 cells at $37°$ under an atmosphere of 95% O_2 and 5% CO_2. (■) Rotenone 35 μM, (▲) CoQ_0 5 μM, (●) CoQ_1 5 μM, (○) CoQ_2 5 μM, (✕) CoQ_1 20 μM, (◆) Control. Cytotoxicity was assessed by trypan blue exclusion, and ATP was measured as described in Materials and Methods. Values are expressed as the mean ± standard error of 3 separate experiments. *Statistically significant from rotenone treatment alone using one-way analysis of variance, followed by Tukey's post hoc test.

Assaying the Cellular Reduction of CoQ_1

A cell suspension consisting of 1 million hepatocytes/ml was isolated by the collagenase perfusion method[37] and incubated in rotating round-bottom flasks at $37°$ under an atmosphere of 95% O_2 and 5% CO_2 (Fig. 3). The cytotoxicity of CoQ_0, CoQ_1 and CoQ_2 limited their study to concentrations below 50 μM in cells. CoQ_1 20 μM dissolved in methanol was added to the hepatocytes and at various time points (10, 20, 30, and 60 minutes) 450 μl aliquots of cell suspension were added to 50 μl of 15% metaphosphoric acid. The extract was centrifuged at 13,500 RPM for 5 min and 100 μl of the supernatant was injected into a 20 μl sample loop. The HPLC was running in isocratic mode with a mobile phase of MeOH, H_2O and CH_3COOH at a ratio of (70:29:1) with 2 mM $C_2H_7NO_2$ running at room temperature with a flow rate of 0.7 ml/min. Under these conditions the CoQ_1 hydroquinone eluted at 10 minutes, while CoQ_1 eluted at 22 minutes.

CoQ_1 hydroquinone was verified by liquid chromatography/mass spectroscopy performed on a PSIEX biomass electrospray ionization mass spectrometer.

---

[37] S. Orrenius, H. Thor, J. Rajs, and M. Berggren, *Forensic Sci.* **8**, 255 (1976).

Measuring ATP Content in Isolated Hepatocytes

A 1-ml sample of cell suspension was added to 1 ml of 0.5% KOH solution and allowed to stand on ice for 3 minutes. An additional aliquot of 1 ml was added to the suspension, which was centrifugally filtered with conical filters (Amicon Inc.) at 2500 rpm. ATP was measured by using a Phenomenex C18 column (250×4.6 mm, 5 um) on a gradient UV-HPLC program consisting of 100% 0.1 M KH_2PO_4 (pH 6) for the first 15 min, followed by a linear gradient to 100% $MeOH:KH_2PO_4$ (10:90; pH 6) for 30 minutes, followed by a 10-minute washout period with 100% 0.1 M KH_2PO_4. The HPLC system consisted of a Shimadzu SCL-6B module controlling two Shimadzu LC-6A pumps connected to a Shimadzu SPD-6AV UV detector set at 254 nm.[38]

Ease of Reduction of CoQ Analogs

All three CoQ analogs were rapidly reduced by isolated rat liver cytosol with $CoQ_0 > CoQ_1 > CoQ_2$. The rates of NADH and NADPH oxidation were almost abolished by the addition of 5 μM dicumarol. Interestingly, the rate of dicumarol-resistant reduction between the analogs was inversely proportional to their rates of reduction in the absence of dicumarol (Table II).

TABLE II

OXIDATION OF NADH AND NADPH MEDIATED BY CYTOSOLIC NAD(P)H QUINONE OXIDOREDUCTASE (NQO_1)

CoQ Analog	NAD(P)H oxidation (nmol/min)					
	NADPH	NADPH with Dicumarol	% NQO_1 indep.	NADH	NADH with Dicumarol	%NQO_1 indep.
CoQ_0	92.2	0.8 ± 0.1	0.8	52 ± 4	0.76 ± 0.04	1.4
CoQ_1	31.8	1.5 ± 0.3	4.7	18.2 ± 0.7	0.93 ± 0.06	5.1
CoQ_2	17.5	1.7 ± 0.3	9.7	15.1 ± 0.3	1.03 ± 0.05	6.8

A 15.4 $\mu g/ml$ sample of rat liver cytosol was added to 0.1 M Tris-HCl (pH 7.4) in a quartz cuvette with or without 5 μM dicumarol. NADPH or NADH (100 μM each) were added to the cuvette. The reaction was started by the addition of CoQ analog at a concentration of 20 μM. NADH and NADPH were followed at 340 nm. NAD(P)H oxidation is represented as the mean $\pm$ the S.E.M. from 3 separate experiments.

[38] V. Stocchi, L. Cucchiarini, M. Magnani, L. Chiarantini, P. Palma, and G. Crescentini, *Anal. Biochem.* **146**, 118 (1985).

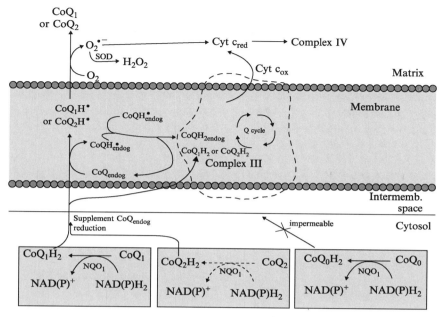

FIG. 5. Interaction of CoQ analogs with mitochondrial respiration in Complex I–inhibited cells.

Over 10% of the 20 μM CoQ$_1$ was rapidly converted to its hydroquinone within 5 minutes of its addition (Fig. 5), however, CoQ$_1$ hydroquinone was gradually oxidized over a period of 20 minutes back to CoQ$_1$, presumably because of the loss in NAD(P)H reducing potential in the cell combined with either the use of CoQ$_1$ for cellular respiration and antioxidant activity or autoxidation (Fig. 5).

Curiously, in cells, CoQ$_1$ was the only analog able to significantly prevent rotenone-induced hepatocyte cytotoxicity throughout a seven-hour exposure. CoQ$_2$ showed low levels of activity after 5 hours that were not statistically significant. CoQ$_0$ showed a complete lack of effect. ATP depletion by rotenone was more than 85% complete 1.5 hours following its administration. Both CoQ$_0$ and CoQ$_2$ showed low but significant prevention of ATP depletion. However, CoQ$_1$ (Table III) was found to be at least 4 times more effective than the other two at preventing ATP depletion. A 4-fold increase in the concentration of CoQ$_1$ slightly elevated the protective activity of the lower concentration of CoQ$_1$.

TABLE III

RESTORATION OF ATP LEVELS IN COMPLEX I–INHIBITED HEPATOCYTES BY CoQ ANALOGS

	% Hepatocyte ATP Relative to the Control	
Treatment	1 hr	2 hr
Control	100	100
+ Rotenone 35 mM	14 ± 3	19 ± 3
+ CoQ$_0$ 5 mM	23 ± 5	22 ± 3
+ CoQ$_1$ 5 mM	$52 \pm 5^*$	$37 \pm 4^*$
+ Dicumarol 15 μM	$15 \pm 4^{**}$	$18 \pm 4^{**}$
+ CoQ$_2$ 5 mM	28 ± 4	25 ± 3
+ CoQ$_1$ 20 mM	$66 \pm 5^*$	$67 \pm 6^*$

Hepatocytes (10^6 cells/ml) were suspended in Krebs-Henseleit buffer at 37° under an atmosphere of 95% O_2, 5% CO_2 were preincubated with 35 μM rotenone for 15 minutes prior to the addition of CoQ analogs. Cells that were NQO$_1$ inhibited were preincubated with dicumarol for 20 minutes prior to the addition of rotenone. Values are expressed as the mean $\pm$ standard error of 3 separate experiments.
  *Statistically significant from the rotenone treatment alone or
  **rotenone + CoQ$_5$ (5 μM) treatment using one-way analysis of variance, followed by Tukey's post hoc test.

Discussion

Low levels of either CoQ$_1$ or menadione have been shown to partially restore mitochondrial function in Complex I inhibited fibroblasts[39] or isolated hepatocytes.[35] The mechanism of this cytoprotective activity has been attributed to the action of NAD(P)H dependent quinone oxidoreductase that reduced menadione or CoQ$_1$ to menadiol or CoQ$_1$ hydroquinone, respectively. These reduced quinones could participate in the mitochondrial electron transport chain, effectively bypassing Complex I inhibition. Also, it was found that CoQ$_1$ hydroquinone acted as an effective antioxidant that prevented rotenone-associated cell death (which could be generated by reductive stress leading to the liberation of free iron and generation of ROS).[35] A non-cytotoxic concentration of the NQO$_1$ inhibitor, dicumarol, was able to inhibit NQO$_1$ activity and abolish CoQ$_1$ cytoprotection.

Short-chain analogs of CoQ such as CoQ$_0$, CoQ$_1$ and CoQ$_2$ were efficiently reduced by cytosolic NQO$_1$ but were likely poor substrates for a cytosolic, dicumarol-insensitive UQR that reduces CoQ$_9$ and CoQ$_{10}$ (described elsewhere).[14] Therefore, cytosolic NQO$_1$ is the major enzyme involved in the reduction of these short chain CoQ analogs to their perspective hydroquinones.

[39] F. A. Wijburg, N. Feller, C. J. de Groot, and R. J. Wanders, *Biochem. Int.* **22,** 303 (1990).

We have conducted a study with lower concentrations of rotenone with a longer incubation time and found that among the CoQ analogs tested in this study, CoQ_1 was optimal for preventing Complex I-mediated cell death. This activity could be due to the high NADH oxidoreductase activity of CoQ_1 through Complex I as compared with both CoQ_0 and CoQ_2.[40] However, this would be an unfavorable hypothesis for several reasons: first, it does not explain why NQO_1 inhibition would have such a dramatic effect on the cytoprotective activity of CoQ_1 in Complex I inhibited cells[35]; second, CoQ_1 Complex I–mediated activity has been shown to be highly sensitive to rotenone inhibition[41]; third, it was unlikely that the CoQ_1 hydroquinone could deliver its electrons directly to Complex III given that the CoQ_1 hydroquinone had a K_{cat}/K_m more than 3 times higher than the CoQ_2 hydroquinone for purified E-coli ubiquinol oxidase,[31] and endogenous CoQ has been shown to compete for CoQ_1 in the physiological active site for complex I[42]; fourth, the CoQ_2 and CoQ_1 hydroquinone forms were found to inhibit CoQ_1-generated Complex I activity.[31]

Short-chain analogs of $CoQH_2$ should be able to reduce endogenous CoQ_9 to its hydroquinone, thereby supplementing proton movement in the mitochondrial inner membrane through Complex III. This would explain why CoQ_1 could not overcome Complex III inhibition.[35] CoQ_0 would therefore be the most-effective CoQ_9 reducer, given that it possessed the greatest rate of reduction by NQO_1; however, hydrophilicity may limit its hydroquinone from interacting with endogenous CoQ_9. The low cytoprotective activity of CoQ_2 could be explained by its poorer reduction by NQO_1 (especially when using NADPH as a cofactor). Studies involving the effects of the hydroquinones of CoQ analogs on Complex I–inhibited MSFs that are depleted of endogenous CoQ would better resolve this question.

The only short-chain analog of CoQ currently in use for the treatment of Parkinson's disease is idebenone (2,3-dimethoxy-5-methyl-6-decan-1-ol-1,4, benzoquinone, IDB).[43] Following its approval for use in Japan, the therapeutic mechanism by which it acts has been extensively investigated. Interestingly, IDB has relatively poor Complex I NADH:ubiquinone oxidoreductase activity and high Complex I NADH oxidase activity,[44] suggesting that it may

[40] G. Lenaz, Biochim. Biophys. Acta 1364, 207 (1998).

[41] Y. Nakashima, K. Shinzawa-Itoh, K. Watanabe, K. Naoki, N. Hano, and S. Yoshikawa, J. Bioenerg. Biomembr. 34, 89 (2002).

[42] R. Fato, S. D. Bernardo, E. Estornell, C. G. Parentic, and G. Lenaz, Mol. Aspects Med. 18(Suppl. S269) (1997).

[43] "Idebenone-Monograph" Altern. Med. Rev. 6, 83 (2001).

[44] M. D. Esposti, A. Ngo, A. Ghelli, B. Benelli, V. Carelli, H. McLennan, and A. W. Linnane, Arch. Biochem. Biophys. 330, 395 (1996).

supplement electron transport through Complex II or through the reduction of cytochrome C by superoxide generated through Complex I. However, given that the reduction of IDB can occur through NQO_1, it may be cytoprotective by serving both bioenergetic and antioxidant roles in the cell through a combination of intracellular reductive pathways.

Role of CoQ and Analogs in the Modulation of Mitochondrial Permeability Transition: Possible Link Between Mitochondrial Permeability Transition and CoQ_1-Mediated Cytoprotection

Recently there has been interest in the role of CoQ and its analogs in the regulation of mitochondrial permeability transition (MPT). It was found that the CoQ analogs CoQ_0 and CoQ_2 were able to effectively inhibit the calcium-induced MPT in skeletal muscle mitochondria. This MPT-inhibitory activity was associated with the ability of these analogs to increase the threshold for causing calcium induced MPT[45,46] through the interaction with a specific site in the mitochondria, whereas CoQ_1 opposed this inhibitory activity, conceivably through the displacement of inhibitory quinones from the site. Since then, a number of CoQ analogs have been shown to have varying effects on modulating MPT caused by calcium exposure in mitochondria.[47] Regrettably, many of these CoQ analogs also inhibit electron transport through Complex I, thus limiting their study to SMFs. Since rotenone has been shown to induce MPT, CoQ_1 could also modulate the sensitivity of Complex I to rotenone-induced MPT.

Conclusions

While the ongoing clarification of the role of CoQ in subcellular fractions and isolated enzymes continues, there is a need to address the role of this biogenic quinone in the cell. These studies could lead to the design of novel molecular pharmaceutical strategies for disease treatment or prevention in particular those with an age-related neurological etiology.

Acknowledgments

This research was funded by the National Science and Engineering Research Council of Canada.

[45] E. Fontaine, F. Ichas, and P. Bernardi, *J. Biol. Chem.* **273,** 25734 (1998).
[46] E. Fontaine, O. Eriksson, F. Ichas, and P. Bernardi, *J. Biol. Chem.* **273,** 12662 (1998).
[47] L. Walter, H. Miyoshi, X. Leverve, P. Bernard, and E. Fontaine, *Free Radic. Res.* **36,** 405 (2002).

[7] Dietary Coenzyme Q_{10} and Mitochondrial Status

By CHING KUANG CHOW

Introduction

Coenzyme Q (ubiquinone) is a lipid-soluble compound composed of a redox active quinoid moiety and a hydrophobic tail. The predominant form of coenzyme Q in humans is coenzyme Q_{10}, which contains 10 isoprenoid units in the tail, while the predominant form in rodents is coenzyme Q_9. Coenzyme Q is an essential cofactor in the mitochondrial electron transport chain, where it accepts electrons from Complexes I and II and plays important roles in energy production in the mitochondria.[1,2] The reduced form of coenzyme Q (ubiquinol) may act as an antioxidant by reducing reactive radicals and forms ubisemiquinone in biological systems.[3,4] Coenzyme Q is considered an important antioxidant because it is synthesized endogenously, is regenerated by intracellular reducing mechanisms, and is present in relatively high concentrations.[3,5] Protection against oxidative damage by coenzyme Q has been demonstrated in liposomes, low-density lipoproteins, biological membranes, proteins and DNA.[6–8] Ubisemiquinone, on the other hand, may react with oxygen to form superoxide and ubiquinone. The latter can be reduced back to ubiquinol in cells by NADPH quinone oxidoreductase (DT diaphorase), or the Complexes I and II of the respiratory chain in mitochondria.[1,9]

Coenzyme Q is synthesized in all cells by enzymes in the endoplasmic reticulum and Golgi membranes and then transported to other cellular organelles. The isoprenyl side-chain is derived from mevalonate, the ring system from tyrosine, and the methyl groups from S-adenosylmethionine.[2,10] All tissues seem to be capable of synthesizing coenzyme Q_{10} in

[1] R. E. Beyer, *Biochim. Cell Biol.* **70,** 390 (1992).
[2] L. Ernster and G. Dallner, *Biochem. Biophys. Acta* **1271,** 195 (1995).
[3] H. Noack, U. Kube, and W. Augustin, *Free Radical Res.* **20,** 375 (1994).
[4] P. Forsmark-Andree, C. P. Lee, G. Dallner, and L. Ernster, *Free Radical Biol. Med.* **22,** 391 (1997).
[5] P. Forsmark-Andree, G. Dallner, and L. Ernster, *Free Rad. Biol. Med.* **19,** 749 (1995).
[6] P. Forsmark, F. Aberg, B. Norling, K. Nordenbrand, G. Dallner, and L. Ernster, *FEBS Lett.* **285,** 39 (1991).
[7] P. Forsmark-Andree and L. Ernster, *Mol. Asp. Med.* **15,** S73 (1994).
[8] R. Stocker, V. W. Bowry, and B. Frei, *Proc. Natl. Acad. Sci. USA* **88,** 1646 (1991).
[9] H. Nohl, L. Gille, K. Schonheit, and Y. Liu, *Free Rad. Biol. Med.* **20,** 207 (1996).
[10] W. A. Maltese and J. R. Aprille, *J. Biol. Chem.* **260,** 11524 (1985).

sufficient amounts under normal conditions.[11] However, decreased levels of coenzyme Q_{10} are found in patients with cardiomyopathies, congestive heart failure, degenerative muscle diseases, and during aging. Coenzyme Q_{10} treatment has been shown to improve both clinical and biochemical lesions in patients with mitochondrial disorders,[12–14] and a coenzyme Q_{10}-rich diet retards the normally observed age-associated decline in overall mitochondrial respiratory function in rat skeletal muscle.[15] Also, patients with primary deficiency of coenzyme Q_{10} caused by a genetic defect in its synthesis, and those with mitochondrial disorders, and those taking medicines that suppress cholesterol synthesis are benefited, and without known adverse effects, from coenzyme Q_{10} supplementation.[16–19] These studies suggest that endogenous and normal dietary sources of coenzyme Q_{10} may be inadequate to maintain optimal health.

Absorption, Tissue Distribution, and Metabolism of Coenzyme Q_{10}

Coenzyme Q is present in small amounts in a wide variety of foods but is relatively high in organ meats such as heart, liver, and kidney as well as in beef, soy oil, sardines, mackerel, and peanuts. The intake of coenzyme Q_{10} from food sources in relation to total amount is relatively small. The average Danish diet, for example, estimated from consumption data and from analysis of food items, contains about 3 to 5 mg coenzyme Q_{10} per day, primarily derived from meat and poultry (64% of the total). Single doses of coenzyme Q_{10} (30 mg/person), administered either as a meal or as capsules, is absorbed to a significant degree.[20]

All tissues, cells and membrane contains coenzyme Q_{10}, but in highly variable amounts.[21] Using ^3H-coenzyme Q_{10}, Bentinger *et al.*,[22] have

[11] P. G. Elmberger, A. Kalen, E. L. Appelkvist, and G. Dallner, *Eur. J. Biochem.* **168,** 1 (1987).

[12] S. A. Mortensen, *Clin. Invest.* **71,** S116 (1993).

[13] A. Kalen, E.-L. Appelkvist, and G. Dallner, *Lipids* **24,** 579 (1989).

[14] M. Battino, A. Gorini, R. F. Villa, M. L. Genova, C. Bovina, S. Sassi, G. L. Littarru, and G. Lenaz, *Mech. Aging Dev.* **78,** 173 (1995).

[15] S. Suguyama, K. Yamada, and T. Ozawa, *Biochem. Mol. Biol. Int.* **37,** 1111 (1995).

[16] N. Bresolin, L. Bet, A. Binda, M. Moggio, G. Comi, F. Nador, C. Ferrante, A. Carenzi, and G. Scarlato, *Neurology* **38,** 892 (1988).

[17] Y. Ihara, R. Namba, S. Kuroda, T. Sato, and T. Shirabe, *J. Neurol. Sci.* **90,** 263 (1989).

[18] P. H. Langsjoen, P. Langsjoen, R. Willis, and K. Folkers, *Mol. Aspects Med.* **15,** S165 (1994).

[19] Y. Nishikawa, M. Takahashi, S. Yorifuji, Y. Nakamura, S. Ueno, S. Tarui, T. Kozuka, and T. Nishimura, *Neurology* **39,** 399 (1989).

[20] C. Weber, A. Bysted, and G. Holmer, *Mol. Aspects Med.* **18,** S2251 (1997).

[21] G. Dallner and P. Sindeler, *Free Rad. Biol. Med.* **29,** 285 (2000).

[22] M. Bentinger, G. Dallner, T. Chojnacki, and E. Swiezewska, *Free Rad. Biol. Med.* **34,** 563 (2003).

recently shown that administration of the radioactive compound intraperitoneally resulted in an efficient uptake into the circulation, with high concentrations found in spleen, liver and white blood cells; lower concentrations in adrenals, ovaries, thymus and heart; and practically no uptake in kidney, muscle and brain. In liver homogenate, most radioactive coenzyme Q_{10} appears in the organelles, but it is also present in the cytosol and transport vesicles. Interestingly, mitochondria, purified on a metrizamide gradient from the mitochondrial fraction, have a very low concentration of ^{3}H-coenzyme Q_{10} (10.5% of the total radioactivity), which is mainly present in the lysosomes (58.8%). Two water-soluble metabolites, which have an unchanged substituted benzoquinone ring with a shortened and carboxylated side chain, are found in all organs that take up ^{3}H-coenzyme Q_{10}, and the majority is conjugated with phosphate and excreted through the kidney and appears in the urine. In addition to some metabolites, intact coenzyme Q_{10} is excreted in feces, likely through the bile.[22] These findings suggest that coenzyme Q_{10} is readily taken up and metabolized by the cells.

Effect of Dietary Coenzyme Q_{10} on Levels of Coenzyme Q_{10} in Tissues and Mitochondria

A unique feature of coenzyme Q_{10} not shared by most endogenously synthesized bioactive compounds is that its content, at least in certain organs, can be greatly augmented by exogenous administration, and that dietary coenzyme Q_{10} does not seem to interfere with the metabolism of endogenous coenzyme Q. Experimental data available suggest that dietary coenzyme Q_{10} is taken up into circulation at a variable degree, and significant uptake occurs only in certain organs. Results obtained from most of the short-term coenzyme Q_{10} supplementation studies have shown a significant increase of coenzyme Q_{10} content in the liver, spleen, and serum but not in other organs.[23–28] Relatively little information concerning the influence of exogenously administered coenzyme Q_{10} on its mitochondrial content is available. Ibrahim et al.,[28] have shown that coenzyme Q_{10} supplementation (500 mg/kg diet) to 12-month-old rats for 2 or 4 weeks resulted in significant increases of coenzyme Q_{10} content in both the

[23] Y. Zhang, F. Aberg, E.-L. Appelkvist, G. Dallner, and L. Ernster, *J. Nutr.* **125,** 446 (1995).
[24] Y. Zhang, M. Turunen, and E.-L. Appelkvist, *J. Nutr.* **126,** 2089 (1996).
[25] T. Yuzuriha, M. Takada, and K. Katayama, *Biochim. Biophys. Acta* **759,** 286 (1983).
[26] S. Reahal and J. Wrigglesworth, *Drug Metab. Dispos.* **20,** 423 (1992).
[27] K. Lonnrot, P. Holm, H. Huhtala, and H. Alho, *Biochem. Mol. Bio. Int.* **44,** 727 (1998).
[28] W. H. Ibrahim, H. N. Bhagavan, R. K. Chopra, and C. K. Chow, *J. Nutr.* **130,** 2343 (2000).

TABLE I

EFFECT OF DIETARY COENZYME Q_{10} AND VITAMIN E ON LEVELS OF COENZYME Q_{10} IN
HOMOGENATE AND MITOCHONDRIAL FRACTION[*]

Feeding period	Tissue	10E	10E+500Q10	110E+500Q10	1320E+500Q10
2 weeks	Liver				
	Homogenate	0.20[**]	5.00	7.85	3.33
	Mitochondria	ND[***]	6.34	14.32	6.54
	H/M[****]	—	1.25	1.82	1.96
	Spleen				
	Homogenate	0.03	0.07	0.11	0.14
	Mitochondria	ND	0.87	1.25	0.85
	H/M	—	12.43	11.36	6.07
	Heart				
	Homogenate	0.15	0.13	0.16	0.14
	Mitochondria	0.45	0.36	0.54	0.54
	H/M	3.00	2.77	3.38	3.85
4 weeks	Liver				
	Homogenate	0.19	6.79	11.95	3.97
	Mitochondria	ND	19.40	26.86	10.45
	H/M	—	2.86	2.22	2.63
	Spleen				
	Homogenate	0.03	0.14	0.19	0.06
	Mitochondria	ND	0.85	2.34	0.41
	H/M	—	6.07	12.32	6.83
	Heart				
	Homogenate	0.12	0.14	0.14	0.13
	Mitochondria	0.31	0.54	0.37	0.45
	H/M	2.58	3.85	2.64	3.46

[*] Twelve-month-old rats were fed a synthetic diet that contained either 10 IU vitamin E (10E), 10E plus 500 mg coenzyme Q10 (500Q10), 110E plus 500Q10 or 1320E plus 500Q10 per kg diet for 2 or 4 weeks.[28]

[**] Mean value in nmoles/mg protein of 4 animals. The data is partly derived from Ibrahim et al.[28]

[***] Not detectable.

[****] Ratio of mitochondrial fraction to homogenate.

mitochondria fraction and homogenate of liver and spleen (Table I). On the other hand, when 24-month-old mice were orally treated with coenzyme Q_{10} at a dose of 123 mg/kg body mass daily for 13 weeks, significantly higher coenzyme Q_{10} concentrations were also found in kidney mitochondria, but not homogenate.[29] In addition to those in the liver, significant increases in mitochondrial coenzyme Q_{10} content are also found in the

[29] A. Lass, M. J. Forster, and R. S. Sohal, *Free Radical Biol. Med.* **26,** 1375 (1999).

kidney, heart, brain and skeletal muscle of 14-month-old rats receiving coenzyme Q_{10} supplementation (150 mg/kg diet) for 4 or 13 weeks.[30] Similarly, Mathews et al.,[31] have shown significant increases of coenzyme Q_{10} concentrations in cerebral cortex and cerebral cortex mitochondria of 12-month-old rats treated orally with 200 mg coenzyme Q_{10}/kg body mass daily for 2 months. Additionally, higher levels of coenzyme Q_9 and coenzyme Q_{10} are found in the liver mitochondria of the old rats (24 to 32 month-old) than in those of the young (7-month-old) rats.[32] These findings suggest that older rodents may retain higher tissue levels of the compound via metabolic adaptation to meet their needs and that treating with high doses of coenzyme Q_{10} for longer periods may enable its uptake into other organs in addition to the liver, spleen, and serum.

Dietary Vitamin E and Coenzyme Q_{10} Uptake and Retention

As a lipophilic redox active compound, the status of coenzyme Q_{10} is likely affected by factors affecting the absorption and oxidative stability of lipid soluble substances. Mataix et al.,[33] for example, have shown that dietary monounsaturated fats increase coenzyme Q_{10} content in mitochondria while polyunsaturated ones decrease the levels. Also, a combination of aerobic exercise with dietary polyunsatured fats, but not monounsaturated ones, also decreases coenzyme Q_{10} content. The findings suggest that coenzyme Q_{10} content is depending on oxidative stress and dietary fat unsaturation.

Vitamin E is functionally interrelated to a number of antioxidants, including ascorbic acid, glutathione, lipoic acid and coenzyme Q.[34,35] High concentrations of both coenzyme Q and vitamin E are found in the inner membranes of mitochondria. The functional interaction between vitamin E and coenzyme Q has long been recognized. A number of studies have examined the interaction between vitamin E and coenzyme Q_{10} *in vivo*

[30] L. K. Kwong, S. Kamzalov, I. Rebrin, A. C. Bayne, C. K. Jana, P. Morris, M. J. Forster, and R. S. Sohal, *Free Rad. Biol. Med.* **33**, 627 (2002).

[31] R. S. Matthews, L. Yang, S. Browne, M. Baik, and F. Beal, *Proc. Natl. Acad. Sci. USA* **95**, 8892 (1998).

[32] T. Armeni, M. Tomasetti, S. Svegliati-Baroni, F. Saccucci, M. Marra, C. Pieri, G. P. Littarru, G. Principato, and M. Battino, *Mol. Aspects Med.* **18**, S247 (1997).

[33] J. Mataix, M. Manas, J. Quiles, M. Battino, M. Cassinello, M. Lopez-Frias, and J. R. Huertas, *Mol. Aspects Med.* **18**, S129 (1997).

[34] D. A. Stoyanovsky, A. N. Osipov, P. J. Quinn, and V. E. Kagan, *Arch. Biochem. Biophys.* **323**, 343–351 (1995).

[35] C. K. Chow, *Free Radical Biol. Med.* **11**, 215 (1991).

with inconsistent findings. For example, Zhang et al.,[23,24] have shown that vitamin E supplementation increased the levels of both endogenous and exogenous coenzyme Q in the liver and plasma, while dietary coenzyme Q_{10} had no effect on tissue vitamin E. Ibrahim et al.,[28] fed 3 different levels (10, 110 and 1320 IU/kg diet) of vitamin E along with a 500 mg coenzyme Q_{10}/kg diet to 12-month-old rats for 2 or 4 weeks and found that the animal group receiving 110 IU vitamin E/kg had significantly higher levels of coenzyme Q_{10} in both the homogenate and mitochondria of liver and spleen, but not heart (and other organs), than those receiving 10 IU vitamin E in the diet. On the other hand, the animal group receiving 1320 IU/kg vitamin E had lower levels when compared with the groups receiving 10 IU or 110 IU vitamin E in the diet (Table I). The results clearly suggest an interaction between exogenously administered vitamin E and coenzyme Q_{10} in terms of uptake and tissue retention. Moderate levels of vitamin E enhance tissue uptake and/or retention of dietary coenzyme Q_{10}, while high levels of vitamin E have an opposite effect.

Measurement of Mitochondrial Levels of Coenzyme Q_{10}

Isolation of Mitochondria

Mitochondrial fraction from heart and skeletal muscle is prepared according to the modified procedure of Bhattacharya et al.[36] Heart and skeletal muscle are first thoroughly minced and incubated with 5 volumes of 0.2 M Tris-HCl buffer containing 29 mg% Nagarse, 100 mM sucrose, 10 mM EDTA, 46 mM KCl, and 0.5% bovine serum albumin (pH 7.4) (buffer A) at room temperature for 5 minutes. After the incubated tissue minces are washed with 10 volumes of the same buffer without Nagarse (buffer B) twice, a 10% homogenate is prepared with buffer B with a Teflon pestle for four strokes by hand in a Potter-Elvehjem-type homogenizer at 4°. After centrifugation at 300g for 20 minutes or 500g for 10 minutes to remove cell debris and nuclei, the supernatant was centrifuged at 12,000 g for 10 minutes. The resulting pellet is resuspended in buffer B. Mitochondrial fraction from liver, spleen, brain and other soft tissues is prepared using the same buffer without incubating with buffer A. The concentration of protein in the homogenate and mitochondrial fraction is determined by using the procedure of Miller (1959).[37]

---

[36] S. K. Bhattacharya, J. H. Thakar, P. L. Johnson, and D. R. Shanklin, Anal. Biochem. **192,** 344 (1991).
[37] G. L. Miller, Anal. Biochem. **31,** 964 (1959).

For sub-fractionation of the mitochondrial fraction, the $300g$ supernatant is centrifuged at $2,800g$ for 20 minutes, and the pellet is washed and resuspended in 57% metrizamide, pH 7.3. The suspension is then layered with 33 to 20% metrizamide solution and centrifuged at $100,000g$ for 3 hours in a swinging bucket rotor.[22,38]

Measurement of Coenzyme Q

A number of good procedures have been reported for measuring coenzyme Q_{10} levels, both the reduced and oxidized forms, in biological samples.[39–41] Because coenzyme Q_{10} is labile to oxidation during sample preparation, it is not practical and of little use to measure the reduced form of the compound in the isolated mitochondria fraction or its subfractions. The procedure adapted from that of Okamoto et al.,[41] described in the following text is relatively simple and highly reproducible for measuring oxidized coenzyme Q_{10} in biological samples. By converting the remaining reduced coenzyme Q to the oxidized form by potassium hexacyanoferrate, measurement of the oxidized form in the sample is equivalent to the measurement of the total coenzyme Q_{10}.

One-half ml of homogenate, mitochondrial fraction or serum/plasma in duplicate is first mixed with 2 ml 95% ethanol and then extracted with 5 ml n-hexane. After drying under nitrogen, the lipid extract (4 ml n-hexane) is dissolved in 50 μl of ethanol. Subsequently, an aliquot of 0.5 ml of 5% potassium hexacyanoferrate solution is added to the ethanol solution, and the mixture is allowed to stand for 10 min at room temperature. The oxidized (or total) coenzyme Q is then extracted with 2 ml ethanol and 5 ml of hexane.[41] After drying under nitrogen, the lipid extract is dissolved in 200 μl methanol. The levels of oxidized (total) coenzyme Q_{10} in each sample are measured twice by HPLC (Beckman 112 Solvent Delivery Module, TosoHaas TSK 6080 Sample processor with a 50 μl sample loop, Waters 490 UV-visible detector and Shimadzu CR6A ChromatopacA). A Waters Nova Pac 3.9×150 mm C-18 reverse-phase column is used as the stationary phase and an isocratic methanol-hexane mixture (70:30) as the mobile phase, and the compound is monitored at a wavelength of 275 nm. With the flow rate of 1.5 ml/min, the retention time is about 6.8 minutes.

[38] R. W. Wattiaux, S. Wattiaux-de Connick, M. F. Ronveaux-Dupal, and F. Dubois, J. Cell Biol. **78,** 349 (1978).

[39] J. Lagendijk, J. B. Ubbink, and W. J. H. Vermaak, J. Lipid Res. **37,** 67 (1996).

[40] M. Podda, C. Weber, M. G. Traber, and L. Packer, J. Lipid Res. **37,** 893 (1996).

[41] T. Okamoto, K. Fukui, M. Nakamoto, and T. Kishi, J. Chromat. **342,** 35 (1985).

Implications of the Effect of Dietary Coenzyme Q_{10} and Vitamin E
on Mitochondrial Coenzyme Q_{10}

Exogenous administration of coenzyme Q_{10} markedly increases uptake
and retention of coenzyme Q_{10} in the homogenate and mitochondria
only in certain organs of the rat.[28] Interestingly, in rodents, coenzyme
Q_{10} supplementation resulted in significantly higher levels of the compound in the liver and spleen, where very low levels of the endogenous
compound are present (Table I). On the other hand, non-responsive organs
such as the heart and kidney have relatively high levels of endogenous
coenzyme Q_{10}.[28] The significance or implication of this finding has yet to
be determined.

Vitamin E and coenzyme Q are essential for maintaining functions and
integrity of mitochondria, and high concentrations of these compounds are
found in its inner membrane. The enhancing effect of moderate levels of
vitamin E on mitochondrial and tissue levels of coenzyme Q_{10} and the suppressing effect of high levels of vitamin E observed[28] are interesting. Since
both vitamin E and coenzyme Q_{10} are lipid soluble, they may have a similar
absorption/transport mechanism. It is possible that a moderate increase in
the dose of vitamin E may facilitate the absorption and/or incorporation of
coenzyme Q, while high levels of vitamin E may compete with coenzyme
Q_{10} and thus suppress its absorption and/or uptake.

In summary, dietary coenzyme Q_{10} increases coenzyme Q_{10} content in
tissue and mitochondria in an organ-dependent manner. Tissue and mitochondrial levels of coenzyme Q_{10} are similarly affected by dietary coenzyme Q_{10}. However, dietary vitamin E is a key factor determining the
uptake and retention of exogenous administrated coenzyme Q_{10} in both
homogenate and mitochondrial fraction.

Section II

Anticancer Quinones and Quinone Oxidoreductases

[8] NAD(P)H:Quinone Oxidoreductase 1 (NQO1, DT-Diaphorase), Functions and Pharmacogenetics

By DAVID ROSS and DAVID SIEGEL

Introduction

NQO1 is a flavoprotein that is known to catalyze two electron reduction of a broad range of substrates.[1–3] As the name of the enzyme suggests, a common group of substrates are quinones, which are reduced via a hydride transfer mechanism to generate the corresponding hydroquinone derivative. Elucidation of the mechanism of catalysis has been facilitated by elucidation of the crystal structure of the enzyme and has been recently summarized.[4–7] The broad substrate specificity has also been explained by structural studies demonstrating the presence of a highly plastic active site that can accommodate a range of structures.[8] Because of the many deleterious effects of quinonoid compounds, including their capacity to arylate nucleophiles and generate aggressive oxygen species via redox cycling mechanisms, removal of a quinone from a biological system by NQO1 has been considered to be a detoxification reaction.[9–12] Two electron reduction of certain antitumor quinones such as mitomycin C, E09, streptonigrin and B-lapachone by NQO1, however, results in bioactivation

[1] L. Ernster, *Methods Enzymol.* **10,** 309 (1967).

[2] C. Lind, E. Cadenas, P. Hochstein, and L. Ernster, *Methods Enzymol.* **186,** 287 (1990).

[3] L. Ernster, *in* "Pathophysiology of lipid peroxides and free radicals" (K. Yagi, ed.), p. 149. Japan Sci. Soc. Press/Karger, Tokyo/Basel, 1998.

[4] R. Li, M. Bianchet, P. Talalay, and L. M. Amzel, *FASEB J.* **9,** A1338 (1995).

[5] M. Faig, M. A. Bianchet, P. Talalay, S. Chen, S. Winski, D. Ross, and A. L. Mario, *Proc. Natl. Acad. Sci. USA* **97,** 3177 (2000).

[6] G. Cavelier and L. M. Amzel, *Proteins* **43,** 420 (2001).

[7] R. Li, M. A. Bianchet, P. Talalay, and L. M. Amzel, *Proc. Natl. Acad. Sci. USA* **92,** 8846 (1995).

[8] M. Faig, M. A. Bianchet, S. Winski, R. Hargreaves, C. J. Moody, A. R. Hudnott, D. Ross, and L. M. Amzel, *Structure. (Camb.)* **9,** 659 (2001).

[9] C. Lind, P. Hochstein, and L. Ernster, *Arch. Biochem. Biophys.* **216,** 178 (1982).

[10] H. Thor, M. T. Smith, P. Hartzell, G. Bellomo, S. A. Jewell, and S. Orrenius, *J. Biol. Chem.* **257,** 12419 (1982).

[11] D. Di Monte, G. Bellomo, H. Thor, P. Nicotera, and S. Orrenius, *Arch. Biochem. Biophys.* **235,** 343 (1984).

[12] D. Di Monte, D. Ross, G. Bellomo, L. Eklow, and S. Orrenius, *Arch. Biochem. Biophys.* **235,** 334 (1984).

of these compounds to more toxic metabolites.[13–16] Since NQO1 is expressed at high levels throughout many human solid tumors,[17,18] compounds efficiently bioactivated by NQO1 have been designed for the therapy of tumors rich in NQO1.[8,19–21] Currently, a new NQO1-targeted aziridinylbenzoquinone, RH1,[20] is undergoing phase 1 clinical trials. The role of NQO1 in chemoprotection and its contrasting role in bioactivation of antitumor quinones[22,23] and both the gene and protein structure of NQO1[24,25] have been recently summarized. The purpose of this review is to introduce the enzyme, to review the possible functions of NQO1, to summarize current information on polymorphic forms of NQO1 and finally, to summarize the relevance of the common NQO1*2 polymorphism for chemoprotection, susceptibility to disease and cancer chemotherapy.

Possible Functions of NQO1

Early Work on Mitochondrial Electron Transport and Vitamin K Metabolism

When NQO1 was first isolated by Ernster,[26,27] there was considerable controversy regarding a potential role for NQO1 in mitochondrial electron transport. This work was summarized in a fascinating historical review of

[13] D. Siegel, N. W. Gibson, P. C. Preusch, and D. Ross, *Cancer Res.* **50,** 7483 (1990).

[14] M. I. Walton, P. J. Smith, and P. Workman, *Cancer Commun.* **3,** 199 (1991).

[15] H. D. Beall, Y. Liu, D. Siegel, E. M. Bolton, N. W. Gibson, and D. Ross, *Biochem. Pharmacol.* **51,** 645 (1996).

[16] J. J. Pink, S. M. Planchon, C. Tagliarino, M. E. Varnes, D. Siegel, and D. A. Boothman, *J. Biol. Chem.* **275,** 5416 (2000).

[17] D. Siegel, W. A. Franklin, and D. Ross, *Clin. Cancer Res.* **4,** 3083 (1998).

[18] D. Siegel and D. Ross, *Free Radic. Biol. Med.* **29,** 246 (2000).

[19] H. D. Beall, A. M. Murphy, D. Siegel, R. H. J. Hargreaves, J. Butler, and D. Ross, *Mol. Pharmacol.* **48,** 499 (1995).

[20] S. Winski, R. H. J. Hargreaves, J. Butler, and D. Ross, *Clin. Cancer Res.* **4,** 3083 (1998).

[21] S. L. Winski, E. Swann, R. H. Hargreaves, D. L. Dehn, J. Butler, C. J. Moody, and D. Ross, *Biochem. Pharmacol.* **61,** 1509 (2001).

[22] D. Ross, *in* "Comprehensive toxicology" (F. P. Guengerich, ed.), Vol. 3, p. 179. Pergamon, New York, 1997.

[23] D. Ross, J. K. Kepa, S. L. Winski, H. D. Beall, A. Anwar, and D. Siegel, *Chem. Biol. Interact.* **129,** 77 (2000).

[24] D. Ross, *in* "Encyclopedia of Molecular Medicine" p. 2208. John Wiley and Sons, New York, 2001.

[25] D. Ross, "Atlas genetics cytogenetics oncology hematology" http://www.infobiogen.fr/services/chromcancer/Genes/NQO1ID375.html, 2002.

[26] L. Ernster and F. Navazio, *Acta Chem. Scand.* **12,** 595 (1958).

[27] L. Ernster, *Fed. Proc.* **17,** 216 (1958).

the enzyme by Ernster in 1987.[28] NQO1 was found not to be a component of the mitochondrial respiratory chain, and attention turned to other possible physiological functions for the enzyme such as a potential role in vitamin K metabolism, suggested in the early 1960s by Martius.[29] The role of NQO1 in generating reduced vitamin K1 cofactor and triggering protein carboxylation critical to prothrombin synthesis was suggested in rat liver,[30] but vitamin K1 was found not to be a substrate for purified NQO1 isolated from rat liver cytosol, suggesting that other enzyme systems may be more important for vitamin K1 reduction.[31]

Detoxification of Quinones via Two Electron Reduction

A considerable body of literature exists supporting the role of NQO1 as a detoxification system. Induction of NQO1 has been demonstrated to protect against the cytotoxicity, mutagenicity and carcinogenicity of many compounds.[32–34] Induction of NQO1 occurs however via the XRE and ARE elements in the NQO1 promoter,[35] and many protective enzymes in addition to NQO1 may also be induced via this mechanism. For example, recent work using t-butylhydroquinone, a common inducer used to elevate NQO1, increased the expression of 63 different genes as indicated by microarray analysis.[36] This observation emphasizes the need to perform specific studies to define the role of a particular gene in detoxification and a good example is work performed with menadione.

Work by two independent groups a few miles apart in Stockholm in the early 1980s using the naphthoquinone menadione[9,10] defined the role of NQO1 as a detoxification enzyme in quinone metabolism. Both menadione-induced arylation of cellular nucleophiles and generation of aggressive oxygen species were diminished in the presence of functional NQO1. Detoxification of quinones by NQO1 has been confirmed by studies in NQO1 knockout mice that demonstrated increased menadione toxicity in NQO1-deficient animals.[37] NQO1 knockout animals also demonstrate

[28] L. Ernster, *Chemica Scripta*, **27A,** 1 (1987).
[29] C. Martius, *in* "The Enzymes" (P. D. Boyer, H. Lardy, and K. Myrback, eds.), Vol. 7, p. 517. Academic Press, New York, 1960.
[30] R. Wallin, S. R. Rannels, and L. F. Martin, *Chemica Scripta* **27A,** 193 (1987).
[31] P. C. Preusch and D. M. Smalley, *Free Rad. Res. Comm.* **8,** 401 (1990).
[32] C. Huggins and R. Fukunishi, *J. Exp. Med.* **119,** 923 (1964).
[33] A. M. Benson, M. J. Hunkler, and P. Talalay, *Proc. Natl. Acad. Sci. USA* **77,** 5216 (1980).
[34] P. Talalay and H. J. Prochaska, *Chemica Scripta* **27A,** 61 (1987).
[35] J. K. Kepa, R. D. Traver, D. Siegel, S. L. Winski, and D. Ross, *Rev. Toxicol.* **1,** 53 (1997).
[36] J. Li, J. M. Lee, and J. A. Johnson, *J. Biol. Chem.* **277,** 388 (2002).
[37] V. Radjendirane, P. Joseph, Y. H. Lee, S. Kimura, A. J. P. Klein-Szanto, F. J. Gonzalez, and A. K. Jaiswal, *J. Biol. Chem.* **273,** 7382 (1998).

increased benzo[a]pyrene and 7,12-dimethylbenzanthracene-induced mouse skin carcinogenesis.[38,39] A role for NQO1 in detoxification is consistent with the distribution of the enzyme in animal and human systems. In mice, rats and humans, NQO1 is mainly localized to epithelial and endothelial tissues,[17,18] which facilitates exposure of compounds entering the body to NQO1. An interesting and puzzling species difference in enzyme distribution is that both rat and mouse liver contain high levels of NQO1, but only trace levels of NQO1 could be detected by immunoblot analysis in five human liver samples.[18] This confirms earlier work demonstrating a low activity of NQO1 in human liver cytosol relative to other species.[30]

Sharpening the Double-Edged Sword: NQO1 in Bioactivation

It is often difficult to generalize whether NQO1 functions as a detoxification enzyme or a toxification system with an individual substrate. The reactions of the hydroquinone generated will determine whether NQO1 mediated reduction results in deleterious or protective effects in a cellular system. Hydroquinones are not necessarily innocuous species and may autoxidize to generate reactive oxygen species or rearrange to produce reactive arylating species.[23] Efficient detoxification of quinones via two electron reduction relies on excretion of the more water soluble hydroquinone or conjugation with glucuronide or sulfate and subsequent excretion.

NQO1 is expressed at high levels throughout many human solid tumors, and a number of compounds clinically used as antitumor agents or in development as experimental antitumor agents can be efficiently bioactivated by NQO1.[23,40–42,43] Apart from the bioactivation of antitumor quinones such as mitosenes, indolequinones, aziridinylbenzoquinones and others, NQO1 can also bioactivate simpler quinones. Even though the naphthoquinone menadione is detoxified by NQO1, closely related naphthoquinones such as 2-hydroxy-1,4-naphthoquinone[44] and β-lapachone can be activated by NQO1.[16] Interestingly, although NQO1 is predominantly a cytosolic enzyme, recent work has demonstrated a significant nuclear pool of

[38] D. J. Long, R. L. Waikel, X. J. Wang, L. Perlaky, D. R. Roop, and A. K. Jaiswal, *Cancer Res.* **60,** 5913 (2000).

[39] D. J. Long, R. L. Waikel, X. J. Wang, D. R. Roop, and A. K. Jaiswal, *J. Natl. Cancer Inst.* **93,** 1166 (2001).

[40] D. Ross, D. Siegel, H. Beall, A. S. Prakash, R. T. Mulcahy, and N. W. Gibson, *Cancer Metastasis Rev.* **12,** 83 (1993).

[41] D. Ross, H. Beall, R. D. Traver, D. Siegel, R. M. Phillips, and N. W. Gibson, *Oncol. Res.* **6,** 493 (1994).

[42] R. J. Riley and P. Workman, *Biochem. Pharmacol.* **43,** 1657 (1992).

[43] H. D. Beall and S. I. Winski, *Front Biosci.* **5,** D639 (2000).

[44] R. Munday, B. L. Smith, and C. M. Munday, *Chem. Biol. Interact.* **117,** 241 (1999).

NQO1 in human tumor cells when using both confocal microscopy and immuno-electron microscopy.[45] A nuclear pool of NQO1 may be of considerable importance for the bioactivation of NQO1-directed antitumor quinones that are designed to crosslink DNA. Whether a similar pool of nuclear NQO1 exists in normal cells is currently unclear.

NQO1 as an Antioxidant Enzyme: Role in Ubiquinone and Vitamin E Metabolism

In addition to activation and deactivation of exogenous compounds, NQO1 has been shown to play a role in the metabolism of endogenous quinones such as ubiquinone and vitamin E quinone. These quinones have very large hydrophobic tails, and in their reduced state they protect cellular membranes against lipid peroxidative injury. The reduction of ubiquinone by NQO1 has been shown to generate ubiquinol, and this compound possesses excellent antioxidant properties.[46,47] Vitamin E quinone is formed during free radical attack on vitamin E and has been shown to undergo reduction by NQO1 to generate vitamin E hydroquinone.[48] Since vitamin E quinone is devoid of antioxidant activity, in this situation NQO1 may extend the antioxidant potential of vitamin E by generation of vitamin E hydroquinone, a compound that has been suggested to have antioxidant properties superior to vitamin E.[49]

NQO1 as a Component of a Stress Response: Stabilization of p53

Studies with proteins typically considered as metabolic enzymes suggest that these proteins may have additional roles outside the range of their normal metabolic functions. For example, glutathione-S-transferase associates with c-Jun N terminal kinase leading to inhibition of kinase activity and modulation of signaling and cellular proliferation.[50–52] Recent studies

[45] S. L. Winski, Y. Koutalos, D. L. Bentley, and D. Ross, *Cancer Research* **62,** 1420 (2002).

[46] R. E. Beyer, J. Segura-Aguilar, S. Di Bernardo, M. Cavazzoni, R. Fato, D. Fiorentini, M. Galli, M. Setti, L. Landi, and G. Lenaz, *Proc. Natl. Acad. Sci. USA* **93,** 2528 (1996).

[47] L. Landi, D. Fiorentini, M. C. Galli, J. Segura-Aguilar, and R. E. Beyer, *Free Radic. Biol. Med.* **22,** 329 (1997).

[48] D. Siegel, E. M. Bolton, J. A. Burr, D. C. Liebler, and D. Ross, *Mol. Pharmacol.* **52,** 300 (1997).

[49] I. Kohar, M. Baca, C. Suarna, R. Stocker, and P. T. Southwell-Keely, *Free Rad. Biol. Med.* **19,** 197 (1995).

[50] T. Wang, P. Arifoglu, Z. Ronai, and K. D. Tew, *J. Biol. Chem.* **276,** 20999 (2001).

[51] V. Adler, Z. Yin, S. Y. Fuchs, M. Benezra, L. Rosario, K. D. Tew, M. R. Pincus, M. Sardana, C. J. Henderson, C. R. Wolf, R. J. Davis, and Z. Ronai, *EMBO J.* **18,** 1321 (1999).

[52] J. E. Ruscoe, L. A. Rosario, T. Wang, L. Gate, P. Arifoglu, C. R. Wolf, C. J. Henderson, Z. Ronai, and K. D. Tew, *J. Pharmacol. Exp. Ther.* **298,** 339 (2001).

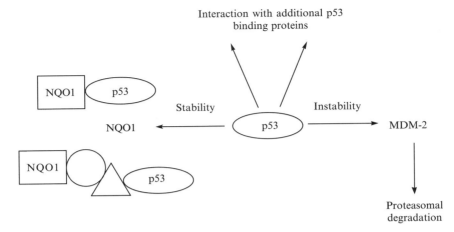

Fig. 1. Proposed mechanism of stabilization of p53 via a protein-protein interaction with NQO1. See reference 56.

have shown that NQO1 may influence the stability of the tumor suppressor protein p53 by inhibiting its degradation.[53–55] In these studies the authors hypothesized that the NQO1-mediated conversion of NADH to NAD^+ promoted stabilization of p53. Our own work has shown a direct physical interaction between p53 and NQO1.[56] In these studies, p53 was found to directly associate with NQO1 in an *in vitro* translation/transcription system and in human cancer and primary cell lines. The association between p53 and NQO1 was not affected by pretreatment with ES936, a mechanism based inhibitor of NQO1, suggesting that the catalytic activity of NQO1 was not needed for a protein-protein interaction of p53 and NQO1 proteins. Anwar *et al.* proposed that the protein-protein interaction of NQO1 and p53 may represent an alternative mechanism of p53 stabilization by NQO1,[56] as shown in Fig. 1.

[53] G. Asher, J. Lotem, B. Cohen, L. Sachs, and Y. Shaul, *Proc. Natl. Acad. Sci. USA* **98**, 1188 (2001).

[54] G. Asher, J. Lotem, R. Kama, L. Sachs, and Y. Shaul, *Proc. Natl. Acad. Sci. USA* **99**, 3099 (2002).

[55] G. Asher, J. Lotem, L. Sachs, C. Kahana, and Y. Shaul, *Proc. Natl. Acad. Sci. USA* **99**, 13125 (2002).

[56] A. Anwar, D. Dehn, D. Siegel, J. K. Kepa, L. J. Tang, J. A. Pietenpol, and D. Ross, *J. Biol. Chem.* **278**, 10368 (2003).

Future Work on Possible Physiological Roles of NQO1

A number of possible functions of NQO1 have been investigated. Many of the pharmacological studies of NQO1 have relied on the use of the inhibitor dicoumarol. The major problem with the use of dicoumarol is its non-specific nature.[57] The recent development of a mechanism-based inhibitor of NQO1[58] offers the promise of more specific targeting of NQO1. Taken together with the application of gene targeting methods[59] and the development of NQO1 knockout animals,[37] these approaches should be able to more precisely define the role of NQO1 in cellular systems.

Pharmacogenetics of NQO1

There are two well-characterized polymorphisms in NQO1 with defined phenotypes and population frequencies—the NQO1*2[60,61] and NQO1*3[62,63] polymorphisms. A recent review by Nebert and coworkers[64] screened the SNP database at the University of Utah and identified 22 SNPs in NQO1. Three of these SNPs had been reported earlier in a study of 84 Japanese volunteers.[65] In addition to the NQO1*2 polymorphism, 5 SNPs were found in the 5′ flanking region of NQO1, 10 SNPs at 9 sites in intron 1, a synonomous mutation in exon 2, 2 SNPs in the 3′ untranslated region and 3 SNPs in the 3′ flanking region of the gene.[64] The NQO1*3 polymorphism was not detected in the SNP database study. Apart from the NQO1*2 and NQO1*3 polymorphisms, the frequency of the additional 21 variant alleles in the population is unknown and their significance, if any, for phenotype remains to be characterized. Further work needs to be performed to determine if any of the SNP's characterized in the SNP database[64] represent true polymorphisms with allele frequencies of greater than 1%.

The two NQO1 variant alleles that have been well-characterized are both coding region alleles that are of sufficient frequency to be

---

[57] P. C. Preusch, D. Siegel, N. W. Gibson, and D. Ross, *Free Rad. Biol. Med.* **11,** 77 (1991).

[58] S. L. Winski, M. Faig, M. A. Bianchet, D. Siegel, E. Swann, K. Fung, M. W. Duncan, C. J. Moody, L. M. Amzel, and D. Ross, *Biochemistry* **40,** 15135 (2001).

[59] T. Yoshida and H. Tsuda, *Biochem. Biophys. Res. Commun.* **214,** 701 (1995).

[60] R. D. Traver, T. Horikoshi, K. D. Danenberg, T. H. W. Stadlbauer, P. V. Danenberg, D. Ross, and N. W. Gibson, *Cancer Res.* **52,** 797 (1992).

[61] R. D. Traver, D. Siegel, H. D. Beall, R. M. Phillips, N. W. Gibson, W. A. Franklin, and D. Ross, *Br. J. Cancer* **75,** 69 (1997).

[62] S. S. Pan, G. L. Forrest, S. A. Akman, and L.-T. Hu, *Cancer Res.* **55,** 330 (1995).

[63] L. T. Hu, J. Stamberb, and S. S. Pan, *Cancer Res.* **56,** 5253 (1996).

[64] D. W. Nebert, A. L. Roe, S. E. Vandale, E. Bingham, and G. G. Oakley, *Genet. Med.* **4,** 62 (2002).

[65] A. Iida, A. Sekine, S. Saito, Y. Kitamura, T. Kitamoto, S. Osawa, C. Mishima, and Y. Nakamura, *J. Hum. Genet.* **46,** 225 (2001).

characterized as polymorphisms (allele frequency > 0.01). The most prevalent variant allele is the NQO1*2 polymorphism,[66,67] which has profound implications for phenotype[68]; individuals carrying the homozygous NQO1*2 allele have no NQO1 activity. The phenotypic implications of the NQO1*3 allele vary according to substrate,[62] and the frequency of the NQO1*3 allele in the population is very low.[66]

Research into the identification and significance of NQO1 polymorphisms is increasing, and a search of the PubMed database in March 2003 resulted in 82 hits when using the keywords NQO1 and polymorphism. Ten additional studies of NQO1 polymorphisms were obvious from additional PubMed searches of NQO1-related publications generating a total of 92 publications related to NQO1 polymorphisms. The majority of work on NQO1 polymorphisms has focused on the NQO1*2 polymorphism, and to give the reader some sense of impact of this polymorphism in different disciplines, we have classified each of these studies into a particular subgroup (Table I).

Polymorphisms in NQO1

NQO1*2 Polymorphism

Genotype-Phenotype Relationships

The NQO1*2 polymorphism was characterized in a collaboration between our own laboratory and that of Dr. Neil Gibson. During investigations of a series of human colon carcinoma cell lines[60] and lung tumor cell lines,[61] a good correlation was observed between NQO1 mRNA levels and NQO1 activity. In each series of tumor cell lines, however, one cell line (BE human colon carcinoma cells and H596 lung cancer cells) demonstrated high mRNA levels but no NQO1 activity. After SSCP analysis and DNA sequencing the same homozygous point mutation—a C to T mutation at position 609 of the cDNA—was identified in each cell line.[60,61] Immunoblot analysis revealed that the reason underlying the lack of NQO1 activity in each cell line was a lack of NQO1 protein. Although the recombinant mutant NQO1*2 protein demonstrated poor activity relative to wild-type protein, virtually no NQO1 protein could be detected in cells,

---

[66] A. Gaedigk, R. F. Tyndale, M. Jurima-Romet, E. M. Sellers, D. M. Grant, and J. S. Leeder, *Pharmacogenetics* **8,** 305 (1998).

[67] K. T. Kelsey, J. K. Wiencke, D. C. Christiani, Z. Zuo, M. R. Spitz, X. Xu, B. K. Lee, B. S. Schwartz, R. D. Traver, and D. Ross, *Br. J. Cancer* **76,** 852 (1997).

[68] D. Siegel, A. Anwar, S. L. Winski, J. K. Kepa, K. L. Zolman, and D. Ross, *Mol. Pharmacol.* **59,** 263 (2001).

TABLE I
NQO1 POLYMORPHISMS, FREQUENCY OF CITATIONS IN DIFFERENT FIELDS[a]

Field	Number of citations
Toxicological	9
Association with specific cancers (apart from leukemia)	29
Association with leukemia	10
Association with Parkinson's disease	2
Association with diabetes	1
Genotype distribution and phenotype	8
Mechanistic studies	15
Reviews	9
Relevance for chemotherapy	6
NQO2[b]	3
Total	92[a]

[a] Eighty-two studies were identified in a PubMed search using the keywords NQO1 and polymorphism and were classified into subgroups. An additional 10 studies were identified by PubMed searches using the key word NQO1. Each study was assigned to one subgroup only and since some studies could be entered in various groups, this classification solely represents the views of the authors. The table describes the wide range of studies of NQO1 polymorphisms in different disciplines. The vast majority of the citations listed are focused on the NQO1*2 polymorphism.
[b] Citations for NQO2 were also retrieved as part of this search.

and, more importantly, in human tissues genotyped as homozygous for the NQO1*2 polymorphism.[17,18,60,61,69] A number of cell lines commonly used in research are homozygous for the NQO1*2 polymorphism (Table II).

The NQO1*2 polymorphism shows a clear gene-dose effect with respect to phenotype. Individuals genotyped as NQO1*1/*1 (or wild type) had the highest levels of NQO1 protein in saliva while individuals carrying the homozygous NQO1*2/*2 polymorphism had no detectable levels of NQO1 protein.[69] Heterozygous individuals (NQO1*1/*2) had intermediate levels of NQO1 protein.[69] In other studies, cell lines and tumor samples homozygous for the NQO1*2 polymorphism had no detectable NQO1 activity.[60,61,70–73] In addition, tumor samples genotyped as NQO1*1/*2 or

[69] D. Siegel, S. M. McGuinness, S. Winski, and D. Ross, Pharmacogenetics 9, 113 (1999).
[70] V. Misra, H. J. Klamut, and A. M. Rauth, Br. J. Cancer 77, 1236 (1998).
[71] P. Eickelmann, T. Ebert, U. Warskulat, W. A. Schulz, and H. Sies, Carcinogenesis 15, 219 (1994).
[72] P. Eickelmann, W. A. Schulz, D. Rohde, B. Schmitz-Dräger, and H. Sies, Biol. Chem. Hoppe Seyler 375, 439 (1994).
[73] W. A. Schulz, A. Krummeck, I. Rosinger, P. Eickelmann, C. Neuhas, T. Ebert, B. Schmitz-Dräger, and H. Sies, Pharmacogenetics 7, 235 (1997).

TABLE II
COMMONLY USED HUMAN CELL LINES WITH THE NQO1*2/*2 GENOTYPE[a]

Cell Line	Tissue
BE	Colon
Caco-2	Colon
NCI-H1570	Lung
NCI-H596	Lung
MDA-MB-231	Breast
MDA-MB-468	Breast

[a] Modified from reference 69.

TABLE III
NQO1 ALLELE FREQUENCIES[a]

	Allele		
Population	*1	*2	*3
Chinese	0.47	0.49	0.04
Inuit	0.54	0.46	<0.01
Native American	0.59	0.40	0.01
Caucasian	0.79	0.16	0.05

[a] Adapted from reference 66.

heterozygous had significantly lower NQO1 activity than samples containing the wild type or concensus sequence.[74]

Ethnic Distribution

The allele frequency of the NQO1*2 polymorphism (Table III) has been reported to be as high as 0.49 in some Asian populations.[66] The prevalence of the NQO1*2/*2 genotype in different ethnic groups is shown in Fig. 2. The NQO1*2/*2 genotype is essentially a null polymorphism and as many as 22% of individuals in some Asian populations lack NQO1 (Fig. 2).

[74] R. A. Fleming, J. Drees, B. W. Loggie, G. B. Russell, K. R. Geisinger, R. T. Morris, D. Sachs, and R. P. McQuellon, *Pharmacogenetics* **12,** 31 (2002).

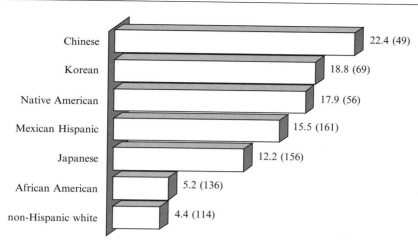

FIG. 2. Percentage of individuals in different populations with the NQO1*2/*2 genotype. Data are from healthy individuals and is primarily taken from reference 67 and unpublished data in a Japanese population (A. Sadler, M. Yano, and D. Ross, unpublished) and a Native American population (A. Sadler and D. Ross, unpublished) adapted from reference 24.

Mechanisms Underlying the Lack of NQO1*2 Protein in Cells and Tissues

Our studies have demonstrated that the NQO1*2 protein is rapidly degraded by the ubiquitin proteasomal system with a half-life of approximately 1.2 h.[68] The mechanisms underlying the structural instability of the NQO1*2 protein are unclear. Although we have generated a crystal structure of human wild type recombinant NQO1,[5] crystallization of the NQO1*2 mutant protein for structural elucidation studies has not been possible. However, the mutation in the NQO1*2 protein is a proline to serine at position 187, which may disrupt the structure of an external loop of the protein. A potential role for the protein chaperone Hsp70 has also been demonstrated in recent studies.[75] NQO1 has two Hsp binding sites and Hsp70 interacts with early immature forms of the wild-type NQO1 protein, but not with the mature protein, presumably to assist in correct protein folding. We were unable to detect any interaction of Hsp70 and the NQO1*2 protein (Fig. 3) and we have hypothesized that this leads to incorrect folding of the NQO1*2 mutant protein followed by ubiquitination and proteasomal degradation.[75]

[75] A. Anwar, D. Siegel, J. K. Kepa, and D. Ross, *J. Biol. Chem.* **16,** 14060 (2002).

A

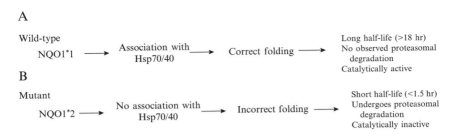

FIG. 3. The potential role of protein chaperones in the folding of the NQO1*1 protein. The protein chaperones HSP 70 and HSP 40 interact with newly-synthesized forms of the NQO1*1 protein but not the NQO1*2 protein.[75]

NQO1*3 Polymorphism: Phenotype and Implications for NQO1 mRNA Alternative Splicing

The NQO1*3 polymorphism was characterized by Pan and colleagues[62,63] and represents a C465T change coding for an arginine to tryptophan change in the protein. The variant protein had similar stability to wild-type protein and the implications of the NQO1*3 polymorphism for phenotype are variable according to the substrate. For example, the NQO1*3 recombinant protein demonstrated a similar ability to reduce menadione relative to wild-type protein but showed a 60% decreased ability to metabolize mitomycin C.[62]

At least three distinct transcripts of NQO1 mRNA (2.7, 1.7 and 1.2 kb) have been reported because of alternative poladenylation sites in the NQO1 gene.[76] A fourth smaller transcript of 1.1 kb was identified in human cancer cell lines as a product of alternative splicing.[77] The truncated transcript corresponds to a protein lacking exon 4 which contains the quinone binding site. Although the truncated protein could be expressed in E. coli, it was not detected in human tumor cells[77] or in Cos7 cells,[63] so it appears that the alternatively spliced mRNA is not translated into protein. The recombinant NQO1*3 protein purified from E. coli had no detectable activity when using a variety of electron acceptors,[77] as would be predicted from a protein lacking the critical quinone binding site. The truncated mRNA could be detected in patients but expressed wide interindividual variability.[78]

[76] A. K. Jaiswal, O. W. McBride, M. Adesnik, and D. W. Nebert, *J. Biol. Chem.* **263**, 13572 (1988).

[77] P. Y. Gasdaska, H. Fisher, and G. Powis, *Cancer Res.* **55**, 2542 (1995).

[78] K.-S. Yao, A. K. Godwin, C. Johnson, and P. J. O'Dwyer, *Cancer Res.* **56**, 1731 (1996).

Recently Pan et al.[79] have reported that in a human colon carcinoma cell line, alternative splicing of NQO1 at the 5′ splice site of intron 4 increased in cells carrying the NQO1*3 polymorphism. The NQO1*3 polymorphism disrupts the binding of small nuclear RNA (snRNA) in spliceosomes and transfection of snRNA constructs compensating for the NQO1*3 polymorphism increased NQO1 protein expression and enzymatic activity. These authors concluded that the NQO1*3 polymorphism was the major cause of increased alternate splicing and decreased expression of NQO1 protein in this particular colon carcinoma cell line.[79] The significance of the NQO1*3 polymorphism may therefore not be limited to the phenotype exhibited by the recombinant NQO1*3 protein but may also involve decreased expression of NQO1 wild type protein.

The frequency of the NQO1*3 polymorphism is low (Table III) and varied from 0 to 0.05 in different ethnic groups.[66]

Relevance of the NQO1*2 Polymorphism

As stated previously, a review of PubMed citations in March 2003 resulted in 92 hits related to NQO1 polymorphisms. The vast majority of these publications focus on the NQO1*2 polymorphism. We cannot discuss all of these studies in detail because of limitations of space, but we have attempted to summarize these studies in Table IV. Three studies on NQO2 were not included in Table IV, resulting in a total of 89 publications summarized. In addition, a few areas of significance of the NQO1 polymorphism are highlighted later.

NQO1*2 Polymorphism and Benzene Toxicity

The metabolism of benzene is complex,[80] but it is clear that metabolism is necessary for the induction of benzene toxicity. A number of reactive metabolites have been proposed as being responsible for benzene toxicity, and the evidence favoring each of these metabolites has recently been critically reviewed.[81] The conclusion of this analysis was that there was not sufficient evidence to define a particular benzene metabolite or group of metabolites as being responsible for benzene toxicity. The greatest weight of evidence, however, favors the generation of polyphenolics via cytochrome P4502E1-mediated metabolism of benzene, accumulation of metabolites such as catechol and hydroquinone in the target organ—the bone marrow—and subsequent oxidation of these metabolites to reactive quinone metabolites via a number of possible pathways.[81]

[79] S. S. Pan, Y. Han, P. Farabaugh, and H. Xia, *Pharmacogenetics* **12**, 479 (2002).
[80] D. Ross, *Eur. J. Haematol.* **57**(Suppl. 60), 111 (1996).
[81] D. Ross, *J. Toxicol. Environ. Health* **61**, 357 (2000).

TABLE IV
STUDIES ON NQO1 POLYMORPHISMS

First Author (Year)	Study	Remarks
Traver (1992)[60]	Characterization of NQO1*2	The C to T substitution at position 609 of the human NQO1 cDNA was identified in cell lines with high NQO1 mRNA expression but no NQO1 catalytic activity.
Eickelmann (1994)[71]	Kidney Tumors	No NQO1 catalytic activity was detected in three separate kidney tumor samples with normal NQO1 mRNA levels.
Eickelmann (1994)[72]	PCR-RFLP/Kidney Tumors	A PCR-RFLP assay is described for the detection of the NQO1*2 polymorphism. No NQO1 catalytic activity was detected in kidney tumor samples from patients genotype as homozygous for the NQO1*2 polymorphism.
Kolesar (1995)[82]	Colorectal Cancer	A C to T substitution at position 609 of the human NQO1 cDNA was detected in a patient with colon cancer.
Rosvold (1995)[83]	Lung Cancer and Smoking	The NQO1 codon 187 (pro→ser) mutation was identified as a polymorphism (NQO1*2). No correlation was observed between this mutant allele and the risk of lung cancer.
Rosvold (1995)[84]	Colorectal Cancer	A letter suggesting that the NQO1 mutation described by Kolesar (1995) was a polymorphism.
Kolesar (1995)[85]	Colorectal Cancer	Response by Kolesar et al.
Ross (1996a)[80]	Benzene	Review.
Kuehl (1995)[86]	Characterization of NQO1*2	Genotype-phenotype studies of the NQO1*2 polymorphism.
Ross (1996b)[87]	Characterization of NQO1*2	A letter indicating that the BE human colon carcinoma cell line previously genotyped by Kuehl et al. as NQO1*1/*2 is actually NQO1*2/*2.
Rebbeck (1996)[88]	Breast Cancer	The NQO1*2 allele was associated with an increased risk of breast cancer.
Traver (1997)[61]	Characterization of NQO1*2/Lung Cancer	The NQO1 protein was not detected in human tumor cell lines with the NQO1*2/*2 genotype. This study also suggested that the NQO1*2 polymorphism may be over represented in lung cancers.
Wiencke (1997)[89]	Lung Cancer	The NQO1*1/*1 genotype was associated with an increased risk of lung cancer.

(continued)

TABLE IV *(continued)*

First Author (Year)	Study	Remarks
Schulz (1997)[73]	Urological Malignancies	The NQO1*2/*2 genotype was associated with an increased risk of renal cell and urothelial carcinomas.
Rothman (1997)[90]	Benzene Poisoning	The NQO1*2/*2 genotype and rapid CYP4502E1 phenotypes were associated with an increased risk of benzene poisoning.
Kelsey (1997)[67]	Anticancer Chemotherapy	This study suggested that the NQO1*2 polymorphism may influence the response of tumor cells to quinone antitumor drugs. Prevalence of the NQO1*2/*2 genotype was reported in different ethnic populations.
Bartsch (1998)[91]	Pancreatic Disease	No correlation was observed between the NQO1*2 allele frequency and risk of pancreatic disease.
Gaedigk (1998)[66]	Mutant Allele Frequency	The NQO1*2 allele frequency was higher in Chinese, Canadian Native Indian and Canadian Inuit populations relative to a Caucasian population.
Kadlubar (1998)[92]	DNA Adducts/Pancreas	No correlation was observed between the NQO1*2 allele frequency and the levels of pancreatic DNA adducts in smokers and non-smokers.
Steiner (1999)[93]	Prostate Disease	The NQO1*2 allele was not associated with an increased risk of prostate disease.
Siegel (1999)[69]	Genotype/Phenotype Studies	A correlation was observed between the amount of NQO1 protein isolated from saliva and the NQO1*2 genotype. Individuals with the NQO1*2/*2 genotype had no detectable NQO1 protein whereas individuals with the NQO1*1/*2 genotype had intermediate NQO1 protein levels.
Ozawa (1999)[94]	Bronchial DNA Adducts	The NQO1*2 polymorphism was not associated with an increased level of bronchial-DNA adducts in smokers and non-smokers.
Longuemaux (1999)[95]	Renal Cell Carcinoma	The NQO1*2 polymorphism was not associated with an increased risk of renal cell carcinoma.
Clairmont (1999)[96]	Basal Cell Carcinoma	The NQO1*2/*2 genotype was associated with an increased risk of basal cell carcinoma.

(continued)

TABLE IV   *(continued)*

First Author (Year)	Study	Remarks
Smith (1999)[97]	Benzene Poisoning	Commentary on Moran *et al.* 1999.
Moran (1999)[98]	Benzene Metabolism	Human bone marrow progenitor cells carrying the NQO1[*]2 allele have decreased NQO1 protein levels following induction by benzene metabolites.
Chen (1999)[99]	Lung Cancer	The NQO1[*]2 allele was protective against lung cancer in a Japanese population. Increased risk of lung cancer was associated with the NQO1[*]1 wild type allele.
Larson (1999)[100]	Myeloid Leukemia	The NQO1[*]2 allele was associated with an increased risk of therapy-related acute myeloid leukemia.
Kristiansen (1999)[101]	Type I Diabetes	No correlation was observed between the NQO1[*]2 polymorphism and the risk of type I diabetes in a Danish population.
Wiemels (1999)[102]	Pediatric Leukemias	The mutant NQO1[*]2 allele was associated with an increased risk of infant leukemias with chromosomal rearrangements.
Kelland (1999)[103]	17-Demethoxygeldana-mycin	The sensitivity of tumor cells to 17-demethoxygeldanamycin was associated with expression of NQO1[*]1 protein.
Zheng (1999)[104]	Manganism	No correlation was observed between the NQO1[*]2 allele frequency and occupational chronic manganism.
Lin (1999)[105]	Lung Cancer	No correlation was observed between the NQO1[*]2 polymorphism and risk of lung cancer in a Taiwan population. Stratification of tumors according to histological subtype suggested that the NQO1[*]1/[*]1 genotype was more common in adenocarcinomas than controls.
Harth (2000)[106]	Colorectal Cancer	No correlation was observed between the NQO1[*]2 polymorphism and risk of colorectal cancer.
Gaedigk (2000)[107]	Drug Metabolism	This study describes interethnic differences in the distribution of polymorphisms including NQO1[*]2 in closely related populations.
Shi (2000)[108]	High-Output Genotyping	The NQO1[*]2 polymorphism was successfully genotyped in a high output assay utilizing Taq poymerase and fluorogenic Taqman probes.

(continued)

TABLE IV   *(continued)*

First Author (Year)	Study	Remarks
Martone (2000)[109]	p53 Mutations in Bladder Cancer	No correlation was observed between the NQO1*2 allele frequency and p53 mutations in human bladder cancers.
Nakajima (2000)[110]	Susceptibility to Industrial Chemicals	Review.
Schelonka (2000)[111]	Eye Disease	The NQO1*2/*2 genotype was associated with the absence of NQO1 protein in human donor eyes.
Lafuente (2000)[112]	Colorectal Cancer	The NQO1*2 polymorphism was associated with a increased risk of colon cancers in association with K-ras codon 12 mutations.
Naoe (2000)[113]	Leukemias	The NQO1*2 polymorphism was associated with an increased risk of therapy-related leukemia and myelodysplastic syndrome.
Ross (2000)[23]	NQO1	Review.
Siegel (2001)[68]	NQO1*2 Protein Stability	The mutant NQO1*2 protein was found to have a greatly reduced protein half-life compared to the NQO1*1 protein as a result of degradation by the proteasomal pathway.
Pavanello (2001)[114]	Genotoxic Risk	No correlation was observed between the NQO1*2 polymorphism and bio-markers of genotoxic exposure.
Peters (2001)[115]	Glioma	No correlation was observed between the NQO1*2 allele frequency and risk of glioma.
Smith (2001)[116]	Leukemias	The NQO1*2 allele was associated with an increased risk of *de novo* acute myeloid leukemia in adults.
Shao (2001)[117]	Parkinson's Disease	The NQO1*2 polymorphism in combination with a polymorphism in mono-amine oxidase B resulted in an additive increased risk of Parkinson's disease.
Xu (2001)[118]	Lung Cancer	No correlation was observed between the NQO1*2 allele frequency and susceptibility to lung cancer. An association, however, was observed between the NQO1*2/2 genotype and risk of lung cancer in heavy former and current smokers.

(continued)

TABLE IV   (continued)

First Author (Year)	Study	Remarks
Iida (2001)[65]	SNPs in Quinone Reductase	A Japanese population was screened for single nucleotide polymorphisms in quinone metabolizing enzymes including NQO1 and NQO2.
Bergamaschi (2001)[119]	Effect of Ozone	The NQO1*1/*1 genotype was associated with both a decrease in pulmonary function and increased serum clara cell protein 16 in patients exposed to ozone.
Winski (2001)[21]	Chemotherapy	A human colon carcinoma cell line genotyped as NQO1*2/*2 (BE) and lacking NQO1 catalytic activity was transfected with the NQO1*1 coding region. Stable clones obtained following transfection demonstrated increased sensitivity to the DNA cross-linking agent RH-1.
Le Marchand (2001)[120]	Genotyping	The NQO1*2 polymorphism was successfully genotyped by PCR-RFLP from buccal DNA obtained in a large, community-based mail-in study.
Gallou (2001)[121]	Renal Cell Carcinoma	No correlation was observed between the NQO1*2 allele frequency and mutations in the von Hippel-Lindau tumor suppressor gene in renal cell carcinoma.
Yin (2001)[122]	Lung Cancer	The NQO1*2 allele was not associated with an increased risk of lung cancer in a Chinese population.
Lewis (2001)[123]	Lung Cancer	The NQO1*2 allele was associated with an increased risk of small cell lung cancer. No correlation was observed between the NQO1*2 allele and non-small cell lung cancer.
Harada (2001)[124]	Parkinson's Disease	The NQO1*2 polymorphism was not associated with an increased risk of Parkinson's disease.
Siegel (2001)[125]	Bone Marrow Expression	NQO1 protein was detected by immunohistochemistry in human bone marrow endothelial cells suggesting a role for NQO1 in the metabolism of xenobiotics in the bone marrow.

(continued)

TABLE IV   *(continued)*

First Author (Year)	Study	Remarks
Phillips (2001)[126]	Response to Mitomycin C	No correlation was observed between the NQO1*2 allele frequency and response to mitomycin C treatment in a panel of human tumor xenografts transplanted into mice.
Verdina (2001)[127]	Urinary Biomarkers/ Benzene Exposure	No correlation was observed between the NQO1*2 allele frequency and the levels of benzene urinary metabolites in policeman exposed to low concentrations of benzene.
Fleming (2002)[74]	Response to Mitomycin C	The NQO1*2 allele was associated with significantly lower NQO1 catalytic activity and reduced survival in patients receiving intraperitoneal hyperthermic chemotherapy with mitomycin C.
Krajinovic (2002)[128]	Leukemias	The NQO1*2 and NQO1*3 alleles in combination with variant alleles in CYP2E1, MPO or GSTM1 were associated with an increased risk of acute lymphoblastic leukemia in children.
Siegelmann (2002)[129]	Breast Cancer	No correlation was observed between the NQO1*2 allele frequency and the risk of breast cancer. The NQO1*1/*1 genotype however, was associated with an increased risk of ductal carcinoma with poor histological grade.
Anwar (2002)[75]	HSP70 Binding	The NQO1*1 protein but not the NQO1*2 protein was found to associate with HSP70 and HSP40 in cellular and cell free systems. The lack of association between NQO1*2 protein and HSP proteins may explain the diminished catalytic activity of the mutant NQO1 protein.
Naoe (2002)[130]	Leukemias	The NQO1*2 polymorphism does not affect survival of patients with AML.
Nebert (2002)[64]	HUGO Review	Review.
Krajinovic (2002)[131]	Leukemias	The NQO1*2 and variant CYP1A1 alleles were associated with a poorer prognosis in children with acute lymphoblastic leukemia.

(continued)

TABLE IV   *(continued)*

First Author (Year)	Study	Remarks
Brockstedt (2002)[132]	DNA Adducts Breast Cancer	No correlation was observed between the NQO1*2 and NQO1*3 allele frequencies and the level of DNA adducts detected in normal human breast tissues.
Tuominen (2002)[133]	PAH Exposure	The NQO1*2 polymorphism may influence the DNA adduct profile in control and aluminum smelter workers.
Hamajima (2002)[134]	PCR/Genotyping	This study describes the use of PCR with confronting two-pair primers as an alternative method for single nucleotide polymorphism analysis. The NQO1*2 polymorphism was utilized as a model.
Hamajima (2002)[135]	Japanese Cancers	The NQO1*2/*2 genotype was not associated with an increased risk of cancer of the stomach, colon, rectum, breast, prostate or malignant lymphoma in a Japanese population. This study suggested that the NQO1*1/*1 genotype was associated with an increased risk of lung cancer. In addition, an increased risk of lung and esophageal cancer in combination with smoking was observed for the NQO1*2/*2 genotype.
Goode (2002)[136]	Breast Cancer	No correlation was observed between the NQO1*2 allele frequency and survival in women with breast cancer.
Krajinovic (2001)[137]	Leukemias	Review.
Morgan (2002)[138]	Leukemia	Review.
Carere (2002)[139]	Urban Pollution	No correlation was observed between the NQO1*2 allele frequency and genetic biomarkers for exposure to urban air pollution in Rome traffic policemen.
Sunaga (2002)[140]	Lung Cancer	The NQO1*1/*1 genotype was associated with increased risk of lung adenocarcinoma. A greater risk was found in combination with the NQO1*1/*1 and GSTT-1 null genotypes. This enhanced risk was more evident in smokers compared to non-smokers.
Hamajima (2002)[141]	Japanese Noncancers	Polymorphisms in drug metabolizing enzymes including NQO1 were analyzed in a Japanese population.

(continued)

TABLE IV   *(continued)*

First Author (Year)	Study	Remarks
Zheng (2002)[142]	Occupational Manganism	No correlation was observed between the NQO1*2 allele frequency and occupational chronic manganism.
Pan (2002)[79]	NQO1*3 Polymorphism	The NQO1*3 (465C→T) mutation was found to be associated with defective NQO1 RNA splicing and decreased NQO1 protein expression.
Asher (2002)[55]	p53 Interaction	The mutant NQO1*2 protein, unlike the wild type NQO1*1 protein, failed to stabilize and prevent degradation of wild type p53.
Kolesar (2002)[143]	Lung Cancer	NQO1*2 in normal tissue and lung tumors. The NQO1*2/*2 genotype was associated with a shorter survival time in patients treated with chemotherapy for stage II/III non-small cell lung cancer.
Seedhouse (2002)[144]	Leukemia	No correlation was observed between the NQO1*2 allele frequency and risk of *de novo* or therapy-related acute myeloblastic leukemia.
Smith (2002)[145]	Leukemias	The NQO1*2 polymorphism was found to be associated with an increased risk of *de novo* leukemias with MLL translocations in infants and children.
Soucek (2002)[146]	Lymphomas	No correlation was observed between the NQO1*2 allele frequency and patients with Hodgkin's and non-Hodgkin's lymphomas.
Sachse (2002)[147]	Colorectal Cancer	No correlation was observed between the NQO1*2 allele frequency and the risk of colorectal cancer.
Blanco (2002)[148]	Leukemia	No correlation was observed between the NQO1*2 allele frequency and the risk of therapy-related acute myeloid leukemia and myelodysplastic syndrome in children treated for ALL.
Wan (2002)[149]	Benzene Poisoning	Study found that Chinese patients with the NQO1*2/*2 genotype were at increased risk of benzene poisoning.
Wu (2002)[150]	Nasopharyngeal Carcinoma	The NQO1*2 polymorphism was found to be associated with genetic susceptibility to nasopharyngeal carcinoma.

(continued)

TABLE IV   *(continued)*

First Author (Year)	Study	Remarks
Takagi (2002)[151]	Colorectal Cancer	A correlation was observed between the NQO1*2/*2 genotype and telomere shortening in colorectal cancer.
Chen (2002)[152]	Benzene Poisoning	Review.

[82] J. M. Kolesar, J. G. Kuhn, and H. A. Burris, III, *JNCI* **87,** 1022 (1995).

[83] E. A. Rosvold, K. A. McGlynn, E. D. Lustbader, and K. H. Buetow, *Pharmacogenetics* **5,** 199 (1995).

[84] E. A. Rosvold, K. A. McGlynn, E. D. Lustbader, and K. H. Buetow, *J. Natl. Cancer Inst.* **87,** 1802 (1995).

[85] J. M. Kolesar, H. A. Burris III, and J. G. Kuhn, *JNCI* **87,** 1803 (1995).

[86] B. L. Kuehl, J. W. E. Paterson, J. W. Peacock, M. C. Paterson, and A. M. Rauth, *Br. J. Cancer* **72,** 555 (1995).

[87] D. Ross, R. D. Traver, D. Siegel, B. L. Kuehl, V. Misra, and A. M. Rauth, *Br. J. Cancer* **74,** 995 (1996).

[88] T. R. Rebbeck, A. K. Godwin, and K. H. Buetow, *Mol. Carcinog.* **17,** 117 (1996).

[89] J. T. Wiencke, M. R. Spitz, A. McMillan, and K. T. Kelsey, *Cancer Epidemiol. Bio. Prev.* **6,** 87 (1997).

[90] N. Rothman, M. T. Smith, R. B. Hayes, R. D. Traver, B. A. Hoener, S. Campleman, G. L. Li, M. Dosemeci, M. Linet, L. P. Zhang, L. Q. Xi, S. Wacholder, W. Lu, K. B. Meyer, N. Titenko-Holland, J. T. Stewart, S. N. Yin, and D. Ross, *Cancer Res.* **57,** 2839 (1997).

[91] H. Bartsch, C. Malaveille, A. B. Lowenfels, P. Maisonneuve, A. Hautefeuille, and P. Boyle, *Eur. J. Cancer Prev.* **7,** 215 (1998).

[92] F. F. Kadlubar, K. E. Anderson, S. Haussermann, N. P. Lang, G. W. Barone, P. A. Thompson, S. L. MacLeod, M. W. Chou, M. Mikhailova, J. Plastaras, L. J. Marnett, J. Nair, I. Velic, and H. Bartsch, *Mutat. Res.* **405,** 125 (1998).

[93] M. Steiner, M. Hillenbrand, M. Borkowsi, H. Seiter, and P. Schuff-Werner, *Cancer Lett.* **135,** 67 (1999).

[94] S. Ozawa, B. Schoket, L. P. McDaniel, Y. M. Tang, C. B. Ambrosone, S. Kostic, I. Vincze, and F. F. Kadlubar, *Carcinogenesis* **20,** 991 (1999).

[95] S. Longuemaux, C. Delomenie, C. Gallou, A. Mejean, M. Vincent-Viry, R. Bouvier, D. Droz, R. Krishnamoorthy, M. M. Galteau, C. Junien, C. Beroud, and J. M. Dupret, *Cancer Res.* **59,** 2903 (1999).

[96] A. Clairmont, H. Sies, S. Ramachandran, J. T. Lear, A. G. Smith, B. Bowers, P. W. Jones, A. A. Fryer, and R. C. Strange, *Carcinogenesis* **20,** 1235 (1999).

[97] M. T. Smith, *Proc. Natl. Acad. Sci. USA* **96,** 7624 (1999).

[98] J. L. Moran, D. Siegel, and D. Ross, *Proc. Natl. Acad. Sci. USA* **96,** 8150 (1999).

[99] H. Chen, A. Lum, A. Seifried, L. R. Wilkens, and L. Le Marchand, *Cancer Res.* **59,** 3045 (1999).

[100] R. A. Larson, Y. Wang, M. Banerjee, J. Wiemels, C. Hartford, M. M. Beau, and M. T. Smith, *Blood* **94,** 803 (1999).

[101] O. P. Kristiansen, Z. M. Larsen, J. Johannesen, J. Nerup, T. Mandrup-Poulsen, and F. Pociot, *Hum. Mutat.* **14,** 67 (1999).

In the mid 1990s, a collaborative effort led by the National Cancer Institute examined the risk of benzene poisoning as a function of metabolic phenotype and genotype of occupationally exposed workers in China.[90] Benzene poisoning was defined by decreases in white blood cell and platelet counts after exposure and was strongly related to the subsequent development of acute nonlymphocytic leukemia and related myelodysplastic

[102] J. L. Wiemels, A. Pagnamenta, G. M. Taylor, O. B. Eden, F. E. Alexander, and M. F. Greaves, *Cancer Res.* **59,** 4095 (1999).

[103] L. R. Kelland, S. Y. Sharp, P. M. Rogers, T. G. Myers, and P. Workman, *J. Natl. Cancer Inst.* **91,** 1940 (1999).

[104] Y. Zheng, F. He, P. Chan, Z. Pan, Z. Wang, J. Pan, and X. Zhou, *Zhonghua Yu Fang Yi. Xue. Za Zhi.* **33,** 78 (1999).

[105] P. Lin, H. J. Wang, H. Lee, H. S. Lee, S. L. Wang, Y. M. Hsueh, K. J. Tsai, and C. Y. Chen, *J. Toxicol. Environ. Health* **58,** 187 (1999).

[106] V. Harth, S. Donat, Y. Ko, J. Abel, H. Vetter, and T. Bruning, *Arch. Toxicol.* **73,** 528 (2000).

[107] A. Gaedigk, *Int. J. Clin. Pharmacol. Ther.* **38,** 61 (2000).

[108] M. M. Shi, S. P. Myrand, M. R. Bleavins, and F. A. de la Iglesia, *Mol. Pathol.* **52,** 295 (1999).

[109] T. Martone, P. Vineis, C. Malaveille, and B. Terracini, *Mutat. Res.* **462,** 303 (2000).

[110] T. Nakajima and T. Aoyama, *Ind. Health* **38,** 143 (2000).

[111] L. P. Schelonka, D. Siegel, M. W. Wilson, A. Meininger, and D. Ross, *Invest. Ophthalmol. Vis. Sci.* **41,** 1617 (2000).

[112] M. J. Lafuente, X. Casterad, M. Trias, C. Ascaso, R. Molina, A. Ballesta, S. Zheng, J. K. Wiencke, and A. Lafuente, *Carcinogenesis* **21,** 1813 (2000).

[113] T. Naoe, K. Takeyama, T. Yokozawa, H. Kiyoi, M. Seto, N. Uike, T. Ino, A. Utsunomiya, A. Maruta, I. Jin-nai, N. Kamada, Y. Kubota, H. Nakamura, C. Shimazaki, S. Horiike, Y. Kodera, H. Saito, R. Ueda, J. Wiemels, and R. Ohno, *Clin. Cancer Res.* **6,** 4091 (2000).

[114] S. Pavanello and E. Clonfero, *Med. Lav.* **91,** 431 (2000).

[115] E. S. Peters, K. T. Kelsey, J. K. Wiencke, S. Park, P. Chen, R. Miike, and M. R. Wrensch, *Cancer Epidemiol. Biomarkers Prev.* **10,** 151 (2001).

[116] M. T. Smith, Y. Wang, E. Kane, S. Rollinson, J. L. Wiemels, E. Roman, P. Roddam, R. Cartwright, and G. Morgan, *Blood* **97,** 1422 (2001).

[117] M. Shao, Z. Liu, E. Tao, and B. Chen, *Zhonghua Yi. Xue. Yi. Chuan Xue. Za Zhi.* **18,** 122 (2001).

[118] L. L. Xu, J. C. Wain, D. P. Miller, S. W. Thurston, L. Su, T. J. Lynch, and D. C. Christiani, *Cancer Epidemiol. Biomarkers Prev.* **10,** 303 (2001).

[119] E. Bergamaschi, G. De Palma, P. Mozzoni, S. Vanni, M. V. Vettori, F. Broeckaert, A. Bernard, and A. Mutti, *Am. J. Respir. Crit Care Med.* **163,** 1426 (2001).

[120] L. Le Marchand, A. Lum-Jones, B. Saltzman, V. Visaya, A. M. Nomura, and L. N. Kolonel, *Cancer Epidemiol. Biomarkers Prev.* **10,** 701 (2001).

[121] C. Gallou, S. Longuemaux, C. Delomenie, A. Mejean, N. Martin, S. Martinet, G. Palais, R. Bouvier, D. Droz, R. Krishnamoorthy, C. Junien, C. Beroud, and J. M. Dupret, *Pharmacogenetics* **11,** 521 (2001).

[122] L. Yin, Y. Pu, T. Y. Liu, Y. H. Tung, K. W. Chen, and P. Lin, *Lung Cancer* **33,** 133 (2001).

[123] S. J. Lewis, N. M. Cherry, R. M. Niven, P. V. Barber, and A. C. Povey, *Lung Cancer* **34,** 177 (2001).

syndromes (relative risk = 70). A case-control study was performed employing 50 cases of benzene poisoning and 50 controls. CYP2E1 phenotyping was assessed by measuring urinary excretion of 6-hydroxy chlorzoxazone over 8 h after administration of chlorzoxazone while NQO1 genotype was assessed by PCR-RFLP analysis. Individuals with a greater capacity for CYP4502E1-mediated metabolism were at a greater risk for benzene poisoning (relative risk = 2.6) as were individuals carrying the

[124] S. Harada, C. Fujii, A. Hayashi, and N. Ohkoshi, *Biochem. Biophys. Res. Commun.* **288**, 887 (2001).

[125] D. Siegel, J. Ryder, and D. Ross, *Toxicol. Lett.* **125**, 93 (2001).

[126] R. M. Phillips, A. M. Burger, H. H. Fiebig, and J. A. Double, *Biochem. Pharmacol.* **62**, 1371 (2001).

[127] A. Verdina, R. Galati, G. Falasca, S. Ghittori, M. Imbriani, F. Tomei, L. Marcellini, A. Zijno, and V. D. Vecchio, *J. Toxicol. Environ. Health* **64**, 607 (2001).

[128] M. Krajinovic, H. Sinnett, C. Richer, D. Labuda, and D. Sinnett, *Int. J. Cancer* **97**, 230 (2002).

[129] N. Siegelmann-Danieli and K. H. Buetow, *Oncology* **62**, 39 (2002).

[130] T. Naoe, Y. Tagawa, H. Kiyoi, Y. Kodera, S. Miyawaki, N. Asou, K. Kuriyama, S. Kusumoto, C. Shimazaki, K. Saito, H. Akiyama, T. Motoji, M. Nishimura, K. Shinagawa, R. Ueda, H. Saito, and R. Ohno, *Leukemia* **16**, 203 (2002).

[131] M. Krajinovic, D. Labuda, G. Mathonnet, M. Labuda, A. Moghrabi, J. Champagne, and D. Sinnett, *Clin. Cancer Res.* **8**, 802 (2002).

[132] U. Brockstedt, M. Krajinovic, C. Richer, G. Mathonnet, D. Sinnett, W. Pfau, and D. Labuda, *Mutat. Res.* **516**, 41 (2002).

[133] R. Tuominen, P. Baranczewski, A. Warholm, L. Hagmar, L. Moller, and A. Rannug, *Arch. Toxicol.* **76**, 178 (2002).

[134] N. Hamajima, T. Saito, K. Matsuo, and K. Tajima, *J. Mol. Diagn.* **4**, 103 (2002).

[135] N. Hamajima, K. Matsuo, H. Iwata, M. Shinoda, Y. Yamamura, T. Kato, S. Hatooka, T. Mitsudomi, M. Suyama, Y. Kagami, M. Ogura, M. Ando, Y. Sugimura, and K. Tajima, *Int. J. Clin. Oncol.* **7**, 103 (2002).

[136] E. L. Goode, A. M. Dunning, B. Kuschel, C. S. Healey, N. E. Day, B. A. Ponder, D. F. Easton, and P. P. Pharoah, *Cancer Res.* **62**, 3052 (2002).

[137] M. Krajinovic, D. Labuda, and D. Sinnett, *Rev. Environ. Health* **16**, 263 (2001).

[138] G. J. Morgan and M. T. Smith, *Am. J. Pharmacogenomics.* **2**, 79 (2002).

[139] A. Carere, C. Andreoli, R. Galati, P. Leopardi, F. Marcon, M. V. Rosati, S. Rossi, F. Tomei, A. Verdina, A. Zijno, and R. Crebelli, *Mutat. Res.* **518**, 215 (2002).

[140] N. Sunaga, T. Kohno, N. Yanagitani, H. Sugimura, H. Kunitoh, T. Tamura, Y. Takei, S. Tsuchiya, R. Saito, and J. Yokota, *Cancer Epidemiol. Biomarkers Prev.* **11**, 730 (2002).

[141] N. Hamajima, T. Saito, K. Matsuo, T. Suzuki, T. Nakamura, A. Matsuura, K. Okuma, and K. Tajima, *J. Epidemiol.* **12**, 229 (2002).

[142] Y. X. Zheng, P. Chan, Z. F. Pan, N. N. Shi, Z. X. Wang, J. Pan, H. M. Liang, Y. Niu, X. R. Zhou, and F. S. He, *Biomarkers* **7**, 337 (2002).

[143] J. M. Kolesar, S. C. Pritchard, K. M. Kerr, K. Kim, M. C. Nicolson, and H. McLeod, *Int. J. Oncol.* **21**, 1119 (2002).

[144] C. Seedhouse, R. Bainton, M. Lewis, A. Harding, N. Russell, and E. Das-Gupta, *Blood* **100**, 3761 (2002).

[145] M. T. Smith, Y. Wang, C. F. Skibola, D. J. Slater, L. L. Nigro, P. C. Nowell, B. J. Lange, and C. A. Felix, *Blood* **100**, 4590 (2002).

homozygous NQO1*2 polymorphism (relative risk = 2.4). When both of these metabolic pathways were combined, rapid CYP2E1 metabolism and a lack of NQO1 caused by the homozygous NQO1*2 polymorphism, the relative risk for benzene poisoning increased to 7.6 fold.[90] This was the first study examining the significance of a sequence of metabolic pathways during the metabolism of benzene in humans, and studies in knockout animals have reinforced the potential importance of both CYP2E1 and NQO1 to benzene toxicity.[153,154] Interestingly, in NQO1 knockout animals, both male and female −/− mice were more susceptible to benzene induced hematotoxicity relative to NQO1 +/+ animals, but only female −/− animals were more susceptible to benzene-induced micronuclei relative to their NQO1 +/+ comparison group.[154] It is possible that different benzene metabolites may be responsible for hematotoxicity and genotoxicity,[154] but these data also suggest that genotoxicity as reflected by micronuclei induction may not be related to hematotoxicity.

The mechanism proposed for the role of CYP2E1 and NQO1 in benzene toxicity is shown in Fig. 4. Increased metabolism of benzene to polyphenolics or inhibited detoxification of benzene-derived reactive quinones caused by the NQO1*2 polymorphism may lead to increased toxicity. One potential problem with this mechanism was that for NQO1 to detoxify benzene-derived quinones, the enzyme would have to be present in human bone marrow, the target organ of benzene toxicity. Our studies, however, did not detect NQO1 in human bone marrow aspirates.[155,98] Two potential

[146] P. Soucek, J. Sarmanova, V. N. Kristensen, M. Apltauerova, and I. Gut, Int. Arch. Occup. Environ. Health 75(Suppl. 1), 86 (2002).
[147] C. Sachse, G. Smith, M. J. Wilkie, J. H. Barrett, R. Waxman, F. Sullivan, D. Forman, D. T. Bishop, and C. R. Wolf, Carcinogenesis 23, 1839 (2002).
[148] J. G. Blanco, M. J. Edick, M. L. Hancock, N. J. Winick, T. Dervieux, M. D. Amylon, R. O. Bash, F. G. Behm, B. M. Camitta, C. H. Pui, S. C. Raimondi, and M. V. Relling, Pharmacogenetics 12, 605 (2002).
[149] J. Wan, J. Shi, L. Hui, D. Wu, X. Jin, N. Zhao, W. Huang, Z. Xia, and G. Hu, Environ. Health Perspect. 110, 1213 (2002).
[150] D. H. Wu, Di Yi. Jun. Yi. Da. Xue. Xue. Bao. 22, 1126 (2002).
[151] S. Takagi, Y. Kinouchi, N. Hiwatashi, M. Hirai, S. Suzuki, S. Takahashi, K. Negoro, N. Obana, and T. Shimosegawa, Anticancer Res. 22, 2749 (2002).
[152] Y. Chen, G. Li, and S. Yin, Wei Sheng Yan. Jiu. 31, 130 (2002).
[153] J. L. Valentine, S. S. T. Lee, M. J. Seaton, B. Asgharian, G. Farris, J. C. Corton, F. J. Gonzalez, and M. A. Medinsky, Toxicol. Appl. Pharmacol. 141, 205 (1996).
[154] A. K. Bauer, B. Faiola, D. J. Abernethy, R. Marchan, L. J. Pluta, V. A. Wong, K. Roberts, A. K. Jaiswal, F. J. Gonzalez, B. E. Butterworth, S. Borghoff, H. Parkinson, J. Everitt, and L. Recio, Cancer Res. 63, 929 (2003).
[155] D. Ross, D. Siegel, D. G. Schattenberg, X. M. M. Sun, and J. L. Moran, Environ. Health Perspect. 104(Suppl. 6), 1177 (1996).

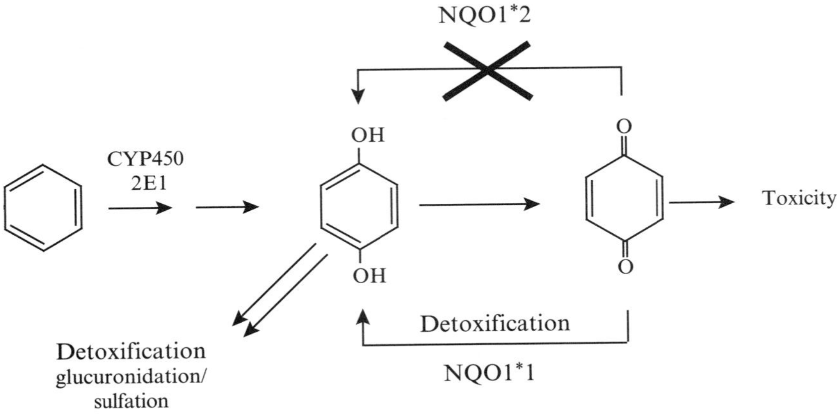

NQO1*2

CYP450
2E1

OH

OH

O

O

Toxicity

Detoxification
glucuronidation/
sulfation

Detoxification

NQO1*1

FIG. 4. The proposed role of cytochrome P4502E1 and NQO1 in benzene toxicity.

mechanisms were proposed to explain this contradiction, and both of these mechanisms have now been confirmed. The first mechanism involved induction of NQO1 by polyphenolic metabolites of benzene. Although NQO1 was not present in either human bone marrow or purified CD34+ human bone marrow cells, exposure of these cells *in vitro* to hydroquinone or catechol led to induction of NQO1 activity but *only* in cells from individuals genotyped as NQO1*1/*1 or wild-type.[98] No induction of NQO1 activity was observed in individuals genotyped as NQO1*2/*2, and intermediate levels of induction were observed in heterozygous or NQO1*1/*2 individuals.[98] Presumably, the lack of induction of NQO1 associated with the NQO1*2 polymorphism reflected the instability of the NQO1*2 protein due to proteasomal degradation[68] rather than any effect on the mechanism and extent of enzyme induction. This mechanism would therefore provide an explanation for increased benzene toxicity associated with the NQO1*2 polymorphism; benzene metabolites would induce NQO1, which could detoxify benzene-derived quinones but not in individuals with the NQO1*2/*2 genotype.

The second hypothesis that we tested was that NQO1 was indeed present in human bone marrow but was contained in cells that were not removed by conventional aspiration techniques. We performed immunohistochemistry on archived bone marrow core samples and found that NQO1 could be detected in human bone marrow but was present predominantly in bone marrow endothelial cells.[125] Endothelial cells are not readily removed from bone marrow by aspiration. The role of bone marrow endothelial cells in benzene toxicity is as yet unclear but is a subject of investigation in our laboratory. Bone marrow endothelial cells participate in the maturation of hematopoietic progenitor cells in a number of ways, including expression of adhesion molecules and secretion of cyto-kines.[156,157] The potential mechanisms underlying a protective role of NQO1 against benzene toxicity in human bone marrow are summarized in Fig. 5. Since additional roles of NQO1 are being uncovered, such as its involvement in antioxidant defense and stabilization of p53 (see previous text), it is feasible that additional mechanisms may be involved in NQO1-mediated protection against benzene toxicity.

NQO1*2 Polymorphism and Leukemia

Since the homozygous NQO1*2 polymorphism is effectively a null polymorphism and heterozygous individuals demonstrate decreased NQO1 protein and activity relative to the wild-type or concensus sequence, the NQO1*2 polymorphism provides a convenient molecular tool to examine the role of NQO1 in disease. Of the many cancers that have been examined for an association with the NQO1*2 polymorphism (Table IV), the most frequent association has been found with leukemias.

Eight separate studies have examined the possible relationship of deficient NQO1 caused by the NQO1*2 polymorphism and leukemia. Two additional studies examined the prognosis of leukemia patients with the NQO1*2 polymorphism after chemotherapy.[130,131] Of the eight studies examining risk of leukemia, six of these found an increased risk of various types of leukemia associated with the NQO1*2 polymorphism while two found no association of therapy related leukemia. An increased risk of leukemia associated with the NQO1*2 polymorphism was found in therapy-related leukemia,[100] therapy-related leukemia/myelodysplastic syndrome,[113] pediatric leukemias (particularly with MLL gene rearrangements[102,145]), childhood acute lymphoblastic leukemia (ALL),[128] and adult de-novo leukemias, both acute myeloid leukemia (AML) and ALL.[116] In the study by Krajinovic et al.,[128] the wild-type NQO1 genotype was found to be protective against childhood ALL, whereas children carrying at least one mutant allele of NQO1 (NQO1*2 or NQO1*3) were at increased risk. Two additional studies found no association of the NQO1*2 polymorphism with therapy related leukemia.[158,159]

[156] C. Voermans, P. M. Rood, P. L. Hordijk, W. R. Gerritsen, and C. E. van der Schoot, Stem Cells 18, 435 (2000).
[157] R. Mohle, M. A. Moore, R. L. Nachman, and S. Rafii, Blood 89, 72 (1997).
[158] D. J. Long, A. Gaikwad, A. Multani, S. Pathak, C. A. Montgomery, F. J. Gonzalez, and A. K. Jaiswal, Cancer Res. 62, 3030 (2002).
[159] K. Mikami, M. Naito, A. Tomida, M. Yamada, T. Sirakusa, and T. Tsuruo, Cancer Res. 56, 2823 (1996).

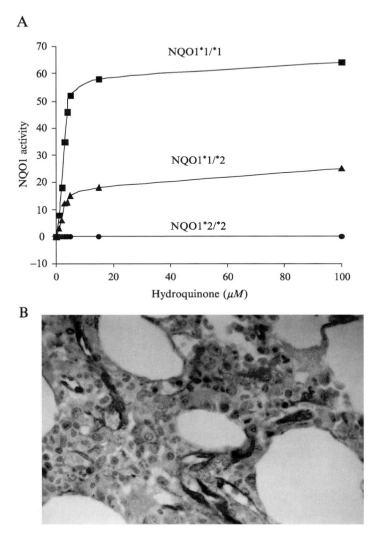

FIG. 5. Potential mechanisms underlying a protective role of NQO1 against benzene toxicity. (A) Induction of NQO1 in human bone marrow mononuclear cells after exposure to hydroquinone depends on NQO1 genotype.[98] Catechol induces a similar induction of NQO1, and induction of NQO1 by hydroquinone was also detected in bone marrow CD34[+] enriched cell populations.[98] The inability of polyphenolic metabolites of benzene to increase NQO1 activity in NQO1*2/*2 cells was presumably related to the instability of the mutant NQO1*2 protein rather than any effect on induction. (B) Presence of NQO1 in bone marrow endothelial cells. Although NQO1 could not be detected in aspirated human bone marrow cells, it could be detected by immunohistochemistry in bone marrow endothelial cells.[125] (See color insert.)

The NQO1*2*2 genotype has also been associated with an increased risk of benzene induced myelotoxicity in occupationally exposed workers in China,[82] and biologically plausible mechanisms can be proposed to account for this increased risk (see previous text). Another important study in NQO1 knockout animals has also recently been published[158] demonstrating that NQO1 knockout animals were at a greater risk of myeloid hyperplasia and increased levels of blood neutrophils, basophils and eosinophils. NQO1 knockout mice also contained decreased levels of p53 in bone marrow relative to wild type animals,[158] reinforcing a possible role for NQO1 in stabilization of p53 (see previous text). Taken together, these data suggest that low activity NQO1 genotypes, and particularly the NQO1*2/*2 genotype, are associated with myeloid abnormalities and increase the risk of leukemias. Assessing the impact of the NQO1*2 polymorphism should be incorporated as a part of the design of any large epidemiological studies of genetic risk factors for the induction of leukemia.

NQO1*2 Polymorphism and Chemotherapy

Since NQO1 can bioactivate certain antitumor quinones such as mitomycin C, E09 streptonigrin, β-lapachone[13–16] and newer experimental agents such as RH1,[20,21] the NQO1*2 polymorphism would be expected to impact therapy using these agents. Pre-clinical studies using gene transfection and gene targeting approaches[20,21,59,159] have clearly demonstrated the role of NQO1 in bioactivation of antitumor quinones and provided proof of principle for this approach. The impact of the NQO1*2 polymorphism has also been demonstrated by using excised human tumors *in vitro;* tumor tissue genotyped as NQO1*2/*2 was more resistant to mitomycin C than tumors genotyped as wild-type.[160] Perhaps the most compelling study to date is a clinical study performed by Fleming *et al.*[74] In this work, patients with disseminated peritoneal cancer received surgical debulking and intraperitoneal hyperthermic perfusions of mitomycin C. In 117 patients genotyped for the NQO1*2 polymorphism, individuals with the low activity NQO1 genotypes (heterozygous or homozygous for the NQO1*2 polymorphism) had reduced survival relative to individuals with the wild type or concensus NQO1 genotype. Genotype-phenotype relationships were also confirmed in this study, and NQO1*1/*2 heterozygotes had significantly less tumor NQO1 activity than individuals with the wild-type NQO1 genotype.[74] This study demonstrates the potential importance of the NQO1*2 polymorphism to therapy using antitumor quinones such as

[160] M. Yano, Y. Akiyama, H. Shiozaki, M. Inoue, Y. Doki, Y. Fujiwara, D. Ross, A. Sadler, and M. Monden, *Proc. Amer. Assoc. Cancer Res.* **41,** 4585 (2000).

mitomycin C. The relevance of the NQO1*2 polymorphism to therapy using other compounds such as E09, RH1 and β-lapachone remains to be determined.

Acknowledgments

The authors are supported by NIH grants CA51210, ES09554 and NS44613.

[9] Structure and Mechanism of NAD[P]H:Quinone Acceptor Oxidoreductases (NQO)

By Mario A. Bianchet, Margarita Faig, and L. Mario Amzel

Introduction

Cytosolic NAD(P)H:quinone acceptor oxidoreductases (NQO) are flavoenzymes that catalyze the obligatory 2-electron reduction of quinones to hydroquinones.[1] This reaction prevents the reduction of quinones by one-electron reductases that would result in the formation of reactive oxygen species (ROS), generated by redox cycling of semiquinones in the presence of molecular oxygen.[2] In addition to its possible role in the detoxification of dietary quinones, the enzyme has been shown to catalyze the reductive activation of quinolic chemotherapeutic compounds such as mitomycins, anthracyclines, and aziridinyl-benzoquinones. NQO type 1 (NQO1; EC 1.6.99.2; also called QR1 and DT-Diaphorase) is a 274-residue enzyme expressed under oxidative or electrophilic stress among a battery of phase II enzymes.[3–6] Also expressed as part of the same response is an iso-enzyme called NQO type 2 (NQO2) that is 43 residues shorter at its C-terminus. NQO2 has a 49% sequence identity with NQO1. Both enzymes can use NAD[P]H as a source of reducing equivalents, but NQO1 uses this cofactor more efficiently. NQO2 has been shown to prefer dihydronicotinamide ribosyl (NRH) as cofactor.[7,8] NQO activity is observed in many solid

[1] L. Ernster and F. Navazio, *Acta Chem. Scand.* **12**, 595–602 (1958).
[2] H. Sies and H. de Groot, *Toxicol. Lett.* **64–65**, 547–551 (1992).
[3] A. Benson, M. Hunkeler, and P. Talalay, *Proc. Natl. Acad. Sci. USA* **77**, 5216–5220 (1980).
[4] H. Wefers, T. Komai, P. Talalay, and H. Sies, *FEBS Lett.* **169**, 63–66 (1984).
[5] H. J. Prochaska, P. Talalay, and H. Sies, *J. Biol. Chem.* **262**, 1931–1934 (1987).
[6] P. Talalay, *Biofactors* **12**, 5–11 (2000).
[7] Q. Zhao, X. Yang, W. Holtzclaw, and P. Talalay, *Proc. Natl. Acad. Sci. USA* **94**, 1669–1674 (1997).

tumors including lung, colon, liver, and breast. This property makes this enzyme an ideal target for development of bio-activatable cytotoxic compounds.

The crystal structures of three mammalian NQO1, rat (rNQO1),[9] human (hNQO1; PDB code 1D4A)[10,11] and mouse (mNQO1; PDB code 1DXQ),[10] and the human NQO2 (PDB code 1QR2),[12] have been determined. In addition, several structures of complexes of NQO1 with substrates, chemotherapeutic prodrugs and inhibitors were recently reported.[13,14] The NQO structure and bio-reduction mechanism are being investigated in part to design better cancer chemotherapeutics.[5]

Primary Structure Analysis

Two-electron reduction of quinones seems to be a fundamental protective mechanism present in all mammalian species. NQO homologs can be identified in non-mammalian systems, but discussion of these enzymes is outside the scope of this review. Suffice it to say that NQO sequences are widely found across living organisms ranging from bacteria to mammals. In the alignment of Fig. 1 it is possible to group the protein into three classes: (1) sequences that show a C-terminal domain (NQO1-like); (2) shorter sequences (NQO2-like) that lack such domain; (3) a subset of the shorter sequences that show an additional deletion of around eighteen residues between the positions 64 and 82 (flavodoxin-like; see the following).

Structure Description

The physiological unit of these quinone reductases is a dimer. NQO1 and NQO2 dimers display a similar fold for the bulk of the structure: two interlocked catalytic domains of 220 residues each with similar topology (Fig. 2). The NQO1 dimer exhibits two additional C-terminal domains of

[8] K. Wu, R. Knox, X. Sun, P. Joseph, A. Jaiswal, D. Zhang, P. Deng, and S. Chen, *Arch. Biochem. Biophys.* **347**, 221–228 (1997).

[9] R. Li, M. Bianchet, P. Talalay, and L. Amzel, *Proc. Natl. Acad. Sci. USA* **92**, 8846–8850 (1995).

[10] M. Faig, M. A. Bianchet, P. Talalay, S. Chen, S. Winski, D. Ross, and L. M. Amzel, *Proc. Natl. Acad. Sci. USA* **97**, 3177–3182 (2000).

[11] J. V. Skelly, M. R. Sanderson, D. A. Suter, U. Baumann, M. A. Read, D. S. Gregory, M. Bennett, S. M. Hobbs, and S. Neidle, *J. Med. Chem.* **42**, 4325–4330 (1999).

[12] C. Foster, M. Bianchet, P. Talalay, Q. Zhao, and L. Amzel, *Biochemistry* **38**, 9881–9886 (1999).

[13] M. Faig, M. A. Bianchet, S. Winski, R. Hargreaves, C. J. Moody, A. R. Hudnott, D. Ross, and L. M. Amzel, *Structure (Camb)* **9**, 659–667 (2001).

[14] S. L. Winski, M. Faig, M. A. Bianchet, D. Siegel, E. Swann, K. Fung, M. W. Duncan, C. J. Moody, L. M. Amzel, and D. Ross, *Biochemistry* **40**, 15135–15142 (2001).

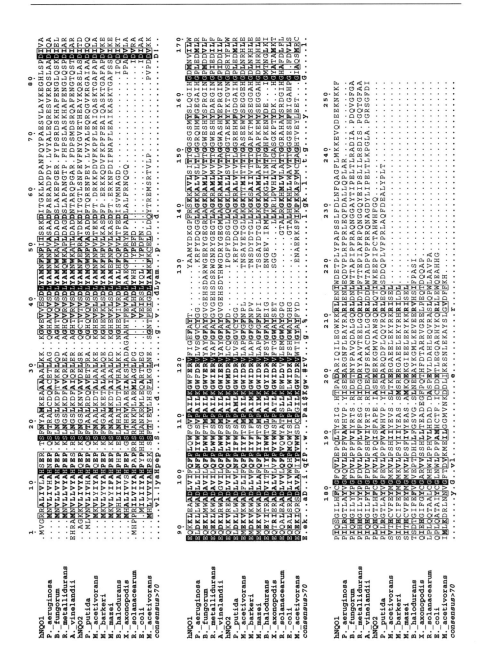

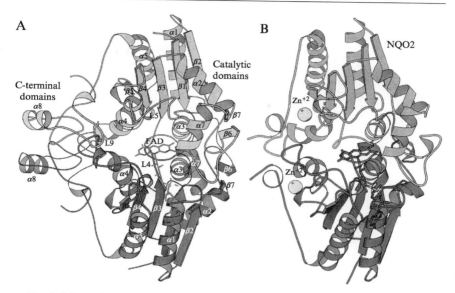

FIG. 2. Schematic representation of the structures of the NQO1 (left) and NQO2 (right) dimers. Both molecules are shown in the same orientation. In both cases the two-fold axis of symmetry is horizontal. The extra C-terminal domain of NQO1 is clearly visible on the left of the molecule. In NQO2 a short sequence and a metal (shown as a sphere) are at the C-terminal of NQO2. This figure, as well as Fig. 3 and Fig. 5B, were prepared with the programs MOLSCRIP[16] and raster3D.[17]

54 residues long. Extensive contact between monomers confers high stability to these dimers. NQO1 and NQO2 dimer bury 5428 Å^2 4004 Å^2 of solvent-accessible area respectively, having 265 monomer-monomer contacts shorter than 4 Å. Most contacts occur between residues belonging to four main regions: residues 153 to 164 are in close contact with residues in the region 235 to 262; residues 103 to 104 contact their equivalent in the other monomer, as do residues 42 to 52. FAD molecules are non-covalently

FIG. 1. Sequences of selected NQO enzymes from diverse sources. The numbers refer to the sequence of hNQO1. Secondary structure elements of mammalian NQO1 enzymes are shown at the top. This sequence alignment figure and the one in Fig. 10 were prepared with the program ESPrint.[15]

[15] P. Gouet, E. Courcelle, D. I. Stuart, and F. Metoz, *Bioinformatics* **15,** 305–308 (1999).
[16] P. Kraulis, *J. Appl. Cryst.* **24,** 946–950 (1991).
[17] Merrit and Bacon, *Methods in Enzymology* **277,** 505–524 (1997).

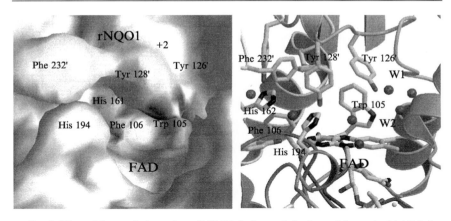

FIG. 3. View of the catalytic pocket of NQO1. Left panel: Surface of the cavity highlighting the importance of Tyr 128 and Phe 232. Residues at the bottom belong to one monomer, those at the top to the other. Right panel: Same view as the left panel in all atom representation. Water molecules that are an integral part of the binding strategy are shown as spheres. The left panel of the figure was drawn using the graphic program Grasp.[18]

bound to each monomer. Their isoalloxazine rings lay at one side of two equivalent crevices present at opposite ends of the monomer-monomer interface. These crevices form two identical catalytic sites that work independently.

Monomer Structure

The first 220 residues of both enzymes form highly similar catalytic domains. The additional 50 residues of NQO1 form a C-terminal domain. The NQO fold is a distinct fold within FAD-containing proteins.[19] The catalytic domain (residues 1–220) is similar to that of *Clostridium* flavodoxin, a FMN-containing protein. The FAD molecule is bound to this domain in a manner similar to that of the riboflavin in flavodoxin (Fig. 3). The bulk of the fold is an open, twisted five-strand parallel β-sheet surrounded by five α-helices that interconnect consecutive strands in an alternating α/β topology. This fold departs from that of flavodoxin by an eighteen-residue crossover between $\beta2$ and $\alpha2$. These crossover residues participate in stabilization and protect the flavin from interacting with other proteins. Loss

[18] A. Nicholls, K. Sharp, and B. Honing, *Proteins, Structure, Function and Genetics* **11**, 281–286 (1991).
[19] O. Dym and D. Eisenberg, *Protein Sci.* **10**, 1712–1728 (2001).

of this stretch of sequence (Fig. 1) results in proteins that resemble more closely electron carriers such as *Clostridium* flavodoxin.

The NQO catalytic domain is a modified Rossman fold in which the third α/β pair is lost or included in the crossover element between the two sections of the parallel β-sheet. This insertion detaches from the bulk making an additional anti-parallel hairpin with an α-helix that participates extensively in the dimer interface ($\alpha2'$).

NQO1 has an additional C-terminal domain that spans from residue 220 to the C-terminus. This domain, an anti-parallel hairpin motif followed by a short helix that ends with two dozen residues with undefined secondary structure, folds against the other monomer catalytic domain. NQO2's shorter C-terminus coordinates a structural Zn^{+2} ion (Fig. 2B).

Catalytic Site

The crystal structure of complexes of NQO1 with $NADP^+$ and with simple electron acceptor substrate (duroquinone; 2,3,5,6 tetramethyl-p-benzoquinone; DQ) identified the catalytic site of the enzyme as a pocket at the dimer interface[9] (Fig. 3). The FAD isoalloxazine *si*-face from one of the monomers catalytic domains makes one side of the catalytic pocket, while the other side is made by residues from the other catalytic domain opposite to the other FAD molecule.

The shape and size of the catalytic site is such that it can accommodate ring-containing compounds as required for enzymes that metabolize a broad range of substrates. In NQO1, the catalytic site is a 360 A^3 pocket lined with aromatic residues. A similar site is observed in NQO2, although three-aminoacid differences (His 161[*] for Asn, Tyr 126 for Phe and Tyr 128 for Ile; Tyr points in and the Ile points out of the pocket) render the site wider over the isoalloxazine ring A, and more polar (Fig. 4).

In the catalytic sites of both enzymes, a triptophan and a phenylalanine residue (Trp 105, and Phe 106) are at the bottom of the pocket. Above the FAD, residues of loop L5 from the opposite side of the other catalytic domain, Tyr 126'[†] (Phe in NQO2) and Tyr 128' (Ile in NQO2), provide interactions with the substrates—apolar contacts in NQO2 but also polar interactions in the case of NQO1. Residue 161, a histidine in NQO1 or an asparagine in NQO2, provides possible polar interaction of mechanistic importance that will be discussed later. The entrance to the catalytic site is limited by glycines 149 and 150 of loop L6, by His 194 and by Pro 68

---

[*] Some publications use numbers for NQO1 that includes one additional residue (Met 1). Numbers in those publications are one greater than those used here.

[†] A prime symbol after the residue number indicates a residue of the other monomer.

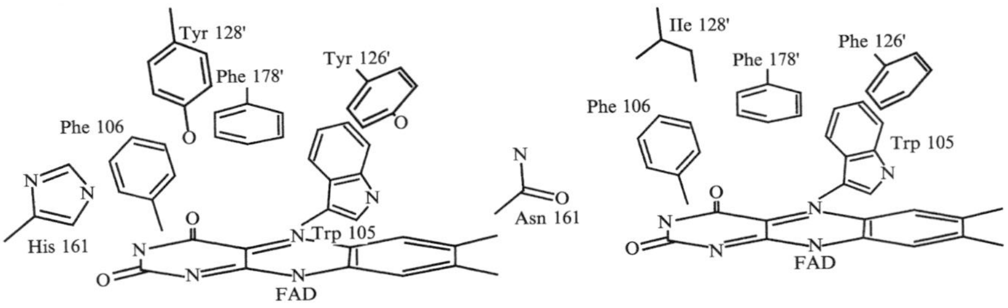

FIG. 4. Comparison of the binding site residues of NQO1 (left) and NQO2 (right).

of the N-terminal of helix $\alpha7$ of the second monomer. In NQO1, Tyr 128 gates the pocket, swinging in and out to protect the isoalloxazine ring C4a-C10a bond, which is sensible to molecular oxygen attack.

FAD Binding Site

Two FAD molecules in an extended conformation are observed bound to the catalytic domain of each monomer. The conformation and location of NQO1- and NQO2-bound flavin are very similar. These prosthetic groups are tightly bound and do not come off the enzymes under native conditions. Most of the interactions between FAD and the protein (Fig. 5) are to one of the catalytic domains of the dimer. The FAD head (dimethylisoalloxazine) sits centered over loop L4, contacting residues of loop L6. Its tail (ribitol, diphosphates and adenosine moieties) is in a cleft between the N-terminal regions of alpha helices $\alpha1$ and $\alpha5$. The isoalloxazine re-side has extensive contacts with loop L4 that forms a bed with main chain atoms of residues 104 to 106 and the aliphatic portions of the side chains of Leu 103 and Gln 104 (Tyr 104 in NQO2). Rings A and B, the hydrophilic part of the isoalloxazine moiety, make seven hydrogen bonds with main chain NH groups and the side chain hydroxyls of Thr 148 and Tyr 155 that anchor this side of the flavin rigidly in position. Ring A makes two hydrogen bonds with the main chain NH groups of loop L6: N1 with Gly 149, and O2 with Gly 150. Tyr 155 and Thr 148 make hydrogen bonds with ring A of the isoalloxazine: two hydrogen bonds from Tyr 155 OH—one to O3, and the other to N3—and one from the Thr 148 Oγ to the O2. The main-chain NH of Trp 105 in L4, provides two hydrogen bonds to atoms of the flavin: to O4 in ring A, and to N5 in the central ring B.

Ring C, a dimethyl benzene ring, is the most hydrophobic portion of the FAD. Its methyl groups are in a hydrophobic pocket, made by the aliphatic parts of Ile 50, Glu 117, and two residues of helix $\alpha7$ of the second monomer,

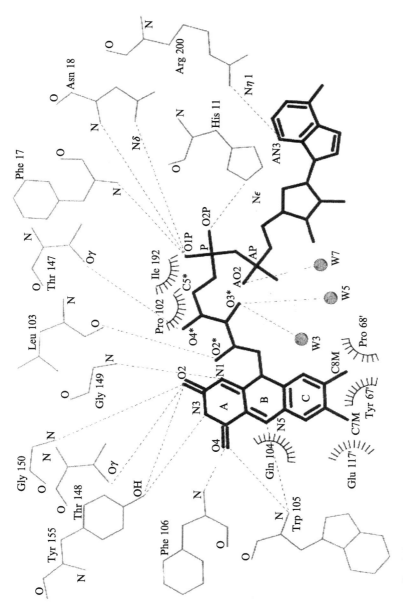

FIG. 5. Schematic representation of the interactions of FAD with NQO1. The FAD is shown with thick lines. Hydrogen bonds are shown as dotted lines and other close contacts as a decorated arc.

Tyr 67 and Pro 68. Carbon atoms C5A and C6 of ring C have contacts from below with the aliphatic portion of the Gln 105 side chain. In NQO2 and rat NQO1, residue 105 is a tyrosine that has a closer contact than the glutamine with ring C. From above, two highly conserved waters (W1 and W2; Fig. 4B) contact the methyl groups of ring C. The ribitol shows two hydrogen bonds with the protein: from its O2 to the carbonyl O of Leu 103, and from the ribitol's O4 to Oγ of Thr 147. The ribitol carbons have two contacts: between Pro 102 and C4 and between Ile 192 and C5. O5, that bridges the ribitol with the di-phosphate, has a hydrogen bond with the side chain Oγ of Thr 147. The phosphate oxygen O1P makes three hydrogen bonds with helix α1: one with the main chain NH of Phe 17, and two with Asn 18—one with the side-chain atom Nδ, and another with the main chain NH. O2P of the first phosphate group makes a hydrogen bond with the side chain Nε of His 11, also in helix α1. The Nε of Gln 66, in the N-terminal of the second monomer helix α7, makes hydrogen bond with O1A of the second phosphate group. The ribose sugar of the adenosine moiety has a short contact with a methylene of Arg 200. Likewise, the adenine lies perpendicular to the aromatic ring of Phe 17, having also a hydrogen bond from its N3A to the guanidinium Nη of Arg 200, in helix α5.

In the flavin binding site of NQO2, four residue differences with respect to NQO1 modify the binding of the ADP moiety of FAD: Ser 21 (Ala) makes an additional hydrogen bond to the adenosine N7, Val 204 (Leu) modifies contacts between the protein and the adenosine, Glu 194 (Gly) makes an additional hydrogen bond to the second phosphate oxygen and Pro 192 (Ile) that modifies the contacts between protein and the diphosphate.

Electron Donor Binding

NQO1

NADP$^+$. NADP$^+$ has fewer specific interactions with the protein than does FAD (Fig. 6A). The nicotinamide ring is in Van der Waals contact with the FAD ring A and with the side chains of Phe 178' and Tyr 128'. In the crystal structure, the nicotinamide ring has been observed to adopt two conformations involving a 180° rotation around the N1 and C4 axis.[9] In one of theses conformations, the carboxyamide moiety makes hydrogen bonds to Tyr 126' and Tyr 128'. One hydroxyl of the ribose makes a hydrogen bond with His 161 and its ring oxygen with the main chain NH of Gly 150. The phosphates have potential hydrogens bonds to His 194. The main chain NH of Phe 232' makes a hydrogen bond with the phosphate of the adenine ribose. The AMP moiety interacts mainly with the C-terminal domain.

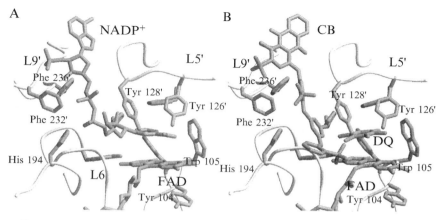

FIG. 6. Binding of NADP$^+$ and simple substrates to NQO1. (A) NADP$^+$ bound to rNQO1. (B) Cibacron blue (CB) and duroquinone (DQ) bound to rNQO1. In the cases of both NADP$^+$ and CB/DQ complexes with rNQO1, loop L9 is in the open conformation.

PHOSPHATE BINDING CLEFT. The phosphate binding site is a cleft that ends in the catalytic site between the hairpin (β8-L9-β9) and L5. During binding of the substrate acceptor, CBQ (Cibacron Blue®, a dye homologous to NAD) or the indolequinone ES936, these loops widen the cleft (Fig. 7). Phe 232' and loop L9 retreat 2.5 Å, opening a connection to the catalytic site. On the other side of the cleft, loop L5 remains unperturbed by the binding. Tyr 133 and Glu 124, the ends of L5, are kept by a number of hydrogen bond interactions in conformations outside the allowed regions of the Ramachandran plot.

NADH. Three polar interactions are expected to be missing when NADH is bound to the enzyme. The hydride transfer is expected to take place from the 4-*pro*-S hydrogen (B-side) of the reduced nicotinamide that corresponds to the minor conformation observed for NADP$^+$. In this conformation, tyrosines 126 and 128 do not provide hydrogen bonds to the nicotinamide. The hydrogen bond involving Phe 232 is also lost in the case of the non-phosphorylated nucleotide.

NQO2

ELECTRON DONOR. While FAD-protein interactions are well conserved in NQO2, there are numerous substitutions among residues involved in NAD(P)H binding. The N-terminal truncation certainly removes many of the residues involved in contacts with the ADP moiety, but aromatic stacking of the nicotinamide ring can still occur. Based on the position of

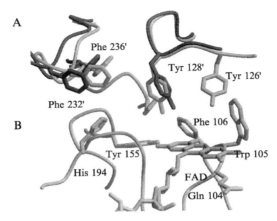

FIG. 7. Conformational changes on cofactor binding. Binding of NAD(P)$^+$ and other large substrates to NQO1 requires opening of loop L9. The superposition shows the two conformations of the loop and other associated changes: open (dark gray), closed (light gray).

the NADP$^+$ nicotinamide in NQO1, it is safe to assume that NRH (dihydroribosyl-nicotinamine) binds with the nicotinamide ring in the NQO2 catalytic pocket, stacking between the isoalloxazine ring and Ile 128.

Substrate Acceptor Binding

Duroquinone Binding to NQO1

Two NQO1-DQ structures (rat, rNQO1 and human, hNQO1) show the catalytic pockets occupied by this small substrate (Figs. 7B and 8A). The total accessible surface area buried by DQ is 427 Å^2. DQ is bound to the active site through a series of contacts involving the flavin and several hydrophobic and hydrophilic residues. The substrate is sandwiched between the rings of Phe 178′ and the isoalloxazine rings A and B. Five aromatic residues—Phe 178′, Trp 106, Phe 106, Tyr 126′, and Tyr 128′—Gly 150 and the central portion of the isoalloxazine of FAD provide most of the contacts. The rat NQO1-DQ structure also includes CB, which certainly affects the DQ binding but does not impede it (Fig. 7B). In hNQO1, a DQ hydrogen bonded water molecule bridges His 161 and Tyr 128′.

Menadione Binding to NQO2

Menadione (vitamin K3) binds to NQO2 in a manner similar to the binding of DQ to rNQO1 (Fig. 8B). The substrate lies with its quinone ring on top of the flavin ring A and with its aromatic ring pointing out of the pocket, contacting loop L6. The methyl group stacks between Trp 105

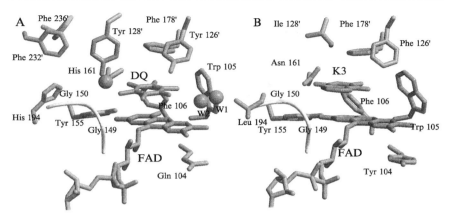

FIG. 8. Binding of simple substrates to hNQO1 and NQO2. (A) Duroquinone (DQ) bound to hNQO1. (B) Menadione (vitamin K3; K3) bound to hNQO2.

and Phe 106 and is sandwiched between Phe 178′ and the isoalloxazine ring A. and one oxygen atom points in direction of Asn 161, but does not make a specific contact—3.6 Å from the Asn 161 Oδ. The quinone oxygen is at 3.5 Å from the flavin N5, ideally suited to receive the hydride during the reduction.

Mechanism

The dimethylisoalloxazine moiety of FAD is very tightly bound to the enzyme. The oxygens (O2 and O4) as well as two of the nitrogens (N1 and N5) are hydrogen bonded to main chain NH groups. Hydrogen bonding to main chain groups is usually associated with very stable rigid binding. The change in the reduction potential from the free flavin (-207 mV at pH 7.0) to the NQO1 bound potential (-159 mV at pH 7.0)[20] should result from tighter binding by the enzyme of the reduced form of the flavin. Enzyme affinity for 1-deaza-FAD or 5-deaza-FAD (K_d 1.7 10^{-5} M and 1.6 10^{-4} M, respectively) are dramatically decreased from that for FAD (K_d 1.8 10^{-7} M).[21]

The obligatory two-electron reduction carried out by these enzymes has been proposed to occur via two direct hydride transfers: one from the electron donor (NAD[P]H in NQO1, or NRH in NQO2) to the FAD, the

[20] G. Tedeschi, S. Chen, and V. Massey, *J. Biol. Chem.* **270,** 1198–1204 (1995).
[21] G. Tedeschi, S. Chen, and V. Massey, *J. Biol. Chem.* **270,** 2512–2516 (1995).

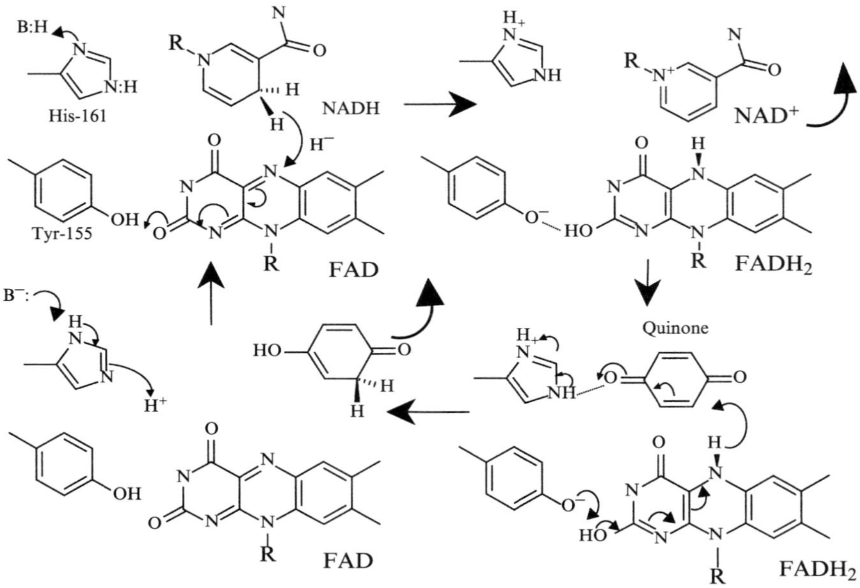

FIG. 9. Proposed mechanism for the reduction of quinones by NQO1. The catalytic cycle can be thought as starting at the lower-left corner and proceeding clockwise. Thick curved arrows initiated in a compound indicate that the compound is released from the enzyme before the ensuing step.

other from the reduced FAD to the substrate electron acceptor. The crystal structures of rNQO1 provide a simple rationale for the observed ping-pong mechanism[22]: NAD(P)H the nicotinamide moiety (donor) and the quinone (acceptor) share the same site. Thus, binding of the substrate acceptor can not proceed without de-binding oxidized $NAD(P)^+$. NAD(P)H binds the enzyme with its 4-pro-S hydrogen (B-side) 3.4 Å from the N5 of the isoalloxazine. This arrangement is ideal for a direct hydride transfer to the FAD.

The mechanism proposed for NQO1 has two hydride transfer steps: first, from the NAD(P)H to the oxidized FAD bound to the enzyme and second, from the reduced $FADH_2$ to the substrate acceptor (Fig. 9).

The hydride transfer results in a negative charge on the flavin that must be stabilized once $NAD(P)^+$ vacates the catalytic site. It is commonly

[22] R. Li, M. A. Bianchet, P. Talalay, and L. M. Amzel, Proc. Natl. Acad. Sci. USA 92, 8846–8850 (1995).

assumed that this negative charge moves to the isoalloxazine N1. But in NQO1, this atom is hydrogen bonded to the NH of Gly 149, making it not a good candidate for charge stabilization. A more likely structure involves a tautomerization to the enolic form, leading to placement of a negative charge on O2. Tyr 155 is hydrogen bonded to this oxygen and is thus in position to transfer its OH proton. In time, His 161 can stabilize the charge of Tyr 155 or perhaps transfer a proton to it. After substrate release, this histidine $N\varepsilon$ becomes solvent exposed, and it is easily protonated. Also, His 161 is poised to change the protonation state of its N_δ, that can compensate for the charge generated by the hydride transfer to and from the FAD isoalloxazine ring. Analysis of the hydrogen bonding pattern shows that His 161 N_δ makes a hydrogen bond with Tyr 132 and with a chain of well-conserved hydrogen bonded groups. This proton wire connects His 161 with the solvent and can easily shuttle protons in and out, changing the protonation state of the histidine. The corresponding role in NQO2 has been proposed to be played by Asn 161 and an associated bound water molecule. Mutation of the two residues proposed to be involved in the mechanism, Y155F and H161Q, have shown values of k_{cat} of 33% and 8% of the wild-type enzyme.[23] The modest effect of Y155F mutation in comparison to the H161Q suggests a larger role for His 161 in charge compensation than Tyr 155. Tyr 155 helps to place the $FADH^-$ charge closer to His 161, shortening the distance from 5.5 Å (charge delocalized between N1 and O2) to 3.5 Å (charge at Tyr 155 OH). Loss of His 161 results in an uncompensated charge. His 161 is ideally located to stabilize this resonance form by donating a proton to the negatively charged oxygen. The electronic events that result in charge stabilization after the hydride transfer were studied by *ab initio* quantum mechanics methods.[24]

The reduction of the substrate acceptor (quinone) by the reduced form of the enzyme can occur via a simple reversal of the steps just described because NAD(P)H and substrate share the same site. The substrate acceptor binds the site in an orientation suited to accept a hydride from the reduced flavin (Fig. 8A). The proton in the O2 of the flavin is transferred back to Tyr 155 and the isoalloxazine returns to the oxidized quinoid form. Transfer of the hydride to the quinone results in an ionized hydroquinone (hydroquinolate). In many of the complexes studied, His 161 $N\varepsilon$ is hydrogen bonded or in close proximity to the substrate quinone oxygen, and it can either transfer a proton to the hydroquinolate or simply stabilize its negative charge. Thus, the net result of this second half of the reaction may

[23] S. Chen, K. Wu, D. Zhang, M. Sherman, R. Knox, and C. Yang, *Mol. Pharmacol.* **56,** 272–278 (1999).
[24] G. Cavelier and L. M. Amzel, *Proteins* **43,** 420–432 (2001).

involve, in addition to the hydride transfer, the transfer of a proton from O2 to the hydroquinolate.

The addition of hydride to one of the quinolic oxygens appears to be a favorable mechanism, because the phenoxy (or phenoxide) formed is a very stable species. An initial hydride transfer to a ring carbon followed by tautomerization (Michael addition) is also a very strong possibility. The resulting enolate can be protonated by His 161 to give 4-hydroxy-3,5-diene-1-cyclohexanone. Although there is no good candidate in the site to be used as a base for the enolization of the ketone, this compound may be released from the enzyme and the quinolization of the enolic form can occur directly in solution. Addition of the hydride to the carbonyl carbon would be less favorable because it would produce an unconjugated and unstable alkoxide.

Structure-Based Mutagenesis

Even before publication of the structures, attempts were made by several groups to identify residues involved in QR catalysis and cofactor binding by using site-directed mutagenesis.[25–27] The residues targeted in these studies (Table I) were selected by a variety of criteria that in most cases identified positions at or close to the catalytic site of the enzyme. Several mutations were carried out in the region 124 to 128. F124L has little or no effect on the activity of the enzyme, in agreement with the observation in the structure that the side chain of residue 124 is not close to the catalytic site. The side chain at position 127 points away from the catalytic site, so one does not expect direct effects from substitutions at this position, in agreement with the results observed in the mutant T127V. The small effects observed on the K_m for NADH ($\times 3.5$) of the mutation T127E are probably due to non-specific effects of the introduction of the negative charge. As mentioned previously, the side chain of Tyr^{128} participates in two interactions with the nicotinamide: a hydrogen bond of the OH with the carboxyamide (in the non-productive conformation) and an interaction of the aromatic ring with the ribose. Although mutations Y128V and Y128F have very little effect on activity, Y128D produces a 10-fold reduction in k_{cat} and an even larger effect on k_{cat}/K_m, probably the result of perturbations of the charge stabilization system of the enzyme by the charge of the aspartate.

[25] Q. Ma, K. Cui, R. Wang, A. Lu, and C. Yang, *Arch. Biochem. Biophys.* **294**, 434–439 (1992).
[26] Q. Ma, K. Cui, F. Xiao, A. Lu, and C. Yang, *J. Biol. Chem.* **267**, 22298–22304 (1992).
[27] H. H. Chen, J. X. Ma, G. L. Forrest, P. S. Deng, P. A. Martino, T. D. Lee, and S. Chen, *Biochem. J.* **284**, 855–860 (1992).

TABLE I
MUTAGENESIS AND KINETIC STUDIES OF QUINONE REDUCTASE

	Activity (%)		Kinetics								
	DCIP		NADH			NADPH			DCIP		
			K_m	k	(k/K_m)	K_m	k	(k/K_m)	K_m	k	(k/K_m)
Enzyme	NADH	NADPH	K_{mo}	k_o	$(k/K_m)_o$	K_{mo}	k_o	$(k/K_m)_o$	K_{mo}	k_o	$(k/K_m)_o$
QR-wt	100	100	1.0	1.0	1.00	1.0	1.0	1.00	1.0	1.0	1.00
QR-rat	77	85	0.9	0.8	0.80	0.7	0.8	1.00	0.7	0.7	1.1
F124L[25]	72	71	1.7	0.8	0.50	0.7	0.7	1.06	1.2	0.9	0.7
T127V	73	47	1.0	0.7	0.75	1.2	0.6	0.47	0.6	0.7	1.3
T127E	61	58	3.5	1.0	0.29	2.3	1.0	0.42	0.8	0.7	0.8
Y128V	57	53	1.5	0.6	0.42	0.8	0.5	0.54	0.7	0.6	0.8
Y128F	73	50	1.9	0.8	0.43	1.5	0.7	0.70	0.5	1.0	1.9
Y128D	6	4	2.0	0.1	0.04	2.7	0.1	0.03	1.3	0.2	0.1
G150F	9	9	12.7	0.2	0.01	7.0	0.3	0.04	1.8	0.1	0.1
G150V	20	10	5.0	0.2	0.04	3.3	0.1	0.03	1.9	0.2	0.1
S151F	20	10	26.4	0.7	0.03	19.0	0.6	0.03	1.0	0.5	0.5
S151A	91	102	1.2	0.9	0.71	0.7	0.9	1.20	1.3	1.0	0.8
Y155D	36	14	19.0	1.0	0.05	11.0	0.4	0.03	1.2	0.8	0.7
Y155V	55	61	0.8	0.5	0.69	0.8	0.8	0.93	0.4	0.7	1.3
K76V[26]	82	93	1.2	0.9	0.8	2.0	1.0	0.5	1.3	1.2	0.9
C179A	88	82	1.2	0.9	0.7	1.0	0.8	0.7	1.0	0.8	0.8
R177A[28]	15–25		0.7	—	—	1.0	—	—	0.6	—	—
R177H	30–40		1.0	—	—	2.3	—	—	1.0	—	—
R177C	15–28		1.0	—	—	1.0	—	—	1.0	—	—
R177L	2–4		1.0	—	—	1.0	—	—	1.0		

Another series of mutations were introduced in the region 150 to 155 (Table I). The main chain NH of Gly 150 makes a hydrogen bond with the isoalloxazine, while the $C\alpha$ is packed very closely against the cofactor. The introduction of a side chain (G150F, and G150V) must produce significant distortions of the catalytic site, in agreement with the observed effects on the kinetic parameters. Ser 151 points toward the outside of the protein; a phenylalanine at this position (S151F) probably destabilizes this region of the protein and will have an effect on the local structure in the area of the catalytic site. On the other hand, the activity of the mutant S151A is normal, as can be expected from structural considerations. The mutants

[28] S. Chen and X. Liu, *Mol. Pharmacol.* **42**, 545–548 (1992).

involving Tyr 155 are very interesting. Tyr 155 makes a hydrogen bond with O2F that in the proposed mechanism is involved in the stabilization of the FADH$_2$ form of the enzyme. The two mutations studied, Y155D and Y155V, have very different effects: while Y155D produces a large increase in K_m for NADH ($\times$19.0), Y155V produces only very modest changes in k_{cat} and K_m. Aspartic acid at this position could partially substitute the function of the tyrosine, but the side chain is too short to reach and make a hydrogen bond with O2F. Model building shows that a water molecule could bind in the position occupied by the tyrosine OH in the wild type. This water molecule would be hydrogen bonded to Asp 155, His 161 and the O2 of the flavin. With this arrangement the charge relay system could work in a manner very similar to that of the wild type. The loss of activity ($\times$12.7 increase in K_m and $\times$10 decrease in k_{cat} for NADH) is what can be expected from the less-optimal fit that results from the replacement and from the introduction of a possible negative charge in the region of the charge relay. In the case of valine, analysis of the structure suggests that a water molecule can become bound at the same position. In this case, binding of the water molecule would be less favorable (hydrogen bonded only to His 161 and the O2 of the flavin), but its proton donor/acceptor properties would be more similar to those of the original tyrosine.

In the region 177 to 179, only Phe 178 interacts with the nicotinamide; Arg 177 and Cys 179 point away from the binding site. Mutations at positions 177 and 179 have, as expected, minor effects on the activity

The results of these mutagenesis experiments deserve a general comment. It is remarkable that mutations that are so well placed all had such small effects on the different activities of the enzyme. These results can be explained because in QR, the two hydride transfers occur directly between the flavin and the NADH or the quinone. Thus, the transfers are direct, uncatalyzed reactions between correctly placed groups. The function of the enzyme is thus two fold: (a) to correctly position the reacting groups, and (b) to minimize the unfavorable effects of charge separation. There are many groups in the enzyme that cooperate to carry out these functions. The net result is that the enzyme possesses high redundancy, and no individual mutation—short of disrupting the structure of the protein—has a very large effect on the activity, unlike the more conventional enzymes such as, for example, serine or aspartate proteases. However, mutation of the residue that appears to be the most important contributor to the charge stabilization, His 161, does have a large effect on enzyme activity.

Lys 140 is conserved in most NQO enzymes from bacteria to mammals (Fig. 1). Mutation of this lysine residue dramatically affects NQO1 catalytic activity.[29] Although it does not participate directly in catalysis, it has an important function: it provides specific polar interactions that anchor loop L5:

the Lys 140 NH$_3$ makes three hydrogen bonds with the main-chain carbonyl oxygens of residues 92, 93, and 95.

Species Differences

Residues around the active site of redox enzymes play important roles in the electron transfer reactions. One of these roles is the modulation of the reduction potential of the redox centers. The three mammalian NQO enzymes show significant differences in their ability to reduce diverse compounds. Most natural substrates (i.e., menadione) are reduced two to four times faster by the rat than by the human enzyme. The rate of reduction of chemotherapeutic compounds by NQO1 from rat is consistently higher than its homologs in mouse and human, despite the higher sequence identity between the rodent enzymes.[29,30] Sequence identity between the three mammalian species for which NQO1 structural data are available are: rat and human, 86%; mouse and rat, 93%; and, between human and mouse, 86%. Sequence alignment (Fig. 10) based on the NQO1 structure suggests that two catalytic site residues, 104 and 130, are the most probable culprits. Measured reaction rates when using human and rat recombinant enzymes with residues 104 and 130 mutated to those of the other species corroborate the importance of these positions.[29] For example, rNQO1 with Tyr 104 changed to Gln displays a kinetic behavior similar to hNQO1. Structurally, this is the most significant difference between human (or mouse) and rat NQO1. Comparison of the human or the mouse structure with the rat structure shows that, as a consequence of changing a bulky aromatic residue (tyrosine) to a smaller amphipathic residue (glutamine), there is more space in the FAD binding pocket of the mouse and human enzymes, allowing the dimethyl benzene ring of the flavin to move 0.5 Å deeper into the cavity (Fig. 11). In solution, the reduced form of FAD adopts a butterfly conformation. As mentioned previously, FAD-enzyme interactions seem to stabilize the flat conformation of the isoalloxazine, even in the reduced form; but an analysis of the FAD contacts suggests that the binding site might accommodate a certain degree of isoalloxazine bend around its central axis, because most contacts are at the center of the isoalloxazine ring, relaxing toward the extremes. Pro 103 limits the available space for the aromatic ring of Tyr 104 that in turn pushes against the bottom of the isoalloxazine ring. These contacts push up ring C and cause differences in

[29] S. Chen, R. Knox, K. Wu, P. Deng, D. Zhou, M. Bianchet, and L. Amzel, *J. Biol. Chem.* **272**, 1437–1439 (1997).

[30] S. Chen, R. Knox, A. Lewis, F. Friedlos, P. Workman, P. Deng, M. Fung, D. Ebenstein, K. Wu, and T. Tsai, *Mol. Pharmacol.* **47**, 934–939 (1995).

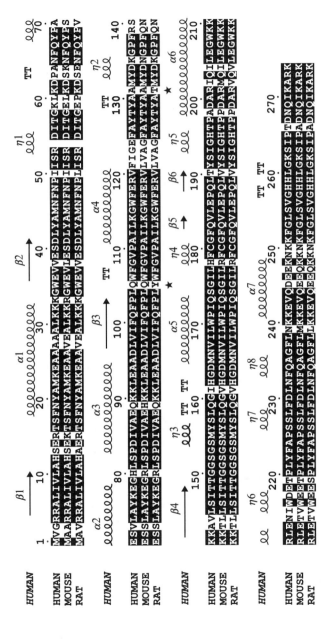

FIG. 10. Alignment of the sequences of human, mouse, and rat NQO1. Identical residues are in black boxes. The secondary structure elements are indicated in the top. The two sequence differences in residues of the catalytic site are indicated with a star.

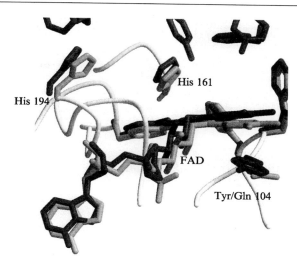

FIG. 11. Position of the isoalloxazine ring of FAD in hNQO1 (dark gray) and rNQO1 (light gray). The double arrow points to the displacement of ring A due to the change of Tyr 104 to Gln. This displacement has important consequences for the catalytic differences between the human (mouse) and the rat enzyme.

the depth of the FAD placement. In the case of the human and mouse enzymes, Gln 104 makes looser contacts with the plane of the flavin than does the Tyr in the rat enzyme. The shortest distance from any atom of Gln 104 and an FAD atom is 3.2 Å (glutamine α-carbon to FAD N5). These distances grow toward the end of the side chain and are up to 4.9 Å at the amide group. In contrast, in rNQO1, the maximal distance from a Tyr 104 atom to an FAD atom is 3.7 Å. Thus, after reduction of the FAD, Tyr 104 opposes any tendency to bend that results from the loss of the electronic conjugation in the central ring of the isoalloxazine and results in the destabilization of the reduced FAD. Thus the reduced FAD in rNQO1 will have a redox potential higher than that of the human and mouse enzymes, widening the spectra of electron acceptors that the enzyme is able to reduce. Also, this increase in the reaction's driving force could increase the rate of hydride transfer to the acceptors, in agreement with the observed increase of the rate of reduction observed with the rat enzyme for almost all electron acceptors. As the substrate (and the nicotinamide portion of NAD) participates in a stacking interaction with the flavin, this change in the position of the flavin in hNQO1 and mNQO1 affects the position of the substrates.

Crystallography Studies of Complexes of hNQO1 with
  Chemotherapeutic Compounds

Several crystallographic structures of NQO and complexes with substrate acceptors and chemotherapeutic compounds are reported in the protein data bank (1QRD, rat NQO1 ternary complex with DQN and CB;
2QR2, hNQO2-K3, 1DXO hNQO1-DQ; 1H66 hNQO1-RH1; 1H69
hNQO1-ARH019; 1GG5 hNQO1-EO9; 1KBQ hNQO1-ES936; 1KBO
hNQO1-ES1340).

Kinetic studies by Massey and coworkers[20] showed that at concentrations of NADPH up to 0.2 mM no saturation is observed in kinetic of
NQO1 reduction. These observations are in agreement with previous determinations that found that the K_m s of NQO1 for both substrates, acceptor
and donor, were in the millimolar range.[31] Therefore, in the preparation
of the crystalline complexes, solutions containing high concentrations of
substrate (>10 mM) were used.

Two families of chemotherapeutic compounds were used by us in crystal
studies of complexes with hNQO1, aziridinylbenzoquinones and indolequinones (Fig. 12). Aziridinylbenzoquinones have undergone clinical trials as
potential antitumor agents.[32] Among these drugs, MeDZQ (Fig. 12A with
X and X′ = methyl) has been shown to be a good substrate for NQO1,
and highly toxic *in vitro* for cell lines expressing high levels of the

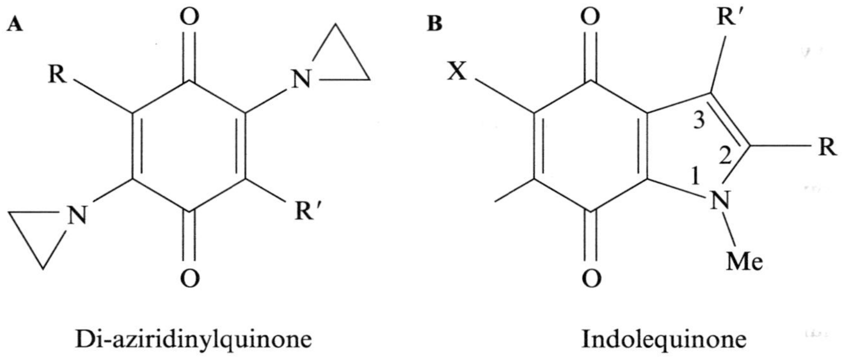

Di-aziridinylquinone   Indolequinone

FIG. 12. General structures of the compounds of two families of chemotherapeutic prodrugs.

[31] S. Chen, P. Deng, J. Bailey, and K. Swiderek, *Protein Sci.* **3,** 51–57 (1994).
[32] S. C. Schold, Jr., H. S. Friedman, T. D. Bjornsson, and J. M. Falletta, *Neurology* **34,** 615–619
  (1984).

enzyme.[33,34] Indolequinones are a family of antitumor agents structurally related to the commonly used drug mitomycin C.

All of the compounds studied share a common pharmacophoric motif: a 1,4-benzoquinone core (Fig. 13). Three of them have in common an aziridinyl group at position 5. (Aziridinyl substituents have been shown to increase the potency and *in vitro* selectivity of quinolic drugs toward NQO1-rich cells under aerobic conditions.[35]) Despite this common feature, these compounds bind NQO1 in a wide variety of ways. These differences in the mode of binding have implications with respect to their kinetics of reduction by NQO1. Good substrates bind the enzyme with the pharmacophore deep in the binding site, in contrast to inhibitors or bad substrates that bind with the benzoquinone core almost outside the pocket.

Binding of these compounds to NQO1 buries between 447 and 585 Å^2 of apolar accessible area of the protein (total buried area varies between 591 and 740 Å^2).

Good Substrates

Aziridinylbenzoquinones

RH1 BINDING. RH1 is an alkyl-substituted analog of MeDZQ that exhibits greater selective toxicity against cells with elevated NQO1 activity than the parent compound.[36] The two RH1 aziridinyl rings bend away from the isoalloxazine ring, allowing the quinone core of the drug to interact fully with rings A and B of the flavin; the 5-aziridinyl group stacks against Trp 105 and the other rests above loop L6 (Fig. 14A), displacing His 194 from its position in the apo structure. Quinone oxygens O1 and O4 make hydrogen bonds with the His 161 N_ε and the Tyr 128' OH. The angle between the planes of the drug and the isoalloxazine is 15°, departing from the exact aromatic ring stacking observed in the rNQO1 complexes with CB/DQ and NADP⁺. The 5-aziridinyl nitrogen is the closest to the FAD N5, the atom of reduced FAD that carries the hydride that is transferred to the drug. Provided there is no change in orientation of the bound substrate when the FAD is reduced, this atom will receive the hydride during the reduction of RH1.

[33] N. Gibson, J. Hartley, J. Butler, D. Siegel, and D. Ross, *Mol. Pharmacol.* **42**, 531–536 (1992).

[34] D. Ross, H. Beall, R. D. Traver, D. Siegel, R. M. Phillips, and N. W. Gibson, *Oncol. Res.* **6**, 493–500 (1994).

[35] R. M. Phillips, M. A. Naylor, M. Jaffar, S. W. Doughty, S. A. Everett, A. G. Breen, G. A. Choudry, and I. J. Stratford, *J. Med. Chem.* **42**, 4071–4080 (1999).

[36] S. Winski, R. Hargreaves, J. Butler, and D. Ross, *Clin. Cancer Res.* **4**, 3083–3088 (1998).

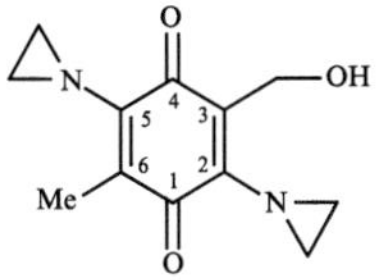

RH1

3-hydroymethyl-6-methyl-2,5 (diaziridinyl)-1,4 benzoquinone

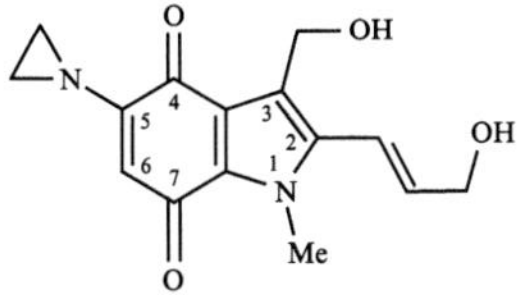

EO9

3-hydroymethyl-5-azirinyl-1-methyl-2(1-H-indole,4,7,-dione)prop-β-en-α-ol

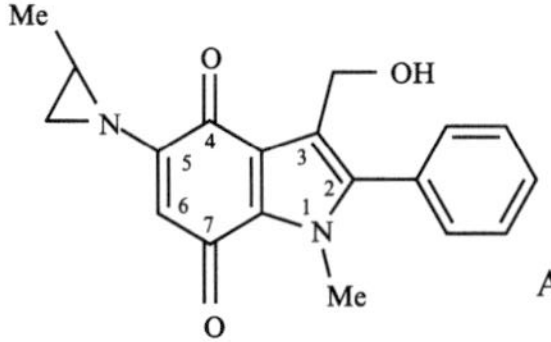

ARH019

3-(Hydroymethyl)-5-(methylaziridin-1-yl)-1-methyl-2-phenyl-indole-4,7-dione

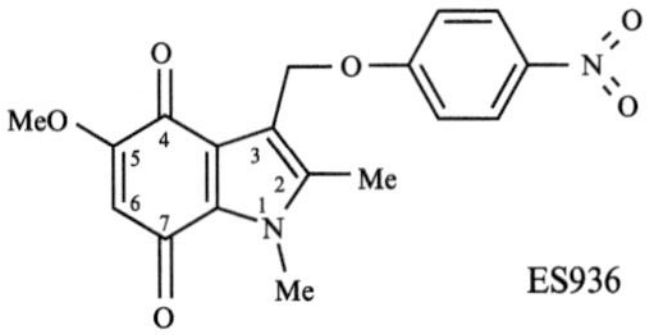

ES936

5-methoxy-1,2-dimethyl-3-[(4-nitrophenoxy)methyl]indole-4,7-dione

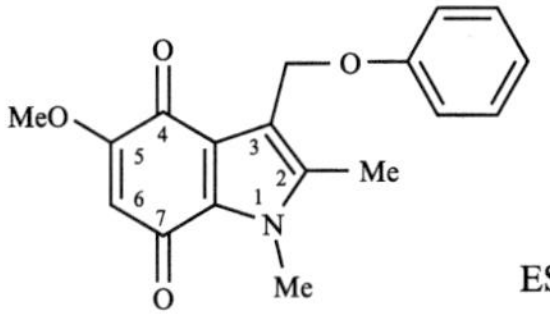

ES1340

5-Methoxy-1,2-dimethyl-3--(phenoxy)methyl-indole-4,7-dione

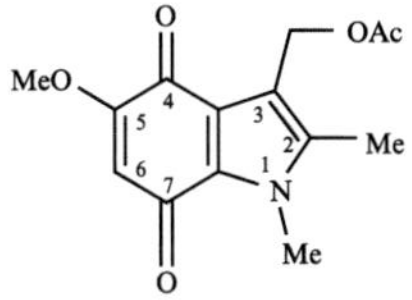

ES1951

3-(Acetoxymethyl)-5-methoxy-1,2-dimethyl-indole-4,7-dione

FIG. 13. Chemical formulas of the prodrugs discussed in the text. All of these compounds have a benzoquinone moiety with a carbon and a nitrogen at adjacent positions of the ring.

Indolequinones

ARH019 BINDING. The ARH019 methyl-aziridinyl group stacks against the Trp-106 indole (Fig. 14B). The 2-phenyl group stacks over Gly 149 and Gly 150 in L6, pointing toward the outside of the active site pocket. Quinone oxygens O7 and O4 are hydrogen bonded to the His 161 N_ε and the Tyr 126′ OH. The 3-hydroxymethyl group further stabilizes the binding by making a hydrogen bond to Tyr 128′ OH. As is the case with RH1, the plane of the ARH019 indolequinone departs from an exact aromatic parallel stacking with the FAD (plane-to-plane angle of 16°). The structure of this complex illustrates how the enzyme's active site is capable of accommodating a quinone with a large aromatic substituent.

EO9 BINDING. EO9, the better known of these prodrugs, is reduced readily by rat NQO1 but 23 times more slowly by human NQO1, with a K_m 5-fold higher than menadione.[37–39] Nonetheless, EO9 is a better NQO1 substrate than the clinically proven MC and shows high toxicity in NQO1-rich cell lines.[37] Despite their chemical similarity, EO9 binds in an orientation opposite to that of ARH019 (Fig. 14C); the positions of the pharmacophoric atoms do not match those described previously. Quinone oxygen atom O4 is hydrogen bonded to Tyr 126′ OH. The distance from the quinone O7 to the His-161 N_ε is 3.3 Å, indicating a weak hydrogen bond interaction. Bound EO9 is more centered over the FAD isoalloxazine, closer to ring B, and resembles DQ in the hNQO1-DQ complex. The drug's plane is almost parallel to the isoalloxazine rings (angle of 3.6° ± 0.8°), with its 5-aziridinyl group in the pocket defined by Phe-106. Indolequinone ring atom C6 is the closest to the FAD N5. EO9 binding buries 176 Å^2 of polar area and 492 Å^2 of apolar area.

Poor Substrates and Inhibitors

ES936 Binding. ES936 is a potent inhibitor of hNQO1.[14] The plane of ES936 stacks parallel to the isoalloxazine ring, with the indole ring placed deep in the catalytic site (Fig. 14D). The 3-nitrophenyl group occupies part of the NADP$^+$ cleft contacting Met 154, His 161 and Met 131′ and stacking with the ring of Tyr 128′. The 4,7-dione ring points out of the pocket with its 5-methoxy stacking against glycines 149 and 150 of loop L6. The 1-methyl stacks against Trp-106, and the indole-2-methyl fits in the pocket

[37] H. Beall, A. Murphy, D. Siegel, R. Hargreaves, J. Butler, and D. Ross, *Mol. Pharmacol.* **48,** 499–504 (1995).

[38] M. Walton, P. Smith, and P. Workman, *Cancer Commun.* **3,** 199–206 (1991).

[39] M. Maliepaard, A. Wolfs, S. M. N. de Groot, and L. Janssen, *Br. J. Cancer* **71,** 836–839 (1995).

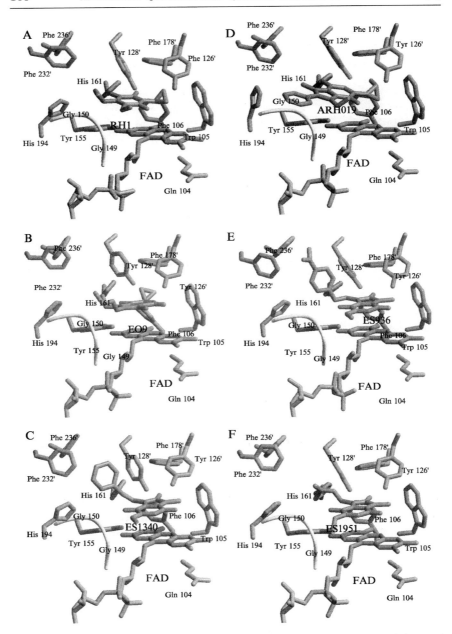

FIG. 14. Binding of six prodrugs to NQO1. (A) RH1; (B) ARH019; (C) E09; (D) ES936; (E) E1340; (F) ES1951. Side chains that make important interactions with the drugs are

defined by Phe 106, Trp 105, Phe 178' and the main chain of Ser 175'. Enzyme-inhibitor interactions are mostly hydrophobic contacts, with only one hydrogen bond between the indolequinone O7 and Tyr 126'. The indo-lyl-1-nitrogen is the closest atom to the hydride at N5 of the reduced FAD (3.6 Å).

To accommodate the nitro group, loop L9' opens the NAD adenosine site, as in the case of the NADP$^+$ and the DQ-CB-rNQO1 complexes. This group has contacts with Phe 232', Phe 236' and with the main chain carbonyl of Tyr 128' (3.2 Å).

ES1340 Binding. ES1340 is an analog of ES936 without the p-nitro group. The phenoxy is not a good leaving group. As a result, ES1340 is not an inhibitor of NQO1 and is actually a substrate, albeit an inefficient one. The hNQO1/ES1340 complex (Fig. 14E) is highly similar to that of ES936, although in this case the NADP$^+$ binding cleft remains closed.

ES1951 Binding. ES1951 binds to hNQO1 in the same manner as ES936 and ES1340[‡] (Fig. 14F). The acetoxy group of ES1951, smaller than the 3-substituents ES1340 or ES1951, interacts with His 161 and Tyr 128'. This group is a good living group, and therefore ES1951 is also an efficient mechanism-based enzyme inhibitor. A mechanism similar to that of ES936 inhibition is expected.

Summary

Compounds bind to the NQO enzymes, stacking their rings between the FAD isoalloxazine ring and the plane formed by polar residues 128 and 178 from the other monomer. Tyr 128 provides polar interactions with the rings and other groups of the compounds. Apolar groups lay at a polar patch above loop L6 or contact Trp 105 and Phe 106. In NQO1, either the Nε of His 161 or the OH of Tyr 126' provides polar interactions with one of the quinone oxygens. Larger substituents could interact with the NAD(P)$^+$ cleft, increasing complementarity with the protein.

RH1, ARH019 and E09 bind with their benzoquinone ring in one side of the pocket, stacked between Tyr 128' and flavin rings A and B. Tyr 128' swings over the substrate, making contacts with the aromatic core of the drugs (Fig. 15). RH1 and ARH019, a benzoquinone and an indolequi-none, bind to NQO1 with a similar spatial arrangement, such that their

---

[‡] Faig *et al.*, manuscript in preparation.

---

shown. All drawings are in the same orientation, highlighting the changes that take place in the binding site to accommodate different substrates. Tyr 128 and Phe 232 experience the largest changes. Gly 149 and Gly 150 are shown as a main chain trace.

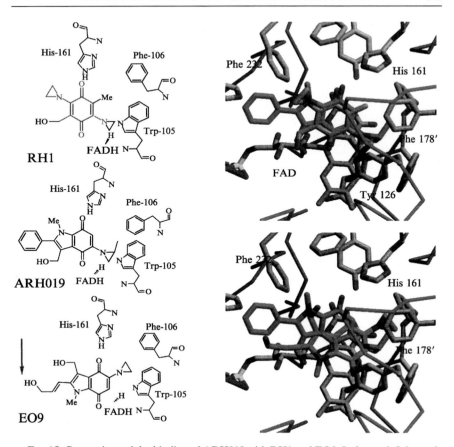

FIG. 15. Comparison of the binding of ARH019 with RH1 and EO9. Left panel: Schematic representation of the residues involved in binding of these prodrugs and their relative positions. Note that ARH019 and EO9, despite being highly similar, bind in opposite orientations. Right panel: Binding site of NQO1 showing the overlap of the structures of RH1 and ARH019 (top) and of EO9 and ARH019 (bottom). (See color insert.)

pharmacophore atoms overlap with a low RMS deviation (≤ 0.6 Å). Surprisingly, this is not the case with the two indolequinones, EO9 and ARH019. Although chemically highly similar (they differ only in their substituents at position 2 and 5), they bind to the enzyme in different orientations.

EO9 and ARH019 bind hNQO1 in dissimilar ways even though both are aziridinylindolequinones. The comparison of the reduction kinetics of a group of closely related compounds, studied by Bailey *et al.*[40] (EO9 homologs) and by Beall *et al.*[41] (ARH019 homologs), illustrate the effect of small differences in the substituents (OMe, aziridinyl, methyl-aziridinyl, open aziridinyl) at the position 5. Analogous compounds have similar relative rate of reduction within one subfamily.[13] Interestingly, species differences have a greater effect on EO9 than on the other prodrugs: EO9 is reduced 27 times faster by rNQO1 than by hNQO1. The binding position of EO9, closer to the center of the active site than the other compounds, apparently makes the drug more susceptible to the change in the FAD position. This compound can serve as a model for other drugs such as the antibiotic streptonigrin, another good substrate for the enzyme.[37]

Reduction rates of drugs by NQO1 similar to those described previously show a strong correlation with the modes of binding suggested by the structures. Aziridinylbenzoquinones (Fig. 12A) tolerate large substitutions only at position 3 or only at position 6. The RH1-hNQO1 complex shows that substituents at one of these two symmetrical positions points toward the outside of the pocket and the other toward the inside. Thus, only one of these positions can accommodate a large substituent. Consequently, drugs with progressively larger symmetrical substitutions have decreasing rates of reduction.[33] In the case of indolequinones, substituents of different shapes and sizes are generally well tolerated at position 2.[35,42] The indolequinone-NQO1 complexes show that substituents at this position point out of the binding pocket. A variety of small substituents at position 5 are tolerated. At position 3, small substituents have a small effect on the reduction rate, with electron-withdrawing groups increasing the rate. Indolequinones such as ES936 and ES1951 that have small substituents at 2- and 3-substituents of the type CH_2R (R = leaving group) inactivate the enzyme.[41] Reduction of these indolequinones in the catalytic site (Fig. 16) places a reactive carbonium at position 3 buried in the catalytic pocket. The resulting indolequinone iminium species is capable of alkylating nearby protein groups, inactivating the enzyme. Mass spectrometry of digested ES936-treated proteins showed that after reduction and loss of p-nitrophenol, the reactive iminium species generated alkylates one of the tyrosines in the binding site (126 or 128).[14]

[40] S. Bailey, N. Suggett, M. Walton, and P. Workman, *Int. J. Radiat. Oncol. Biol. Phys.* **22,** 649–653 (1992).
[41] H. Beall, S. Winski, E. Swann, A. Hudnott, A. Cotterill, S. N. O. S. Green, R. Bien, D. Siegel, D. Ross, and C. Moody, *J. Med. Chem.* **41,** 4755–4766 (1998).
[42] C. J. Moody and E. Swann, *Farmaco* **52,** 271–279 (1997).

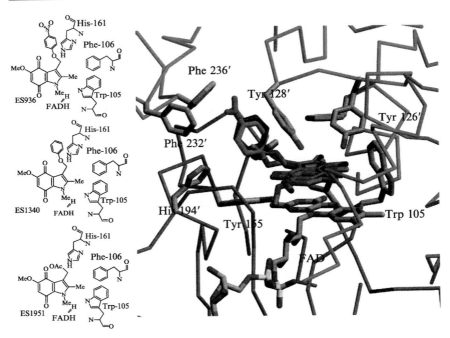

FIG. 16. Binding of inhibitors and poor substrates to hNQO1. Left: Schematic representation of the residues involved in binding ES936 (top), ES1340 (middle), and ES1951 (bottom). Right: Overlap of the positions of the three drugs in the binding site of hNQO1. The three compounds bind in highly similar positions. What determines whether they are a poor substrate or an inhibitor is the nature of the group at the 3-position: good leaving groups make compounds inactivate the enzyme. (See color insert.)

Computer Modeling of Other Complexes of NQO1

CB1954

CB1954 (5-[aziridin-1-yl]-2,4-dinitrobenzamide), when activated by NQO (or other enzymes with nitroreductase activity), becomes a cytotoxic agent able to cross-link DNA. The size and shape of CB1954 are similar to those of RH1. Complexes of NQO1 with this substrate have been computer-modeled on the basis of mutagenesis and crystallographic data.[43,44] Chen et al. have studied the effect on the reduction rate of selected substitutions

[43] S. Chen, K. Wu, D. Zhang, M. Sherman, R. Knox, and C. S. Yang, *Mol. Pharmacol.* **56,** 272–278 (1999).
[44] J. Skelly, M. Sanderson, D. Suter, U. Baumann, M. Read, D. Gregory, M. Bennett, S. Hobbs, and S. Neidle, *J. Med. Chem.* **42,** 4325–4330 (1999).

of catalytic site residues such as Y128D, G150V, H194D and Y155F. The reduction kinetics of CB1954 did not show significant changes in K_m with these mutations. Most of these substitutions show modest variations in V_{max}, ranging between 36% and 72% of the wild-type enzyme rate. Only the G150V substitution showed a larger decrease in V_{max} to 4% of the wild-type rate. Based on this and on the structure of the ternary complex rNQO1-CB-DQ, they proposed a model represented schematically in Fig. 17A.

Skelly *et al.* have also modeled a CB1954-rNQO1 complex (Fig. 17B). In addition, based on the crystallographic differences between human and rat enzymes, they also predict species differences in CB1954 binding. (CB1954 is shifted toward the central ring in the human case.) Both CB1954 models have placed the compound over the FAD rings A, but they differ on the placement of the hydrophobic aziridinyl group. In Skelly's model, CB1954 binding resembles the RH1 binding observed in the crystallographic complex.[13] This model also seems to agree with the mutagenesis data.

Inhibitors

Inhibitors such as dicoumarol, chrysin and phenidone have shown significant changes in their K_i in response to two particular mutations, Y128D and G150V. In particular, Y128D suggests a direct interaction

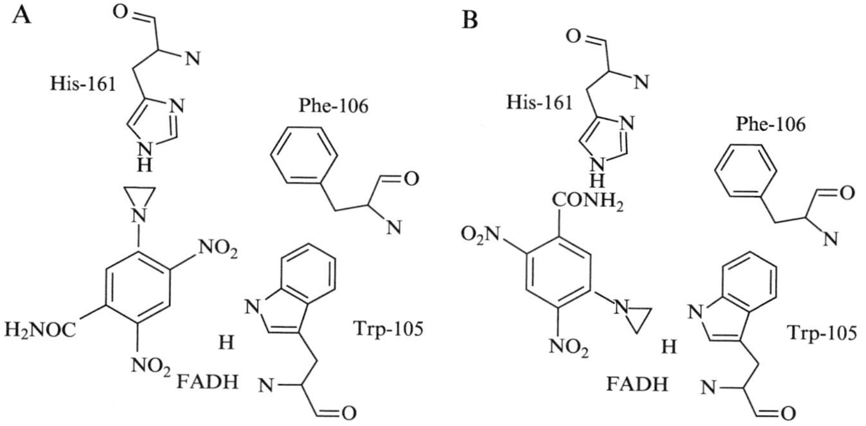

FIG. 17. Proposed models for binding of CB1954 to NQO1. The model on the left is that proposed by Chen and coworkers and that on the right by Skelley *et al.* These models were proposed before the structures of the prodrug complexes were reported. Skelley's model has more features in common with the experimental structures of the prodrugs. However, the proposed position of the carboxamide in this model is opposite to that observed in the complexes with NAD(P)⁺.

between the phenol of Tyr 128 and an inhibitor ring. Chen and coworkers proposed models for binding of dicoumarol, chrysin and phenidone that display π–π stacking of the aromatic rings of Tyr 128 and those of the inhibitors. Crysin and phenidone have been modeled with their rings stacking against Tyr 128 in a plane almost perpendicular to the isoalloxazine ring. Direct interaction between Tyr 128 and inhibitors were *a posteriori* observed in crystallographic experiments.[14] However, the parallel stacking was always between the inhibitors' ring and the isoalloxazine. Only the modeled dicoumarol complex resembles the one observed in similar crystal complexes of the indolequinone inhibitors ES936 and ES1340.

The poor success of these predictions illustrates the difficulty of modeling in cases such as that of NQO1, an enzyme with a broad specificity that binds substrate with few specific protein-ligand interactions and shows a high plasticity in its binding site.

Acknowledgments

We thank Paul Talalay for introducing us to this exciting system and for a long and fruitful collaboration. Our studies on the chemotherapeutic prodrugs were done in collaboration with David Ross. We thank Sandra Gabelli for help with this manuscript. This work was supported by Grant GM45540 from the National Institute of General Medical Sciences.

[10] Diaziridinylbenzoquinones

By Angela Maria Di Francesco, Timothy H. Ward, and John Butler

Introduction

One of the first compounds screened by the NCI in the 1950s was a simple benzoquinone. Since this time, hundreds of quinones have been screened and many, including mitomycin C, mitozantrone and doxorubicin are now in routine clinical use. Quinones have become one of the largest group of clinically used anticancer drugs in the United States.

The diaziridinylbenzoquinones (DBQs) are highly cytotoxic quinone-based alkylating agents that have also been studied since the early 1950s. These compounds are relatively easy to synthesize and hence numerous interesting compounds have been screened. Several DBQs have been clinically tested, but they have generally not been very successful.

The interest in the DBQs has increased over the last few years since it was discovered that the two-electron reducing enzyme, human

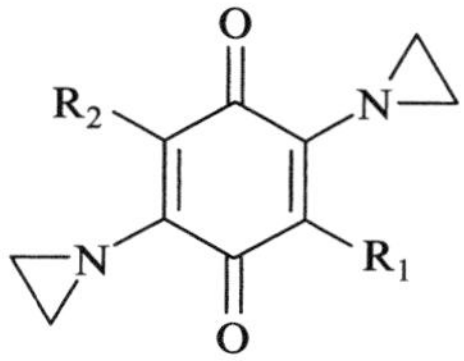

FIG. 1. General structure of diaziridinylbenzoquinones.

DT-Diaphorase (DTD), is over-expressed in many different types of tumors. Some DBQs are excellent substrates for this enzyme and are extremely cytotoxic toward human tumor cells both *in vitro* and *in vivo*.

This chapter concentrates on the known mechanisms underlying the cytotoxicities of the DBQs and highlights how these mechanisms might be applied in the design of more-successful antitumor quinones.

History

Some of the earliest chemical studies on aziridinylquinones were carried out by the groups of Holzer and Petersen over 50 years ago.[1-3] Many novel synthetic routes were discovered during these times, and several clinically interesting quinones were discovered. Among these were DZQ (Fig. 1; $R_1 = R_2 = H$) and triaziquone (Fig. 1; $R_1 = H$, $R_2 = $ aziridinyl). Trenimon (also known as triaziquone) was found to be highly cytotoxic toward different cancer cell lines including those derived from Ehrlich ascites and lymphatic leukemias.[4,5] This quinone underwent clinical trials in Europe for the treatment of breast cancer, leukemia, Hodgkin's disease, ovarian carcinoma, bronchial carcinoma and cervical carcinoma.[6-8] One of the main side effects of the drug was myelosuppresion.

The next major series of clinical studies on aziridinylquinones was carried out in Japan by Nakao and co-workers. It would appear that their

[1] H. Holzer, *Biochem. Z.* **330**, 59 (1958).
[2] S. Petersen, W. Gauss, and E. Urbschat, *Angew Chem.* **67**, 217 (1955).
[3] G. Domagk, S. Petersen, and W. Gauss, *Z. Krebsforsch* **59**, 617 (1954).
[4] D. Kummer and H. D. Ochs, *Z. Krebsforsch Klin.* **73**, 315 (1970).
[5] M. G. Ihlenfeldt, M. Gantner, B. Harrer, H. Puschendorf, and H. Grunicke, *Cancer Res.* **41**, 289 (1981).
[6] F. Gasparri, P. Periti, G. Scarselli, T. Mazzei, L. Savino, F. Branconi, and R. G. Dancygier, *Chemioterapia* **5**, 44 (1986).
[7] J. Wilde, *Z. Erkr. Atmungsorgane* **142**, 101 (1975).
[8] W. Krafft, H. Behling, A. Schirmer, D. Bruckmann, W. Preibsch, J. Kademann, F. Marzotko, and G. Schmeisser, *Zentralbl. Gynakol.* **101**, 468 (1979).

main purpose was to synthesize a series of benzoquinones to mimic the alkylating ability of mitomycin C. More than 40 novel benzoquinones were synthesized by this group, and all of them were initially screened for activity against mice bearing L1210 tumors.[9] As a consequence of these studies, carboquone (Fig. 1; R_1 = Me, R_2 = CH[OCH$_3$]CH$_2$OCONH$_2$, also known as carbazilquinone) was further investigated by using different animal models including those derived from lung, urothelial and pancreatic cancers.[10–12] This DBQ has been used in combination therapy in Japan (with cisplatinum and adriamycin) for the treatment of prostate cancer[13] and in Finland (with cyclophosphamide and adriamycin) for the treatment of ovarian cancer.[14]

In the late 1970s, Driscoll and coworkers synthesized a series of DBQs to find compounds that could target tumors within the central nervous system. Essentially, these quinones were designed to be able to cross the blood-brain barrier because of their high lipid solubility and low ionization.[15,16] The most active compound identified was AZQ (Fig. 1; R_1 = R_2 = NHCO$_2$Et, also known as diaziquone). This DBQ has subsequently been used to treat numerous different cancer types including those of the brain,[17,18] the pancreas[19] and the skin.[20] It has more recently been used in combination therapy for the treatment of acute myeloid leukemia.[21]

[9] H. Nakao, M. Arakawa, T. Nakamura, and M. Fukushima, *Chem. Pharm. Bull.* **20**, 1968 (1972).

[10] T. Mitsudomi, S. Kaneko, M. Tateishi, T. Yano, T. Ishida, S. Kohnoe, Y. Maehara, and K. Sugimachi, *Anticancer Res.* **10**, 987 (1990).

[11] K. Naito, H. Hisazumi, S. Mihara, T. Asari, K. Kobashi, T. Amano, and T. Uchibayashi, *Cancer Chemother. Pharmacol.* **20**(Suppl. S1) (1987).

[12] S. Imai, S, Y. Nio, T. Shiraishi, T. Manabe, and T. Tobe, *Anticancer Res.* **11**, 657 (1991).

[13] H. Ito, *Nippon Ika Daigaku Zasshi* **62**, 456 (1995).

[14] J. U. Maenpaa, E. Heinonen, S. M. Hinkka, P. Karnani, P. J. Klemi, T. A. Korpijaakko, T. A. Kuoppala, A. M. Laine, M. A. Lahde, E. K. Nuoranne et al., *Gynecol. Oncol.* **57**, 294 (1995).

[15] A. H. Khan and J. S. Driscoll, *J. Med. Chem.* **19**, 313 (1976).

[16] F. Chou, A. H. Khan, and J. S. Driscoll, *J. Med. Chem.* **19**, 1302 (1976).

[17] L. J. Ettinger, N. Ru, M. Krailo, K. S. Ruccione, W. Krivit, and G. D. Hammond, *J. Neurooncol.* **9**, 69 (1990).

[18] R. P. Castleberry, A. H. Ragab, C. P. Steuber, B. Kamen, S. Toledano, K. Starling, D. Norris, P. Burger, and J. P. Krischer, *Invest New Drugs* **8**, 401 (1990).

[19] E. J. Tilchen, T. Fleming, G. Mills, N. Oishi, J. D. Bonnett, R. B. Natale, G. Harker, and J. Coltman, *Cancer Treat. Rep.* **71**, 1309 (1987).

[20] H. Host, R. Joss, H. Pinedo, U. Bruntsch, F. Cavalli, G. Renard, G. M. van Glabbeke, and M. Rozencweig, *Eur. J. Cancer Clin. Oncol.* **19**, 295 (1983).

[21] E. J. Lee, S. L. George, P. C. Amrein, P. A. Paciucci, S. L. Allen, and C. A. Schiffer, *Leukemia* **12**, 139 (1998).

A more water-soluble analog, BZQ (Fig. 1; $R_1 = R_2 = -NHC_2H_4OH$), was also identified by Driscoll *et al.* This quinone underwent phase I/II clinical trials in the United Kingdom but did not proceed any further.[22]

It is interesting to note that trenimon, AZQ, BZQ and carboquone entered clinical trials even though relatively few mechanistic studies were published. It was shown that the aziridinylquinones were capable of alkylating DNA, but the actual mechanisms of activation and their interactions with the bases of DNA were not investigated. Indeed, if these studies had been carried out, they would have shown that all of these quinones behave differently.

A more-mechanistic approach toward developing antitumor diaziridinylquinones began a few years ago after it was recognized that DTD is over-expressed in many tumors[23-25] and that some diaziridinylquinones are excellent substrates for this enzyme.[26-28] The lead compound in these studies was MeDZQ (Fig. 1: $R_1 = R_2 = CH_3$), which is an exceptional substrate for the enzyme, and the resulting hydroquinone efficiently cross links DNA.[26,27,29] RH1 (Fig. 1; $R_1 = -CH_3$, $R_2 = -CH_2OH$) is a water-soluble analog of MeDZQ that is scheduled for phase I/II clinical trials in the United Kingdom.

Mechanisms

Figure 2 shows the known cytotoxic pathways for the diaziridinylquinones.

pKs

The scheme in Fig. 2 shows that the alkylating ability of the aziridines comes about as a consequence of protonation. The extent of protonation, the potential alkylating ability and the cytotoxocity of an aziridinyl quinone is strongly dependent on its structure.

The pK of a simple aliphatic aziridine is around 8.0, which is lower than that of ammonia (9.24) and significantly lower than the 4-membered ring

[22] R. A. Betteridge, A. G. Bosanquet, and E. D. Gilby, *Eur. J. Cancer* **26,** 107 (1990).

[23] J. J. Schlager and G. Powis, *Int. J. Cancer* **15,** 403 (1990).

[24] M. Belinsky and A. K. Jaiswal, *Cancer Metastasis Rev.* **12,** 103 (1993).

[25] A. M. Malkinson, D. Seigel, G. L. Forresr, A. F. Gazdar, H. K. Oie, D. C. Chan, P. A. Bunn, M. Mabry, D. J. Dykes, S. D. Harrison, and D. Ross, *Cancer Res.* **52,** 4752 (1992).

[26] N. W. Gibson, J. A. Hartley, J. Butler, D. Siegel, and D. Ross, *Mol. Pharmacol.* **42,** 531 (1992).

[27] R. H. Hargreaves, C. C. O'Hare, J. A. Hartley, D. Ross, and J. Butler, *J. Med. Chem.* **42,** 2245 (1999).

[28] A. M. Di Francesco, R. H. Hargreaves, T. W. Wallace, S. P. Mayalarp, A. Hazrati, J. A. Hartley, and J. Butler, *Anticancer Drug Des.* **15,** 347 (2000).

[29] H. D. Beall, A. M. Murphy, D. Siegel, R. H. J. Hargreaves, J. Butler, and D. Ross, *Mol. Pharmacol.* **48,** 499 (1995).

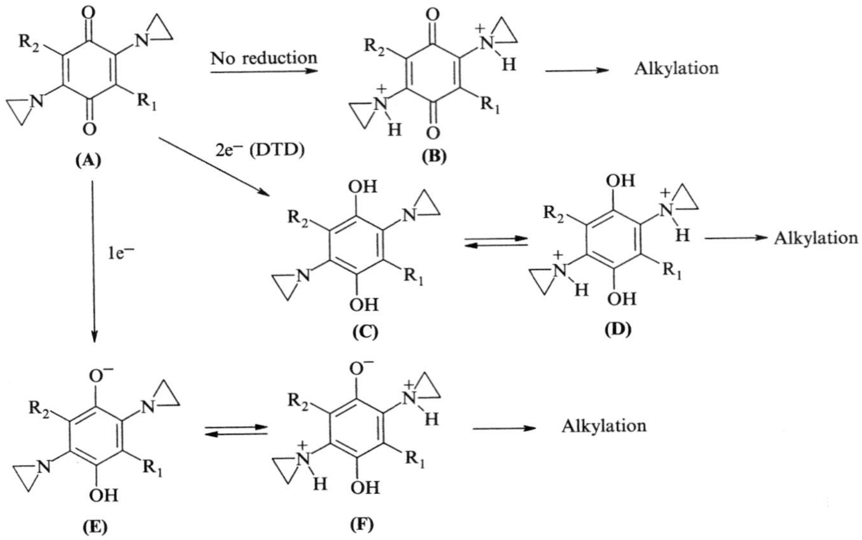

No reduction

Alkylation

(A)

2e⁻ (DTD)

(B)

1e⁻

Alkylation

(C)

(D)

Alkylation

(E)

(F)

FIG. 2. Mechanisms of activation of diaziridinylbenzoquinones.

azetidine (11.29). The lower pK of aziridine is a consequence of the strain in the three-membered ring that leads to the lone pair on the nitrogen atom, being in an orbital with lesser p character than in a normal sp^3 N atom. Essentially this means that the orbital has more s character and the electrons are held more tightly to the nucleus. In the case of azetidine, the strain is much less because the ring is bigger and so the inductive effect (+I) of the cyclic ring (this is a secondary amine) will predominate, resulting in an increase in the pK.[30] It is for these reasons that the azetidine analogs are relatively stable and are orders of magnitude less toxic than the corresponding DBQs.[31]

Because of the instability of the aziridinylquinones, very few pK measurements have been made. However, it has been shown that the pK for DZQ (Fig. 1; $R_1 = R_2 = -H$) is between 3.8 and 4.0,[32,33] whereas BZQ

[30] S. Searles, M. Tamres, F. Block, and L. A. Quarterman, *J. Amer. Chem. Soc.* **78**, 4917 (1956).

[31] S. P. Mayalarp, R. H. Hargreaves, J. Butler, C. C. O'Hare, and J. A. Hartley, *J. Med. Chem.* **39**, 531 (1996).

[32] E. B. Skibo and C. Xing, *Biochemistry* **37**, 15199 (1998).

[33] R. J. Driebergen, J. J. Holthuis, A. Hulshoff, S. J. Postma-Kelder, W. Verboom, D. N. Reinhoudt, and P. Lelieveld, *Anticancer Res.* **6**, 605 (1986).

(Fig. 1; $R_1 = R_2 = -NHCH_2CH_2OH$) has a pK of around 4.5 (H. Shah-bakhti, 1998, unpublished result). These pKs are significantly lower than that of simple aziridine, if only because the electron withdrawing $C = O$ groups of the quinone reduce the availability of the donating electrons on the nitrogen. In a similar manner, the electron withdrawing ability of the two $-NHCH_2CH_2OH$ groups in BZQ will result in a higher pK than DZQ.[†] The pK_2 values for the DBQs (i.e., for the protonation of the second aziridine) have not been measured.

The pKs of semiquinones derived from simple benzoquinones are typically in the range 4.0 to 5.1,[34] and there is no reason to propose that they should be dramatically different for the DBQs. Similarly, the pKs of the $-OH$ groups in simple benzhydroquinones are all >9.9.[35] Thus the states of protonation of the $-OH$ groups of the semiquinones and hydroquinones of the DBQs at physiologically relevant pH are as shown in Fig. 2.

When the aziridinylquinones are reduced to aromatic semiquinones or hydroquinones, it can be predicted that the pKs should change because the conjugated aromatic ring should stabilize the free amine as compared with the protonated form. However, this will be compensated for by the inductive effect of the $-O^{•-}$ and/or the $-OH$ groups. Moreover, these groups could also stabilize the protonated form of the aziridine by $-H$ bonding. The pKs are therefore expected to be higher than those of the parent quinones. Once again, relatively few measurements have been made. However, apparent pK (aziridine) values of between 6 and 8 have been measured after the electrochemical reduction of diaziridinylquinones.[33,36] The pKs of the semiquinones of AZQ and DZQ have more recently been reported to be 6.3–6.6 and 5.0, respectively.[37] These values are consistent with the pH dependence of the alkylation of DNA by the electrochemically reduced AZQ and DZQ.[38]

The Reductions

Thermodynamically, the ease of reduction of quinones can be expressed in terms of the reduction potentials:

[†] It has been proposed that the strong dependence of the pK values on quinone substituents indicates that the oxygen on the quinone is protonated rather than the nitrogen on the aziridine.[32]

[34] A. J. Swallow, *in* "Function of Quinones in Energy Conserving Systems" (B. L. Trumpower, ed.), pp. 59–72. Academic Press, London, 1982.

[35] L. A. Bishop and L. K. J. Tong, *J. Amer. Chem. Soc.* **87,** 501 (1967).

[36] R. J. Driebergen, Ph.D. Thesis, Utrecht (1987).

[37] C. Xing and E. B. Skibo, *Biochemistry* **39,** 10770 (2000).

[38] K. J. Lusthof, N. J. de Mol, L. H. M. Janssen, W. Verboom, and D. N. Reinhoudt, *Chem-Biol. Interact.* **70,** 249 (1989).

$$Q + e^- \rightleftarrows Q^{\bullet -} \qquad\qquad E[Q/Q^{\bullet -}] \tag{1}$$

$$Q + 2e^- + 2H^+ \rightleftarrows QH_2 \qquad E[Q, 2H^+/QH_2] \tag{2}$$

The semiquinones and hydroquinones are in equilibrium:

$$2Q^{\bullet -} + 2H^+ \rightleftarrows QH_2 + Q \tag{3}$$

The reduction potentials of the two reduced forms are related according to:

$$E[Q^{\bullet -}, 2H^+/QH_2] = 2E[Q2H^+/QH_2] - E[Q/Q^{\bullet -}] \tag{4}$$

As mentioned previously, pKs of semiquinones are typically 4 to 5, and the pKs of hydroquinones are >9.9, and hence $Q^{\bullet -}$ and QH_2 represent the correct states of protonation of the quinone moieties at physiological pH.

The $E[Q/Q^{\bullet -}]$ values for benzoquinones vary dramatically, depending on the electron donating/withdrawing abilities of the substituents. Hence, for example, 1,4-benzoquinone has a $E[Q/Q^{\bullet -}]$ value of +78 mV, whereas for BZQ, the value is −370 mV.[39,40] The two-electron potentials, $E[Q, 2H^+/QH_2]$ are typically between +390 and +500 mV for the simple benzoquinones[35] and should be about the same values for the DBQs. All of these potentials are quoted as the standard reduction potentials (NHE).

The reduction potentials give useful information about the chemistry of the substances and can be used to predict the positions of equilibria. Thus, for example, as mentioned previously, the one-electron reduction potentials of 1,4-benzoquinone (BQ) and BZQ are 78 mV and −370 mV, respectively. Therefore if an equilibrium could set up between the two quinones,

$$BZQ^{\bullet -} + BQ \rightleftarrows BZQ + BQ^{\bullet -} \tag{5}$$

because the reduction potential of the BZQ/BZQ$^{\bullet -}$ couple is much more negative than that of the BQ/BQ$^{\bullet -}$ couple, the equilibrium will be way over to the right-hand side.

The reduction potentials can also partially explain why some diaziridinylquinones are not very selective at targeting cells that over-express DTD. Thus, for example, although DZQ (Fig. 1; $R_1 = R_2 = -H$) and PDZQ (Fig. 1; $R_1 = $ phenyl, $R_2 = -H$) are highly cytotoxic and are excellent substrates for DTD, they produce very poor cytotoxicity differentials between DTD-rich and -deficient cell lines.[24,29,31] These DBQs have relatively high (more positive) $E[Q/Q^{\bullet -}]$ and $E[Q, 2H^+/QH_2]$ values, and hence they can also be reduced and activated by other cellular components such as glutathione (see following text), ascorbate and even NAD(P)H.[41,42]

[39] P. Wardman, *J. Phys. Chem. Ref. Data* **18**, 1637 (1989).
[40] J. Butler, B. M. Hoey, and J. S. Lea, *Biochim. Biophys. Acta* **925**, 144 (1987).

The reduction potentials of the quinones also influence the stability of the semiquinones and hydroquinones in the presence of oxygen. The one- and two-electron reduction potentials of oxygen are $E[O_2/O_2^{\bullet -}] = -155$ mV, and the $[O_2, 2H^+/H_2O_2]$ couple has a potential of $+300$ mV at pH $7^{39,43}$ (molar concentrations of oxygen). Hence, the semiquinones that have a reduction potential, $E[Q/Q^{\bullet -}]$, less than -155 mV will be readily oxidized by oxygen:

$$Q^{\bullet -} + O_2 \rightleftarrows Q + O_2^{\bullet -} \qquad (6)$$

It should be stressed that as (6) is an equilibrium, most semiquinones, even those with positive $E[Q/Q^{\bullet -}]$ values, will eventually be re-oxidized in air.

The corresponding oxidation of hydroquinones can be represented as:

$$QH_2 + O_2 \rightleftarrows Q + H_2O_2 \qquad (7)$$

Because $[O_2, 2H^+/H_2O_2]$ is $+300$ mV and $E[Q, 2H^+/QH_2]$ values are around $+390$ mV, the oxidation of most hydroquinones (reaction [7]) is thermodynamically favorable. However, the reaction does not readily occur, and most hydroquinones react very slowly with oxygen in aqueous solution. This is because for the reactions to occur, two electrons would have to be transferred simultaneously to the antibonding orbitals of oxygen. This process is spin forbidden.[44] There are a few examples in the literature wherein the oxidation is more rapid.[45,46] In these cases, the reactions probably occur via the reaction of oxygen with the semiquinones that are in equilibrium with the hydroquinone:

$$QH_2 + Q \rightleftarrows 2Q^{\bullet -} + 2H^+ \qquad (8)$$

$$2Q^{\bullet -} + O_2 \rightleftarrows 2Q + 2O_2^{\bullet -} \qquad (9)$$

$$2O_2^{\bullet -} + 2H^+ \rightarrow H_2O_2 + O_2 \qquad (10)$$

$$\text{Net: } QH_2 + O_2 \rightleftarrows H_2O_2 + Q \qquad (11)$$

[41] C-S. Lee, J. A. Hartley, M. D. Berardini, J. Butler, D. Seigel, D. Ross, and N. W. Gibson, *Biochemistry* **31**, 3019 (1992).
[42] J. Butler and B. M. Hoey, *Biochim. Biophys. Acta* **1161,** 73 (1993).
[43] W. H. Koppenol and J. Butler, *Adv. Free Rad. Biol. Med.* **1,** 91 (1985).
[44] W. H. Koppenol and J. Butler, *FEBS Lett.* **83,** 1 (1977).
[45] E. Cadenas, D. Mira, A. Brunmark, C. Lind, J. Segura-Aguilar, and L. Ernster, *Free Rad. Biol. Med.* **5,** 71 (1988).
[46] J. Butler, V. Spanswick, and J. Cummings, *Free Rad. Res.* **25,** 141 (1996).

The rate of autoxidation of hydroquinones can also be increased in the presence of superoxide dismutase.[47,48] Consistent with these mechanisms, the autoxidation of AZQ hydroquinone has been shown to generate superoxide radicals.[49]

The Enzymes

Quinones can be reduced by a variety of one-electron reducing enzymes including NADPH:cytochrome P450 reductase,[42,49,50] NADH: cytochrome b_5 reductase[51] and xanthine oxidase.[40,52] They are also reduced by the two electron reducing enzymes, DTD[26–29] and xanthine dehydrogenase.[53] Some DBQs and mitomycin C may also be activated by HAP1, an AP-endonuclease that has a redox function.[54] However, it is now generally accepted that the most important enzymes involved in quinone reduction are NADPH:cytochrome P450 reductase and DT-diaphorase.

The main function of the NADPH:cytochrome P450 reductases is to reduce the numerous forms of cytochrome P450. However, transfection studies have now confirmed that the P450 reductases can reduce and activate several different types of xenobiotics including nitroimidazoles,[55] N-oxides[56] and quinones.[57]

Studies on the rates of reduction of different types of compounds by cytochrome P450 reductase have shown that there is a correlation between the rates and the one-electron reduction potentials of the substrates. Essentially, as $E[Q/Q^{•-}]$ becomes more positive, the rates of reduction increase. Furthermore, the rates are relatively insensitive to the different structures of the substrates.[42,50] This is consistent with an outer-sphere electron transfer process and implies that there is little or no binding of the substrates to the enzyme.

[47] E. Cadenas, *Biochem. Pharmacol.* **49,** 127 (1995).
[48] R. Jarabak and J. Jarabak, *Arch. Biochem. Biophys.* **318,** 418 (1995).
[49] G. R. Fisher and P. L. Gutierrez, *Free Rad. Biol. Med.* **11,** 597 (1991).
[50] N. Cenas, Z. Anusevicius, D. Bironaite, G. L. Bachmanova, A. I. Archakov, and K. Ollinger, *Arch. Biochem. Biophys.* **315,** 400 (1994).
[51] M. F. Belcourt, W. F. Hodnick, S. Rockwell, and A. C. Sartorelli, *J. Biol. Chem.* **173,** 8875 (1998).
[52] K. J. Lusthof, W. Richter, N. J. de Mol, L. H. Janssen, W. Verboom, and D. N. Reinhoudt, *Arch. Biochem. Biophys.* **277,** 137 (1990).
[53] C. A. Pritos and D. L. Gustafson, *Oncol. Res.* **6,** 477 (1994).
[54] M. J. Prieto-Alamo and F. Laval, *Carcinogenesis* **20,** 415 (1999).
[55] A. V. Patterson, K. J. Williams, R. L. Cowen, M. Jaffar, B. A. Telfer, M. Saunders, R. Airley, D. Honess, A. J. van der Kogel, C. R. Wolf, and I. J. Stratford, *Gene Ther.* **9,** 946 (2002).
[56] Y. Jounaidi and D. J. Waxman, *Cancer Res.* **60,** 3761 (2000).
[57] P. Joseph, Y. Xu, and A. K. Jaiswal, *Int. J. Cancer* **65,** 263 (1996).

The National Cancer Institute database shows that although cytochrome P450 reductases are widely distributed in their panel of 60 tumor cell lines, they do not appear to be overexpressed in any particular tumor type.[58] However, the cytochrome P450 reductases play a role in chemotherapy because they can selectively form reactive radicals under hypoxic conditions. Essentially, if the one-electron potential of a damaging radical is more negative than that of oxygen ($E[O_2/O_2^{\bullet-}] = -155$ mV), the radical will be inactivated in normal, fully oxygenated cells (via reaction [6]). On the other hand, if the radical is produced in the hypoxic regions of a tumor, it can be around long enough to damage the tumor cells. This concept is also very relevant to the selective targeting of DTD-rich tumor cells by alkylating quinones. It would be expected that if the $E[Q/Q^{\bullet-}]$ values of the quinone are more negative than -155 mV, reduction by the one-electron reducing enzymes would result in futile redox cycling. Thus the quinones would alkylate only as a consequence of two-electron reduction by DTD (assuming that the autoxidation of the hydroquinone is slow). Alternatively, the hypoxic tumor cells may also be alkylated as a consequence of activation of the semiquinones. The selective targeting of hypoxic cells could be even further enhanced because DTD may be over-expressed in hypoxia.[59,60]

There have been numerous studies on the kinetics of reduction of diaziridinylquinones by DT-diaphorase and values of V_{max}, K_m and K_{cat} have been determined.[26,29,61] However, many diaziridinylquinones are reduced intracellularly and are cytotoxic at nanomole concentrations—that is, at concentrations that are apparently orders of magnitude lower than the reported V_{max} and K_m values. It would therefore appear that if the quinones were to follow simple Michaelis-Menten kinetics within the cell, relatively little substrate-enzyme binding must occur. The ability of a diaziridinylquinone to be reduced within a cell may therefore be controlled simply by its ease of reduction and some structural factors that could impede or assist the approach to the site of electron transfer within the enzyme.

The two-electron reduction potentials of the quinone $E[Q,2H^+/QH_2]$ are relevant as they are related to the overall free energy of the reactions. Thus, regardless of the mechanism, if the two-electron potentials are more

[58] The NCI database is available on: http://dtp.nci.nih.gov/mtargets/mt_index.html

[59] K.-S. Yao, S. Xanthoudakis, T. Curran, and P. J. O'Dwyer, *Mol. Cell. Biol.* **14,** 5997 (1994).

[60] P. J. O'Dwyer, K.-S. Yao, P. Ford, A. K. Godwin, and M. Clayton, *Cancer Res.* **54,** 3082 (1994).

[61] B. Prins, A. S. Koster, W. Verboom, and D. N. Reinhoudt, *Biochem. Pharmacol.* **38,** 3753 (1989).

negative than that for the $NAD(P)^+/NAD(P)H$ couples, the electron transfer cannot readily take place. However, the relationships between the initial rates or V_{max} values and the reduction potentials are not as simple as observed for the cytochrome P450 reductase reductions. This is illustrated by considering the rates of reduction of AZQ and DZQ. These DBQs have similar $E[Q/Q^{\bullet-}]$ values and are expected to have similar $E[Q, 2H^+/QH_2]$ values.[27,40,62] However, AZQ is a poor substrate for DTD, whereas DZQ is an excellent substrate.[26,27] Similarly, although MeDZQ is excellent substrate for the enzyme, it is thermodynamically more difficult to reduce than AZQ.[27,40]

A recent study on the rates of reduction of a series of different quinones (benzoquinones, naphthoquinones and anthraquinones) by DT-diaphorase has shown that the rates have some dependence on both the reduction potentials and the van der Waal volumes.[63] These results are consistent with an earlier study on alkyl substituted diaziridinylbenzoquinones, where it was demonstrated that as the size of the alkyl groups were increased, the rates with DTD and the cytotoxicities were decreased.[27] This previous study also showed that fast rates of reduction could be achieved if a methyl group was maintained in the structure, and this led to the speculation that only one side of the quinone has to interact with the sites of reduction in the enzyme. This idea can explain why some apparently more bulky quinones (e.g., certain indoloquinones such as EO9 and streptonigrin) can be good substrates for the enzyme, whereas adriamycin and other anthraquinones are poor substrates.[29,64–66] Some diaziridinylquinones with very bulky groups in the 3 position (e.g., Fig. 1; $R_1 = H$, $R_2 = $ phenyl-esters or acridines) can be substrates for the enzyme.[28,67]

The structure of human DT-diaphorase has now been published,[68,69] and it appears that the unoccupied potential binding site of the protein is quite large (10 Å wide, 9 Å deep and 4 Å high). This relatively large

[62] A. Dzielendziak, J. Butler, B. M. Hoey, J. S. Lea, and T. H. Ward, *Cancer Res.* **50,** 2003 (1990).

[63] Z. Anusevicius, J. Sarlauskas, and N. Cenas, *Arch. Biochem. Biophys.* **15,** 254 (2002).

[64] H. W. Beall, S. Winski, E. Swann, A. R. Hudnott, A. S. Cottrill, N. O'Sullivan, S. J. Green, R. Bien, D. Siegel, D. Ross, and C. J. Moody, *J. Med. Chem.* **41,** 4755 (1998).

[65] M. Jaffar. M. A. Naylor, N. Robertson, S. D. Lockyer, R. M. Phillips, S. A. Everett, G. E. Adams, and I. J. Stratford, *Anticancer Drug Des.* **13,** 105 (1998).

[66] W. Wallin, *Cancer Lett.* **30,** 97 (1986).

[67] A. M. Di Francesco, S. P. Mayalarp, J. E. Simpson, S. Kim, A. E. Scott, J. Butler, and M. Lee (submitted).

[68] M. Faig, M. A. Bianchet, S. Winski, R. Hargreaves, C. J. Moody, A. R. Hudnott, D. Ross, and L. M. Amzel, *Struct. Fold. Des.* **9,** 659 (2001).

[69] M. Faig, M. A. Bianchet, P. Talalay, S. Chen, S. Winski, D. Ross, and L. M. Amzel, *Proc. Natl. Acad. Sci. USA* **97,** 3177 (2000).

volume means that in principle, many different types of quinone substrates can be accommodated. Interestingly, the high-resolution structures of the enzyme complexed with RH1 and with two aziridinylindolequinones have now shown that these quinones can bind in different orientations. It was proposed that the binding cavity is highly plastic and can change to accommodate the binding of different substrates. Nonetheless, these structures clearly show that small side-groups on the DBQs can be accommodated inside the cavity while much larger groups on the opposite side can fit outside this cavity. This implies that more complex, possibly more DNA-sequence-selective DBQs, could be activated by DTD.

DNA Interactions

Several investigations have been carried out on the interactions of AZQ with DNA in model systems and intact cells.[41,70–72] All of these studies show that AZQ can form interstrand cross-links as a consequence of reduction. Because AZQ is a relatively poor substrate for DTD, it is probable that these cross-links are formed via the one-electron reducing enzymes or from other cellular reducing agents (see previous text). The semiquinone radicals of AZQ are expected to be relatively stable in air[40] and hence should be around long enough to interact with the bases of DNA. One study has shown that for a series of AZQ analogs, there is a direct relationship between the individual $E[Q/Q^{\bullet-}]$ values, the cross-linking efficiencies and the strengths of the e.s.r semiquinone radical signals from treated intact human cells.[62]

The products of the alkylation reaction of chemically reduced AZQ with calf thymus DNA and a synthetic oligodeoxynucleotide have been studied in detail, and numerous products including monoadducts with both dG and dA and two main di-adducts have been identified.[73] The predominant interstrand cross-links were between dG residues separated by two intervening base pairs. ^{32}P-postlabeling studies have also shown that when AZQ is reacted with DNA, two major (22% and 40%) and at least eight minor products can be detected. Once again, one of the major products was identified as a guanine adduct.[74]

[70] L. Szmigiero, L. C. Erickson, R. A. Ewig, and K. W. Kohn, *Cancer Res.* **44**, 4447 (1984).
[71] T. H. Ward, J. Butler, H. Shahbakhti, and J. T. Richards, *Biochem. Pharmacol.* **53**, 1115 (1997).
[72] C. L. King, W. N. Hittelman, and T. L. Loo, *Cancer Res.* **44**, 5634 (1984).
[73] S. C Alley and P. B. Hopkins, *Chem. Res. Toxicol.* **7**, 666 (1994).
[74] R. C. Gupta, A. Garg, K. Earley, S. C. Agarwal, G. R. Lambert, and S. Nesnow, *Cancer Res.* **51**, 5198 (1991).

Interestingly, AZQ also produces extensive DNA single strand breaks in mammalian cells.[70–72] These strand breaks may be produced via the formation of superoxide radicals (equilibrium [6]) or as a consequence of DNA repair.[71,75] There have been many studies that show that DBQs can redox cycle. However, the production of DNA single strand breaks via superoxide production must play a relatively minor role in the cytotoxicity of diaziridinylquinones. The evidence for this comes from the actual cytotoxicity data. These simple toxicity or growth inhibition tests use IC_{50} values that are defined as the concentration of compound necessary to inhibit the growth of a population of cells by 50%. Simple redox active quinones such as 1,4-benzoquinone and duroquinone have typical ID_{50} values of around 10 to 50 μM in most tumor cell lines. However, if DNA-alkylating aziridines are incorporated in the quinones, the redox capabilities of these compounds can be very similar to the simple benzoquinones, but the cytotoxicity values decrease dramatically. Thus, for example, the mono alkylating aziridinylbenzoquinone 2-aziridinyl-3,5-dimethyl-1,4-benzoquinone has ID_{50} values of around $\sim$0.2 μM, and the bis alkylating diaziridinylquinone MeDZQ has even lower ID_{50} values of around 1 to 2 nM.[27,31] It has also been proposed from measurements of intracellular pyridine nucleotide pools that the DBQs damage cells via redox cycling only at relatively high concentrations. The predominant damaging event at much lower concentrations is attributed to alkylation of DNA.[76]

The superoxide/hydrogen peroxide generating abilities of the diaziridinylquinones are less important to cytotoxicity because most mammalian cells have very efficient radical scavenging systems.[77] Furthermore, DNA single strand breaks can be efficiently repaired. Indeed, it has been demonstrated that the cellular DNA single strand breaks formed from high concentrations of AZQ are repaired over a period of a few hours after treatment.[71,72] This is also consistent with an alkaline elution study that showed that the cytotoxicity of AZQ in different cell lines could be correlated with the extent of cross-links but not with the strand breaks.[70]

In one study, it was shown that although AZQ readily produces e.s.r detectable semiquinone radicals in cells, no such radicals could be detected in BZQ-treated cells. This is consistent with the one-electron reduction potentials.[40,78] BZQ has subsequently been shown to produce DNA

[75] C. S. Lee, *Mol. Cell*, **10**, 723 (2001).
[76] W. A. Morgan, B. Prins, and A. S. Koster, *Arch. Toxicol.* **71**, 582 (1997).
[77] B. Halliwell and J. M. C. Gutteridge, *in* "Free Radicals in Biology and Medicine" (3rd Ed.). Clarendon Press, 1999.
[78] J. Butler, A. Dzielendziak, J. S. Lea, T. H. Ward, and B. M. Hoey, *Free Rad. Res. Comms.* **8**, 231 (1989).

cross-links at pH 7 in the absence of reduction.[79] Similarly, it has been shown that the cytotoxicity of BZQ is independent of the presence of the DTD inhibitor dicoumarol.[26]

Carboquone can form DNA cross-links and strand breaks in mammalian cells *in vitro*.[80,81] There have been relatively few studies on the mechanisms of activation. Surprisingly (see MeDZQ, below), carboquone appears to be a relatively poor substrate for DT-diaphorase and the one-electron reducing enzymes and does not efficiently alkylate DNA.[61,76,38]

There have been numerous studies on the interaction of DZQ with DNA that show that DZQ produces DNA strand breaks and alkylation.[41,61,79,82] It has now also been shown that the DZQ-DNA adducts could be repaired by glycosylases.[75] However, one of the most interesting findings is that DZQ and presumably other diaziridinylquinones can also alkylate the DNA phosphate backbone.[32] It would be expected that these alkylated phosphates will be unstable and hence produce strand breaks. This might explain why DZQ induces DNA strand breaks under the same conditions in which other diaziridinylquinones, such as MeDZQ, produce DNA cross-links.[26]

Reduced DZQ, trenimon and the other mono-substituted DBQs (i.e., Fig. 1; $R_1 = -H$, $R_2 = -X$) alkylate DNA at 5'-TGC sequences. This is in contrast to the bis-substituted DBQs, which appear to react randomly.[41,79,83,84] The 5'-GC selectivity occurs only after reduction and takes place because the $-H$ group is small enough to allow the hydroquinones to fit between the bases. In this position, the $-OH$ groups of the hydroquinone hydrogen bond with O2 and the $C4-NH_2$ groups of cytosine, and the protonated aziridine group can associate with the N7 group of guanine (Fig. 3). The TGC selectivity can be explained by the three-center hydrogen bond between the N6-H of the adenine and the O4 of the thymine. This means that the space between the guanines and cytosines in a TGC sequence is relatively large.[85] Interestingly, this sequence selectivity allows both of the aziridines to be in close contact with two guanines on opposite

[79] J. A. Hartley, M. Berardini, M. Ponti, N. W. Gibson, A. S. Thompson, D. E. Thurston, B. M. Hoey, and J. Butler, *Biochemistry* **30**, 11719 (1991).

[80] Y. Maehara, H. Anai, Y. Sakaguchi. T. Kusumoto, Y. Emi, and K. Sugimachi, *Oncology* **47**, 282 (1990).

[81] R. Kanamaru, M. Asamura, Y. Hayashi, H. Sato, and T. Saito, *Tohoku J. Exp. Med.* **124**, 331 (1978).

[82] M. Yuki and I. S. Haworth, *Anticancer Drug Des.* **8**, 269 (1993).

[83] S. C. Alley and P. B. Hopkins, *Chem. Res. Toxicol.* **7**, 666 (1994).

[84] R. H. J. Hargreaves, S. P. Mayalarp, J. Butler, C. C. O'Hare, and J. A. Hartley, *J. Med. Chem.* **40**, 357 (1997).

[85] R. E. Dickerson, *Methods Enzymol.* **211**, 67 (1992).

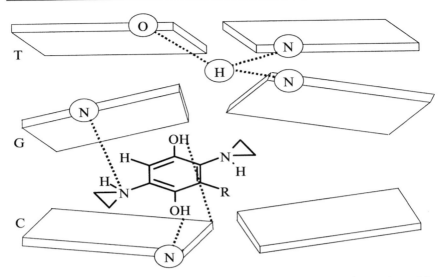

Fig. 3. Schematic representation of a DNA TGC sequence and the interactions with a reduced diaziridinylbenzoquinone. The three-center hydrogen bond between the N6-H of the adenine and the O4 of the thymine and N4 of cytosine is indicated by the dashed line. Adapted with permission from reference 84. Copyright 1997 American Chemical Society.

strands of the DNA. These types of compounds are therefore very efficient at cross-linking DNA and thus are more cytotoxic.[84]

It was noted in the previous studies that MeDZQ (Fig. 1; $R_1 = R_2 = -CH_3$) did not follow the same trend. This quinone was extremely efficient at cross-linking DNA and was as toxic as DZQ in a leukemia cell line. However, consistent with the proposed mechanism, MeDZQ did not alkylate DNA at TGC sequences. Subsequent studies showed[27,86] that this quinone and other alkyl substituted DBQs can efficiently cross-link DNA at 5'-GNC sequences after reduction. Essentially, the hydroquinone of MeDZQ interacts face on with DNA to form hydrogen bonds between the two hydroquinone OH groups and the two guanine O-6 s. The protonated aziridines also associate with the two guanine N-7 positions (Fig. 4). Once again, the sequence selectivity allows the two aziridines to come in close contact with the two guanines, and hence these compounds are very efficient at cross-linking DNA. If both of the methyl groups are replaced by ethyl or propyl, the cross-linking efficiency decreases significantly as these

[86] M. D. Berardini, R. L. Souhami, C.-S. Lee, N. W. Gibson, J. Butler, and J. A. Hartley, *Biochemistry* **32**, 3306 (1993).

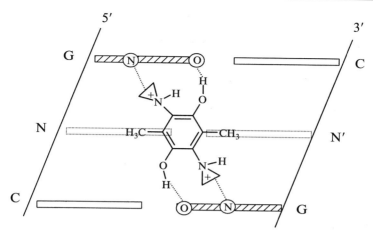

FIG. 4. Schematic representation of a DNA GNC sequence and the interactions with reduced MeDZQ. Adapted with permission from reference 27. Copyright 1999 American Chemical Society.

larger alkyl groups sterically hinder the hydrogen bonding interactions. The sequence selectivity of the alkylating diaziridinylquinones has been extended by using triplex-directed oligonucleotides conjugated to the quinone.[87]

Other Targets for Diaziridinylbenzoquinones

There is little doubt that the cytotoxicity of most DBQs can be explained by the direct interaction of the aziridines with DNA. However, even simple benzoquinones can undergo complex reactions with other cellular complements. Reactions of these types can be relevant to the cytotoxicities of DBQs.

The reactions of the diaziridinylquinones with glutathione are complex, and many different products have been reported. Interestingly, it appears that although some reactions with glutathione result in an activation of the aziridines, other reactions result in a deactivation or loss of aziridine. Because mammalian cells have millimolar concentrations of glutathione, the overall cytotoxicity of a particular diaziridinylbenzoquinone and its ability to selectively target different cell types must be affected by glutathione reactions.

[87] M. W. Reed, A. Wald, and R. B. Meyer, *J. Am. Chem. Soc.* **120,** 9729 (1998).

Some of proposed reactions[88–91] are:

$$GS^- + Q \rightleftarrows GS^{\bullet} + Q^{\bullet -} \tag{12}$$

$$GS^- + GS^{\bullet} \rightarrow GSSG^{\bullet -} \tag{13}$$

$$GSSG^{\bullet -} + Q \rightarrow GSSG + Q^{\bullet -} \tag{14}$$

$$GSSG^{\bullet -} + O_2 \rightarrow GSSG + O_2^{\bullet -} \tag{15}$$

$$2GSH + Q \rightarrow QH_2 + GSSH \tag{16}$$

$$Q + O_2^{\bullet -} \rightleftarrows Q^{\bullet -} + O_2 \tag{17}$$

$$2Q^{\bullet -} + 2H^+ \rightleftarrows QH_2 + Q \tag{18}$$

The potential of the couple $[RS^{\cdot}, H^+/RSH]$ is around $+900$ mV at pH 7 for cysteine and should be the same for gluathione.[92] The $GS^{\cdot}$ radicals are therefore very strongly oxidizing, whereas semiquinone radicals are reducing. This means that the equilibrium (12) should be way over to the left-hand side. However, it is possible that the equilibrium is driven over to the right-hand side because of the formation of the fast reacting radicals $GS^{\bullet}$ and $Q^{\bullet -}$.

Because hydroquinones and semiquinones can be formed in these reactions, the aziridine can become activated and react with excess glutathione (Fig. 5). Products of this type have been observed with AZQ.[89,93]

Glutathione can also undergo addition reactions with halogen-substituted DBQs (Fig. 1; R1 = R2 = Cl or F) and with simple 1,4-benzoquinones to eventually produce glutathionyl-quinone conjugates,[89,90,94] (Fig. 6; products 1 and 2). One study has shown that glutathione adducts of this type are much less efficient at alkylating DNA than the parent quinone.[95]

[88] C. Giulivi, A, Forlin, S. Bellin, and E. Cadenas, *Chemico-Biol. Inter.* **108,** 137 (1998).
[89] P. Gutierrez and S. Silva, *Chem. Res. Toxicol.* **8,** 455 (1995).
[90] J. Butler and B. M. Hoey, *Free Rad. Biol. Med.* **12,** 337 (1992).
[91] N. Takahashi, J. Schreiber, V. Fisher, and R. P. Mason, *Arch. Biochem. Biphys.* **252,** 41 (1987).
[92] P. S. Surdhar and D. A. Armstrong, *J. Phys. Chem.* **91,** 6532 (1987).
[93] D. Ross, D. Siegel, N. W. Gibson, D. Pacheco, D. J. Thomas, M. Reasor, and D. Wierda, *Free Rad. Res. Comm.* **8,** 373 (1990).
[94] J. Goin, C. Giulivi, J. Butler, and E. Cadenas, *Free Rad. Biol. Med.* **18,** 525 (1995).
[95] K. J. Lusthof, N. J. de Mol, L. H. Janssen, B. Prins, W. Verboom, and D. N. Reinhoudt, *Anticancer Drug Des.* **5,** 283 (1990).

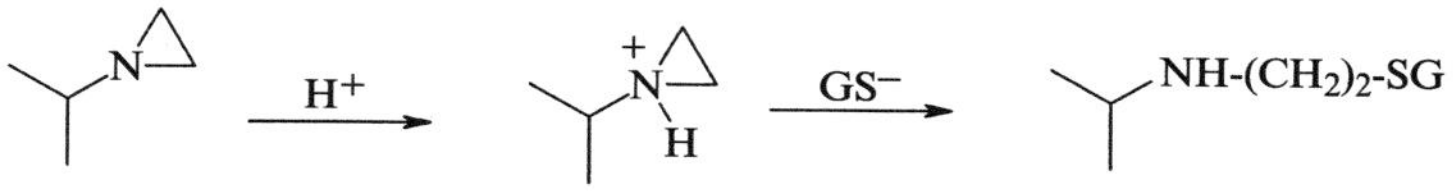

FIG. 5. The reaction between a hydroquinone or semiquinone with excess glutathione.

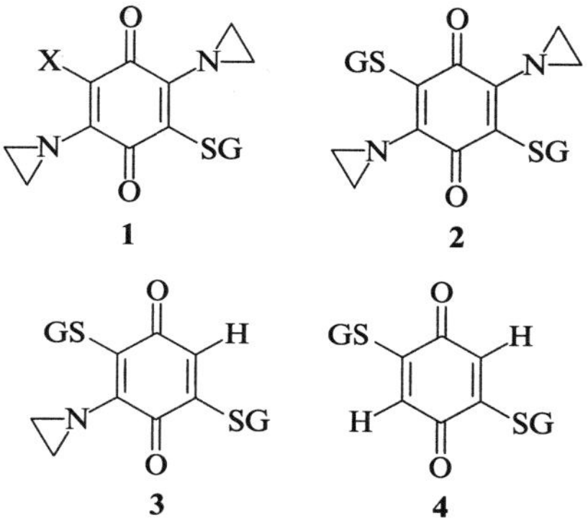

FIG. 6. Products from the interactions of diaziridinylbenzoquinones with glutathione.

Interestingly, two independent groups[88,89] have also shown that when the reactions are carried out at high GSH:quinone ratios, the quinone aziridines can be replaced by glutathione (Fig. 6; products 3 and 4). It is likely that the leaving group is the protonated aziridine. It has been proposed[32] that in some instances, protonated aziridines may be displaced directly from the quinone even in the absence of glutathione or reducing agents.

There is relatively little information on the direct interactions of diaziridinylbenzoquinones with amino acids, either free in solution or in proteins. However, an early study on trenimon showed that it inhibited the template activity of chromatin in an RNA polymerization system, and this inhibition was not simply due to DNA alkylation. It was suggested that trenimon may be able to react with the nucleoproteins.[96] Trenimon also covalently binds to cysteines in hemoglobin and bovine gammaglobulin.[97] Similarly, it has been shown that ^{14}C-labeled DZQ covalently binds to bovine serum albumin (BSA) and to some proteins in *E. coli*. The extent of binding to BSA actually *decreased* as the pH was lowered. This means that the binding was not due to acid-assisted aziridine activation. It was

[96] B. Puschendorf, H. Wolf, and H. Grunicke, *Biochem. Pharmacol.* **20,** 3039 (1971).
[97] J. H. Linford, *Chem. Biol. Interact.* **6,** 149 (1973).

shown that the main sites of attack were at cysteine residues.[98] It is therefore probable that all of these interactions with proteins are similar to those with glutathione.

It is difficult to assess the extent to which the reactions with proteins contribute to the relatively high cytotoxicities of DBQs. The situation is further complicated because many simple benzoquinones can also initiate reactions 12 through 16 and undergo addition reactions (Fig. 6; products 1 and 2) with cysteines and proteins.[90,99–101] Furthermore, these simple benzoquinones can also undergo redox cycling.[102,103] Nonetheless, it is probable that the reactions between the DBQs and proteins that are involved in the synthesis and maintenance of DNA will have the largest effect on the cytotoxicities. Hence, for example, it has been shown that AZQ inhibits thioredoxin reductase and that this inhibition is not affected by the presence of radical scavengers or superoxide dismutase or catalase. It has also been shown that there is a correlation between the AZQ inhibition of thioredoxin reductase and the cytotoxicity of AZQ in a human sarcoma cell line.[104,105] Unfortunately, simple non-alkylating benzoquinones were not screened.

All of the DBQs can damage DNA as a consequence of redox cycling and/or as a consequence of alkylation. These quinones can therefore trigger the onset of apoptosis. AZQ, DZQ and MeDZQ induces p53 in a MCF-7 human breast cancer cell line.[106] Interestingly, the induction was reduced in the presence of dicoumarol, a DTD inhibitor and catalase. This work was also significant because it showed that although 1,4-benzoquinone and menadione can also increase p53 levels, the induction is much less than with the DBQs. The main conclusions from this study are that bioreductive activation of AZQ is necessary to induce p53 and that the aziridines are necessary.

In a similar manner to the preceding, it has been demonstrated that AZQ and DZQ can induce p21, an inhibitor of cyclin-dependent kinases. Interestingly, BZQ has little effect on p21 expression.[107] It was also

[98] K. J. Lusthof, N. J. de Mol, L. H. Janssen, R. J. Egberink, W. Verboom, and R. N. Reinhoudt, *Chem. Biol. Interact.* **76**, 193 (1990).

[99] J. J. Miranda, *Biochem. Biophys. Res. Commun.* **275**, 517 (2000).

[100] Y. A. Kang, O. N. Bae, M. Y. Lee, S. M. Chung, J. Y. Lee, and J. H. Chung, *J. Toxicol. Environ. Health* **65**, 1367 (2002).

[101] R. J. Boatman, J. C. English, L. G. Perry, and L. A. Fiorica, *Chem. Res. Toxicol.* **13**, 853 (2000).

[102] G. Powis, *Free Rad. Biol. Med.* **6**, 63 (1989).

[103] J. O'Brien, *Chem. Biol. Interact.* **80**, 1 (1991).

[104] B-L. Mau and G. Powis, *Biochem. Pharmacol.* **43**, 1613 (1992).

[105] B-L. Mau and G. Powis, *Biochem. Pharmacol.* **43**, 1621 (1992).

[106] E. O. Ngo, L. M. Nutter, T. Sura, and P. L. Gutierrez, *Chem. Res. Toxicol.* **11**, 260 (1998).

[107] X. Qiu X, H. J. Forman, A. H. Schonthal, and E. Cadenas, *J. Biol. Chem.* **271**, 31915 (1996).

shown that the extent of induction correlates with the ability of the quinones to be substrates for DTD in different cell lines.[108] Subsequent studies have confirmed that the induction of p21 by DZQ is regulated at the transcriptional level and requires the activation of p53.[109] MeDZQ has been shown to induce apoptosis in BE and HT29 human colon adenocarcinoma cell lines.[110]

The Future

At the present time, RH1 is undergoing phase I/II clinical trials in the United Kingdom. This quinone has been shown to be an excellent substrate for DTD and is selectively toxic toward DTD-rich and -deficient cell lines.[111] RH1 has a reduction potential similar to that of MeDZQ[27] ($[Q/Q^{\bullet-}]$ -220 mV vs NHE at pH 7 as opposed to -240 mV for MeDZQ), and hence the semiquinone is relatively unstable in fully oxygenated cells (see previous section titled The Enzymes). Similarly, as the 2, 3, 5 and 6 positions of the quinone are blocked, RH1 should not readily undergo Michael additions with cysteines in proteins and with glutathione (see previous section titled Other Targets for Diaziridinylbenzoquinones). Interestingly, although RH1 has properties similar to those of MeDZQ, it is a better substrate for DTD and is also more than 10 times more cytotoxic in DTD-rich cells. RH1 also produces much higher cytotoxicity ratios between different DTD-rich and -deficient cell lines.[111] Pharmacology studies on RH1 have indicated that the half-life in the systemic circulation is relatively long[112,113] and has reduced kidney metabolism when compared with EO9, another bioreductive quinone.[113]

Despite the fact that several DBQs have already been clinically tested and have not been very successful, there is a clear role for this class of compounds in cancer chemotherapy. RH1 was the first quinone to be specifically designed to be activated by DTD, and all of the previous attributes indicate the drug should have a promising future as a DTD-targeting anti-cancer agent. The outcome of the present clinical trial is eagerly awaited.

[108] X. B. Qiu and E. Cadenas, *Arch. Biochem. Biophys.* **346,** 241 (1997).

[109] R. C. Wu, A. Hohenstein, J. M. Park, X. Qiu, S. Mueller, E. Cadenas, and A. H. Schonthal, *Oncogene* **17,** (1998).

[110] X. Sun and D. Ross, *Chem. Biol. Interact.* **6,** 267 (1996).

[111] S. L. Winski, R. H. J. Hargreaves, J. Butler, and D. Ross, *Clin. Cancer Res.* **4,** 3083 (1998).

[112] P. Khan, S. Abbas, R. H. Hargreaves, R. Caffrey, V. Megram, and A. McGown, *J. Chromatogr. B Biomed. Sci. App.* **729,** 287 (1999).

[113] P. M. Loadman, P. M. Philips, L. E. Lim, and M. C. Bibby, *Biochem. Pharmacol.* **59,** 831 (2000).

[11] Quinone Reductase–Mediated Nitro-Reduction: Clinical Applications

By RICHARD J. KNOX and SHIUAN CHEN

Introduction

The quinone reductase enzyme NQO1 catalyzes the reduction of quinones (e.g., menadione) to their hydroquinone form. However, NQO1 has a number of unusual properties. It can use either NADPH or NADH with equal activity as a cofactor,[1] and it has been shown that simple reduced nicotinamide derivatives can also act as cofactors for this enzyme.[2] NQO1 can also simultaneously transfer two electrons to a substrate, and this gives rise to the accepted role of NQO1 in the detoxification of quinones[3–5] (as suggested by Ernster[6]). This is because the aerobic toxicity of quinones arises from one-electron reduction reactions. However the two-electron reduction of quinones to hydroquinones by NQO1 represents a detoxification pathway, because hydroquinones are stable with respect to autoxidation and therefore bypass the radical–generating semiquinone. Hydroquinones may then undergo conjugation with glucuronate and other water-soluble ligands and be excreted from the cell.[3,7,8] Thus, under aerobic conditions, NQO1 would appear to play an important role in the cellular defense against oxygen stress caused by quinones. However, not all quinones are detoxified in this manner, and cytotoxic anti-tumor quinones—the mitomycins, aziridinylbenzoquinones and anthrocyclins—are bioactivated by NQO1 (and other enzyme systems) to cytotoxic compounds.[9–11]

Another unusual property of NQO1 is that it has activity against a remarkable range of substrate types. As well as quinones, NQO1 reduces

[1] L. Ernster, L. Danielson, and M. Ljunggren, *Biochim. Biophys. Acta* **58**, 171 (1962).
[2] F. Friedlos, M. Jarman, L. C. Davies, M. P. Boland, and R. J. Knox, *Biochem. Pharmacol.* **44**, 25 (1992).
[3] C. Lind, H. Vadi, and L. Ernster, *Arch. Biochem. Biophys.* **190**, 97 (1978).
[4] C. Lind, P. Hochstein, and L. Ernster, *Arch. Biochem. Biophys.* **216**, 178 (1982).
[5] H. Thor, M. T. Smith, P. Hartzell, G. Bellomo, S. A. Jewell, and S. Orrenius, *J. Biol. Chem.* **257**, 12419 (1982).
[6] L. Ernster, *Methods Enzymol.* **10**, 309 (1967).
[7] C. Lind, *Biochem. Pharmacol.* **34**, 895 (1985).
[8] C. Lind, *Arch. Biochem. Biophys.* **240**, 226 (1985).
[9] D. Siegel, N. W. Gibson, P. C. Preusch, and D. Ross, *Cancer Res.* **50**, 7293 (1990).
[10] D. Siegel, N. W. Gibson, P. C. Preusch, and D. Ross, *Cancer Res.* **50**, 7483 (1990).
[11] M. I. Walton, P. J. Smith, and P. Workman, *Cancer Commun.* **3**, 199 (1991).

other types of substrates such as benzotriazine di-(N)-oxides,[12] azo compounds,[13–15] chromium VI compounds[16] and nitro compounds.[17,18]

This chapter will examine the clinical applications of the nitroreductase activity of three quinone reductase enzymes, NQO1, *E. coli* B nitroreductase and—in particular—NQO2. The nitroreductase properties of these enzymes are particularly relevant to the treatment of cancer, and the rationale for this approach will be described. With NQO2, an endogenous human enzyme has been discovered with significant nitroreductase activity that can bioactivate a proven anti-tumor prodrug designated CB 1954.

Prodrugs Activated by Nitro-Reduction in Cancer Therapy

Prodrugs: An Introduction

The majority of the agents that are now used in cancer chemotherapy lack any intrinsic anti-tumor selectivity. They mostly act by an anti-proliferative mechanism, and their action is on cells that are in cycle, or in some cases, on a specific phase of the cell cycle, rather than by a specific toxicity directed toward a particular type of cancer cell. Thus, the limiting toxicity of the majority of anti-cancer agents is a result of a toxic effect on the normal host tissues that are the most rapidly dividing such as bone marrow, gut mucosa and the lymphatic system. Further, most human solid cancers do not have a high proportion of cells that are rapidly proliferating, and they are therefore not particularly sensitive to this class of agent. Thus, because of host toxicity, treatment has to be discontinued at dose levels that are well below the dose that would be required to kill all viable tumor stem cells. This intrinsic poor selectivity of anti-cancer agents has been recognized for a long time, and attempts to improve selectivity and allow greater doses to be administered have been numerous.

A sufficiently high degree of selectivity might be obtained by the use of prodrugs. Prodrugs are chemicals that are toxicologically and pharmacodynamically inert but may be converted *in vivo* to active products.

---

[12] R. J. Riley and P. Workman, *Biochem. Pharmacol.* **43,** 167 (1992).

[13] M. Huang, G. T. Miwa, N. Cronheim, and A. Y. H. Lu, *J. Biol. Chem.* **254,** 11223 (1979).

[14] M. Huang, G. T. Miwa, and A. Y. H. Lu, *J. Biol. Chem.* **254,** 3930 (1979).

[15] M. T. Huang, R. C. Smart, P. E. Thomas, C. B. Pickett, and A. Y. H. Lu, *Chemica Scripta* **27A,** 49 (1987).

[16] S. De Flora, A. Morelli, C. Basso, M. Romano, D. Serra, and F. A. De, *Cancer Res.* **45,** 3188 (1985).

[17] T. Sugimura, K. Okabe, and M. Nagao, *Cancer Res.* **26,** 1717 (1966).

[18] R. J. Knox, M. P. Boland, F. Friedlos, B. Coles, C. Southan, and J. J. Roberts, *Biochem. Pharmacol.* **37,** 4671 (1988).

Conversion of the prodrug to the active form can take place by a number of mechanisms such as changes of either pH, oxygen tension, temperature or salt concentration or by spontaneous decomposition of the drug, internal ring opening or cyclization.[19–21] However, the major approach in prodrug design for cancer chemotherapy is the synthesis of inert compounds that are converted to an active drug by enzyme action.[19] Thus, in cancer chemotherapy, the prodrug would be inert but converted *in vivo* into a highly toxic metabolite by an enzyme present at high levels in the cancer cells but not in other cells. A classic example of the high degree of selectivity that can be attained with prodrugs is the activation by the enzyme β-glucuronidase of a relatively non-toxic alkylating agent to an extremely reactive and toxic metabolite that can cure cancers in experimental models.[22] The quinone reductase enzyme NQO1 would appear to be a good target as a prodrug-activating enzyme. NQO1 activity is present in a variety of tissues and cell lines of human origin. High NQO1 activity has been reported in human tumor cell lines of breast,[23–25] brain,[26,27] colon,[23,24,28–30] lung[23–25] and liver origin.[23,24,30–32] There is a marked increase in the level of NQO1 detected in solid human tumors from thyroid, adrenal, breast, ovarian, colon, and cornea and in non-small cell lung.[30,33,34] Further, NQO1 levels in bone marrow, a tissue sensitive to conventional cytotoxic chemotherapy, are low,[35–37] directing toxicity away from tissues that are usually sensitive to

[19] T. A. Connors, *Xenobiotica* **16**, 975 (1986).

[20] T. A. Connors, Bioactivation and Cytotoxicity, *in* "Progress in Drug Metabolism" (J. W. Bridges and L. F. Chasseaud, eds.), p. 41. Wiley, London, 1976.

[21] T. A. Connors, *Biochimie* **60**, 979 (1978).

[22] T. A. Connors and M. E. Whisson, *Nature* **210**, 866 (1966).

[23] J. J. Schlager and G. Powis, *Int. J. Cancer* **45**, 403 (1990).

[24] M. Belinsky and A. K. Jaiswal, *Cancer Metastasis Rev.* **12**, 103 (1993).

[25] H. D. Beall, A. M. Murphy, D. Siegel, R. H. Hargreaves, J. Butler, and D. Ross, *Mol. Pharmacol.* **48**, 499 (1995).

[26] M. S. Berger, R. E. Talcott, M. L. Rosenblum, M. Silva, F. Ali-Osman, and M. T. Smith, *J. Toxicol. Environ. Health* **16**, 713 (1985).

[27] T. Okamura, K. Kurisu, W. Yamamoto, H. Takano, and M. Nishiyama, *Int. J. Oncol.* **16**, 295 (2000).

[28] R. M. Phillips, A. de la Cruz, R. D. Traver, and N. W. Gibson, *Cancer Res.* **54**, 3766 (1994).

[29] S. S. Pan, G. L. Forrest, S. A. Akman, and L. T. Hu, *Cancer Res.* **55**, 330 (1995).

[30] A. K. Jaiswal, *Free Radic. Biol. Med.* **29**, 254 (2000).

[31] T. Cresteil and A. K. Jaiswal, *Biochem. Pharmacol.* **42**, 1021 (1991).

[32] P. Joseph, T. Xie, Y. Xu, and A. K. Jaiswal, *Oncol. Res.* **6**, 525 (1994).

[33] N. A. Schor and C. J. Cornelisse, *Cancer Res.* **43**, 4850 (1983).

[34] D. Siegel and D. Ross, *Free Radic. Biol. Med.* **29**, 246 (2000).

[35] F. Friedlos, P. J. Biggs, J. A. Abrahamson, and R. J. Knox, *Biochem. Pharmacol.* **44**, 1739 (1992).

[36] R. J. Riley and P. Workman, *Biochem. Pharmacol.* **43**, 1657 (1992).

[37] D. Siegel, J. Ryder, and D. Ross, *Toxicol. Lett.* **125**, 93 (2001).

conventional cytotoxic chemotherapy. Thus NQO1 is currently being exploited as a target in the development of anti-cancer prodrugs, and cells expressing high levels of NQO1 are also sensitive to drug treatment by the cytotoxic anti-tumor quinones, the mitomycins and anthracyclines.[9–11] However, these agents can also be activated by other enzyme systems such that the tumor selectivity conferred by the raised expression of NQO1 can be lost here or masked.[12,38–40]

NQO1 as a Nitroreductase

Although predominantly a quinone reductase, NQO1 can also reduce certain nitrocompounds aerobically, and this gives a potential for highly selective activation by NQO1. Reduction of nitro groups is of central importance in the toxicity of nitro compounds.[41] It is apparent that the reduction of nitro compounds can involve the formation of three reduction products corresponding to 2-, 4- or 6-electron reductions, respectively (Fig. 1). These products can be formed by a series of single electron transfers followed by dis- and co-proportionation reactions. The first step leads to the formation of the nitro radical anion. Nitro groups cannot be reduced in air by conventional one-electron reductases because the product of this reduction, the nitro radical anion, reacts very rapidly with oxygen to regenerate the parent compound (and superoxide) (Fig. 1). Because NQO1 can transfer two-electron equivalents to its substrate, it can bypass this fundamental block to aerobic nitroreduction. However, the nitroreductase properties of NQO1 are much weaker than its quinone reductase ability and are often discounted (Table I). Nevertheless, there is a specific example in which aerobic nitroreduction by NQO1 leads to the formation of a very potent anti-tumor compound in rats.

CB 1954 (5-[Aziridin-1-yl]-2,4-Dinitrobenzamide)

The dinitrobenzamide CB 1954 (5-[aziridin-1-yl]-2,4-dinitrobenzamide) (Fig. 2) represents one of the very few examples of an agent that effects a genuine and demonstrable anti-tumor selectivity where this constitutes a vital requirement for any chemotherapeutic anti-cancer agent.[42] Although structurally only a weak monofunctional alkylating agent toward nucleophiles (by virtue of its single aziridine ring substituent), CB 1954

[38] P. Workman, M. I. Walton, G. Powis, and J. J. Schlager, *Br. J. Cancer* **60,** 800 (1989).
[39] S. M. Bailey, M. D. Wyatt, F. Friedlos, J. A. Hartley, R. J. Knox, A. D. Lewis, and P. Workman, *Br. J. Cancer* **76,** 1596 (1997).
[40] S. Pan, P. A. Andrews, and C. J. Glover, *J. Biol. Chem.* **259,** 959 (1984).
[41] R. P. Mason and J. L. Holtzman, *Biochem.* **14,** 1626 (1975).
[42] D. E. Thurston, *Br. J. Cancer* **80** (Suppl. 1), 65 (1999).

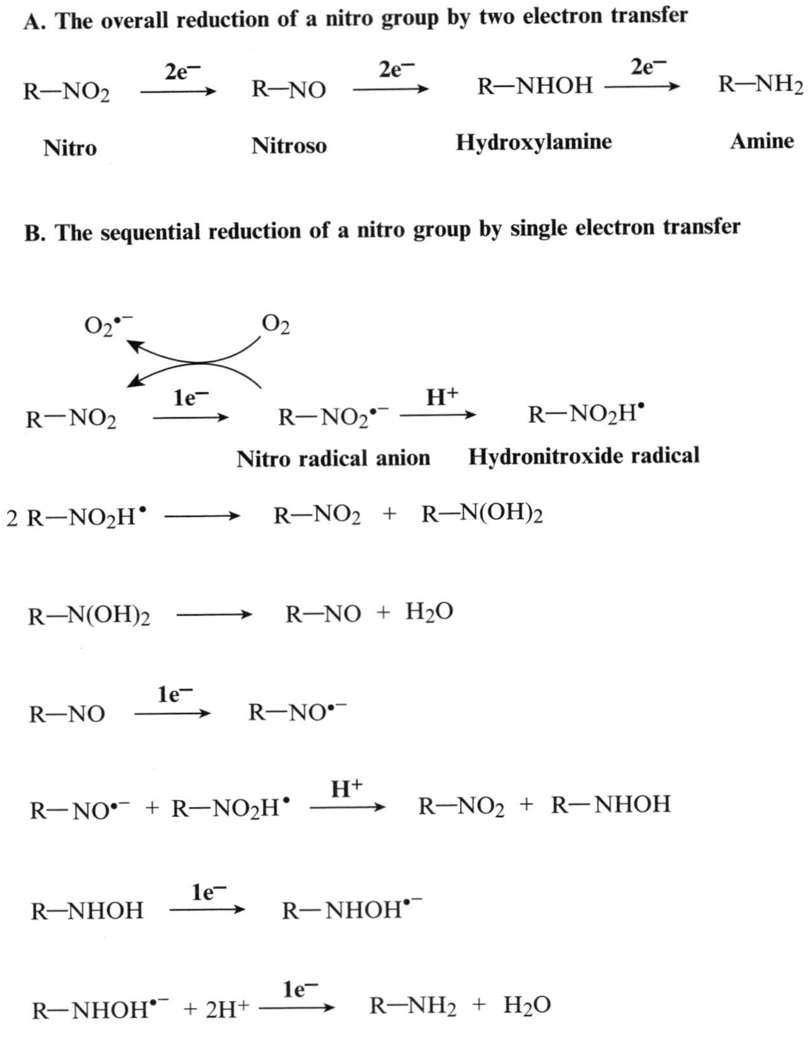

A. The overall reduction of a nitro group by two electron transfer

$$R-NO_2 \xrightarrow{2e^-} R-NO \xrightarrow{2e^-} R-NHOH \xrightarrow{2e^-} R-NH_2$$

Nitro   Nitroso   Hydroxylamine   Amine

B. The sequential reduction of a nitro group by single electron transfer

$$R-NO_2 \xrightarrow{1e^-} R-NO_2^{\bullet -} \xrightarrow{H^+} R-NO_2H^{\bullet}$$

(with $O_2^{\bullet -}$ / O_2 crossover shown above)

Nitro radical anion   Hydronitroxide radical

$$2\ R-NO_2H^{\bullet} \longrightarrow R-NO_2 + R-N(OH)_2$$

$$R-N(OH)_2 \longrightarrow R-NO + H_2O$$

$$R-NO \xrightarrow{1e^-} R-NO^{\bullet -}$$

$$R-NO^{\bullet -} + R-NO_2H^{\bullet} \xrightarrow{H^+} R-NO_2 + R-NHOH$$

$$R-NHOH \xrightarrow{1e^-} R-NHOH^{\bullet -}$$

$$R-NHOH^{\bullet -} + 2H^+ \xrightarrow{1e^-} R-NH_2 + H_2O$$

FIG. 1. Scheme for the reduction of nitro-containing aromatic compounds.

showed a dramatic and highly selective activity against the rat Walker 256 tumor and could actually cure this tumor. Such selectivity was unprecedented for any monofunctional-alkylating agent, and it was evident that the sensitivity of the Walker tumor toward CB 1954 pointed to a unique biochemical feature. The prospect that a human tumor could also be found

TABLE I

COMPARISON OF KINETIC CONSTANTS

	Substrate			
	Menadione		CB 1954	
Enzyme	K_m (μM)	k_{cat} (s^{-1})	K_m (μM)	k_{cat} (s^{-1})
Human NQO2	2.3 ± 0.3	38.3 ± 0.03	263 ± 13	6.01
E. coli NTR	80	700	862 ± 145	6.0
Rat NQO1	2.5 ± 0.1	1200 ± 150	840 ± 40	0.07
Mouse NQO1	4.3 ± 0.4	900 ± 100	1280 ± 80	0.01
Human NQO1	2.7 ± 0.3	550 ± 10	1370 ± 70	0.01

Comparison of the kinetic constants of the quinone reductase enzymes that can also bioactivate CB 1954. NRH was used as a co-substrate for NQO2, while the values for the other enzymes were determined by using NADH.

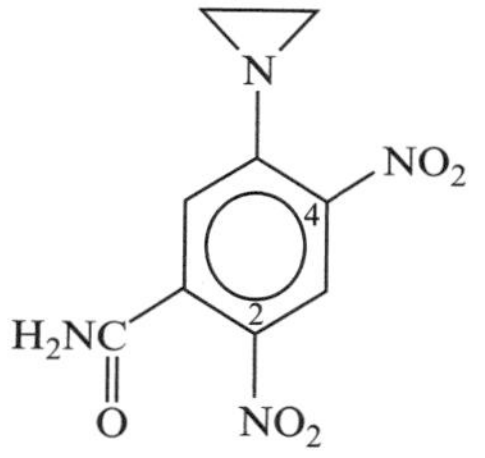

FIG. 2. The structure of CB 1954 (5-[aziridin-1-yl]-2,4-dinitrobenzamide).

that shared the sensitivity of the Walker tumor made the mechanism of action of CB 1954 the subject of continual interest for more than 30 years. Indeed, CB 1954 has been described as "a drug in search of a human tumor to treat."[43]

The Bioactivation of CB 1954 by NQO1

The reason CB 1954 is so selective is that it is a prodrug that is enzymatically activated to form a difunctional agent that can form DNA–DNA interstrand cross-links. The bioactivation of CB 1954 in Walker and other rat cells involves the reduction of its 4-nitro group to form the 4-hydroxylamino derivative 5-(aziridin-1-yl)-4-hydroxylamino-2-nitrobenzamide. This

---

[43] P. Workman, J. E. Morgan, K. Talbot, K. A. Wright, J. Donaldson, and P. R. Twentyman, *Cancer Chemother. Pharmacol.* **16,** 9 (1986).

nitroreductase activity occurs in air in the presence of either NADH or NADPH and was shown to be the enzyme NQO1 (Fig. 3). The dose of CB 1954 required to achieve the same degree of cytotoxicity is about 10,000 times less in cells able to perform this conversion than in cells that cannot, because 5-(aziridin-1-yl)-4-hydroxylamino-2-nitrobenzamide is highly cytotoxic (even to those cells resistant to CB 1954) and can form interstrand cross-links in their DNA. It is the formation of this compound that accounts for the sensitivity of Walker cells toward CB 1954, because all cell lines have a comparable sensitivity toward the reduced 4-hydroxylamino derivative.[44]

Although 5-(aziridin-1-yl)-4-hydroxylamino-2-nitrobenzamide can produce DNA–DNA interstrand cross-links in cells, it cannot form these lesions in naked DNA.[45] There is a further activation step that converts 5-(aziridin-1-yl)-4-hydroxylamino-2-nitrobenzamide to the proximal, DNA cross-linking, cytotoxic species. An enzymatic esterification and activation of the hydroxylamine, analogous to that formed by metabolism of 4-nitroquinoline-N-oxide and N-acetylaminofluorene, was proposed.[45] In fact, 5-(aziridin-1-yl)-4-hydroxylamino-2-nitrobenzamide can be activated non-enzymatically, to a form capable of reacting with naked DNA to produce interstrand cross-links, by a direct chemical reaction with acetyl-coenzyme A and other thioesters[45] (Fig. 3). The ultimate, DNA-reactive derivative of CB 1954 is probably 4-(N-acetoxy)-5-(aziridin-1-yl)-2-nitrobenzamide (Fig. 3). Another product of this reaction with thioesters is 4-amino-5-(aziridin-1-yl)-2-nitrobenzamide, which is the major urinary metabolite of CB 1954 in the rat. Formation of this product is actually in competition with the activation reaction (Fig. 3). The 4-hydroxylamino derivative can react spontaneously with oxygen, and in aqueous solution it yields 5-(aziridin-1-yl)-2-nitro-4-nitrosobenzamide and hydrogen peroxide. Mild biological reducing agents such as NAD(P)H, reduced thiols, and ascorbic acid rapidly re-reduced the nitroso compound to the hydroxylamine (Fig. 3).[46] Both compounds were equally efficient at inducing cytotoxicity and DNA interstrand cross-linking in cells *in vitro*. However the 4-nitroso compound does not react with thioesters. The nitroso compound is acting as a prodrug for the hydroxylamine and needs to be reduced to this compound to exert its cytotoxic effects.[46] Importantly, *in vivo*, neither compound shows any anti-tumor activity.[46] Serum proteins deactivate the 4-hydroxylamine rapidly, and the 4-nitroso is very quickly reduced to the

[44] M. P. Boland, R. J. Knox, and J. J. Roberts, *Biochem. Pharmacol.* **41**, 867 (1991).

[45] R. J. Knox, F. Friedlos, T. Marchbank, and J. J. Roberts, *Biochem. Pharmacol.* **42**, 1691 (1991).

[46] R. J. Knox, F. Friedlos, P. J. Biggs, W. D. Flitter, M. Gaskell, P. Goddard, L. Davies, and M. Jarman, *Biochem. Pharmacol.* **46**, 797 (1993).

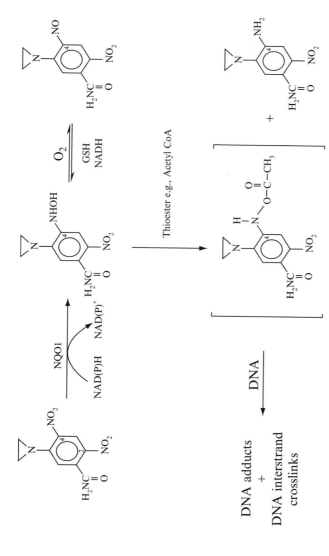

Fig. 3. The bioactivation of CB 1954. The initial step is the reduction of CB 1954 by the enzyme NQO1 to form 5-(aziridin-1-yl)-4-hydroxylamino-2-nitrobenzamide. This hydroxylamine derivative can react with thioesters to produce a DNA reactive species. It is postulated that this is the N-acetoxy derivative. Formation of the amine is in competition with reactions with DNA.

hydroxylamine in the circulation. These factors emphasize the potential value of *in situ* activation of CB 1954. The CB 1954 induced DNA cross-link is also formed with a very high frequency (up to 70%) and is poorly repaired. Therefore the bioactivation results in a compound that is intrinsically more cytotoxic than other difunctional agents.[47]

Given the levels of NQO1 expression in Walker and other rat tumor and tumor-cell lines, these observations explained why they were sensitive toward CB 1954. However, NQO1 is also present in human cells, and it was thus difficult to explain why no CB 1954–sensitive tumors or tumor-cell lines had been described.

Method 1. Identification of the Products From the Reduction of CB 1954 and Their Cytotoxicity

1. Initially, a mixture of [U-3H]CB 1954 (100 μM; 140 $\mu Ci/mM$), NADH (500 μM) and the purified enzyme (50 ng) was incubated in PBS (1 ml) at 37°. After 2 h, a sample (200 μl) was removed and injected onto an ODS-5 reverse-phase HPLC column (Bio-Rad 250 × 4.5 mm) and eluted (1 ml/min) with a methanol gradient (0 to 30% linear over 30 min, 30 to 100% linear over 10 min) in 10 mM sodium acetate buffer (pH 4.5). The eluate was continually monitored for absorbance at 325 nm and absorbance spectra of eluting peaks determined by a PDA detector. Fractions (0.5 minutes) were collected and the tritium activity of each was determined by liquid scintillation counting. After 24 h, a further aliquot was analyzed as above, to assess the effects of oxidation on the sample.

2. Reduction products were identified by incorporation of tritium activity and retention time, and absorbance spectra were compared to synthetic standards. All six reduction products corresponding to 2-, 4- or 6-electron reductions at either the 2- or 4-nitro groups of CB 1954 have been chemically synthesized.[46,48,49] Elution times of the standards are: CB 1954, 23.1 min; 5-(aziridin-1-yl)-2-hydroxylamino-4-nitrobenzamide, 27.2 min; 5-(aziridin-1-yl)-4-hydroxylamino-2-nitrobenzamide, 10.8 min, 2-amino-5-(aziridin-1-yl)-4-nitrobenzamide, 25.4 min; 4-amino-5-(aziridin-1-yl)-2-nitrobenzamide, 13.2 min and 5-(aziridin-1-yl)-2-nitro-4-nitrosobenzamide 29.7 min. Representative absorbtion spectra are shown in Fig. 4.

3. To assess the cytotoxicity of the enzymatic reduction products, CB 1954 (200 μM), NADH (5 mM) and enzyme (100 ng) were incubated in 10 mM sodium phosphate buffer (pH 7) at 37°. After 2 h, an aliquot (500 ul) was injected onto the ODS-5 HPLC column and eluted as

[47] R. J. Knox, F. Friedlos, and M. P. Boland, *Cancer Metastasis Rev.* **12**, 195 (1993).
[48] M. Jarman, D. H. Melzack, and W. C. J. Ross, *Biochem. Pharmacol.* **25**, 2475 (1976).
[49] R. J. Knox, F. Friedlos, M. Jarman, and J. J. Roberts, *Biochem. Pharmacol.* **37**, 4661 (1988).

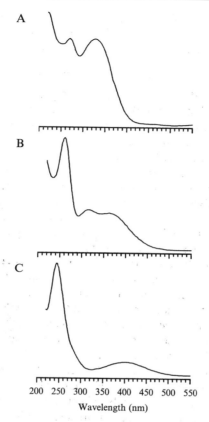

Fɪɢ. 4. The absorbance spectra of (A) CB 1954 and its two major reduction products, (B) the 4-hydroxylamine derivative and (C) the 2-hydroxylamine derivative.

described previously. Samples (1 ml) were collected, individually sterilized by passage through 0.2 um filters, and a proportion (500 μl) added to a culture of rat Walker 256 cells or Chinese hamster V79 (10 ml/2 × 10^5 per ml) which, after a 2-h incubation at 37° were assayed for colony-forming ability. After 24 h, a further aliquot was analyzed as shown previously to assess the effects of oxidation on the sample.

4. For routine assays for CB 1954 reduction and 2- and 4-hydroxylamine formation, aliquots (10 μl) were taken every 6 min and assayed immediately by HPLC (Partisil 10 SCX [4.2 × 150 mm] [Whatman, Maidstone, Kent, U.K.] eluted isocratically with 0.13 M sodium phosphate [pH 5] at 1.5 ml/min). The concentration of CB 1954 and formed products was

determined in each sample by reference to the corresponding peak area with an external standard, quantified by absorbance at 325 nm. Initial rates were calculated by curve fitting (FigP, Biosoft, Cambridge, U.K.).

Human NQO1 Does Not Readily Activate CB 1954

The discovery of NQO1 as the enzyme responsible for the bioactivation of CB 1954 rekindled the possibility of finding comparably sensitive human tumors that also expressed this enzyme. As discussed previously, high NQO1 activity has been reported in human tumor cell lines of various origins and in human tumor biopsy samples. Both the rat and human forms of NQO1 have been cloned and sequenced, and human NQO1 cDNA and proteins are 83% and 85% homologous with the rat liver cytosolic cDNA and protein, respectively.[50,51] Both are inducible cytosolic flavoproteins encoded by a single gene, and the human protein is biochemically very similar to the rat protein, with only small differences between K_m values for the substrates menadione and NADH having been reported at this time.[52] Thus it might be predicted that human or rat NQO1 would metabolize CB 1954 in a manner similar to the protein from Walker cells and that the cytotoxicity resulting from the bioactivation of CB 1954 might be observed in human tumors expressing significant levels of this enzyme.

A number of human cell lines were indeed shown to contain NQO1 levels comparable to those found in the Walker and some other rat cell lines.[44] The rat cell lines were all sensitive to CB 1954, and the resulting cell kill approached that obtained in Walker cells. Thus Walker cells are not uniquely sensitive toward CB 1954. Human cell lines were, on the other hand, all dramatically less toxically affected by CB 1954 with between 500-fold and 5000-fold higher doses of the agent being required to produce a comparable cytotoxic response to that obtained in cells of rat origin.[44] In contrast to the large difference in their cytotoxic response toward CB 1954, the rat and human cell lines were similarly affected by the 4-hydroxylamino derivative of CB 1954.[44] The fact that human cells were sensitive to 5-(aziridin-1-yl)-4-hydroxylamino-2-nitrobenzamide suggested that their resistance toward CB 1954 was not due to any failure to further activate the hydroxylamine nor to an intrinsic resistance to the DNA adducts formed. It appeared that the human form of NQO1 reduced CB 1954 differently to the rat form and the human form of NQO1 was purified to

[50] A. K. Jaiswal, O. W. McBride, M. Adesnik, and D. W. Nebert, *J. Biol. Chem.* **263**, 13572 (1988).
[51] S. Chen, R. Knox, A. D. Lewis, F. Friedlos, P. Workman, P. S. Deng, M. Fung, D. Ebenstein, K. Wu, and T. M. Tsai, *Mol. Pharmacol.* **47**, 934 (1995).
[52] D. Smith, L. F. Martin, and R. Wallin, *Cancer Lett.* **42**, 103 (1988).

homogeneity from Hep G2 cells.[44] Although NQO1 had been extensively studied, the most common source of the enzyme was rat liver, and information regarding the human protein was limited at that time. However, as might be predicted from the large degree of homology between the rat and human forms of NQO1, the biochemical properties of the Hep G2 and Walker forms of the enzyme, with respect to cofactors and the reduction of menadione, were very similar.[44] In contrast, significant differences were observed in the ability of NQO1 isolated from either Hep G2 or Walker cells to reduce CB 1954 to the active 5-(aziridin-1-yl)-4-hydroxyl-amino-2-nitrobenzamide derivative. Although both forms of the enzyme produced the 4-hydroxylamino derivative as the single product, the human Hep G2 form of the enzyme was intrinsically less able to carry out this reduction, and the k_{cat} value is over six-fold higher for the Walker cell form of the enzyme (4.1 min-1) than for the human NQO1 (0.64 min-1).[44] The intrinsic inability of human NQO1 to produce the required cytotoxic species from CB 1954 accounts for the lack of sensitivity of human cells towards this agent. It is interesting to note that it has been shown that the rat enzyme is more effective in activating two important anti-tumor quinones—diaziquone and mitomycin C—than the human enzyme.[9,10] These results indicate that the available quinone drugs are better substrates for the rat enzyme and not the most ideal compounds to treat human cancer.

The Molecular Basis for the Catalytic Difference Between Rat and Human NQO1

By using Escherichia coli–expressed (recombinant) forms of NQO1 and evaluating them under identical conditions, the above findings, concerning the catalytic difference between rat and human NQO1, were confirmed[51] (Table I). Interestingly, although the amino acid sequence of mouse quinone reductase is more homologous to that of the rat enzyme (Fig. 5), the mouse enzyme behaves similarly to the human enzyme in its ability to reduce these compounds and to generate drug-induced DNA damage (Table I).[51] To determine the region of quinone reductase that is responsible for the catalytic differences, two mouse-rat chimeric enzymes were generated. MR-P, a chimeric enzyme that has mouse amino-terminal and rat carboxy-terminal segments of quinone reductase, was shown to have catalytic properties resembling those of rat quinone reductase, and RM-P, a chimeric enzyme that has rat amino-terminal and mouse carboxyl-terminal segments of quinone reductase, was shown to have catalytic properties resembling those of mouse quinone reductase. Thus it was proposed that the carboxyl-terminal portion of the enzyme plays an important role in the reduction of cytotoxic drugs and the binding of flavones.[51] By using site-directed mutagenesis to replace residues in the rat enzyme with

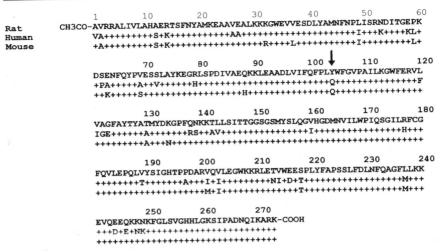

```
              1        10        20        30        40        50        60
Rat     CH3CO-AVRRALIVLAHAERTSFNYAMKEAAVEALKKKGWEVVESDLYAMNFNPLISRNDITGEPK
Human         VA++++++++++S+K++++++++++++AA+++++++++++++++++++++++I+++K++++KL+
Mouse         +A++++++++++S+K+++++++++++++++++R++++L+++++++++++++I+++++++++L+

                       70        80        90       100 ↓     110       120
        DSENFQYPVESSLAYKEGRLSPDIVAEQKKLEAADLVIFQFPLYWFGVPAILKGWFERVL
        +PA+++++A++V++++++H++++++++++++++++++++++Q+++++++++++++++++++F
        ++K+++++S+++++++++++++++++++H++++++++++++++Q+++++++++++++++++

                      130       140       150       160       170       180
        VAGFAYTYATMYDKGPFQNKKTLLSITTGGSGSMYSLQGVHGDMNVILWPIQSGILRFCG
        IGE++++++A+++++++RS++AV++++++++++++++++I+++++++++++++++++H+++
        +++++++++A+++N++++++++++++++++++++++++++++++++++++++++++++++

                      190       200       210       220       230       240
        FQVLEPQLVYSIGHTPPDARVQVLEGWKKRLETVWEESPLYFAPSSLFDLNFQAGFLLKK
        +++++++++T++++++A+++I+I++++++++++NI+D+T++++++++++++++++M+++
        ++++++++++++++++++M+I++++++++++++++T++++++++++++++++++M+++

                      250       260       270
        EVQEEQKKNKFGLSVGHHLGKSIPADNQIKARK-COOH
        +++D+E+NK++++++++++++++++++++++++
        +++++++++++++++++++++++++++++++++
```

FIG. 5. The amino acid sequences of NQO1 from rat, human and mouse origin. For the human and mouse sequences, only those residues that differ from the rat are shown. The impotant residue at position 104 (tyrosine in the rat, glutamine in human and mouse) is indicated.

the human sequences and residues in the human enzyme with the rat sequences, residue 104 (tyrosine in the rat enzyme and glutamine in the human and mouse enzymes) (Fig. 5) was found to be the important residue responsible for the catalytic differences between the rat and the human (and mouse) enzymes. With an exchange of a single amino acid, the rat mutant Y104Q behaved like the wild-type human enzyme, and the human mutant Q104Y behaved like the wild-type rat enzyme in their ability to reductively activate the cytotoxic drug CB 1954 both *in vitro*[53] and *in vivo*.[54] The x-ray structure of rat NQO1 reveals that the phenolic ring of Tyr-104, together with the main chain carbonyl of Trp-103, provides the main interactions with the "bottom face" of the isoalloxazine ring (the face opposite to the one that interacts with substrate and nicotinamide) of FAD.[55] In addition, a water molecule hydrogen bonded to the OH of Tyr-104 is hydrogen bonded with the O3′ and with a phosphate oxygen of FAD. For mouse or human NQO1, replacing Tyr-104 by a Gln

[53] S. Chen, R. Knox, K. Wu, P. S. K. Deng, D. Zhou, M. A. Bianchet, and L. M. Amzel, *J. Biol. Chem.* **272**, 1437 (1997).

[54] K. Wu, E. Eng, R. Knox, and S. Chen, *Arch. Biochem. Biophys.* **385**, 203 (2001).

[55] R. Li, M. A. Bianchet, P. Talalay, and L. M. Amzel, *Proc. Natl. Acad. Sci. USA* **92**, 8846 (1995).

manifests itself predominantly through effects in the positioning of the iso-alloxazine ring.[56] The change in size of the side chain allows the flavin to move deeper into the protein (by approximately 0.7 Å). Such a change may modify the rate of electron transfer between FAD and the substrate CB 1954. These studies indicate that through site-directed mutagenesis, a human NQO1 mutant can be generated (by changing one amino acid) that is as effective as the rat enzyme in the reduction of CB 1954.

To investigate whether the resistance of human tumor cells to CB 1954 could be fully accounted for by the properties of human NQO1, a cell-line panel consisting of V79 cells that had been engineered to express either the human or rat forms of NQO1 was made. Chinese hamster V79 cells have practically no measurable NQO1 activity[57] and thus provide a suitable null background. A control cell line was also constructed with the empty ex-pression vector. Thus the panel consisted of cell lines expressing various levels of either rat or human NQO1 in an identical cellular background. The sensitivity of these various cell lines toward CB 1954 was determined. On the basis of IC_{50} values, the cytotoxic effect of CB 1954 is proportional to the activity of either the rat or human enzyme.[58] The cells expressing the rat enzyme were more sensitive than cells expressing the human enzyme at a comparable level of quinone reductase (NADH-dependant menadione oxidoreductase [NMOR] activity). At the high levels of NMOR activity that have been measured in tumor cell lines (about 20,000 U/mg cytosolic protein),[44] there is a 10,000-fold difference in the concentration of CB 1954 required to produce the same cytotoxic response in cells expressing the rat as opposed to the human form of NQO1 (Fig. 6).[58] These results demon-strate that the resistance of human tumors to CB 1954 can be accounted for solely by the kinetic properties of the enzyme for this prodrug.

The Use of CB 1954 in Targeted Therapies

The intrinsic poor reduction rate of CB 1954 by the human form of NQO1 would seem to exclude CB 1954 from a role in the treatment of humans. However, because human cells do not readily activate CB 1954 but CB 1954 has proven anti-tumor activity once activated, CB 1954 would appear to be an ideal prodrug in targeted cancer therapies that involve introducing a prodrug-activating enzyme at a tumor site. For these applica-tions, an *E. coli* B nitroreductase (NTR) is used that can activate CB 1954 about 100-fold more rapidly than rat NQO1 (Table I). Like NQO1, NTR is

[56] M. Faig, M. A. Bianchet, P. Talalay, S. Chen, S. Winski, D. Ross, and L. M. Amzel, *Proc. Natl. Acad. Sci. USA* **97,** 3177 (2000).

[57] F. Friedlos, J. Quinn, R. J. Knox, and J. J. Roberts, *Biochem. Pharmacol.* **43,** 1249 (1992).

[58] L. K. Mehta, S. Hobbs, S. Chen, R. J. Knox, and J. Parrick, *Anticancer Drugs* **10,** 777 (1999).

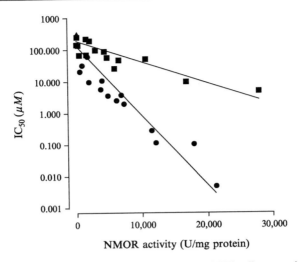

FIG. 6. The sensitivities of individual clones of transfected V79 cells expressing either rat or human DT-diaphorase to CB 1954. Cells expressing human DT-diaphorase (■) or rat DT-diaphorase (●). NQO1 activity is measured by menadione reduction (NMOR) activity as units of nmoles cytochrome C reduced per minute. IC$_{50}$ values are for a 72 h exposure to CB 1954.

also capable of reducing CB 1954 in air to 5-(aziridin-1-yl)-4-hydroxylamino-2-nitrobenzamide by using either NADH or NADPH as co-substrate. In contrast to NQO1, which can only reduce the 4-nitro group of CB 1954, the E. coli nitroreductase (NTR) can reduce either (but not both) nitro groups of CB 1954 to the corresponding hydroxylamino species. The two hydroxylamino species are formed in equal proportions and at the same rates.[59] However, no products are formed in which both nitro groups have been reduced. Thus, once one nitro group has been reduced, NTR cannot then reduce the other nitro group. 5-(Aziridin-1-yl)-4-hydroxylamino-2-nitrobenzamide is a potent cytotoxic agent capable of producing DNA-DNA interstrand crosslinks in cells. In contrast the 2-hydroxylamino species is less cytotoxic and cannot produce interstrand cross-links.[49] As well as being able to reduce CB 1954, NTR shares some other biochemical properties with NQO1. Like NQO1, NTR is also a quinone reductase (Table I) and utilizes either NADH or NADPH and other reduced nicotinamide analogs as cofactors.[60,61] However, it is a much smaller protein

[59] R. J. Knox, F. Friedlos, R. F. Sherwood, R. G. Melton, and G. M. Anlezark, Biochem. Pharmacol. 44, 2297 (1992).
[60] G. M. Anlezark, R. G. Melton, R. F. Sherwood, B. Coles, F. Friedlos, and R. J. Knox, Biochem. Pharmacol. 44, 2289 (1992).

(24 kD) than NQO1 (33.5 kD), and there is no obvious sequence homology between the two enzymes.[60] Further, NTR contains a FMN prosthetic group while NQO1 contains FAD.

The major application for this enzyme is in prodrug gene therapy commonly referred to as virus-directed enzyme prodrug therapy (VDEPT) or GDEPT (gene-directed enzyme prodrug therapy) and is based on the premise that a large therapeutic benefit can be gained by transferring a drug susceptibility gene into tumor cells by using gene therapy technology. The gene encodes an enzyme that can catalyze the activation of a prodrug to its cytotoxic form. However, in contrast to cytotoxic gene therapy approaches that involve expression of a toxic product (for example, diphtheria toxin), the enzyme itself is not toxic and cytotoxicity only results after administration of the prodrug.

Human tumor cells transduced to express NTR are very sensitive to CB 1954.[62–68] Thus, engineered delivery of the NTR gene into tumor cells, with some degree of specificity, should result in a differential cancer cell kill after administration of CB 1954. Importantly, a strong bystander effect operated, and with only 5% of cells transfected, high levels of total cell kill (>90%) were observed.[66] However, success for this technique requires not only a choice of an enzyme/prodrug system but also a delivery system by which the gene encoding for the enzyme can be delivered efficiently and accurately to a human tumor. Gene delivery can be achieved by using viruses, particularly adenovirus. *In vivo* models demonstrated useful efficacy.[66,68] In a model of peritoneal carcinomatosis using the pancreatic SUIT-2 cell line in nude mice, NTR delivered by a replication–deficient adenovirus followed by CB 1954, both administered by the IP route,

[61] R. J. Knox, F. Friedlos, M. Jarman, L. C. Davies, P. Goddard, G. M. Anlezark, R. G. Melton, and R. F. Sherwood, *Biochem. Pharmacol.* **49,** 1641 (1995).

[62] J. A. Bridgewater, C. J. Springer, R. J. Knox, N. P. Minton, N. P. Michael, and M. K. Collins, *Eur. J. Cancer* **31a,** 2362 (1995).

[63] S. M. Bailey, R. J. Knox, S. M. Hobbs, T. C. Jenkins, A. B. Mauger, R. G. Melton, P. J. Burke, T. A. Connors, and I. R. Hart, *Gene Ther.* **3,** 1143 (1996).

[64] J. A. Bridgewater, R. J. Knox, J. D. Pitts, M. K. Collins, and C. J. Springer, *Hum. Gene Ther.* **8,** 709 (1997).

[65] N. K. Green, D. J. Youngs, J. P. Neoptolemos, F. Friedlos, R. J. Knox, C. J. Springer, G. M. Anlezark, N. P. Michael, R. G. Melton, M. J. Ford, L. S. Young, D. J. Kerr, and P. F. Searle, *Cancer Gene Ther.* **4,** 229 (1997).

[66] I. A. McNeish, N. K. Green, M. G. Gilligan, M. J. Ford, V. Mautner, L. S. Young, D. J. Kerr, and P. F. Searle, *Gene Ther.* **5,** 1061 (1998).

[67] J. A. Plumb, A. Bilsland, R. Kakani, J. Zhao, R. M. Glasspool, R. J. Knox, T. R. Evans, and W. N. Keith, *Oncogene* **20,** 7797 (2001).

[68] A. E. Bilsland, C. J. Anderson, A. J. Fletcher-Monaghan, F. McGregor, T. R. Jeffry Evans, I. Ganly, R. J. Knox, J. A. Plumb, and W. Nicol Keith, *Oncogene* **22,** 370 (2003).

doubled median survival of treated mice.[69] The notion of using a virus to deliver the gene encoding an enzyme that can convert a relatively inactive prodrug to a potent anticancer agent is not new and was pioneered by the use of the enzymes thymidine kinase or cytosine deaminase to activate the prodrugs ganciclovir and 5-fluorocytosine, respectively.[70] However, the mechanism of action of activated CB 1954 makes its combination with NTR a potentially powerful addition to the VDEPT stable.[71]

This system has entered phase I clinical trial, initially with the virus and CB 1954 being administered as single agents, but with an intention to combine them when adequate tumor expression of NTR was documented after intra-tumoral injection. The phase I and pharmacokinetic study of CB 1954 has been completed and the results are summarized later.

An interesting use for NTR and CB 1954 is in selective cell ablation in transgenic animals. Conditional targeted ablation of specific cell populations in living transgenic animals is a very powerful strategy to determine cell functions *in vivo*. Targeted ablation is achieved by constructing a transgene incorporating NTR and appropriate tissue-specific promoters and injecting this into the fertilized eggs of the animal under study—normally mice. After birth, genomic integration of the transgene is confirmed and founder mice bred to establish the transgenic lines. The animals are treated with CB 1954 at various stages of their development to assess the effect of the specific ablation of the cell population being studied. Cell ablation occurs very rapidly, starting as early as 7 h after administration of the prodrug, and appears to be independent of a functional p53.[72] Examples of the use of this system include the luminal cells of the mammary gland of transgenic mice. Treatment of NTR-expressing animals resulted in a rapid and selective killing of this population of cells, whereas the closely associated myoepithelial cells were unaffected. Other examples of selective ablation with this system have been observed in adipocytes,[73] astrocytes[74] and neurones.[75,76]

[69] S. J. Weedon, N. K. Green, I. A. McNeish, M. G. Gilligan, V. Mautner, C. J. Wrighton, A. Mountain, L. S. Young, D. J. Kerr, and P. F. Searle, *Int. J. Cancer* **86**, 848 (2000).

[70] R. J. Knox, Gene-directed Enzyme Prodrug Therapy (GDEPT) of Cancer, *in* "Enzyme Prodrug Strategies for Cancer Therapy" (R. G. Melton and R. J. Knox, eds.), p. 209. Kluwer Academic/Plenum Press, New York, 1999.

[71] T. A. Connors, *Gene Therapy* **2**, 1 (1995).

[72] W. Cui, B. Gusterson, and A. J. Clark, *Gene Ther.* **6**, 764 (1999).

[73] R. Felmer, W. Cui, and A. J. Clark, *J. Endocrinol.* **175**, 487 (2002).

[74] W. Cui, N. D. Allen, M. Skynner, B. Gusterson, and A. J. Clark, *Glia* **34**, 272 (2001).

[75] A. R. Isles, D. Ma, C. Milsom, M. J. Skynner, W. Cui, J. Clark, E. B. Keverne, and N. D. Allen, *J. Neurobiol.* **47**, 183 (2001).

[76] D. Ma, N. D. Allen, Y. C. Van Bergen, C. M. Jones, M. J. Baum, E. B. Keverne, and P. A. Brennan, *Eur. J. Neurosci.* **16**, 2317 (2002).

CB 1954 in the Clinic

CB 1954 was entered into an experimental trial in humans in 1970 owing to its high therapeutic index when tested against the Walker tumor *in vivo*, despite its limited effectiveness against some other experimental tumors. No tumor regression was observed in any of the treated patients. The most severe side effect of treatment was diarrhea, with no bone marrow toxicity or liver dysfunction being observed. It is now clear from previous work that CB 1954 will never be effective against human tumors if activation by endogenous NQO1 is relied on. However, the clinical results have shown that CB 1954 can be safely administered to humans.

As part of a gene therapy trial, a formal dose-escalation phase I trial of single agent CB 1954 has recently been completed with identification of a maximum tolerated dose of 37.5 mg/m^2 and dose limiting hepatotoxicity.[77] Intriguingly, there were signs of anticancer activity in one patient. The hepatotoxicity observed was an asymptomatic reversible transaminitis that was possibly associated with the drug. The link is not definite because those patients had underlying progressive metastatic hepatic cancer. In this trial, CB 1954 was administered by intravenous injection every 3 weeks, or intraperitoneally followed by 3-weekly intravenous injections, toward a maximum of six cycles. Thirty patients, 23 to 78 years of age (median 62 years), with predominantly gastrointestinal malignancies, were treated. The dose was escalated from 3 to 37.5 mg/m^2. No significant toxicity was seen until 24 mg/m^2. Dose-limiting toxicities (DLT) were diarrhea and hepatic toxicity, seen at 37.5 mg/m^2. There was no alopecia, marrow suppression or nephrotoxicity. Less than 5% of CB 1954 was renally excreted. It was concluded that CB 1954 is well tolerated and that sufficient serum levels are achieved for an enzyme-prodrug approach to be feasible.

NAD(P)H Quinone Oxidoreductase 2 (NQO2)

There is an additional endogenous CB 1954-reducing enzyme in human tumor cells, and its activity is much greater than that attributable to NQO1.[78,79] However, this activity is latent and only detectable in the presence of dihydronicotinamide riboside (NRH)[79] (Fig. 7) and not in the

[77] G. Chung-Faye, D. Palmer, D. Anderson, J. Clark, M. Downes, J. Baddeley, S. Hussain, P. I. Murray, P. Searle, L. Seymour, P. A. Harris, D. Ferry, and D. J. Kerr, *Clin. Cancer Res.* **7**, 2662 (2001).

[78] K. Wu, R. Knox, X. Z. Sun, P. Joseph, A. K. Jaiswal, D. Zhang, P. S. Deng, and S. Chen, *Arch. Biochem. Biophys.* **347**, 221 (1997).

[79] R. J. Knox, T. C. Jenkins, S. M. Hobbs, S. Chen, R. G. Melton, and P. J. Burke, *Cancer Res.* **60**, 4179 (2000).

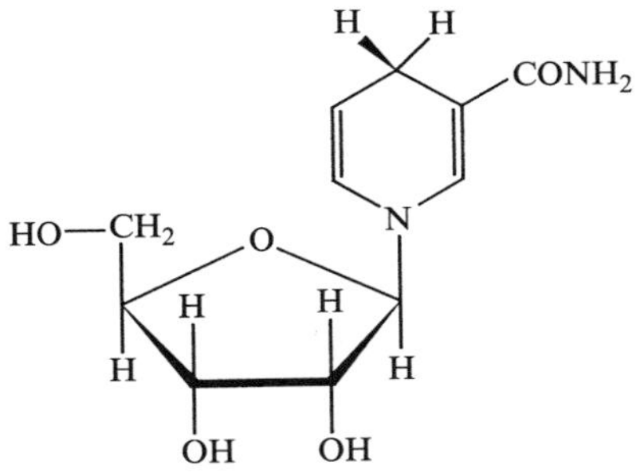

FIG. 7. Dihydronicotinamide riboside (NRH), a co-substrate for human NAD(P)H quinone oxidoreductase2 (NQO2). The biogenic cofactors NADH and NAD(P)H do not act with this enzyme.

presence of either NADH or NADPH. The enzyme responsible for this activity is human NAD(P)H quinone oxidoreductase2 (NQO2).[78,79] NQO2 was originally identified by its homology to NQO1 (NQO1).[80] The last exon in the NQO2 gene is 1603 base pairs shorter than the last exon of the NQO1 gene and encodes for 58 amino acids as compared to 101 amino acids encoded by the NQO1 gene. This makes the NQO2 protein 43 amino acids shorter than the NQO1 protein (Fig. 8). The high degree of conservation between NQO2 and NQO1 gene organization and sequence confirmed that the NQO2 gene encoded for a second member of the NQO gene family in humans. However, it lacked the quinone-reductase activity characteristic of NQO1 and appeared to have little enzymatic activity.[80] This apparent lack of activity is because NQO2 uses dihydronicotinamide riboside (NRH), not NAD(P)H, as an electron donor—a novel and unique property. In the presence of NRH, NQO2 can catalyze the two-electron reduction of quinones and the four-electron nitroreduction of CB 1954.[78] However, even in the presence of NRH, the quinone reductase activity of NQO2 is much less than that of NQO1 (Table I), and NQO2 can be considered a human NRH-dependent nitroreductase. Between human NQO1 and NQO2, the region of amino acid residues 94 to 115 is highly conserved (Fig. 8). The only difference in this region is a Gln at residue 104 in human NQO1 and a Tyr in human NQO2. Residue 104 of rat NQO1 is Tyr, like human NQO2. As indicated previously, residue 104 has been shown to play a critical role in the reduction of CB 1954. By a comparison of the k_{cat}/K_m values, human NQO2 was found to be 3000 times more effective than human NQO1 in reducing CB 1954, supporting the role of residue 104

[80] A. K. Jaiswal, J. Biol. Chem. 269, 14502 (1994).

```
              1        10        20        30        40        50        60
NQO1  CH3CO-VGRRALIVLAHSERTSFNYAMKEAAAAALKKKGWEVVESDLYAMNFNPIISRKDITGKLK
NQO2        AGKKVLIVYAHQEPKSFNGSLKNVAVDELSRQGCTVTVSDLYAMNFEPRATDKDITGTLS


              70        80        90       100       110       120
      DPANFQYPAESVLAYKEGHLSPDIVAEQKKLEAADLVIFQFPLQWFGVPAILKGWFERVF
      NPEVFNYGVETHEAYKQRSLASDITDEQKKVREADLVIFQFPLYWFSVPAILKGWMDRVL


             130       140       150       160       170       180
      IGEFAYTYAAMYDKGPFRSKKAVLSITTGGSGSMYSLQGIHGDMNVILWPIQSGILHFCG
      CQGFAFDIPGFYDSGLLQGKLALLSVTTGGTAEMYTKTGVNGDSRYFLWPLQHGTLHFCG


             190       200       210       220       230       240
      FQVLEPQLTYSIGHTPADARIQILEGWKKRLENIWDETPLYFAPSSLFDLNFQAGFMLKK
      FKVLAPQISFAPEIASEEERKGMVAAWSQRLQTIWKEEPIPCTAHWHFGQ


             250       260       270
      EVQDEEKNKKFGLSVGHHLGKSIPADNQIKARK-COOH
```

FIG. 8. Amino acid sequences of human NQO1 and NQO2. A very conserved region (residues 94–115) between the two enzymes is underlined.

The Physiological Role of NQO2

The physiological role of NQO2, particularly in humans, is not known. An NRH-metabolizing activity described in bovine kidney in the early 1960s[82,83] has now been ascribed to NQO2,[84] and it was postulated that the enzyme functioned in a salvage pathway oxidizing unusable NRH into nicotinamide. In theory, NRH can be generated in cells by reaction of NADH or NADPH with phosphodiesterases and phosphatases. This reaction does occur in serum,[85] but NRH is not detected in cells. Further, in cell lines, there was no effect on CB 1954 cytotoxicity unless a co-substrate was also administered. These data confirm that endogenous co-substrates for the NQO2-catalyzed reduction of CB 1954 are not present. It is possible that both the physiological electron donor and substrate for this enzyme

in the reduction of CB 1954. In this respect, NQO2 resembles NTR (Table I) but, like NQO1, it selectively generates only the 4-hydroxylamine derivative.[81]

[81] R. J. Knox, R. G. Melton, S. K. Sharma, and P. J. Burke, *Br. J. Cancer* **83** (Suppl. 1), 70 (2000).
[82] S. Liao, J. T. Dulaney, and H. G. Williams-Ashman, *J. Biol. Chem.* **237**, 2981 (1962).
[83] S. Liao and H. G. Williams-Ashman, *Biochem. Biophys. Res. Commun.* **4**, 208 (1961).
[84] Q. Zhao, X. L. Yang, W. D. Holtzclaw, and P. Talalay, *Proc. Natl. Acad. Sci. USA* **94**, 1669 (1997).
[85] F. Friedlos and R. J. Knox, *Biochem. Pharmacol.* **44**, 631 (1992).

have yet to be identified or that NQO2 does not function physiologically as an oxidoreductase. In this respect, porcine NQO2 has been identified as a puromycin aminonucleoside-binding protein.[86] A melatonin receptor (MT [3]) from Syrian hamster kidney was identified as the hamster homolog of the human NQO2.[87,88] Melatonin is a hormone synthesized at night in the pineal gland that regulates the circadian rhythm. However, melatonin does not inhibit the CB 1954–reducing activity of NQO2 (Knox, unpublished data), and thus the two activities are separate.

To examine the *in vivo* role of NQO2, NQO2-null mice were generated by using targeted gene disruption.[89] Mice lacking NQO2 gene expression showed no detectable developmental abnormalities and were indistinguishable from wild-type mice. However, they exhibited myeloid hyperplasia of the bone marrow and increased neutrophils, basophils, eosinophils and platelets in the peripheral blood, probably because of decreased apoptosis of bone marrow cells. NQO2-null mice also demonstrated decreased toxicity when exposed to menadione, or menadione with NRH,[89] contrary to the perceived function of a quinone reductase.

Thus the only established role for NQO2 is in protection against myelogenous hyperplasia and in metabolic activation of menadione, leading to hepatic toxicity.

Functional Characterization of Truncated NQO1 and NQO1/NQO2 Chimeric Enzyme Preparations

The carboxyl-terminal portion of NQO1 can be cleaved by various proteases, whereas the amino-terminal portion is not. This proteolytic digestion of the enzyme can be blocked by the prosthetic group FAD, substrates NAD(P)H and menadione, and the inhibitors dicoumarol and phenidone.[90] The results indicate that the subunit of NQO1 has a two-domain structure—that is, an amino-terminal compact domain and a carboxyl-terminal flexible domain (∼43 amino acids) and that the binding of substrates involves an interaction between two structural domains.[90] NQO2 is 43 amino acids shorter than NQO1. Residues situated in the

[86] T. Kodama, H. Wakui, A. Komatsuda, H. Imai, A. B. Miura, and Y. Tashima, *Nephrol. Dial. Transplant* **12,** 1453 (1997).

[87] O. Nosjean, M. Ferro, F. Coge, P. Beauverger, J. M. Henlin, F. Lefoulon, J. L. Fauchere, P. Delagrange, E. Canet, and J. A. Boutin, *J. Biol. Chem.* **275,** 31311 (2000).

[88] O. Nosjean, J. P. Nicolas, F. Klupsch, P. Delagrange, E. Canet, and J. A. Boutin, *Biochem. Pharmacol.* **61,** 1369 (2001).

[89] I. D. Long, K. Iskander, A. Gaikwad, M. Arin, D. R. Roop, R. Knox, R. Barrios, and A. K. Jaiswal, *J. Biol. Chem.* **277,** 46131 (2002).

[90] S. Chen, P. S. Deng, J. M. Bailey, and K. M. Swiderek, *Protein Sci.* **3,** 51 (1994).

carboxyl terminus of NQO1 appear to be involved in the binding of NADH, and x-ray analysis of rat NQO1 revealed that the ribose of the AMP moiety in NAD(P)H is in contact with Phe-232 and Phe-236 that are situated in the carboxy terminus (Fig. 8).[55] To better understand the basis for the catalytic differences between NQO2 and NQO1, a human NQO2 with an additional 43 amino acids from the carboxyl terminus of human NQO1 (i.e., hNQO2-hDT43) was prepared. However, HNQO2-hDT43 still used NRH as an electron donor.[91] Also, human NQO1 without the 43 amino acids from the carboxyl terminus (Δ43-NQO1) could still use NADH as the electron donor, although at a much lower rate.[91] Kinetic analysis of these enzyme forms indicates that the C-terminal domain of NQO1 is important for enzyme catalysis and is more important for NADH oxidation than NRH oxidation. Deletion of the C-terminal domain (i.e., Δ43-NQO1) reduced the binding affinity for both NADH and NRH (as indicated by an increase of K_m values), and the K_m value of NADH for Δ43-NQO1 was found to be similar to that for NQO2. Deletion of the C-terminal domain reduced the rate of NADH oxidation more than NRH oxidation. HNQO2-hDT43 was found to have relatively similar NRH reductase activity to NQO2. Furthermore, the K_m value of NADH for hNQO2-hDT43 increased significantly more than the K_m value of NRH when compared with the values for NQO2, suggesting that an additional region(s) is required for the binding of NADH.[91] However, although the C-terminal domain is important for enzyme catalysis, an additional region(s) is probably involved in differentiating the binding of NADH vs. NRH to NQO2.[91]

Bioactivation of CB 1954 by NQO2

Chinese Hamster V79 cells were transfected with a bicistronic vector coding for the NQO2 protein and puromycin resistance. Seven puromycin-resistant clones (designated TM1 to TM7) were selected at random and their sensitivity to CB 1954 determined by the SRB assay for cell growth in the presence or absence of 100 μM NRH. This concentration of NRH has been shown to be totally non-cytotoxic. Non-transfected V79 cells were used as a control. Expression of NQO2 had no effect on the cytotoxicity of CB 1954 alone against these cell lines.[81] All the transfected cell lines had a similar sensitivity to non-transfected V79 cells, with an IC_{50} value of about 250 μM after a 72-h continuous exposure to CB 1954. The addition of NRH did not increase the sensitivity of non-transfected V79 cells to CB 1954, but all the transfected cell lines showed a very large

[91] S. Chen, K. Wu, and R. Knox, *Free Radic. Biol. Med.* **29**, 276 (2000).

increase in sensitivity toward this prodrug (Fig. 9). Based on IC_{50} values, this increased sensitivity ranged from 100-fold (TM7) to >3000-fold (TM6).[81] Such data show the latent nature of NQO2 and its potential to activate CB 1954 in the presence of NRH.

Activation of NQO2 by NRH produces a dramatic increase in the cytotoxicity of CB 1954 in a number of non-transfected human tumor cell lines (Fig. 10).[81] The cytotoxicity data indicate that NQO2 is expressed in human tumor cells and its cellular activity can be measured biochemically.[81]

Although giving excellent experimental results, NRH is expensive and time-consuming to synthesize. However, simple reduced pyridinium derivatives can, like NRH, act as co-substrates for NQO2. The simplest quaternary (and therefore reducible) derivative of nicotinamide, 1-methylnicotinamide, was, in its reduced form, a co-substrate for NQO2 with a specific activity about 30% that of NRH. Increased chain length and bulk at the 1-position of the nicotinamide ring increased specific activity, and compounds more active than NRH were found.[81] However, there was a limit to this effect, and little activity was seen with either the phenyl or benzyl derivatives. These findings suggest that the co-substrate binding site of NQO2 is

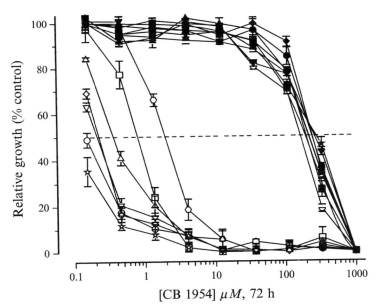

FIG. 9. The sensitivity of transfected V79 cell lines expressing NQO2 to either CB 1954 (solid symbols) or CB 1954 and 100 μM NRH (open symbols). Untransfected V79 (▲); clone TM1 (■); TM2 (●); TM3 (▲); TM4 (◆); TM5 (▼); TM6 (★); TM7 (●); drug exposure was 72 h and relative growth was measured by the SRB assay.

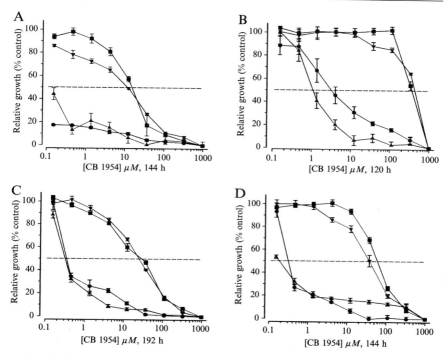

FIG. 10. The sensitivity of human tumor cell lines to either CB 1954 (■); CB 1954 and 100 μM NRH (▲); CB 1954 and 100 μM EP-0152R (●). CB 1954 and 100 μM EP-1017R (▼). (A) PC-3; (B) U87-MG; (C) U373-MG; (D) T98G. Drug exposures were as indicated on each panel. There is no cytotoxicity with either NRH or EP-0152R alone (not shown). Relative growth was measured by the SRB assay.

sterically constrained. This would explain why neither NADH nor NADPH are effective co-substrates for this enzyme. Little enzyme activity was seen with any nicotinic acid derivatives. This would suggest that a negative charge at the 3-position on the pyridine ring is also poorly tolerated in the enzyme's binding site. Because of its stability, ease of synthesis and lack of acute toxicity, 1-carbamoylmethyl-3-carbamoyl-1,4-dihydropyridine (now designated EP-0152R) (Fig. 11) has been selected for further development as an NRH substitute.[81] Results show EP-0152R potentiated the effect of CB 1954 in human cancer cell lines to about the same extent as NRH (Fig. 10). This can be compared with another NQO2 co-substrate, the 1-(3-sulfonatopropyl)-derivative, that had no significant effect on the cytotoxicity of CB 1954 (Fig. 10). The compound is negatively charged at pH7, and this charge prevents it from entering cells. Lack of cellular permeability of this co-substrate probably accounts for its inability to potentiate the cytotoxicity of CB 1954,

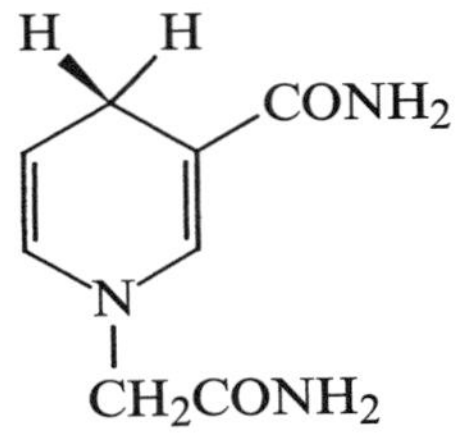

Fig. 11. The structure of EP-0152R (1-carbamoylmethyl-3-carbamoyl-1,4-dihydropyridine). This compound is a co-substrate for the enzyme NQO2.

because both NRH and EP-0152R are neutral at pH7 and are able to cross membranes into cells.[79] EP-0152R was not cytotoxic to any cell line up to 10 mM and no acute toxicity occurred when it was administered to mice. CB 1954 is an effective anti-tumor agent when given with EP-0152R *in vivo* against experimental human tumor xenografts.[81] The use of CB 1954 and EP-0152R is not a true combination therapy, because the co-substrate is present only to act as an electron donor for NQO2. In cell culture there is no advantage in escalating the co-substrate concentration past about 50 μM, because this concentration of cofactor is sufficient for NQO2 to reduce enough CB 1954 to kill the cell. In a patient, this could mean there is no requirement to dose escalate past this threshold and no need to go to the MTD of the co-substrate.

Biodistribution of NQO2 in Humans

The previous data illustrate the potential cytotoxicity resulting from the activation of CB 1954 by NQO2. However, for this to have a clinical application in cancer therapy, there would need to be a favorable biodistribution of NQO2 in humans such that CB 1954 is activated predominately in tumor tissue and not by sensitive normal tissues.

NQO2 is significantly raised in clinical samples obtained from patients suffering from hepatoma and colorectal tumors. Importantly, known chemosensitive tissues have relatively low levels of the enzyme. This would suggest that NQO2-based therapies could have a very favorable therapeutic index. The biochemical assay for the enzyme that measures reduction of CB 1954 to the cytotoxic hydroxylamine has demonstrated that NQO2 levels are very much (6–10x) higher in hepatoma samples obtained from patients when compared with either their matched normal liver or normal donor liver samples (Fig. 12). In a panel of matched human colorectal samples, 50% of the tumor samples had a significantly higher NQO2 level (>3x) than the matched normal tissue. Enzyme levels in hepatic

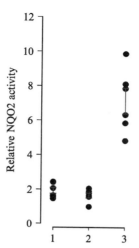

FIG. 12. The activity of the enzyme NQO2 in clinical samples of human liver. (1) Normal donor liver; (2) normal patient liver; (3) matched hepatoma. Enzyme activity was determined by a biochemical assay measuring the reduction of CB 1954 to its 4-hydroxylamine derivative.

metastases of colorectal origin were uniformly high. NQO2 levels in bone marrow stem cells were low, as were levels in white, red, or leukemia blood cells.[92] The actual amount of enzyme present is substantial, and in hepatoma samples it can represent about 0.1% of the total cytosolic protein.

Given the provenance of CB 1954, the apparently favorable distribution of NQO2 in certain tumor types and the lack of acute toxicity of EP-0152R, NQO2 represents a novel target for anti-tumor prodrug therapy. It has a unique activation mechanism relying on a synthetic co-substrate to activate an apparently latent enzyme. A phase I clinical trial combining CB 1954 with EP-0152R is due to start in 2004 under the auspices of Cancer Research UK.

Method 2. Determination of NQO2 in Human Tumor and Tissue Samples

1. Excised tissue or tumors were immediately weighed, then stored at $-80°$ until required.

2. A sample ($\sim$1g) was defrosted into 1 ml of lysis buffer (1% NP-40, 1% aprotinin in PBS) per g of tissue on ice over 60 min. The sample of tumor or normal tissue was disrupted with 4 to 5 strokes of a homogenizer and the homogenate cleared by centrifugation at 13,000g (3 min).

---

[92] R. J. Knox, P. J. Burke, S. Chen, and D. J. Kerr, *Curr. Pharm. Design.* **9,** 2091 (2003).

3. The supernatant was taken, and 100 μl was assayed for NQO2 activity as below.

4. Protein concentration was determined by the Bradford method after appropriate dilution, usually 1/1000 and 1/10,000 in PBS.

5. The assay was started by the addition of 100 μl supernatant to a mixture of CB 1954 (100 μM) and NRH (500 μM) in sodium phosphate buffer, pH 7, to give a final volume of 1 ml. The mixture was incubated at 37°, and aliquots (10 μl) were taken every 6 min and assayed immediately by HPLC (Partisil 10 SCX [4.2 × 150 mm] [Whatman, Maidstone, Kent, U.K.] eluted isocratically with 0.13 M sodium phosphate [pH 5] at 1.5 ml/min). The concentration of CB 1954 was determined in each sample by reference to the corresponding peak area with an external standard, quantified by absorbance at 325 nm. Initial rates were calculated by curve fitting (FigP, Biosoft, Cambridge, U.K.). As a control for NQO1 and nonspecific enzyme activity, NADH was substituted for the NRH and the assay repeated. This rate was subtracted from the rate obtained with NRH to calculate NQO2 activity. One unit of NQO2 enzyme will convert 1 μmole of CB 1954 to its 4-hydroxylamine per minute under these conditions. The limit of sensitivity was 10^{-4} U.

Conclusions

A number of quinone reductase enzymes also have a nitroreductase activity. This activity can be used clinically as a cancer therapy by bioactivating nitro-prodrugs and in particular CB 1954. The advantage of using the nitroreductase activity rather than the quinone reductase activity of these enzymes is that the bioactivation is very selective for the enzyme. Historically, CB 1954 is curative in rat models and is specifically bioactivated by rat NQO1. However, the human form of NQO1 metabolizes CB 1954 much less efficiently than rat NQO1. Thus, even cells that are high in human NQO1 are insensitive to CB 1954. This catalytic difference between the two forms of the enzyme is mainly accounted for by a single amino acid change at residue 104 (tyrosine in the rat enzyme and glutamine in the human enzyme).

In consideration of the proven success of CB 1954 in the rat system, it would be highly desirable to re-create its anti-tumor activity in humans. In this respect, a gene therapy–based approach for targeting cancer cells and making them sensitive to CB 1954 has been proposed. VDEPT (virus-directed enzyme prodrug therapy) has been used to express an *E. coli* quinone reductase in tumor cells. As a nitroreductase, this enzyme can bioactivate CB 1954 much more efficiently than even rat NQO1, and human tumor cells transduced to express this enzyme are very sensitive to CB 1954. This application is in clinical trial.

However, it is also possible to use CB 1954 directly for prodrug therapy of human tumors. There is an additional CB 1954-reducing activity in human tumor cells that is much greater than that attributable to NQO1. However, this activity is latent and only detectable in the presence of dihydronicotinamide riboside (NRH). The enzyme responsible for this activity is NAD(P)H quinone oxidoreductase 2 (NQO2), originally identified by its homology to NQO1. NQO2 uses dihydronicotinamide riboside (NRH), not NAD(P)H, as an electron donor—a novel and unique property. NQO2 can catalyze a two-electron reduction of quinones and the four-electron nitroreduction of CB 1954. NQO2 is 3000-fold more effective than human NQO1 in the reduction of CB 1954. In this respect, NQO2 resembles the *E. coli* nitroreductase but, like NQO1, it selectively generates only the 4-hydroxylamine derivative. NQO2 can be considered a human NRH-dependent nitroreductase, and NRH produces a dramatic increase in the cytotoxicity of CB 1954 in human tumor cell lines both *in vitro* and *in vivo*. NRH can be substituted for by some reduced pyridinium derivatives such as EP-0152R (1-carbamoylmethyl-3-carbamoyl-1,4-dihydropyridine) that, like NRH, act as co-substrates for NQO2. The target enzyme, NQO2, has been shown to be greatly elevated in some human cancers, and this suggests that activation of CB 1954 could be achieved by administration of an NQO2 co-substrate to "switch on" this enzyme. Although it may not be applicable to all tumor types, the activation of CB 1954 by using endogenous NQO2 rekindles the concept of using CB 1954 as a simple chemotherapeutic agent, and this is scheduled to enter a clinical trial in 2004.

In summary, the nitroreductase activities of quinone reductase enzymes such as NQO2 have clinical applications particularly in the area of selective cancer chemotherapy.

[12] Bioactivation and Resistance to Mitomycin C

By HELEN A. SEOW, PHILIP G. PENKETH,
RAYMOND P. BAUMANN, and ALAN C. SARTORELLI

Introduction

The hypoxic state in solid tumors permits the development of populations of malignant cells that pose an obstacle to the successful eradication of these tumors. While current paradigms for the treatment of certain malignant diseases consist of the use of combinations of chemotherapeutic agents or of radiotherapy and are capable of killing tumor cells in well-oxygenated regions of the tumor, such therapies in general have little or no effect on malignant stem cells found in hypoxic regions. Furthermore, it is often difficult to achieve cytotoxic levels of chemotherapeutic agents in areas of hypoxia because of inadequate blood supplies to tumors that result in both poor drug delivery and poor oxygen delivery.[1] Moreover, since hypoxic cells are slowly moving through the cell cycle or are blocked in their progression through the cell cycle, they are resistant to cycle-active drugs. Additionally, many drugs are rapidly taken up by cells in aerobic regions as these agents diffuse through the tumor mass, and therefore, cells in hypoxic regions are often exposed to sublethal concentrations of these agents.[2] Radiotherapy is also ineffective in hypoxic areas because the low oxygen tension of hypoxia does not result in the production of the highly reactive oxygen radicals necessary for cytotoxicity.[3] The hypoxic state, however, can be considered to be a site of vulnerability in solid tumors,[4] with an alkylating agent such as mitomycin C (MC) having preferential cytotoxicity to hypoxic malignant cells because its requirement for bioreductive activation exploits the hypoxic environment, which favors reductive processes.

MC as a Prototypic Bioreductive Agent

MC is a naturally occurring antibiotic that was isolated originally from the microorganism *Streptomyces caspitosus*.[5] MC exerts its antitumor activity primarily by damaging DNA through both monofunctional and bifunctional alkylations (cross-links).[6] Numerous studies have employed HPLC separation methods to identify specific MC-DNA lesions associated with MC treatment, and these have been well described.[7–16] Monoalkylations initially occur through the C1 position of MC to the N2 position of a

[1] P. W. Vaupel, S. Frinak, and H. I. Bicher, *Cancer Res.* **41,** 2008 (1981).

[2] I. F. Tannock, *Br. J. Cancer* **22,** 258 (1968).

[3] J. E. Moulder and S. Rockwell, *Cancer Metastasis Rev.* **5,** 313 (1987).

[4] A. J. Lin, R. S. Pardini, L. A. Cosby, B. J. Lillis, C. W. Shansky, and A. C. Sartorelli, *J. Med. Chem.* **16,** 1268 (1973).

[5] S. Wakaki, H. Marumo, K. Tomioka, G. Shimizu, E. Kato, H. Kamada, S. Kudo, and Y. Fujimoto, *Antibiotics and Chemotherapy* **8,** 228 (1958).

[6] V. Iyer and W. Szybalski, *Proc. Natl. Acad. Sci. USA* **50,** 355 (1963).

[7] M. Tomasz, R. Lipman, D. Chowdary, J. Pawlak, G. L. Verdine, and K. Nakanishi, *Science* **235,** 1204 (1987).

[8] R. Bizanek, D. Chowdary, H. Arai, M. Kasai, C. S. Hughes, A. C. Sartorelli, S. Rockwell, and M. Tomasz, *Cancer Res.* **53,** 5127 (1993).

[9] A. C. Sartorelli, M. Tomasz, and S. Rockwell, *Adv. Enzyme Regul.* **33,** 3 (1993).

[10] D. Gargiulo, G. S. Kumar, S. S. Musser, and M. Tomasz, *Nucleic Acids Symp. Ser.* **34,** 169 (1995).

[11] G. Kumar, R. Lipman, J. Cummings, and M. Tomasz, *Biochemistry* **36,** 14128 (1997).

[12] M. Tomasz and Y. Palom, *Pharmacol. Ther.* **76,** 73 (1997).

guanine base in DNA and may proceed to form a DNA cross-link through the C10 position of MC to an adjacent DNA guanine at its N2 position.[12] MC-induced cross-links are believed to be primarily responsible for cell death. Thus, a single cross-link per genome has been reported to be sufficient to cause the death of a bacterial cell.[17] Furthermore, a direct relationship between the quantity of DNA interstrand cross-links and the degree of cell kill occurs under aerobic and hypoxic conditions in mammalian cells, with only a few cross-links being responsible for cell death, strongly implicating MC-DNA cross-links as lethal lesions.[18]

The salient feature of the molecular mechanism action of MC is that this agent exists as a prodrug, and both its DNA cross-linking and alkylating activities require the reduction of the quinone ring that transforms MC into a highly reactive alkylating species.[17] It is this property that endows MC with the capacity to kill hypoxic malignant cells preferentially. The list of enzymes capable of reducing MC to an intermediate capable of alkylating DNA is a relatively large one. These bioreductive enzymes that require a reduced pyridine nucleotide include NADPH:cytochrome P450 oxidoreductase (EC 1.6.2.4; FpT), NADH:cytochrome b_5 oxidoreductase (EC 1.6.2.2; FpD), DT-diaphorase (EC 1.6.99.2; DTD), xanthine:oxygen oxidoreductase (EC 1.1.3.23; XOD), xanthine:NAD$^+$ oxidoreductase (EC 1.1.1.204; XDH) (for appropriate references to methodology see reference 19), nitric oxide synthase (EC 1.14.13.39),[20] and NAPDH-ferrodoxin (EC 1.18.1.2).[21] However, only five of these enzymes—FpT, FpD, DTD, XOD, and XDH—have been extensively studied for the potential to sensitize tumor cells to the cytotoxic effects of MC.[22] Of particular note, FpT, FpD and DTD appear to be major MC-activating enzymes in tumor cells;

[13] Y. Palom, M. F. Belcourt, G. S. Kumar, H. Arai, M. Kasai, A. C. Sartorelli, S. Rockwell, and M. Tomasz, *Oncol. Res.* **10**, 509 (1998).

[14] L. A. Ramos, R. Lipman, M. Tomasz, and A. K. Basu, *Chem. Res. Toxicol.* **11**, 64 (1998).

[15] Y. Palom, M. F. Belcourt, S. M. Musser, A. C. Sartorelli, S. Rockwell, and M. Tomasz, *Chem. Res. Toxicol.* **13**, 479 (2000).

[16] Y. Palom, M. F. Belcourt, L. Q. Tang, S. S Mehta, A. C. Sartorelli, C. A. Pritsos, K. L. Pritsos, S. Rockwell, and M. Tomasz, *Biochem. Pharmacol.* **61**, 1517 (2001).

[17] V. Iyer and W. Szybalski, *Science* **145**, 55 (1964).

[18] S. R. Keyes, R. Loomis, M. P. DiGiovanna, C. A. Pritsos, S. Rockwell, and A. C. Sartorelli, *Cancer Commun.* **3**, 351 (1991).

[19] A. C. Sartorelli, W. F. Hodnick, M. F. Belcourt, M. Tomasz, B. Haffty, J. J. Fischer, and S. Rockwell, *Oncol. Res.* **6**, 501 (1994).

[20] H. B. Jiang, M. Ichikawa, A. Furukawa, S. Tomita, and Y. Ichikawa, *Biochem. Pharmacol.* **60**, 571 (2000).

[21] H. B. Jiang, M. Ichikawa, A. Furukawa, S. Tomita, T. Ohnishi, and Y. Ichikawa, *Life Sci.* **68**, 1677 (2001).

[22] J. Cummings, V. J. Spanswick, M. Tomasz, and J. F. Smyth, *Biochem. Pharmacol.* **56**, 405 (1998).

XOD and XDH, although present in normal cells and therefore possibly of significance in the cytotoxicity of MC to normal cells, are not normally present in rapidly dividing malignant tumor cells. Furthermore, pH studies have demonstrated that the metabolism of MC by DTD increased as the pH of the reaction buffer decreased from 7.8 to 5.8[23] and, likewise, DNA cross-linking events exhibit the same pH-dependent effects,[24,25] suggesting that DTD may contribute significantly to the cytotoxic effects of MC in hypoxic regions of tumors that can exist as acidic environments. In contrast, at alkaline values of pH, MC inactivates DTD through the alkylation of the enzyme.[24]

Mechanism of the Reductive Activation of MC

The mechanism of bioactivation of MC is depicted in Fig. 1. MC can be enzymatically reduced by enzymes through either of two processes, one- or two-electron additions, to produce MC semiquinone anion radical ($MC^{\bullet-}$) or MC hydroquinone (MCH_2), respectively. Both of these reduced species are theoretically capable of alkylating DNA. FpT[26] and FpD[27] catalyze one-electron reductions of MC by using NADPH or NADH, respectively, as the reductant to generate $MC^{\bullet-}$. $MC^{\bullet-}$ is a highly reactive species with a calculated $T_{1/2}$ in the presence of physiological oxygen concentrations of less than 0.1 ms at 37°.[28] Thus, under aerobic conditions, $MC^{\bullet-}$ reacts rapidly with O_2 at near diffusion controlled rates ($k = 10^9\ M^{-1}\ s^{-1}$) to regenerate the non-toxic parent prodrug, MC, and superoxide radicals as a byproduct. Therefore, in intact cells, our expectation is that under aerobic conditions $MC^{\bullet-}$ does not contribute meaningfully to the alkylation of DNA. However, under hypoxic conditions, the $T_{1/2}$ of $MC^{\bullet-}$ is extended, allowing it to participate in several nucleophilic reactions. Thus $MC^{\bullet-}$ can react directly with surrounding nucleophiles with first-order kinetics, with an estimated rate constant of $k = 7 \times 10^{-2}\ s^{-1}$ calculated from reference.[29] Based on kinetic parameters, a more likely scenario predicts

[23] D. Siegel, N. W. Gibson, P. C. Preusch, and D. Ross, *Cancer Res.* **50**, 7483 (1990).
[24] D. Siegel, H. Beall, M. Kasai, H. Arai, N. W. Gibson, and D. Ross, *Mol. Pharmacol.* **44**, 1128 (1993).
[25] D. Ross, D. Siegel, H. Beall, A. S. Prakash, R. T. Mulcahy, and N. W. Gibson, *Cancer Metastasis Rev.* **12**, 83 (1993).
[26] N. R. Bachur, S. L. Gordon, M. V. Gee, and H. Kon, *Proc. Natl. Acad. Sci. USA* **76**, 954 (1979).
[27] W. F. Hodnick and A. C. Sartorelli, *Cancer Res.* **53**, 4907 (1993).
[28] P. G. Penketh, W. F. Hodnick, M. F. Belcourt, K. Shyam, D. H. Sherman, and A. C. Sartorelli, *J. Biol. Chem.* **276**, 34445 (2001).
[29] C. Nagata and A. Matsuyama, in "Progress in Antimicrobial and Anticancer Chemotherapy," Vol. 2, p. 423. University Park Press, Baltimore, MD, 1970.

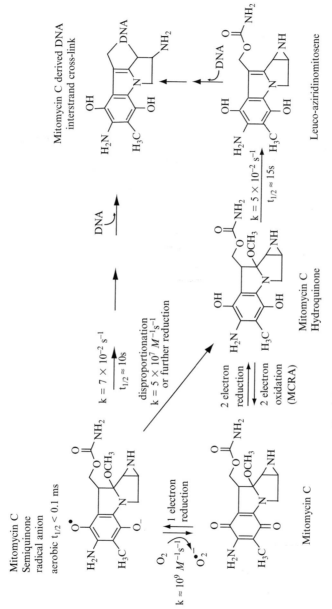

Fig. 1. Reduction/oxidation pathways of mitomycin C.

that $MC^{\bullet-}$ undergoes disproportionation ($k = 5 \times 10^7 \, M^{-1} \, s^{-1}$)[30] or further reduction by other biomolecules to produce MCH_2. Under hypoxic conditions, the flux of a direct two-electron reduction is also likely to increase as the redox potential of the $NADH/NAD^+$ couple becomes more reducing, as NADH accumulates in the absence of electron transfer to O_2.

Unlike $MC^{\bullet-}$, MCH_2 is a relatively long-lived species under aerobic conditions with a $T_{1/2}$ estimated to be 15 s at 37° and pH 7.4 *in vitro,*[28] and like $MC^{\bullet-}$, it can be subject to back oxidization to MC. Thus, in *Streptomyces lavendulae*, which produces MC and is resistant to this prodrug, this process occurs enzymatically through an oxygen-requiring enzyme known as MC resistance protein A (MCRA).[31–34]

In contrast to one-electron reductants, DTD activates MC by a two-electron transfer process by using either NADPH or NADH as a cofactor to directly produce MCH_2. The two-electron addition results in resonance in the quinone ring structure. Subsequently, the methoxy group is lost to produce the leuco-aziridinomitosene, the aziridine ring structure opens and the carbamyl group leaves, resulting in DNA-reactive sites at the C1 and C10 positions of MC. The formation of C1- and C10-reactive sites allows MC to act as a bifunctional cross-linking agent[17] (for a review, see reference 12).

The difference in the production of each reactive species, either MCH_2 or $MC^{\bullet-}$, gives rise to the phenotypic separation of the survival curves that is observed for cells treated with MC under aerobic and hypoxic conditions. This separation, known as the "aerobic/hypoxic differential," is reflected in the cytotoxicity profiles for CHO-K1/dhfr$^-$ cell transfectants overexpressing FpT [35] or FpD[36,37] but not in DTD transfectants.[38] Under aerobic conditions, one-electron reducing enzymes participate in a futile redox cycle in which the production of the $MC^{\bullet-}$ from MC is quickly back-oxidized to its parent compound, MC, at near diffusion-limited rates. Therefore, under

[30] B. M. Hoey, J. Butlerand, and A. J. Swallow, *Biochemistry* **27,** 2608 (1988).
[31] P. R. August, M. C. Flickinger, and D. H. Sherman, *J. Bacteriol.* **176,** 4448 (1994).
[32] P. R. August, J. A. Rahn, M. C. Flickinger, and D. H. Sherman, *Gene* **175,** 261 (1996).
[33] M. F. Belcourt, P. G. Penketh, W. F. Hodnick, D. A. Johnson, D. H. Sherman, S. Rockwell, and A. C. Sartorelli, *Proc. Natl. Acad. Sci. USA* **96,** 10489 (1999).
[34] R. P. Baumann, W. F. Hodnick, H. A. Seow, M. F. Belcourt, S. Rockwell, D. H. Sherman, and A. C. Sartorelli, *Cancer Res.* **61,** 7770 (2001).
[35] M. F. Belcourt, W. F. Hodnick, S. Rockwell, and A. C. Sartorelli, *Proc. Natl. Acad. Sci. USA* **93,** 456 (1996).
[36] M. F. Belcourt, W. F. Hodnick, S. Rockwell, and A. C. Sartorelli, *J. Biol. Chem.* **273,** 8875 (1998).
[37] K. M. Holtz, S. Rockwell, M. Tomasz, and A. C. Sartorelli, *J. Biol. Chem.* **278,** 5029 (2003).
[38] M. F. Belcourt, W. F. Hodnick, S. Rockwell, and A. C. Sartorelli, *Biochem. Pharmacol.* **51,** 1669 (1996).

aerobic conditions only the two-electron reducing pathway is involved in generating the toxic MCH_2 species, while in contrast, the cell kill observed under hypoxic conditions is attributable to the formation of both the MCH_2 and $MC^{\bullet-}$, which undergo a series of spontaneous rearrangements and subsequently alkylate DNA. The combined flux from both one- and two-electron reducing pathways increases the level of MC species capable of alkylating DNA and this is reflected in increased cellular toxicity associated with MC treatment under hypoxia.

MC Bioreduction

Since bioreduction of MC occurs by an enzymatic process, it is not surprising that changes in cellular toxicity associated with drug treatment reflect perturbations in enzyme expression and/or cellular location. Thus, in studies where CHO-K1/dhfr⁻ cells are transfected with various cDNAs to overexpress bioreductive enzymes such as FpT[35] or DTD,[38] sensitivity to MC is enhanced, thereby demonstrating a role for these enzymes in the cytotoxicity of this agent. Overexpression of FpT by 19-fold in CHO-K1/dhfr⁻ cells increases the sensitivity to 10 μM MC by about 3-fold and 60-fold under aerobic and hypoxic conditions, respectively.[35] Approximately a 45-fold increase in sensitivity to 10 μM MC is observed under both aerobic and hypoxic conditions in cells that overexpress DTD by 128-fold.[38] Interestingly, similar findings are not found in cells transfected with the cDNA for FpD.[36,37] However, overexpression of FpD by 9-fold results in a 5-fold decrease in sensitivity to 10 μM MC under aerobic conditions and equivalent sensitivity under hypoxic conditions.[36] In other studies, identical toxicities for transfected cells that overexpress FpD by 5-fold and parental cells occurs when they are treated with 10 μM MC regardless of the state of oxygenation.[37] Studies in which FpD is solublized by removing the membrane anchor (soluble-FpD) and is overexpressed by 9- to 12-fold have demonstrated equivalent aerobic toxicities for both the soluble-FpD CHO-K1/dhfr⁻ transfectant and parental cells and a 25-fold increase in toxicity under hypoxia,[36] thereby restoring the sensitivity of these cells to MC. Similarly, when FpD is overexpressed by 3-fold as a fusion protein containing the nuclear localization signal of the SV40 large T antigen, sensitization is more pronounced under both aerobic and hypoxic conditions as compared with the cells that overexpress FpD in the mitochondrial/endoplasmic reticulum compartments.[37] Furthermore, in the case where overexpressed FpD is localized in the nucleus, enhanced sensitization to MC treatment occurs under hypoxic conditions, and this phenomenon is correlated with increases in the number of DNA adducts.[37] These findings indicate that subcellular distribution of an activating enzyme affects MC cytotoxicity and that bioactivation close to the DNA target results in greater cytotoxicity. Enhanced cytotoxicity is also observed in CHO-K1/dhfr⁻ cells that overexpress cDNA fusion contructs that permit nuclear localization of either FpT[39] or DTD[40] as compared with cells that overexpress these enzymes in their normal cellular localization.

MC Resistance

Classically, it is thought that the rate of MC reduction governs the sensitivity of cells to the cytotoxic effects of this agent and, therefore, theoretically, decreases in the expression of bioreductive activating enzymes[41–45] or, alternately, repair of DNA lesions[46–48] can result in resistance to this agent. However, the finding that *S. lavendulae* produces novel proteins that afford protection from MC through an oxygen-sensitive MCH_2 back-oxidation mechanism and a second mechanism involving a high-affinity MC binding protein coupled with a transport system that presumably produces the dissociation of MC from the high-affinity binding protein as it effluxes MC from the microbial cell[28,31–34,49–51] has sparked interest in determining whether mammalian cells possess similar mechanisms of resistance.

MC Resistance Protein A (MCRA)

A number of genes coding for MC resistance proteins in *S. lavendulae* have been cloned and expressed.[31,32,49–51] One of these genes, *mcrA*, codes for a 54-Kd protein known as MC resistance protein A (MCRA) that has been proposed to function as a MC hydroquinone oxidase.[31,33]

[39] M. F. Belcourt, W. F. Hodnick, S. Rockwell, and A. C. Sartorelli, *Proc. Am. Assoc. Cancer Res.* **39,** 599 (1998).

[40] H. A. Seow, P. G. Penketh, M. F. Belcourt, M. Tomasz, S. Rockwell, and A. C. Sartorelli, *Proc. Am. Assoc. Cancer Res.* **45,** 1088 (2002).

[41] A. M. Dulhanty and G. F. Whitmore, *Cancer Res.* **51,** 1860 (1991).

[42] R. S. Marshall, M. C. Paterson, and A. M. Rauth, *Carcinogenesis* **12,** 1175 (1991).

[43] R. S. Marshall, M. C. Paterson, and A. M. Rauth, *Biochem. Pharmacol.* **41,** 1351 (1991).

[44] S. S. Pan, S. A. Akman, G. L. Forrest, C. Hipsher, and R. Johnson, *Cancer Chemother. Pharmacol.* **31,** 23 (1992).

[45] S. S. Pan, G. L. Forrest, S. A. Akman, and L. T. Hu, *Cancer Res.* **55,** 330 (1995).

[46] P. M. Fracasso and A. C. Sartorelli, *Cancer Res.* **46,** 3939 (1986).

[47] C. S. Hughes, C. G. Irvin, and S. Rockwell, *Cancer Commun.* **3,** 29 (1991).

[48] T. J. Begley and L. D. Samson, *Trends Biochem. Sci.* **1,** 2 (2003).

[49] P. J. Sheldon, D. A. Johnson, P. R. August, H. W. Liu, and D. H. Sherman, *J. Bacteriol.* **179,** 1796 (1997).

[50] P. J. Sheldon, Y. Mao, M. He, and D. H. Sherman, *J. Bacteriol.* **181,** 2507 (1999).

[51] M. He, P. J. Sheldon, and D. H. Sherman, *Proc. Natl. Acad. Sci. USA* **98,** 926 (2001).

Expression of the cDNA for the bacterial resistance protein MCRA in CHO-K1/dhfr⁻ and CHO/AA8 cells results in profound resistance to MC in these cells under aerobic conditions, with little change in sensitivity to MC under hypoxia.[33,34] In fact, MCRA is capable of protecting cells from concentrations of MC up to 500 μM, whereas a 5 log kill of parental CHO cells is observed when treatment is done with 50 μM MC.[33]

MCRA as a Selection Marker

The *mcrA* gene is so effective at conferring MC resistance to mammalian cells that it can be utilized directly as a marker to derive MC-resistant cells after transfection and selection by the antibiotic.[52] Thus, CHO cells, transfected with a plasmid construct containing *mcrA* driven by the cytomegalovirus (CMV) promoter and the neomycin resistance marker, can be expanded as colonies in the presence of neomycin and assessed by immunoblot analysis to determine whether MCRA is expressed. Cell clones that express MCRA are highly resistant to MC when compared with clones transfected with empty vector, as measured by a clonogenic assay.

Mammalian MCRA Functional Homolog

The observations that expression of the bacterial protein MCRA selectively protects mammalian cells from the cytotoxic effects of MC under aerobic conditions, thereby increasing the magnitude of the aerobic/hypoxic differential, and the analogous findings with MC-resistant clones produced by selection, brought into question whether a functional homolog of MCRA is present in MC-resistant neoplastic cells. Thus, MC-resistant CHO cell lines derived by using traditional methods of step-wise increases in MC concentration and drug selection over time produced cell clones that are almost as resistant to MC under aerobic conditions as those expressing the bacterial *mcrA* resistance gene under aerobic conditions.[34] One of these cell clones, designated CHO/HRM, displays a high level of resistance (>23-fold) to 50 μM MC as compared with its parental counterpart and no measurable difference in DTD activity, although it exhibits a modest increase in FpT activity (<4-fold). Both CHO/HRM cells and cells transfected with MCRA cDNA (CHO/MCRA) demonstrate similar patterns of resistance to MC under aerobic and hypoxic conditions—that is, high levels of resistance under aerobic conditions with little decrease in sensitivity under hypoxia. In all other reported studies in which resistance to MC was evaluated under both aerobic and hypoxic conditions, this pattern of

[52] R. P. Baumann, D. H. Sherman, and A. C. Sartorelli, *Biotechniques* **32**, 1030 (2002).

enhanced resistance occurred only under oxygenated conditions (see reference 34 for appropriate references).

Reversal of MC Resistance

The introduction of the cDNAs of bioreductive enzymes into CHO/MCRA and CHO/HRM cells also suggests that a functional homolog of MCRA exists in MC-resistant mammalian cells. Overexpression of DTD in both CHO/MCRA and CHO/HRM cell lines resulted in increased sensitivity to MC, which in the case of the highest DTD overexpressing cell lines approached wild-type cell levels of sensitivity to MC. This result indicates that these resistant cell lines required a much greater MCH_2 flux to produce lethality. Furthermore, these findings suggest that a functional homolog of the MCRA protein, which is capable of oxidizing MCH_2 back to MC and/or other innocuous components, may be responsible for resistance to MC in the drug selected resistant cell lines. Similarly, overexpression of FpT in both CHO/MCRA and CHO/HRM cell lines resulted in increased sensitivity to MC under aerobic conditions[34]; this enhanced kill under aerobic conditions may be attributable to the generation of new lesions as a result of the increased production of hydrogen peroxide and oxygen radicals.

Rapid Screening for DTD Activity by Using a Microtiter Assay

To quickly identify cell clones that highly express DTD, qualitative enzyme activity determinations can be performed in a microtiter plate based on the methodology of Ernster.[53] In each well of a 96-well tissue culture vessel, 10 μl of cell sonicate (1×10^6 cells/ml) is assayed in 100 μl of a reagent mixture consisting of 0.04 M potassium phosphate buffer (pH 7.5), 0.04 mM 2,6-dichlorophenolindophenol (DCPIP), 0.07% bovine serum albumin, and 0.01 mM NADPH. The disappearance of DCPIP is measured by monitoring absorbance at 600 nm at 0, 5, and 10 min after the addition of the reagent mixture. Cell clones with very high levels of DTD activity often become colorless by 10 min.

Search for Oxygen-Sensitive Resistance Mechanisms

The discovery of an oxygen-sensitive resistance profile present in CHO/HRM cells that lack MCRA has fueled a search for the identity of a resistance mechanism based on back-oxidation of MCH_2 to MC. Since many

---

[53] L. Ernster, *Methods Enzymol.* **10**, 309 (1967).

heme-based peroxidases have high affinities for hydroxylated aromatic compounds and hydroquinones as the oxidizable substrate, the possibility that peroxidases may be able to offer protection from MCH_2 must be considered. Numerous peroxidases have been evaluated for their ability to protect cells against MC. The instability of MCH_2 under normal physiological conditions precludes the addition of MCH_2 directly to reaction systems to assay for the presence of molecules capable of oxidizing or metabolizing this MC intermediate. Fortunately, MCH_2 can be "pulse" generated in solution by using chemical reductants such as $NaBH_4$,[28] and then MCH_2 interactions can be assessed by measuring alkylation of target molecules such as 4-(4-nitrobenzyl)pyridine or by the cross-linking of test DNA molecules to study the influence of MCH_2 oxidizing enzymes. In this model system, peroxidases have been found to behave similarly to MCRA in that they have the ability to oxidize MCH_2 and to inhibit DNA cross-linking. The oxidation of hydroquinones by utilizing horseradish peroxidase (HRP) appears to be a two-electron process, but a one-electron process oxidation would be expected to protect in the presence of oxygen and possibly be superior since such a system would further generate H_2O_2 to fuel the peroxidase as a consequence of the oxidation of the semiquinone by O_2. Since *S. lavendulae* generates MC to attack surrounding bacterial competitors, it is advantageous for MCRA to regenerate the prodrug, which can be reactivated in potential competitors, and therefore high yields of MC are produced by MCRA. However, the peroxidases give lower yields of MC from the reduced material. Since the regeneration of MC from the activated toxic species is not advantageous in mammalian systems, this reduced yield of MC as a consequence of the generation of MC species that cannot be reductively reactivated to cross-linking and alkylating intermediates may be advantageous. Although the existence of enzymes capable of oxidizing MCH_2 back to the MC prodrug or to an inactive form(s), and the apparent facile nature of this oxidation gives credibility to MCH_2 oxidation as a possible MC resistance mechanism, the expression of peroxidases such as HRP, lactoperoxidase, or myeloperoxidase in mammalian cells does not produce significant MC resistance. The failure of these peroxidases to protect cells from the cytotoxic effects of MC can be the consequence of several different factors. First, competing substrates that are preferred over MCH_2 as the oxidized substrate for the peroxidase may be present in these cells. Second, the production of H_2O_2 in the cells may be insufficient to supply the H_2O_2 substrate needs of the peroxidase. Theoretically, the presence of elevated levels of MC one-electron reducing systems could alleviate this problem. Third, the level of peroxidase expression achieved in these cells may be inadequate to produce an effect, since they are expressed at very low levels as compared with that found in neutrophils where myeloperoxidase represents 2% of the dry weight of these cells.[54] Support for the concept that a peroxidase could mediate MC resistance arises from studies with Fanconi anemia (FA) lymphoblastoid cells from which a gene has been cloned based on the ability of its protein product to block the hypersensitivity of FA lymphoblastoid cells to MC.[55] Sequence analysis of this gene product has tentatively identified it as a peroxidase.[56] Moreover, its expression is induced by H_2O_2.

An oxygen-dependent alkB protein DNA repair system based on the oxidative N-demethylation of bases that may be involved in resistance to MC has also been recently described in both bacterial and mammalian cells.[48] Increased DNA repair has been documented as a cellular resistance mechanism to a number of alkylating agents and, although it has never been demonstrated to be of major importance for MC resistance, by analogy to other alkylating agent resistance systems, it may potentially contribute to the resistance of cell lines such as CHO/HRM. The alkB protein removes methyl groups from the N1 position of adenine and the N3 position of cytosine by oxidative N-dealkylation (hydroxylation of the α carbon followed by aldehyde loss) and requires O_2. At least three human homologs of the *alkB* gene product have been identified.[57] The major MC interstrand DNA cross-link involves two N2 guanine linkages of which one contains a methylene bridge[7]; hydroxylation at this methylene carbon could result in N-dealkylation of one of the guanine moieties in the cross-link, converting this highly cytotoxic cross-link into a less-toxic single-strand adduct that can be readily repaired. This *alkB*-type repair mechanism, if it were to occur, would produce an O_2-sensitive phenotype similar to that observed in CHO/MCRA cells. However, in an *alkB*-type mechanism, the oxidation reaction would occur after the reduced drug interacts with DNA rather than before the alkylation step, as expected from an MCH_2 oxidase-type resistance mechanism.

An oxygen-sensitive MC resistance mechanism can theoretically also arise as a consequence of an impaired energy-dependent repair/protection mechanism under low-oxygen conditions, since any repair or transport processes that require energy are likely to be impeded. In fact, p-glycoprotein has been reported to have the ability to transport MC[58] and, therefore, can be considered to be a resistance mechanism. However, the contribution of p-glycoprotein is not believed to be a major contributor to the MC

[54] K. Agner, *Acta Physiol. Scand.* **2**, 1 (1941).
[55] I. S. Mian and M. J. Moser, *Mol. Genet. Metab.* **63**, 230 (1998).
[56] J. Ren and H. Youssoufian, *Mol. Genet. Metab.* **72**, 54 (2001).
[57] P. A. Aas, M. Otterlei, P. O. Falnes, C. B. Vagbo, F. Skorpen, M. Akbari, O. Sundheim, M. Bjoras, G. Slupphaug, E. Seeberg, and H. E. Krokan, *Nature* **421**, 859 (2003).
[58] R. Giavazzi, N. Kartner, and I. R. Hart, *Cancer Chemother. Pharmacol.* **13**, 145 (1984).

resistance profile, because p-glycoprotein has poor substrate specificity for MC.

Finally, the selective down-regulation of the two-electron reducing pathways that contribute significantly to the cytotoxicity of MC observed under aerobic conditions could also produce an oxygen-sensitive MC resistance profile. Such a mechanism requires that the contribution of the two-electron reducing enzymes to the total toxicity under hypoxia to be small if selective resistance is only observed under aerobic conditions.

Methodology for the Indirect Determination of MC Activation

Reductive activation of MC transforms this prodrug into a highly reactive species that can interact with a wide array of cellular intermediates and macromolecules. Although it would be difficult to follow MC activation by measuring the numerous products that are formed, one can indirectly determine the degree of metabolic activation of MC in cells treated with this agent by extracting untransformed MC from cellular suspensions and quantifying the amount of metabolized MC substrate. This can be accomplished in cell systems in the following manner. Samples of MC treated cells are collected and mixed with an equal volume of acetonitrile to extract MC at various times after exposure to the drug. Cellular debris and precipitated proteins are removed by centrifugation, and the supernatant is analyzed by separation on a 5 micron 220 × 4.6 mm C-18 reverse phase column (Applied Biosystems RP-18) by elution with a 3 to 27% acetonitrile gradient in 0.03 M KH_2PO_4 (pH 5.4) at a flow rate of 0.8 ml/min. Untransformed MC is eluted as a single peak at approximately 25 min, and absorbance is measured at 360 nm by using a Beckman 168 UV/vis spectrophotometer. The area under the curve, calculated by using Beckman Ultima Gold software, directly correlates with the concentration of MC in the sample. Thus the activation of MC to a reactive species can be calculated by quantifying the amount of remaining MC.

Measurements of the net loss of MC (total two-electron reduction flux minus any back oxidation to parental drug) under aerobic conditions did not reveal a significant net consumption of drug by both MCRA transfected and CHO/HRM selected cells as compared with CHO/AA8 parental cell lines, implying that either an effective re-oxidation process is present, or less likely, DTD is not a major contributor to the aerobic reduction of MC in CHO/HRM cells, DTD does not have access to the MC—possibly because of the activation of an efflux pump—or a combination of the previous mechanisms exists.

[13]   NAD(P)H:Quinone Oxidoreductase 1 Expression, Hydrogen Peroxide Levels, and Growth Phase in HeLa Cells

By Rosario I. Bello, Consuelo Gómez-Díaz, Plácido Navas, and José M. Villalba

Introduction

NAD(P)H:(quinone acceptor) oxidoreductase 1 (NQO1, DT-diaphorase, EC 1.6.99.2) is a cytosolic flavoenzyme ubiquitously present in tissues of nearly all animal species.[1] NQO1 catalyzes the obligatory two-electron reduction of several quinones to their corresponding hydroquinones by using both NADH and NADPH as donor with almost equal affinities. Another prominent feature of NQO1 is its extreme sensitivity to inhibition by the anticoagulant dicoumarol in the micromolar range.[2] A significant part of NQO1 research has been focused on its role as a detoxificating enzyme that prevents the formation of highly reactive quinone metabolites and its role in the bioactivation of some nontoxic quinones that become toxic to the cells once reduced by NQO1.[1,3] In addition, great interest has been paid to the study of NQO1 gene expression, which is regulated by a number of transcription factors such as c-Jun, Jun-B, Jun-D, c-Fos, Fra1, Nrf1 and Nrf2. These bind to several cis-elements of the NQO1 gene promoter, including antioxidant response element (ARE) that contains AP-1 and AP-1-like elements and a basal element. NQO1 expression is coordinately induced with other genes by 3-methylcholantrene, dioxin, transstilbene, phenobarbital, azo dyes, aromatic diamines, aminophenols and phenolic antioxidants.[1,4,5] Activation of the human ARE by phenolic antioxidants is mediated by phosphatidylinositol 3-kinase.[6] Because of its obligatory two-electron reaction mechanism, NQO1 generates relatively stable hydroquinones and thus avoids spurious generation of reactive oxygen species (ROS).[2,3] Moreover, a number of reports have documented a role for NQO1 in the maintenance of the reduced states of ubiquinones, α-tocopherolquinone and α-tocopherol, thereby promoting their antioxidant

[1] A. K. Jaiswal, Free Radic. Biol. Med. 29, 254 (2000).

[2] C. Lind, E. Cadenas, P. Hochstein, and L. Ernster, Methods Enzymol. 186, 287 (1990).

[3] E. Cadenas, Biochem. Pharmacol. 49, 127 (1995).

[4] S. Dhakshinamoorthy and A. K. Jaiswal, J. Biol. Chem. 275, 40134 (2000).

[5] Y. Li and A. K. Jaiswal, J. Biol. Chem. 267, 15097 (1992).

[6] J.-M. Lee, J. M. Hanson, W. A. Chu, and J. A. Johnson, J. Biol. Chem. 276, 20011 (2001).

function in membranes.[7] Taken together, these facts point out the importance of NQO1 as an antioxidant enzyme.

During the last ten years, NQO1 has received renewed interest because of the demonstration of novel functions that include the regulation of the intracellular redox state by controlling the NAD(P)H:NAD(P)$^+$ ratio.[8] A balance in the cytosolic redox state is required to maintain the appropriate cell environment permissive for signaling, and both the inhibition of NQO1 by dicoumarol and treatment of cells with high concentrations of hydroquinone cause the blockade of stress-activated protein kinase/c-Jun NH$_2$-terminal kinase and NFκB pathways and potentiate apoptosis induced by tumor necrosis factor-α (TNF-α).[9] Furthermore, TNFα sensitivity is also increased by overexpression of NQO1 in adenocarcinoma-derived MCF-7 cells.[10] NQO1 also plays an important role in the regulation of p53-mediated apoptotic cell death by increasing p53 stabilization, particularly under oxidative stress conditions.[11] However, p53 stabilization by NQO1 does not appear to require NQO1 activity, but this function relies on a direct interaction between both proteins.[12]

Expression of NQO1 is also related to growth phase in both normal and tumor adherent cells. A significant increase in NQO1 activity at high cellular densities has been demonstrated in normal BALB/c 3T3,[13] human osteoblastic,[14] and HeLa cells.[15] NQO1 expression and activity are also significantly elevated in confluent cell cultures and spheroids of human colon carcinoma HT-29 cells.[16] Changes in NQO1 as a function of cell density have important implications in the area of bioreductive drug metabolism, because plateau phase cultures of tumor cells are valuable models that mimic many characteristics of the tumor microenvironment.[16]

[7] R. E. Beyer, J. Segura-Aguilar, D. di Bernardo, M. Cavazzoni, R. Fato, D. Fiorentini, M. C. Galli, M. Setti, L. Landi, and G. Lenaz, *Proc. Natl. Acad. Sci. USA* **93**, 2528 (1996).

[8] A. Gaikwad, D. J. Long, J. L. Stringer, and A. K. Jaiswal, *J. Biol. Chem.* **276**, 22559 (2001).

[9] J. V. Cross, J. C. Deak, E. A. Rich, Y. Qian, M. Lewis, L. A. Parrot, K. Mochida, D. Gustafson, S. Vande Pol, and D. J. Templeton, *J. Biol. Chem.* **274**, 31150 (1999).

[10] L. M. Siemankowski, J. Morreale, B. D. Butts, and M. M. Briehl, *Cancer Res.* **60**, 3638 (2000).

[11] G. Asher, J. Lotem, R. Kama, L. Sachs, and Y. Shaul, *Proc. Natl. Acad. Sci. USA* **99**, 3099 (2002).

[12] A. Anwar, D. Dehn, D. Siegel, J. K. Kepa, L. J. Tang, J. A. Pietenpol, and D. Ross, *J. Biol. Chem.* **278**, 10368 (2003).

[13] J. J. Schlager, B. J. Hoerl, J. Riebow, D. P. Scott, P. Gasdaska, R. E. Scott, and G. Powis, *Cancer Res.* **53**, 1338 (1993).

[14] P. Collin, A. Lomri, and P. J. Marie, *Bone* **28**, 9 (2001).

[15] R. I. Bello, C. Gómez-Díaz, F. Navarro, F. J. Alcaín, and J. M. Villalba, *J. Biol. Chem.* **276**, 44379 (2001).

[16] R. M. Phillips, A. de la Cruz, R. D. Traver, and N. W. Gibson, *Cancer Res.* **54**, 3766 (1994).

Accordingly, higher levels of NQO1 gene expression have also been observed in many tumors when compared with normal tissues of the same origin.[17]

ROS are endogenously generated by many metabolic reactions and released constitutively by tumor cells.[18] Levels of endogenous ROS (both superoxide and peroxide) are significantly elevated in low-density HeLa cells as compared with cells grown to confluence.[19] We have shown that endogenously generated H_2O_2 is a factor that regulates density-related increases in NQO1 expression in HeLa cells. Treatment of HeLa cells with 10 mM sodium pyruvate, a well-established H_2O_2 scavenger that has been used to decrease intracellular levels of H_2O_2,[20] partially prevents a cell-density-related increase in NQO1 activity. In addition, when added exogenously, H_2O_2 also enhances NQO1 activity in HeLa cells cultured at low but not high cell density.[15] A role for ROS in the regulation of NQO1 expression and activity as a function of cell density is not surprising, because ROS such as superoxide radicals and H_2O_2 can activate NQO1 expression through the ARE.[21]

NQO1 activity is required to sustain the viability of HeLa cells at high density, because inhibition of the reductase with dicoumarol strongly increases cell death.[15] NQO1 activity is also required to promote the survival of HL-60 cells,[22] a promyelocytic cell line that grows in suspension. In the following section we describe the methods we have found most useful for demonstrating the role of hydrogen peroxide and cell density as factors that regulate cell density-mediated modulation of NQO1 activity and for studying the requirement for NQO1 activity to maintain the viability of HeLa cells grown at high density.

Methods

Cell Cultures

HeLa cells can be conveniently maintained at the laboratory in culture flasks in Dulbecco's modified Eagle's medium (DMEM) supplemented with 10% fetal calf serum (FCS), 100 units/ml penicillin, 100 mg/ml

[17] T. Cresteil and A. K. Jaiswal, *Biochem. Pharmacol.* **42,** 1021 (1991).

[18] R. H. Burdon, *Free Radic. Biol. Med.* **18,** 775 (1995).

[19] R. I. Bello, F. J. Alcaín, C. Gómez-Díaz, G. López-Lluch, P. Navas, and J. M. Villalba, *J. Bioenerg. Biomembr.* **35,** 169 (2003).

[20] J.-J. Li, L. W. Oberley, M. Fan, and N. H. Colburn, *FASEB J.* **12,** 1713 (1998).

[21] T. H. Rushmore, M. Morton, and C. B. Pickett, *J. Biol. Chem.* **266,** 11632 (1991).

[22] N. Forthoffer, S. F. Martín, C. Gómez-Díaz, R. I. Bello, M. I. Burón, J. C. Rodríguez-Aguilera, P. Navas, and J. M. Villalba, *J. Bioenerg. Biomembr.* **34,** 209 (2002).

streptomycin and 2.5 mg/ml amphotericin B. Cell cultures are carried out at $37°$ in a humidified atmosphere of 5% CO_2 and 95% air. Prior to experiments involving H_2O_2, cells are changed to Minimum Essential Medium Eagle (MEM) supplemented as described previously. Since DMEM contains both iron and pyruvate, this change avoids the generation of hydroxyl radicals by Fenton chemistry in experiments using H_2O_2.[23] On the other hand, pyruvate contained in DMEM could alter levels of ROS endogenously generated by cell metabolism.[20] To obtain sparse and dense cultures, HeLa cells are detached from culture flasks by incubation for 5 to 10 min with a Ca^{2+}- and Mg^{2+}-free non enzymatic detaching solution, and the number of viable cells is counted by the Trypan blue exclusion method. Cells are then seeded in supplemented MEM at either 1500 or 25,000 viable cells/cm^2 in 50 cm^2 culture plates. Cells are allowed to grow for three days to reach densities of about 8000 (sparse cultures) and 100,000 (dense cultures) cells/cm^2, respectively. A more detailed study of NQO1 activity as a function of cell density can be carried out by culturing cells initially at 1500 cells/cm^2 and then following the culture for 3 to 10 days. The culture medium is changed every two days until cells reach the densities required for each experiment.

Measurement of Endogenous ROS as a Function of Cell Density

Endogenous levels of peroxide and superoxide in cells are quantified by using the probes 2',7'-dichlorodihydrofluorescein diacetate (DCFH-DA) and hydroethidine (Het), respectively. Cells are incubated in the dark with the corresponding probe (final concentrations: 10 μg/ml DCFH-DA and 4 μM Het) for 30 min at $37°$. Stock solutions of probes (10 μM Het and 10 mg/ml DCFH-DA, respectively) are prepared in DMSO. After incubation, cells are washed and then detached from cultured plates. One milliliter of cell suspension (containing 2.5×10^5 to 1×10^6 cells/ml) is transferred to a tube for determination of fluorescence by flow cytometry. Excitation of fluorescent probes is carried out with an Argon laser at 488 nm, and emission fluorescence is determined at 525 nm (FL1) for fluorescein and 620 nm (FL3) for ethidium.[19]

Preparation of Cell Extracts From HeLa Cells for Determination of NQO1 Activity

All procedures are carried out at $4°$. Cells (2 to 10×10^6) are separated for culture plates with the aid of a Ca^{2+}- and Mg^{2+}-free non enzymatic detaching solution and then concentrated by centrifugation at 500g for 5 min

[23] M. Meyer, R. Schreck, and P. A. Baeuerle, *EMBO J.* **12**, 2005 (1993).

and washed with cold 130 mM Tris-HCl pH 7.6, containing 1 mM EDTA, 0.1 mM DTT and 1 mM PMSF. After the final centrifugation step, the cell pellet is resuspended in 1 ml of hypotonic lysis buffer (10 mM Tris-HCl [pH 7.6] containing 1 mM EDTA, 0.1 mM DTT, 1 mM PMSF and 20 μg/μl each of chymostatin, leupeptin, antipain and pepstatin A [CLAP]). Homogenization of cells is carried out for 5 min on ice with the aid of a hand-held glass-glass homogenizer, and then for 5 sec with a mechanical cell homogenizer (Ultra-Turrax T 25, Ika-Labortectnick, Staufen i. Br., Germany) set at the maximal speed. After disruption of cells, the concentration of the lysis buffer is raised to 130 mM Tris by adding enough volume of 250 mM Tris buffer, pH 7.6, containing 1 mM EDTA, 0.1 mM DTT, 1 mM PMSF and CLAP. Unbroken cells and debris are separated by centrifugation at 500g for 5 min, and the supernatant is saved. Cytosolic fractions, containing NQO1 activity, are separated from the membrane residue by ultracentrifugation at 100,000g for 30 min. The sample (typically 0.5 to 1 mg of protein per milliliter) is saved on ice for assay of NQO1 activity and protein determination.

Assay of NQO1 Activity in Cytosolic Fractions From HeLa Cells

NQO1 activity is calculated from the NADH and menadione-dependent, dicoumarol-inhibitable reduction of cytochome c as described by Lind *et al.*[2] Assays are carried at 37° in a stirred spectrophotometer cuvette. The reaction mixture is made by sequentially adding distilled water (enough volume to reach a final volume of 1 ml), 100 μl 0.5 M Tris-HCl (pH 7.5), 100 μl 0.8% Triton X-100, 70 to 140 μl of protein sample (containing about 70 μg of cytosolic protein), 5 μl of menadione stock solution (prepared at 2 mM in ethanol) and 38.5 μl of 2 mM cytochrome c. Assays are carried out either in the absence or in the presence of 10 μM dicoumarol. In the latter case, 5 μl of dicoumarol stock solution (prepared at 2 mM in 6 mM NaOH) is added after protein sampling. Reactions are initiated by adding 50 μl of 10 mM NADH, and the absorbance is then continuously monitored at 550 nm for 1 min in a recording spectrophotometer. Amounts of cytochrome c reduced are calculated from the linear slope of this line, and NQO1 activity is calculated from the difference in reaction rates obtained with and without dicoumarol. An extinction coefficient of 18.5 mM^{-1} cm^{-1} is used in calculations of specific activities.

By using this method, it is found that cytosols obtained from HeLa cells in sparse cultures (about 8000 cells/cm^2) contain very little NQO1 activity (11.67 $\pm$ 6.7 nmoles min^{-1} mg^{-1}), but this activity is dramatically increased up to 40-fold (484.2 $\pm$ 108.7 nmoles min^{-1} mg^{-1}) activity in confluent cultures (about 100,000 cells/cm^2). However, a major portion of this increase

(10 to 15-fold increase) takes place after only one day of culture, when the density of cells increases from about 8000 to 15,000 cells/cm^2. This indicates that some diffusible agent(s), rather than cell-to-cell contacts, account for the observed increment in NQO1 activity.

Role of Hydrogen Peroxide on Cell Density-Mediated Increase in NQO1 Activity

The role of endogenous H_2O_2 produced by cell metabolism in the cell-density-mediated increase in NQO1 activity is assessed by testing the effect of H_2O_2 scavenging in cells by adding sodium pyruvate to the culture medium. Pyruvate is added to form a sterile 100 mM stock solution to a final concentration of 10 mM. Experiments are initiated at about 8000 cells/cm^2, and then the effect of pyruvate addition on cell growth and NQO1 activity is followed for 3 days.

Before the effect of exogenous H_2O_2 on NQO1 activity is estimated, the concentration range of H_2O_2 where no significant losses of cell viability is observed with respect to untreated cells, both in sparse and in confluent cultures, must be determined. This previous determination is necessary because sparse and dense cultures of HeLa cells differ significantly in their resistance to oxidative stimuli.[15] Stock solutions of H_2O_2 are freshly prepared daily in water from commercial concentrated (30%, stored at 4°) H_2O_2 and kept on ice before they are added to the cells. Since H_2O_2 is progressively degraded with time during storage, actual concentrations of concentrated H_2O_2 solutions can be estimated by direct spectrophotometric reading at 240 nm with an extinction coefficient of 43.6 mM^{-1} cm^{-1}. For these experiments, cells are treated with H_2O_2 (0 to 100 μM) for 8 h, and then viability is calculated by the Trypan blue exclusion method. By using this method, it is demonstrated that viability of HeLa cells at high density is not affected significantly by H_2O_2 at concentrations up to 75 μM, and a slight effect is observed only at 100 μM. However, whereas low concentrations of H_2O_2 (12.5 μM) slightly increase the viability of sparse cells, increasing the concentration of H_2O_2 produces a significant decline in the number of viable cells, and at 75 to 100 μM H_2O_2, nearly all cells are killed.

Once concentrations that do not significantly affect the viability of sparse and dense cells have been determined (0 to 24 μM in the case of HeLa cells), these are tested for stimulation of NQO1 activity. Cells are exposed to various sublethal concentrations of H_2O_2 for 13 h. After this incubation period, cells are detached from culture plates and the cytosolic fractions are obtained for determination of NQO1 activity. Exogenous H_2O_2 significantly increases NQO1 activity in sparse but not in confluent HeLa cells. In addition, decreasing endogenous H_2O_2 with pyruvate results in a significant prevention of NQO1 increase (particularly in sparse cultures), but the activity is still stimulated about 13-fold with relation to low-density cultures.[15]

Use of Dicoumarol to Study the Role of NQO1 Activity: Effect of Serum Albumin

A frequent approach to study the consequences of a lack of NQO1 function in cells is to test the effect of NQO1 inhibition with dicoumarol. The inhibitor is added to cultured cells from aqueous stock solutions that are prepared by adding enough NaOH to allow for dissolving concentrated dicoumarol. The same amount of vehicle is added to control cultures. HeLa cells are treated for 24 h either with dicoumarol or with an equivalent amount of NaOH and then are detached from culture plates and their viability estimated by the Trypan blue exclusion method.

A factor that should be carefully controlled when assessing the role of NQO1 in a cellular system by inhibiting the enzyme with dicoumarol is the presence of serum albumin during dicoumarol treatment. As shown in Fig. 1, no effect on the viability of confluent HeLa cells is observed when treatments with dicoumarol are carried out at concentrations up to 100 μM in whole medium containing 10% FCS. Viability is progressively decreased when cultures are exposed to dicoumarol up to 400 μM, and then a sharp decline is observed at 600 μM. However, when dicoumarol treatments are carried out in serum-free medium, the maximal concentration at which no effect on cells is observed is drastically reduced to 2.5 μM. The viability of confluent HeLa cells is then significantly decreased at 20 and 100 μM (Fig. 1). A major part of the protective effect of serum decreasing the effectiveness of dicoumarol on cells can be attributed to serum albumin. When cells are treated with 5 μM dicoumarol, cytotoxicity is completely prevented by either 10% FCS or 5 mg/ml BSA. However, inhibition of NQO1 activity itself by dicoumarol in the *in vitro* enzymatic assay is affected by serum or BSA only when very low concentrations of dicoumarol (0.1 μM) are tested, but the activity can be fully inhibited by dicoumarol at 5 μM even in the presence of 10% FCS or 5 mg/ml BSA (Fig. 2).

Conclusions

It is now firmly established that the quinone reductase NQO1 plays an important role in the regulation of signaling pathways that control cell growth and death. NQO1 activity has been related to both potentiation and inhibition of cell death. Thus, both dicoumarol inhibition of NQO1 in HeLa cells and NQO1 overexpression in human breast adenocarcinoma

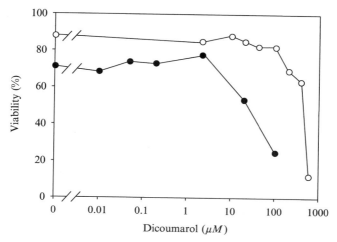

Fig. 1. Effect of dicoumarol on the viability of confluent HeLa cells. Protective effect of serum. HeLa cells were cultured at a density of 25,000 cells/cm^2 and then allowed to grow for three days to reach a density of 100,000 cells/cm^2. Cells were then treated for 24 hr with increasing concentrations of dicoumarol either in whole or in serum-free culture medium (MEM). After incubation, cells were detached and the viability of the culture estimated by the Trypan blue exclusion method. Controls (0% dicoumarol) were also carried out by adding to cultures the same amount of vehicle (NaOH) used in plates containing dicoumarol. No effect of NaOH addition was found in any case. Open circles: 10% FCS-containing MEM. Closed circles: serum-free MEM. Depicted data are representative of two independent experiments.

MCF-7 cells potentiate cell death triggered by TNF-α.[9,10] NQO1 inhibition also potentiates cell death in the absence of serum, both in HeLa and HL-60 cells.[15,22] NQO1 stabilizes the tumor suppressor p53 and participates in the regulation of p53-mediated apoptosis,[11,12] but it is also important in cell growth and death regulation through p53-independent mechanisms.[22]

NQO1 activity is also related to the growth phase in cultured adherent cells. Increased NQO1 activity[13] and a decrease in ROS levels[24] may play roles in density-dependent inhibition of growth in normal adherent cells. However, an increase in NQO1 activity and a decrease in ROS levels can also occur in dense cultures of some tumor cells.[15,16,19] Increased activities of antioxidant enzymes correlate with enhanced resistance of confluent cells to oxidative stress.[15] Hydrogen peroxide is a factor that mediates density-related increases in NQO1 activity in HeLa cells, but additional

[24] G. Pani, R. Colavitti, B. Bedogni, R. Avenzino, S. Borrello, and T. Galeotti, *J. Biol. Chem.* **275**, 38891 (2000).

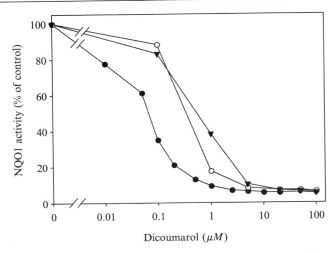

FIG. 2. NQO1 activity in cytosolic fractions from confluent HeLa cells. Effect of albumin and serum on dicoumarol inhibition. Cytosolic fractions were obtained from confluent cells obtained as described in Fig. 1, and NQO1 activity was assayed at the concentrations of dicoumarol indicated. Assays were carried out under standard assay conditions (without serum or BSA, closed circles), in the presence of 10% fetal calf serum (open circles), or in the presence of 5 mg/ml BSA (closed triangles). Depicted data are representative of two independent experiments.

factors related to cell-to-cell interactions and/or growth conditions at high cellular densities should also be considered. A combination of factors, rather than a single triggering stimulus, have also been proposed to lead to elevated NQO1 mRNA levels in plateau phase cultures of adenocarcinoma HT-29 cells.[16]

The role of NQO1 in dense cell cultures can be tested by inhibiting the enzyme with dicoumarol. The concentration of dicoumarol used in these assays should be carefully controlled to ensure its specificity on NQO1 and to avoid inhibition of other reductases that may lead to unclear results that are difficult to interpret.[25] Another factor that must be considered in experiments using dicoumarol is the presence of albumin in experiments carried out with serum. A decrease in cell viability at concentrations of dicoumarol that are in the same range to those required to inhibit the enzyme is observed in the absence of albumin, but the presence of this protein significantly increases the amount of dicoumarol required to affect cells. Cell

[25] P. C. Preusch, D. Siegel, N. W. Gibson, and D. Ross, *Free Radic. Biol. Med.* **11,** 77 (1991).

uptake of anticoagulants such as dicoumarol and warfarin is prevented by their complexation with albumin,[26] and a higher concentration of dicoumarol (200 to 400 μM) must be used. Thus, the effective concentration of inhibitor acting on cells is difficult to control. It is expected that the use of irreversible inhibitors of NQO1 activity, such as the newly developed ES936,[12] and genetic methods that induce a decrease in NQO1 gene expression will help to elucidate the particular role played by NQO1 in the control of cell growth and death.

Acknowledgments

Supported by grants BMC2002–01078 and BMC2002–01602 (Spanish Ministerio de Educación y Cultura and Ministerio de Ciencia y Tecnología) and CVI-276 (Junta de Andalucía). R. I. Bello acknowledges financial support by the Spanish Ministerio de Educación y Cultura. C. Gómez-Díaz was supported by CVI-276.

[26] W. D. Wosilait, M. P. Ryan, and K. H. Byington, *Drug Metab. Dispos.* **9,** 80 (1981).

[14] The "Prochaska" Microtiter Plate Bioassay for Inducers of NQO1

By Jed W. Fahey, Albena T. Dinkova-Kostova, Katherine K. Stephenson, and Paul Talalay

Introduction

The microtiter plate bioassay for NQO1 was first developed in this laboratory by Hans Prochaska as a rapid and direct assay of quinone reductase (NQO1; QR; DT-diaphorase) activity in cultured cells, suitable for identifying, purifying and determining the potency of inducers of this detoxication enzyme.[1,2] De Long *et al.*[3] were the first to suggest exploiting the specific activities of NQO1 in Hepa 1c1c7 murine hepatoma cells for bioassay of the inducer potencies of phase 2 enzymes. These cells mimic animal tissues in responding to a wide variety of chemoprotective agents.[3] Hepa 1c1c7 cells were originally established from a transplantable murine

[1] H. J. Prochaska and A. B. Santamaria, *Anal. Biochem.* **169,** 328 (1988).
[2] H. J. Prochaska, A. B. Santamaria, and P. Talalay, *Proc. Natl. Acad. Sci. USA* **89,** 2394 (1992).
[3] M. J. De Long, H. J. Prochaska, and P. Talalay, *Proc. Natl. Acad. Sci. USA* **83,** 787 (1986).

hepatoma of the C57L/J mouse and the 1c1c7 clone was selected for its highly inducible aryl hydrocarbon hydroxylase (AHH), making this line important in the analysis of the functional characteristics of the *Ah* (*A*ryl *h*ydrocarbon) receptor.[4] The Hepa 1c1c7 line was chosen for the NQO1 assay because: (a) these cells have many characteristics of normal tissues, in particular the capacity for carcinogen activation and xenobiotic metabolism; (b) these cells are amenable to strict control of environmental, nutritional, and hormonal factors; (c) mutants defective in *Ah* receptor or its gene product are available – this is not true for all cells in culture; and (d) these cells have a highly inducible AHH as well as other inducible cytochromes P450, cytochrome P450 reductase, and epoxide hydrolase, thus facilitating metabolic activation of xenobiotics.[3,5]

As adapted for use in microtiter plates, the cells were grown for 24 h, followed by exposure to inducing agents for an additional 24–48 h, and then lysed with digitonin. Cell lysates were assayed for NQO1 activity using a reaction mixture in which both the NADPH and the quinone are regenerated, thus avoiding reagent depletion (Fig. 1). Cell density, as a proxy for total cellular protein, was assayed in replicate plates by vital staining. This assay has proven to be exceptionally useful for its originally envisioned purpose, as well as for the characterization of phase 2 enzyme inducers in a variety of materials. Although other microtiter plate assays have been proposed,[6] none are as responsive or as versatile as the Prochaska assay. The dynamic response of the assay is very wide, permitting the evaluation of compounds over a concentration range of approximately 5 orders of magnitude (Fig. 2).[7]

We have made certain modifications to the assay in order to make it more efficient, and considerable data are now available from laboratories worldwide that support its utility in cancer chemoprevention studies. For example, we have accumulated more than 10 years of baseline data with a single standard NQO1 inducer, β-naphthoflavone (BNF), that has been used in hundreds of separate assays. We have also made observations on the use of this basic assay: (a) with other cell lines and passage numbers, (b) with various common solvent systems, (c) with a wide array of samples, (d) with multiple researchers, (e) as a versatile tool for screening crude plant extracts, (f) as a powerful aid in the directed synthesis of biologically active compounds, and (g) as a dynamic system to monitor conversion of

[4] J. P. Whitlock, *Ann. Rev. Pharm. Toxicol.* **39,** 103 (1999).
[5] M. J. De Long, A. B. Santamaria, and P. Talalay, *Carcinogenesis* **8,** 1549 (1987).
[6] M. Zhu and W. E. Fahl, *Anal. Biochem.* **287,** 210 (2000).
[7] T. Prestera, Y. Zhang, S. R. Spencer, C. A. Wilczak, and P. Talalay, *Adv. Enzyme Regul.* **33,** 281 (1993).

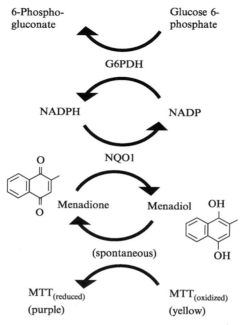

FIG. 1. Principle of the NQO1 assay: Glucose 6-phosphate and glucose-6-phosphate dehydrogenase (G6PDH) generate NADPH continually. This NADPH is used by NQO1 to transfer electrons to menadione. The menadiol thus formed reduces MTT (yellow when oxidized), spontaneously, to the purple formazan which can be measured over a broad range of wavelengths (490 to 640 nm). Both NADPH and menadione are regenerated, which obviates problems encountered with substrate depletion.

inactive precursors to active products. The use of this assay has led to the discovery of a number of potent chemoprotective agents from plants, to their use in controlled animal and human clinical studies, and to the development of similarly active compounds by synthetic means. These findings are reviewed and discussed herein.

Prochaska Bioassay Protocol

As initially described, the induction of NQO1 was measured in Hepa 1c1c7 murine hepatoma cells grown in 96-well microtiter plates containing α-MEM culture medium supplemented with 10% fetal calf serum.[1,2] Approximately 10,000 cells were plated into flat-bottomed, untreated plastic microtiter plates and incubated for 24 h at 37° in a 5% CO_2 atmosphere,

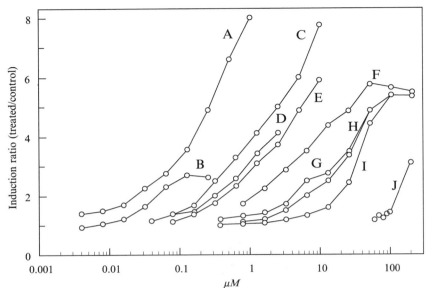

FIG. 2. Dynamic range of the Prochaska Assay. Induction profiles in Hepa 1c1c7 cells are shown for (A) BNF, (B) phenylarsine oxide, (C) sulforaphane, (D) mercury chloride, (E) 1-nitro-1-cyclohexane, (F) 1,3-dithiole-3-thione, (G) benzyl isothiocyanate, (H) 1-butylhydroquinone, (I) BAL, and (J) 1-butylhydroperoxide which had CD's of about 0.02, 0.05, 0.2, 0.3, 0.4, 1.5, 4, 5, 20, and 110 μM, respectively.

during which time they became adherent. Culture medium was then replaced with 150 μL of fresh medium containing antibiotics in order to permit the introduction of non-sterile test compounds (e.g., crude plant extracts). Cells were exposed to inducers for an additional 24 h. Typically the solution to be assayed in DMSO was diluted with cell culture medium and 2-fold serial dilutions were introduced into 8-well columns on the microtiter plates with a multichannel pipettor. The final concentration of DMSO was 0.1% by volume. The plates were then incubated for 24 h, and the medium was discarded. Cells were then lysed with digitonin in the presence of EDTA for 20 min and NQO1 was assayed by addition of an NADPH-generating system that maintained a constant NADPH concentration (glucose-6-phosphate, glucose-6-phosphate dehydrogenase, NADP), FAD, menadione (2-methyl-1,4-naphthoquinone, a quinone that is reduced to menadiol by NQO1 in the presence of NADPH), and MTT (3-[4,5-dimethylthiazo-2-yl]-2,5-diphenyltetrazolium bromide; a tetrazolium dye that is reduced non-enzymatically by menadiol). NQO1 activity was stopped after 5 min by the addition of dicumarol (a specific and very potent inhibitor of NQO1), and the reduced formazan dye was measured spectrophotometrically in a 96-well plate reader at 610 nm. Since all wells have negligible absorbance at zero time, the absolute absorbance at 5 min accurately reflects the change in absorbance during the incubation period. Cytotoxicity and cell density were assessed by crystal violet staining of a duplicate set of plates, followed by scanning at 610 nm[1] or 490 nm.[2] One unit of inducer activity is defined as the *Concentration* that *Doubles* the NQO1 specific activity in a microtiter well containing 150 μL of medium. This concentration has been designated the "CD" value. Hence a compound with a CD of 1.0 μM has a potency of 6667 units of inducer activity per μmol.

Technical Improvements of the Assay

The original procedure has been modified in several ways:

1. The fetal calf serum is treated with activated charcoal (1 g/100 ml) for 90 min at 55°,[8] thereby lowering the basal NQO1 levels.

2. Protein content is measured in an aliquot of cell lysates, thereby providing a more rapid and reliable assessment of cell density and eliminating the need for a duplicate set of plates. Protein concentration is measured in a 20-μL aliquot of the digitonin cell lysate in a separate 96-well microtiter plate. Bicinchoninic acid reagent (300 μL)[9] is added and the product is measured spectrophotometrically (550 nm) after 30 min incubation at 37.5°.[10]

3. The exposure period has been increased from 24 to 48 h which magnifies the amplitude of NQO1 induction for many inducers in this cell line.[2,8,10]

4. An automated microtiter plate washer has been used to remove growth medium and rinse cells just before conducting the assay, and the MTT concentration is monitored at 490 nm.[10]

5. We have introduced the use of various solvents in addition to DMSO. Acetonitrile,[2] and a "triple solvent" composed of a mixture of equal volumes of acetonitrile, DMSO, and dimethyl formamide,[10] are both extremely effective solvents for extraction of plant tissues, and they can be diluted into cell culture medium to final concentrations of 1% and 0.5%,

[8] S. R. Spencer, C. A. Wilczak, and P. Talalay, *Cancer Res.* **50,** 7871 (1990).

[9] P. K. Smith, R. I. Krohn, G. T. Hermanson, A. K. Mallia, F. H. Gartner, M. D. Provenzano, E. K. Fujimoto, N. M. Goeke, B. J. Olson, and D. C. Klenk, *Anal. Biochem.* **150,** 76 (1985).

[10] J. W. Fahey, Y. Zhang, and P. Talalay, *Proc. Natl. Acad. Sci. USA* **94,** 10367 (1997).

respectively, without affecting cell viability. Other solvents such as tetrahydrofuran and methanol[11] can be used for more lipophilic compounds, but these solvents must be used at less than 0.1% and 0.33%, respectively, in order to avoid cytotoxicity.

6. We have typically expressed the inducer potency of plant extracts in units/g fr. wt. (fresh weight) or dry wt. Cytotoxicity is of concern if it occurs at or near the concentration range that results in NQO1 induction. Pezzuto and colleagues[12,13] have proposed using a *Chemoprotective Index* (CI; CI = IC_{50}/CD) which provides a useful metric of the window between activity and toxicity.

7. The assay has been adapted by plant scientists[14–16] for rapid screening of individual plants, by introducing leaf disc punches as sources of NQO1 inducer (glucosinolates/isothiocyanates), placed in individual wells of microtiter plates and then removed prior to assay.

Novel Findings Made with The Prochaska Bioassay

The Prochaska microtiter plate bioassay has been exceptionally useful for identifying and isolating inducers of NQO1 from natural sources, and in guiding the synthesis of more potent analogs of isothiocyanates,[12,17,18] dithiolethiones,[19–21] and curcuminoids.[22,23] It has also been used to survey,

[11] F. Khachik, J. S. Bertram, M.-T. Huang, J. W. Fahey, and P. Talalay, *in* "Proceedings of the International Symposium on Antioxidant Food Supplements in Human Health" (L. Packer, M. Hiramatsu, and T. Yoshikawa, eds.), pp. 203–229. Academic Press, New York, 1999.

[12] C. Gerhäuser, M. You, J. Liu, R. M. Moriarity, M. Hawthorne, R. G. Mehta, R. C. Moon, and J. M. Pezzuto, *Cancer Res.* **57,** 272 (1997).

[13] C. Gerhäuser, K. Klimo, E. Heiss, I. Neumann, A. Gamal-Eldeen, J. Knauft, G. Y. Liu, S. Sitthimonchai, and N. Frank, *Mutat. Res.* **523–524,** 163 (2003).

[14] H. B. Gross, T. Dalebout, C. D. Grubb, and S. Abel, *Plant Sci.* **159,** 265 (2000).

[15] Q. Wang, C. D. Grubb, and S. Abel, *Phytochem. Anal.* **13,** 152 (2002).

[16] C. D. Grubb, H. B. Gross, D. L. Chen, and S. Abel, *Plant Sci.* **162,** 143 (2002).

[17] G. H. Posner, C. G. Cho, J. V. Green, Y. Zhang, and P. Talalay, *J. Med. Chem.* **37,** 170 (1994).

[18] Y. Zhang, T. W. Kensler, C. G. Cho, G. H. Posner, and P. Talalay, *Proc. Natl. Acad. Sci. USA* **91,** 3147 (1994).

[19] P. A. Egner, T. W. Kensler, T. Prestera, P. Talalay, A. H. Libby, H. H. Joyner, and T. J. Curphey, *Carcinogenesis* **15,** 177 (1994).

[20] T. W. Kensler, *Environ. Health Perspect.* **4**(Suppl. 105), 965 (1997).

[21] T. W. Kensler, T. J. Curphey, Y. Maxiutenko, and B. D. Roebuck, *Drug Metabol. Drug Interact.* **17,** 3 (2000).

[22] A. T. Dinkova-Kostova and P. Talalay, *Carcinogenesis* **20,** 911 (1999).

[23] K. Singletary, C. MacDonald, M. Iovinelli, C. Fisher, and M. Wallig, *Carcinogenesis* **19,** 1039 (1998).

screen, and compare the potency of various plant sources in both plant breeding and molecular genetics applications. Highlights of some of these findings are described herein.

Many of the 9 recognized classes of phase 2 enzyme inducers were originally identified using the Prochaska assay.[7,11,24–26] These inducers include (see review[27]): (i) oxidizable *ortho*- and *para*-diphenols, their cognate quinones, and other Michael reaction acceptors; (ii) isothiocyanates[28] – highly electrophilic compounds that are widely consumed as their dietary precursor glucosinolates which are hydrolyzed to isothiocyanates by the enzyme myrosinase, which co-exists in plant cells and is present in the human gastrointestinal tract[29,30]; (iii) dithiocarbamates, which are formed metabolically by conjugation of GSH with isothiocyanates. Synthetic dithiocarbamates are widely used industrial and agricultural chemicals; (iv) 1,2 dithiole-3-thione derivatives such as oltipraz[19]; (v) trivalent arsenic derivatives such as sodium arsenite and phenylarsine oxide; (vi) divalent heavy metal derivatives; (vii) hydroperoxides such as *tert*-butyl hydroperoxide; (viii) polyenes such as carotenoid metabolites[11]; and (ix) vicinal dimercaptans such as 2,3-dimercaptopropanol. Synergism between compounds in certain of these classes has been demonstrated,[31] and whereas many of the inducers that have been isolated from natural products fit into one of these classes, others appear to involve novel chemistry. Compounds identified as inducers by means of the Prochaska assay include withanolides,[32,33] a bicyclic diarylheptanoid,[34] flavonoids,[35–38] and curcumin derivatives and other phenyl propenoids.[22,39–42] Furthermore, unpublished

[24] P. Talalay, M. J. De Long, and H. J. Prochaska, *Proc. Natl. Acad. Sci. USA* **85,** 8261 (1988).

[25] T. Prestera, W. D. Holtzclaw, Y. Zhang, and P. Talalay, *Proc. Natl. Acad. Sci. USA* **90,** 2965 (1993).

[26] P. Talalay, *Proc. Amer. Phil. Soc.* **143,** 52 (1999).

[27] P. Talalay, A. T. Dinkova-Kostova, and W. D. Holtzclaw, *Adv. Enzyme Regul.* **43,** 121 (2003).

[28] Y. Zhang, P. Talalay, C. G. Cho, and G. H. Posner, *Proc. Natl. Acad. Sci. USA* **89,** 2399 (1992).

[29] J. W. Fahey, A. T. Zalcmann, and P. Talalay, *Phytochemistry* **56,** 5 (2001); corrigendum *Phytochemistry* **59,** 237.

[30] T. A. Shapiro, J. W. Fahey, K. L. Wade, K. K. Stephenson, and P. Talalay, *Cancer Epidemiol. Biomarkers Prev.* **10,** 501 (2001).

[31] R. R. Putzer, Y. Zhang, T. Prestera, W. D. Holtzclaw, K. L. Wade, and P. Talalay, *Chem. Res. Toxicol.* **8,** 103 (1995).

[32] R. I. Misico, L. L. Song, A. S. Veleiro, A. M. Cirigliano, M. C. Tettamanzi, G. Burton, G. M. Bonetto, V. E. Nicotra, G. L. Silva, R. R. Gil, J. C. Oberti, A. D. Kinghorn, and J. M. Pezzuto, *J. Nat. Prod.* **65,** 677 (2002).

[33] E. J. Kennelly, C. Gerhäuser, L. L. Song, J. G. Graham, C. W. W. Beecher, J. M. Pezzuto, and A. D. Kinghorn, *J. Agric. Food Chem.* **45,** 3771 (1997).

work from our lab, and recent surveys[13,17,27] reveal a wide spectrum of inducers that includes both plant components and pharmaceutical agents. With curcumin, a naturally occurring spice component, Dinkova-Kostova[22,41] has used the Prochaska bioassay to elucidate which portions of the molecule are essential for inducer activity: synthetic as well as the naturally-occurring structural analogs were examined. These studies showed that the hydroxyl group had to be present at the *ortho*-position on the phenolic rings, and that the rings had to be bridged by a β-diketone moiety in order to be strong inducers. In studies with our colleague G.H. Posner, a number of synthetic norbornyl structural analogs of sulforaphane were examined, and one had equivalent inducer potency to sulforaphane in Hepa 1c1c7 cells.[17,18] Subsequently, Pezzuto and colleagues developed sulforamate, an analog of sulforaphane with comparable inducer potency and somewhat less toxicity.[12] This group has also developed 4′-bromoflavone as a derivative of naturally occurring flavonoids, guided by the Prochaska bioassay.[39]

Numerous reports now document the inducer potency of various plants, plant constituents, and complex dietary ingredients[10,28,30,35,43–64]

[34] D. S. Jang, E. J. Park, M. E. Hawthorne, J. S. Vigo, J. G. Graham, F. Cabieses, B. D. Santarsiero, A. D. Mesecar, H. H. Fong, R. G. Mehta, J. M. Pezzuto, and A. D. Kinghorn, *J. Agric. Food Chem.* **50**, 6330 (2002).

[35] J. W. Fahey and K. K. Stephenson, *J. Agric. Food Chem.* **50**, 7472 (2002).

[36] S. Yannai, A. J. Day, G. Williamson, and M. J. Rhodes, *Food Chem. Toxicol.* **36**, 623 (1998).

[37] Y. Uda, K. R. Price, G. Williamson, and M. J. Rhodes, *Cancer Lett.* **120**, 213 (1997).

[38] C. L. Miranda, G. L. Aponso, J. F. Stevens, M. L. Deinzer, and D. R. Buhler, *Cancer Lett.* **149**, 21 (2000).

[39] L. L. Song, J. W. Kosmeder, S. K. Lee, C. Gerhäuser, D. Lantvit, R. C. Moon, R. M. Moriarty, and J. M. Pezzuto, *Cancer Res.* **59**, 578 (1999).

[40] A. T. Dinkova-Kostova, C. Abeygunawardana, and P. Talalay, *J. Med. Chem.* **41**, 5287 (1998).

[41] A. T. Dinkova-Kostova, M. A. Massiah, R. E. Bozak, R. J. Hicks, and P. Talalay, *Proc. Natl. Acad. Sci. USA* **98**, 3404 (2001).

[42] A. T. Dinkova-Kostova and P. Talalay, *Free Radic. Biol. Med.* **29**, 231 (2000).

[43] N. Tawfiq, S. Wanigatunga, R. K. Heaney, S. R. Musk, G. Williamson, and G. R. Fenwick, *Eur. J. Cancer Prev.* **3**, 285 (1994).

[44] N. Tawfiq, R. K. Heaney, J. A. Plumb, G. R. Fenwick, S. R. Musk, and G. Williamson, *Carcinogenesis* **16**, 1191 (1995).

[45] T. A. Shapiro, J. W. Fahey, K. L. Wade, K. K. Stephenson, and P. Talalay, *Cancer Epidemiol. Biomarkers Prev.* **7**, 1091 (1998).

[46] K. Hashimoto, S. Kawamata, N. Usui, A. Tanaka, and Y. Uda, *Cancer Lett.* **180**, 1 (2002).

[47] M. W. Farnham, J. W. Fahey, and K. K. Stephenson, *J. Amer. Soc. Hort. Sci.* **125**, 482 (2000).

[48] M. W. Farnham, P. E. Wilson, K. K. Stephenson, and J. W. Fahey, *Plant Breeding* (in press).

[49] C. Y. Zhu and S. Loft, *Food Chem. Toxicol.* **41**, 455 (2003).

[50] F. M. Pereira, E. Rosa, J. W. Fahey, K. K. Stephenson, R. Carvalho, and A. Aires, *J. Agric. Food Chem.* **50**, 6239 (2002).

(see recent reviews by Pezzuto[65] and by Talalay[26]). Application of this assay to plant breeding, biochemistry, physiology, and molecular biology suggests that it may facilitate advances in the development of "tailored" vegetables and fruit that are rich inducers.[14–16,47,48,66,67]

Versatility and Limitations of The Prochaska Bioassay

Test Compound Matrix

The Prochaska bioassay, originally described for pure compounds, has been adapted to a wide range of materials including plants and plant extracts, urine, food (crude homogenates of dietary components), honey, and wine.

The application of this bioassay to plant extracts may now represent its most widespread use. In the original studies on plants in our laboratory, a wide variety of vegetables were tested for NQO1 inducer potency.[2] Those of the *Brassica* (e.g., broccoli) and *Allium* (e.g., onion) families showed inducer potency and various others had little or none.[2] This information led to the isolation of sulforaphane from broccoli,[28] and broccoli sprouts.[10] Subsequently, the assay has been used to guide the isolation of resveratrol from grapes,[54] withanolides from tomatillos and other plants,[32,33] isothiocyanates closely related to sulforaphane from wasabi and watercress,[61,66,67]

[51] J. W. Fahey and K. K. Stephenson, *HortScience* **34**, 1159 (1999).

[52] J. Bomser, D. L. Madhavi, K. Singletary, and M. A. Smith, *Planta Med.* **62**, 212 (1996).

[53] L. C. Chang, C. Gerhäuser, L. Song, N. R. Farnsworth, J. M. Pezzuto, and A. D. Kinghorn, *J. Nat. Prod.* **60**, 869 (1997).

[54] M. Jang, L. Cai, G. O. Udeani, K. V. Slowing, C. F. Thomas, C. W. Beecher, H. H. Fong, N. R. Farnsworth, A. D. Kinghorn, R. G. Mehta, R. C. Moon, and J. M. Pezzuto, *Science* **275**, 218 (1997).

[55] H. J. Park, Y. W. Lee, H. H. Park, Y. S. Lee, I. B. Kwon, and J. H. Yu, *Eur. J. Cancer Prev.* **7**, 465 (1998).

[56] K. J. Hintze, A. S. Keck, J. W. Finley, and E. H. Jeffery, *J. Nutr. Biochem.* **14**, 173 (2003).

[57] A. S. Keck, Q. Qiao, and E. H. Jeffery, *J. Agric. Food Chem.* **51**, 3320 (2003).

[58] N. V. Matusheski and E. H. Jeffery, *J. Agric. Food Chem.* **49**, 5743 (2001).

[59] C. Matito, F. Mastorakou, J. J. Centelles, J. L. Torres, and M. Cascante, *Eur. J. Nutr.* **42**, 43 (2003).

[60] S. K. Lee, L. Song, E. Mata-Greenwood, G. J. Kelloff, V. E. Steele, and J. M. Pezzuto, *Anticancer Res.* **19**, 35 (1999).

[61] P. Rose, K. Faulkner, G. Williamson, and R. Mithen, *Carcinogenesis* **21**, 1983 (2000).

[62] R. Mithen, K. Faulkner, R. Magrath, P. Rose, G. Williamson, and J. Marquez, *Theor. Appl. Genet.* **106**, 727 (2003).

[63] K. Faulkner, R. Mithen, and G. Williamson, *Carcinogenesis* **19**, 605 (1998).

[64] Y. H. Shon and K. S. Nam, *J. Ethnopharmacol.* **77**, 103 (2001).

[65] J. M. Pezzuto, *Biochem. Pharmacol.* **53**, 121 (1997).

various proanthocyanidins from blueberries,[52] and flavonoids from *Tephrosia* sp.,[53] onions,[68] buckwheat,[35] and Thai ginger.[35] Numerous other laboratories have reported on the inducer activity of various uncharacterized plant and fungal extracts including aqueous extracts of Brussels sprouts[49] and solvent extracts of 45 different plants.[46] The extraction protocol is critically important, and is addressed elsewhere in this chapter. Chemical structures of NQO1 inducers are reviewed in Chapter 23 of this volume.[69]

One novel dosing protocol utilizes plant leaf disc punches which are incubated in microtiter plates prepared essentially as described above. After a suitable incubation (leaching) period, the discs are removed, and the assay is performed.[14–16] The response is thus proportional to the amount of inducer leaching out into culture medium, and this, in turn, is proportional to cut surface area/volume. In this manner, several thousand chemically mutagenized *Arabidopsis* plants were screened for altered leaf NQO1 inducer potency and bioassay results were used as a guide for selective HPLC analysis of progeny from the putative mutants.[14–16]

We have used the Prochaska assay as a tool to assist us in verifying dietary compliance in clinical trials. For example, when human volunteers were fed various levels of glucosinolates and/or isothiocyanates following or preceding a period of prescribed abstinence from cruciferous vegetables (a source of NQO1-inducing isothiocyanates), we were able to verify compliance by bioassaying their urines for inducer activity during the control period. Likewise, we could confirm compliance with dietary intake of cruciferous vegetables. These results were entirely congruent with both the dietary record, and with an independent chemical test for the presence of isothiocyanate metabolites.[30] Also, in these clinical trials we first used the Prochaska bioassay to monitor dietary ingredients for unexpected phase 2 enzyme induction potency. Had there been significant levels of inducers in, for example, white bread, CocaCola™, green beans or tomatoes, we would have avoided administering these food products to subjects as part of a baseline diet. We were able to test these and all other dietary ingredients by appropriate dilution and homogenization of dietary components in order to develop a "non-inducing" baseline diet. In this case, there were no independent chemical tests that would have ruled out all possible sources of inducers.

[66] D. X. Hou, M. Fukuda, M. Fujii, and Y. Fuke, *Cancer Lett.* **161**, 195 (2000).
[67] D. X. Hou, M. Fukuda, M. Fujii, and Y. Fuke, *Int. J. Mol. Med.* **6**, 441 (2000).
[68] G. Williamson, G. W. Plumb, Y. Uda, K. R. Price, and M. J. Rhodes, *Carcinogenesis* **17**, 2385 (1996).
[69] A. T. Dinkova-Kostova, J. W. Fahey, and P. Talalay, *Meth. Enzymol.* **382,** 423 (2004).

We have also used this bioassay to compare the NQO1 inducing potential of various honeys and molasses.[35] We found that the induction potentials of honey and molasses were directly proportional to their color intensity (darkness), and it appeared that a number of flavones common to honey may have been responsible for this induction. Further analysis of the pure flavones then led us to identify very rich plant sources of these compounds.

Dosing

For assay of either pure chemicals, or plant extracts, a number of solvents including DMSO, dimethyl formamide, acetonitrile, "triple solvent" (a mixture of equal volumes of DMSO, dimethyl formamide and acetonitrile), tetrahydrofuran, ethanol, and methanol can be used to prepare the initial stock solution. In order to avoid cytotoxicity, the maximum tolerated concentration of these solvents for a 48-h induction period is 0.1% for tetrahydrofuran, 0.25% for dimethyl formamide, 0.33% for methanol and DMSO, 0.5% for triple solvent, and 1% for acetonitrile and ethanol. If examining the NQO1 inducer potential of urine or other aqueous solutions, one can add as much as about 10% of the volume of the microtiter plate well fluid.

The Prochaska bioassay presents unique opportunities for examining the kinetics of isothiocyanate effects on enzyme induction. Y. Zhang has shown that isothiocyanates are very rapidly absorbed by cultured cells,[70–74] and in a number of clinical studies they were shown to be rapidly metabolized.[30,45] Furthermore, some isothiocyanates (e.g., allyl isothiocyanate) are quite volatile, whereas others (e.g., sulforaphane) are not. Vapor phase transfer of inducers between nearby wells can complicate assays, even for compounds which are not usually considered to be especially volatile. An interesting example of vicinity effects among microtiter plate wells was encountered with dimethyl fumarate, which has been used previously as an inducer of NQO1.[24,75] This compound has very low volatility, yet when introduced into microtiter plate wells, cells in adjacent wells become induced. The mechanism of this effect is not understood.

[70] Y. Zhang and P. Talalay, *Cancer Res.* **58,** 4632 (1998).
[71] Y. Zhang, *Carcinogenesis* **21,** 1175 (2000).
[72] Y. Zhang, *Carcinogenesis* **22,** 425 (2001).
[73] L. Ye and Y. Zhang, *Carcinogenesis* **22,** 1987 (2001).
[74] Y. Zhang and E. C. Callaway, *Biochem. J.* **364,** 301 (2002).
[75] A. Begleiter, K. Sivananthan, T. J. Curphey, and R. P. Bird, *Cancer Epidemiol. Biomarkers Prev.* **12,** 566 (2003).

Equimolar concentrations of isothiocyanates compared to glucosinolates with added myrosinase (thioglucoside glucohydrolase; EC 3.2.3.1; the enzyme that catalyzes the conversion of glucosinolates to isothiocyanates) do not induce equally in the Prochaska bioassay. When an excess of highly purified myrosinase is added to each well (in the presence of ascorbate), complete hydrolysis of glucosinolates to their cognate isothiocyanates occurs *in-situ,* during the assay.[10] Interestingly, enzymatic release of isothiocyanates from glucosinolates during the assay gives higher inducer potencies than direct addition of the isothiocyanate alone. We have reproduced this situation with a number of glucosinolate/isothiocyanate pairs, using highly purified myrosinase.[76] (Neither the glucosinolate alone, nor myrosinase alone, induced NQO1). In all cases examined (data not shown), enzymatic release of isothiocyanate from its cognate glucosinolate was clearly more effective in inducing NQO1 than was direct addition of the isothiocyanate product.

Advantages of the Hepa 1c1c7 Cell Line

We have examined a variety of rodent, insect and human cell lines and found that the large amplitude of NQO1 response produced by Hepa 1c1c7 cells makes this line ideal for use in such a bioassay. Additionally, these cells maintain their responsiveness through a large number of passages, making standardization of the assay more reliable. Some inducers elevate NQO1 levels in Hepa 1c1c7 cells as high as 8- or 10-fold over the levels found in untreated control cells (Fig. 3; which also provides an example of the passage effect). NQO1 induction by some or all of these compounds, as well as by other standard inducers such as sulforaphane, have also been examined in the following: human gastric epithelial, AGS (Fig. 4); human laryngeal epithelial, HEp-2 (Fig. 4); human adult retinal pigmented epithelial, APRE-19 (Gao *et al.,*[77] and Fig. 4); human breast cancer, MCF7[41]; human skin keratinocyte, HaCaT[41,75]; murine leukemia L1210[77]; and murine keratinocyte, PE,[77] cell lines. In general, most cell lines were less sensitive to inducers than Hepa 1c1c7 cells. The commonly used invertebrate cell lines Sf9 (from *Spodoptera frugiperda*), Sf21 (also from *S. frugiperda*), and Tni (from *Trichoplusia ni*) did not respond even at the highest levels of several compounds tested (data not shown). Others have evaluated the use of cell lines such as a Chinese hamster ovary

[76] M. Shikita, J. W. Fahey, T. R. Golden, W. D. Holtzclaw, and P. Talalay, *Biochem. J.* **341,** 725 (1999).

[77] X. Gao, A. T. Dinkova-Kostova, and P. Talalay, *Proc. Natl. Acad. Sci. USA* **98,** 15221 (2001).

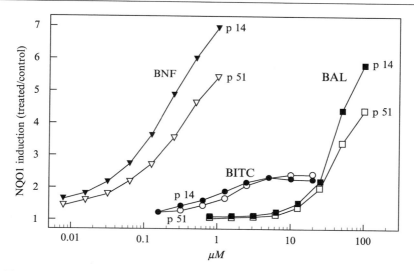

FIG. 3. Induction of NQO1 (treated/control NQO1 activity) following dosing with BNF (▼,▽), benzyl isothiocyanate (BITC; ●,○), or 2,3-dimercapto-1-propanol (British anti-Lewisite or BAL; ■,□). Filled symbols represent Hepa 1c1c7 cell line passage 14 and open symbols represent passage number 51. CD's for the early and late passages of BNF, BITC, and BAL, are 0.022, 0.044, 1.5, 2.2, 20, and 24 μM, respectively.

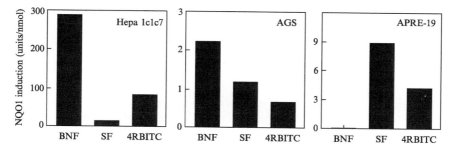

FIG. 4. Induction of NQO1 in Hepa 1c1c7, AGS, and APRE-19 cells treated for 48 h with BNF, sulforaphane (SF), or 4-(rhamnopyranosyloxy)benzyl isothiocyanate (4RBITC). Dosing of Hep-2 cells with the same three inducers yielded no induction with BNF, and produced substantial cytotoxicity with the 2 isothiocyanates such that a CD could not be measured.

(CHO) line which was transfected with high levels of human NQO1 in order to achieve over-expression of the enzyme.[78] Recently published work from Jiang and colleagues[79] has compared induction in Hepa 1c1c7 cells to induction in 6 other mammalian cell lines (MCF7, MDA-MB-231, HeLa, HT-29, HepG2, LNCaP) by using BNF and sulforaphane as standard inducers, and they have come to much the same conclusion – that the Hepa 1c1c7 cells "were the most robust and sensitive cells, which had higher basal as well as up-regulated enzymatic activities."

Because NQO1 induction in cells involves uptake, metabolism, and transcriptional events, it is not surprising that different cell types produce radically different induction profiles when presented with the same array of inducers. For example, human gastric and retinal epithelial cell lines respond very differently to treatment with two dissimilar isothiocyanates, than does the murine hepatoma cell line Hepa 1c1c7 (Fig. 4). Whether this is due to differences in uptake, metabolism, or signal processing, is not known, but comparative uptake kinetics of these and other isothiocyanates can now be examined. There is clearly a very broad variation of potencies among cell types. Thus NQO1 induction in Hepa 1c1c7 cells is as much as two orders of magnitude greater than in AGS cells. The relative NQO1 inducer potencies of two isothiocyanates were compared: 4-methylsulfinylbutyl isothiocyanate, (sulforaphane, from *Brassica oleracea* var. *italica* and 4-(rhamnopyranosyloxy)benzyl isothiocyanate from *Moringa oleifera*. In Hepa 1c1c7 cells, the NQO1 inducer potency of the *Moringa* isothiocyanate was many-fold higher than that of sulforaphane yet with both human cell types tested its potency was considerably lower than that sulforaphane (Fig. 4).

And finally, in Hepa 1c1c7 cells bifunctional inducers such as BNF,[24] induce phase 1 enzymes (e.g., cytochromes P-450), as well as NQO1, but they require activation by the cytochromes P-450 in order to exert inducer activity. Such activation is probably lacking in APRE-19 cells since BNF is not an inducer. This compound is one of the most potent inducers of phase 2 enzymes in Hepa 1c1c7 cells. The behavior of mutant Hepa 1c1c7 cell lines that are defective in either aryl hydrocarbon *(Ah)* receptor function (BP^rc1 cells) or aryl hydrocarbon hydroxylase expression (c1 cells), can be used to determine whether compounds are mono- or bifunctional inducers.[80] For example two isothiocyanates, sulforaphane and progoitrin

[78] L. H. De Haan, A. M. Boerboom, I. M. Rietjens, D. Van Capelle, A. J. De Ruijter, A. K. Jaiswal, and J. M. Aarts, *Biochem. Pharmacol.* **64**, 1597 (2002).
[79] Z. Q. Jiang, C. Chen, B. Yang, V. Hebbar, and A. N. Kong, *Life Sci.* **72**, 2243 (2003).
[80] H. J. Prochaska and P. Talalay, *Cancer Res.* **48**, 4776 (1988).

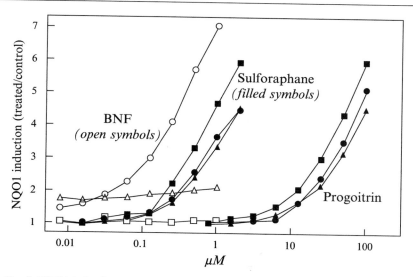

FIG. 5. NQO1 induction in Hepa 1c1c7 cells (●,○) and their mutants BP^{r}c1, defective in aryl hydrocarbon receptor function (▲,△), and c1, defective in aryl hydrocarbon hydroxylase expression (■,□). Cells were treated for 48 h with BNF, sulforaphane, or progoitrin. Only BNF required metabolic activation to induce NQO1.

(2-hydroxybut-3-enyl isothiocyanate), are fully active as inducers in wild type and mutant cell lines and are thus termed "monofunctional," whereas when BNF is tested against the same 3 cell types it does not induce NQO1 in cells lacking the cytochrome P450 (Fig. 5). These mutants have also been used to characterize flavonoids as mono- and bifunctional NQO1 inducers.[35,36]

Bioassay Variability

As is the case with all bioassays, and by the very nature of cell culture, there is inherent and unexplained variability in this bioassay. Certain investigators continue to use the 24 h dosing or exposure period originally suggested.[1] Most laboratories, however, use the 48 h induction period described in Prochaska *et al.*[2] and subsequent publications. Although cells should not be used after a certain number of passages (we usually retire sub-lines after 30 passages, but have gone as high as 45 or more on occasion), we have not detected an overall trend towards reduced sensitivity to inducers, although the absolute amplitude of the response (ratio of NQO1 levels in treated vs. control cells) may decline (Fig. 3). On close examination of the data obtained using BNF as a standard in all assays over a period of 10 years in our laboratory (CD mean, 0.028 μM; SD, 0.015 μM; maximum, 0.082 μM; minimum, 0.015 μM; data from 253 assays performed by 5 different operators between 1993 and 2001), we can find no systematic pattern of variations relating to operator, operator training, cell history, season, culture medium components (source or lot), or any other identifiable extrinsic factor. Careful attention must be paid to timing, conditions, and cell culture status. Assay variability should be tracked and controlled by use of the same standard compound in all experiments (e.g., BNF).

Acknowledgments

We appreciate the patience and care of the following persons who, in addition to the authors, performed some of the bioassays described in this paper: Xiangqun Gao, Kristina Wade, Paula Wilson, Mark Wrona, and Yuesheng Zhang. We pay special tribute to Hans Prochaska [deceased], who along with Annette Santamaria, and Mary De Long, first developed this versatile assay. Hepa1c1c7 cells were originally provided by J. P. Whitlock, Jr.

[15] Structure-Activity Relationships in Two-Electron Reduction of Quinones

By Narimantas Čėnas, Žilvinas Anusevičius, Henrikas Nivinskas, Lina Misevičienė, and Jonas Šarlauskas

Introduction

Quinones may accept electrons from various flavoenzymes, iron-sulfur proteins and photosynthetic reaction centers. The energetics of the quinone reduction have been studied extensively by pulse-radiolysis, electron spin resonance and electrochemical techniques. The studies of Yamazaki and coworkers in the early 1970s have shown that flavoenzymes may reduce quinones in single-electron, mixed single- and two-electron, and two-electron methods.[1,2] The single-electron reduction of quinones by flavoenzyme dehydrogenases-electrontransferases may be treated according to an "outer-sphere electron transfer" model[3–7] and is relatively well

[1] T. Iyanagi and I. Yamazaki, *Biochim. Biophys. Acta* **172,** 370 (1969).
[2] T. Iyanagi and I. Yamazaki, *Biochim. Biophys. Acta* **216,** 282 (1970).
[3] R. Marcus and N. Sutin, *Biochim. Biophys. Acta* **811,** 265 (1985).

understood. In general, the reaction rates increase with an increase in quinone single-electron reduction potential, E^1. In contrast, the mechanisms of two-electron reduction of quinones by flavoenzymes are insufficiently understood. The reactivity of quinones toward a most thoroughly studied two-electron transferring NAD(P)H:quinone oxidoreductase (NQO1, EC 1.6.99.2) does not depend on their potential, pointing to enzyme specificity related to their certain structures.[8] In this chapter, we demonstrate that multiparameter regression analysis, by using redox potential and several simple structural parameters of quinones, may provide important information on the mechanisms of two-electron enzymatic reduction. In view of the simplicity of the single-electron reduction mechanism and the presumable involvement of single-electron transfers in two-electron reduction, the single-electron reduction of quinones will be analyzed first.

Single-Electron Reduction of Quinones by Flavoenzymes

Outer-Sphere Electron Transfer Model in Single-Electron Reduction of Quinones

The single-electron reduction of quinones is frequently treated according to an "outer-sphere electron transfer" model, which describes an electron transfer with weak electronic coupling between the reactants.[3] In the simplest form, the rate constant of single electron transfer between reagents (k_{12}) depends on the electron self-exchange constants of reagents (k_{11} and k_{22}) and equilibrium constant of reaction (K) (log $K = \Delta E^1$ (V)/ 0.059, where ΔE^1 is the difference in the standard single-electron transfer potential of reactants):

$$k_{12} = (k_{11} \times k_{22} \times K \times f)^{1/2} \tag{1}$$

and

$$\log f = (\log K)^2 / 4 \log (k_{11} \times k_{22}/Z^2) \tag{2}$$

[4] G. Powis and P. L. Appel, *Biochem. Pharmacol.* **29,** 2567 (1980).

[5] J. Butler and B. M. Hoey, *Biochim. Biophys. Acta* **1161,** 73 (1993).

[6] N. Čėnas, Ž. Anusevičius, D. Bironaitė, G. I. Bachmanova, A. I. Archakov, and K. Ollinger, *Arch. Biochem. Biophys.* **315,** 400 (1994).

[7] Ž. Anusevičius, M. Martinez-Julvez, C. G. Genzor, H. Nivinskas, C. Gomez-Moreno, and N. Čėnas, *Biochim. Biophys. Acta* **1320,** 247 (1997).

[8] C. Lind, E. Cadenas, P. Hochstein, and L. Ernster, *Methods Enzymol.* **186,** 287 (1990).

where Z is a frequency factor ($10^{11}\ M^{-1}\,\mathrm{s}^{-1}$). According to Eq. (1, 2), in the reaction of electron donor with a series of homologous electron acceptors (k_{22} = constant), log k_{12} will exhibit parabolic (square) dependence on ΔE^1 with a slope $\Delta\log k/\Delta\Delta E^1 = 8.45\ \mathrm{V}^{-1}$ at $\Delta E^1 = \pm 0.15$ V. At $\Delta E^1 = 0$, $k_{12} = (k_{11} \times k_{22})^{1/2}$. For quinoidal compounds and free flavins, the self-exchange constants are around $10^8\ M^{-1}\,\mathrm{s}^{-1}$.[9] The electron exchange between quinone and protonated semiquinone at acidic pH:

$$Q + QH^{\bullet} \rightarrow QH^{\bullet} + Q \tag{3}$$

proceeds separately — that is, the rate-limiting electron transfer is followed by fast ($k \sim 10^{10}\ M^{-1}\mathrm{s}^{-1}$) proton transfer from solvent.[9] In the reactions of redox proteins, the protein k_{11} values may be used to evaluate the orientational distance (in nm) of electron transfer (R_p)[10]:

$$R_p = 0.63 - 0.035 \ln k_{11} \tag{4}$$

The potentials of single-electron reduction of quinones at pH 7.0 (E_7^1) and pK_a of semiquinones (pK_a [$QH^{\bullet}$]) are given in Table I, including the data on antitumor aziridinylbenzoquinones AZQ, DZQ, MeDZQ and BZQ. Obviously, the electron donating substituents and increase in a number of aromatic rings decrease the E_7^1 of quinones (Table I). The E_7^1 for several aziridinylbenzoquinones are unavailable; however, their orientational values may be easily calculated (Table I). The E_7^1 and $pK_a(QH^{\bullet})$ values are usually obtained from the pulse-radiolysis data caused by the rapid dismutation of quinone radicals and should not be confused with the standard potentials of quinones (E_7^0) and pK_a of hydroquinones that are obtained by electrochemical methods and refer to the energetics of the quinone/hydroquinone (Q/QH_2) couple. The $pK_a(QH^{\bullet})$ for alkylsubstituted 1,4-benzoquinones increase with a decrease in their E_7^1; however, for a whole set of compounds examined (Table I), the correlation is absent.

Structure-Activity Relationships in Single-Electron Reduction of Quinones by NADPH:Cytochrome P-450 Reductase and Ferredoxin:NADP⁺ Reductase

Flavoenzymes NADPH:cytochrome P-450 reductase (P-450R, EC 1.6. 4.2), and ferredoxin:NADPH⁺ reductase (FNR, EC 1.18.1.2) perform the single-electron reduction of quinones.[6,7] P-450R and FNR belong to the class of dehydrogenases-electron-transferases, which transform two-electron (hydride) transfer into two subsequent single-electron transfers

---

[9] G. Grampp and W. Jaenicke, *J. Electroanal. Chem.* **229**, 297 (1987), and references therein.
[10] A. G. Mauk, R. Scott, and H. B. Gray, *J. Am. Chem. Soc.* **102**, 4360 (1980).

TABLE I

SINGLE-ELECTRON REDUCTION POTENTIALS OF QUINONES AND RELATED COMPOUNDS

No. Compound	E_7^1 (V)	pK_a(OH$^\bullet$)	k_{cat}/K_m (M^{-1} s^{-1})		$VdWvol$ (Å^3)
			P-450R	FNR	
1. 1,4-BQ	0.09	4.10	1.5×10^7	3.5×10^7	−124.6
2. 2-CH$_3$-1, 4-BQ	0.01	4.45	4.3×10^7	2.0×10^6	142.5
3. 2,3-Cl$_2$-1,4-NQ	−0.035	—	1.1×10^7	2.9×10^5	206.5
4. 2,5-(AZ)$_2$-1,4-BQ (DZQ)	−0.054	5.00	7.5×10^6	1.0×10^5	199.2
5. 2,5-(Az)$_2$-3,6-(NHCOOC$_2$H$_5$)$_2$-1,4-BQ (AZQ)	−0.07	6.30			360.6
6. 2,5-(Az)$_2$-3,6-(NHCOOCH$_3$)$_2$-1,4-BQ	−0.044				324.4
7. 2-CH$_3$-5-Az-1,4-BQ	−0.07^b		9.5×10^6	3.0×10^5	179.9
8. 2,5-(CH$_3$)$_2$-1,4-BQ	−0.07	4.60	8.6×10^6		160.6
9. 2,6-(CH$_3$)$_2$-1,4-BQ	−0.08	4.75			160.6
10. 5-OH-1,4-NQ	−0.09	3.65	4.6×10^7	3.2×10^5	179.7
11. Methylene Blue	−0.10			1.2×10^6	326.2
12. 5,8-(OH)$_2$-1,4-NQ	−0.11	2.80		9.2×10^5	192.3
13. 9,10-PQ	−0.12		7.0×10^7	1.3×10^5	218.3
14. 1,4-NQ	−0.15	4.10	3.1×10^7	2.2×10^6	168.1
15. 2-CH$_3$-5-OH-1,4-NQ	−0.16		3.5×10^7	6.0×10^5	199.6
16. (CH$_3$)$_3$-1,4-BQ	−0.17	5.00	1.9×10^7	6.7×10^5	178.5
17. 2-CH$_3$-1,4-NQ	−0.20	4.50			187.2
18. 2-CH$_3$-3,6-(Az)$_2$-5-CH$_2$OH-1,4-BQ (RH1)	−0.20^b		1.1×10^7	4.6×10^5	246.9
19. 2,5-(CH$_3$)$_2$-3,6-(Az)$_2$-1,4-BQ (MeDZQ)	−0.23^c		6.7×10^6		237.7
20. 9,10-AQ-2,6-(SO$_3^-$)$_2$	−0.25	3.00	2.6×10^6		314.5
21. 2,3,5-(CH$_3$)$_3$-6-Az-1,4-BQ	−0.25^b				213.7
22. (CH$_3$)$_4$-1,4-BQ	−0.26		4.2×10^6	2.0×10^5	197.5
23. Mitomycin C	−0.31	5.10	5.0×10^6		313.9
24. Riboflavin	−0.32	8.50	4.3×10^5	2.4×10^5	422.7

(continued)

TABLE I (continued)

No. Compound	E_7^1 (V)	pK_a(QH$^{\bullet}$)	k_{cat}/K_m (M^{-1} s^{-1})		$VdWvol$ (Å^3)
			P-450R	FNR	
25. 1,8-(OH)$_2$-9,10-AQ	-0.325	3.95	1.7×10^5	2.5×10^5	245.5
26. Adriamycin	-0.34	2.80	5.0×10^4	6.3×10^4	424.5
27. 2,5-(Az)$_2$-3,6-(NHC$_2$H$_4$OH)$_2$-1,4-BQ	-0.38		4.8×10^5		327.2
28. 2,5-(Az)$_2$-3,6-(NHC$_2$H$_5$)$_2$-1,4-BQ (BZQ)	-0.38^b		4.0×10^5		304.2
29. 9,10-AQ-2-SO$_3^-$	-0.38	3.25			266.4
30. 1-OH-9,10-AQ	-0.385	4.60			231.5
31. 2-OH-1,4-NQ	-0.41	4.70	4.4×10^4	6.7×10^3	179.7
32. 2-CH$_3$-3-OH-1,4-NQ	-0.46^b				199.6

Single-electron reduction potentials of quinones and related compounds at pH 7.0 (E_7^1),[a] pK_a of their semiquinones (pK_a[QH$^{\bullet}$]),[a] their bimolecular rate constants of reduction (k_{cat}/K_m) by pig liver NADPH:cytochrome P-450 reductase (P-450R), by Anabaena ferredoxin:NADP$^+$ reductase (FNR) at pH 7.0 and 25°, and their Van der Waals volumes (VdWvol). Abbreviations: AQ, anthraquinone; Az, aziridine; BQ, benzoquinone; NQ, naphthoquinone; PQ, phenanthrene quinone.

[a] P. R. Rich and D. S. Bendall, Biochim. Biophys. Acta 592, 506 (1980); A. J. Swallow, in: "Functions of Quinones in Energy Conserving Systems" (B. L. Trumpower, ed.), p. 59. Academic Press, New York, San Francisco, London (1982); T. Mukherjee, E. J. Land, A. J. Swallow, P. M. Guyan, and J. M. Bruce, J. Chem. Soc. Perkin Trans. I 84, 2855 (1988); P. Wardman, J. Phys. Chem. Ref. Data 18, 1637 (1988), and references therein; T. Mukherjee, E. J. Land, A. J. Swallow, and J. M. Bruce, Arch. Biochem. Biophys. 275, 450 (1989); J. Lind and G. Merenyi, Photochem. Photobiol. 51, 21 (1990); A. Dzielendziak, J. Butler, B. M. Hoey, J. S. Lea and T. H. Ward, Cancer Res. 50, 2003 (1990); J. A. Hardley, M. Berardini, M. Ponti, N. W. Gibson, A. S. Thompson, A. S. Thurston, B. M. Hoey, and J. Butler, Biochemistry 30, 11719 (1991); P. J. O'Brien, Chem. Biol. Interact. 80, 1 (1991); M. C. Rath, H. Pal, and T. Mukherjee, Radiat. Phys. Chem. 47, 221 (1996); E. B. Xing and E. B. Skibo, Biochemistry 39, 10770 (2000).

[b] E_7^1 calculated assuming that (i) aziridine and methyl groups, and ethylamino and 2-ethanolamino groups may similarly influence the E_7^1 value of benzoquinone, (ii) the substitution of methyl group by hydroxymethyl may increase the E_7^1 of benzoquinone by 0.05 V; and (iii) 3-methyl group may decrease the potential of naphthoquinone by 0.05 V.

[c] J. Butler, personal communication.

to single-electron acceptor and possess stabilized neutral (blue) flavin semi-quinone.[11,12] P-450R contains FAD and FMN in the active center, and transfers redox equivalents in a sequence NADPH → FAD → FMN → cytochrome P-450 (cytochrome c).[11] Like cytochrome c, quinones are reduced via reduced FMN (E_7[FMNH•/reduced FMN] = −0.27 V), whereas FMNH• does not transfer electrons to quinones, probably because of a higher potential of FMN/FMNH• couple, −0.11 V.[11] During the reaction of two-electron reduced FNR from *Anabaena* with quinones, the oxidation of FADH$^-$ to FADH• (E_7[FADH$^-$/FADH•] = −0.312 V[12]) is faster than that of FADH•,[7] since E_7(FAD/FADH•) = −0.279 V[12] (Eq. [5]):

$$\begin{array}{ccccc}
Q & Q^{\bullet-} & Q & Q^{\bullet-}+H^+ & \\
\overbrace{\quad\quad} & & \overbrace{\quad\quad} & & '' \\
\text{E-FADH}^- \xrightarrow{\quad\quad} & \text{E-FADH}^\bullet & \xrightarrow{\quad\quad} & \text{E-FAD} &
\end{array} \qquad (5)$$

The rates of P-450R- and FNR-catalyzed reactions are determined spectrophotometrically, monitoring the rate of NADPH oxidation ($\Delta\varepsilon_{340}$ = 6.22 mM^{-1} cm^{-1}). Typically, the saturating NADPH concentrations—100 μM for P-450R and 250 μM for FNR—may be used. In the case of high quinone absorbance at 340 nm, the reaction rate may be determined at 370 nm ($\Delta\varepsilon_{370}$ = 2.66 mM^{-1} cm^{-1}). Under aerobic conditions, the quinone radicals formed are rapidly reoxidized by oxygen, and the reaction rates usually are not corrected for the changes in quinone absorbance. The catalytic constants (k_{cat}) and the bimolecular constants (k_{cat}/K_m) of quinone reduction correspond to the reciprocal intercepts and slopes of plots [E]/v vs. 1/[Q], where [E] and [Q] are the enzyme and quinone concentrations, respectively. Typically, 6 to 9 quinone concentrations may be used. The [E] in the assay may vary between 1.0 and 100 nM, depending on the quinone activity. Enzyme concentrations are determined spectrophotometrically, ε_{460} = 22 mM^{-1} cm^{-1} (P-450R), and ε_{459} = 9.4 mM^{-1} cm^{-1} (FNR).

In the analysis, the redox potential, steric and, possibly, proton transfer energetics effects should be considered, as detailed in the following sections.

Redox Potential Effects

In the analysis, one should use a series of quinones with a broad range of E_7^1, typically from 0.09 V to −0.41 V (Table I). The use of quinones with higher redox potentials is not recommended because of a high rate of direct (nonenzymatic) oxidation of NADPH. The k_{cat}/K_m of quinones in

---

[11] H. Matsuda, S. Kimura, and T. Iyanagi, *Biochim. Biophys. Acta* **1459**, 106 (2000), and references therein.
[12] M. Faro, C. Gomez-Moreno, M. Stankovich, and M. Medina, *Eur. J. Biochem.* **269**, 2656 (2002), and references therein.

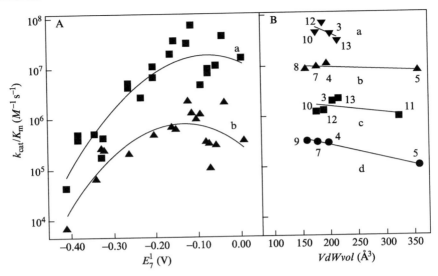

FIG. 1. Structure-activity relationships in single-electron reduction of quinones by pig liver NADPH:cytochrome P-450 reductase (P-450R) and *Anabaena* ferredoxin:NADP$^+$ reductase (FNR). (A) Dependence of quinone k_{cat}/K_m on their single-electron reduction potential at pH 7.0 (E_7^1) in P-450R-(a) and FNR (b)-catalyzed reactions. (B) Dependence of k_{cat}/K_m of benzoquinones (b,d), and polycyclic quinones (a,c) with $E_7^1 = -0.054$--0.12 V in P-450R-(a,b) and FNR (c,d)-catalyzed reactions on their Van der Waals volume (*VdWvol*). The numbers of compounds, their E_7^1, *VdWvol* and k_{cat}/K_m values are taken from Table I.

P-450R- and FNR-catalyzed reactions, partly determined in our previous studies,[7,13,14] are given in Table I. The log k_{cat}/K_m exhibit scattered parabolic dependences on E_7^1, similar to those predicted by an "outer-sphere" electron transfer model (Fig. 1A). The k_{cat} of P-450R and FNR in the reactions with quinones of $E_7^1 = 0.01$ V $-$ -0.26 V are almost potential-independent and were not used in the analysis. Analogous log k_{cat}/K_m vs. E_7^1 relationships are also characteristic for other dehydrogenases-electrontransferases, NADPH:adrenodoxin reductase (EC 1.18.1.2),[15] NADH:cytochrome b$_5$ reductase (EC 1.6.2.2),[4] and mitochondrial NADH:ubiquinone reductase (Complex I, EC 1.6.5.3).[16] It is unclear

[13] P. Grellier, J. Šarlauskas, Ž. Anusevičius, A. Marozienė, C. Houee-Levin, J. Schrevel, and N. Čėnas, *Arch. Biochem. Biophys.* **393**, 199 (2001).

[14] A. Nemeikaitė-Čėnienė, E. Sergedienė, H. Nivinskas, and N. Čėnas, *Z. Naturforsch.* **57c**, 822 (2002).

[15] N. K. Čėnas, J. A. Marcinkevičienė, J. J. Kulys, and S. A. Usanov, *Biokhimiya (in Russian)* **52**, 643 (1987).

[16] D. A. Bironaitė, N. K. Čėnas, and J. J. Kulys, *Biochim. Biophys. Acta* **1060**, 203 (1991).

whether the nonlinearity of plots (Fig. 1A) is a general pattern predicted by an outer-sphere electron-transfer model[3] or is caused by the limitation of k_{cat}/K_m by substrate diffusion. Nevertheless, the plots may be approximated by the second order polynomial regression using E_7^1 and $(E_7^1)^2$ as variables, giving $r^2 = 0.8412$ for P-450R, and $r^2 = 0.6517$ for FNR (Fig. 1A).

Steric Effects

As the simplest structural parameters of quinones, one may use their Van der Waals volumes *(VdWvol)* (Table I) or the number of their aromatic rings, N. The use of *VdWvol* as an additional parameter in the regression, or the use of *VdWvol* + *(VdWvol)*,[2] expecting the existence of an optimal quinone volume for electron transfer, did not improve the correlations (data not shown). It is evident that the reactivity of benzoquinones are less active than policyclic quinones with similar $E_7^1 = -0.054$ V--0.07 V vaguely depends on their *VdWvol*, and benzoquinones are less active than policyclic quinones with similar E_7^1 and *VdWvol* (Fig. 1B). The use of $N = 1–3$ for compounds analyzed (Table I) as an additional parameter did not improve the correlation for P-450 R ($r^2 = 0.8433$) but markedly improved the correlation for FNR:

$$\log k_{cat}/K_m = 4.9579 \pm 0.2186 - (3.9551 \pm 2.5253)E_7^1$$

$$-(20.8006 \pm 5.6292)(E_7^1)^2 + (0.3982 \pm 0.1378)N,$$

$$(r^2 = 0.8492) \tag{6}$$

On the other hand, the use of N^2 as the additional parameter markedly improved the correlation for P-450R:

$$\log k_{cat}/K_m = 5.7669 \pm 0.60211 - (3.7700 \pm 2.4598)E_7^1$$

$$- (24.4922 \pm 5.8146)(E_7^1)^2 + (1.8295 \pm 0.6657)N$$

$$-(0.4960 \pm 0.1725)N^2, \quad (r^2 = 0.9481) \tag{7}$$

but did not improve the correlation for FNR ($r^2 = 0.8500$). Thus it seems that P-450R exhibits certain specificity towards naphthoquinone structure ($N = 2$, Eq. [7]), whereas FNR prefers a general increase in a number of aromatic rings (Eq. [6]).

Proton Transfer Energetics

The oxidation of neutral flavin semiquinone (Eq. [3]) involves the proton dissociation from N-5 position of isoalloxazine, thus formally being a net hydrogen atom transfer. Since at pH 7.0 quinone radicals are anionic

(Table I), the proton from flavin is finally transferred to solvent. However, the possibility of a proton-accompanied electron transfer—that is, the transient formation of QH•—should be considered. In this case, the reactivity of quinones may deviate from an outer-sphere model and increase with an increase in $pK_a(QH•)$. However, the use of $pK_a(QH•)$ (Table I) as an additional parameter in the analysis of reactions of P-450R and FNR did not show an increase in quinone reactivity with an increase in their $pK_a(QH•)$. This argues against the possible formation of QH⁻ in single-electron quinone reduction and is in line with the separate transfer of e^- and H^+.[9]

Finally, the k_{11} values of P-450R and FNR ($10^4 \, M^1 \, s^{-1}$ and $10^2 \, M^1 \, s^{-1}$, respectively), and their electron-transfer distances in quinone reduction (0.31 nm and 0.47 nm, respectively) were calculated by using Eq. (1, 2, 4). The R_p values, being close to the doubled Van der Waals radius of aromatic carbon, 0.36 nm, are consistent with a good solvent accessibility of the flavin cofactors of P-450R and FNR.[12,17]

Summing up, the multiparameter regression analysis supported the view on the proton-unassisted electron transfer following an outer-sphere transfer model in the reduction of quinones by P-450R and FNR. Besides, it provided additional information on the enzyme specificity for policyclic quinones. In the next section, this approach will be extended for the analysis of two-electron reduction of quinones.

Two-Electron Reduction of Quinones by Flavoenzymes

Mechanisms of Two-Electron (Hydride) Transfer

Typically, the two-electron reduction of quinones at pH 7.0 is accompanied by the transfer of two protons and may be considered a net hydride (H^-) and H^+ transfer. The midpoint potentials of quinone/hydroquinone couples (or standard potentials) at pH 7.0 (E_7^0) are given in Table II. The values of E_7^0 were used in the analysis of two-electron reduction of quinones by glucose oxidase, giving separate linear log k_{cat}/K_m vs. E_7^0 relationships for o- and p-quinones.[18] Similar results were obtained in the nonenzymatic oxidation of NADH by quinones[19,20];

[17] M. Wang, D. L. Roberts, R. Paschke, T. M. Shea, B. S. S. Masters, and J.-J. P. Kim, *Proc. Natl. Acad. Sci. USA* **94,** 8411 (1997).

[18] J. J. Kulys and N. K. Čėnas, *Biochim. Biophys. Acta* **744,** 57 (1983).

[19] N. K. Čėnas, J. J. Kanapieniené, and J. J. Kulys, *Biochim. Biophys. Acta* **767,** 108 (1984).

[20] B. W. Carlson and L. L. Miller, *J. Am. Chem. Soc.* **107,** 479 (1985).

TABLE II

STANDARD (TWO-ELECTRON) REDUCTION POTENTIALS OF QUINONES AND RELATED COMPOUNDS

No. Compound	E_7^0 (V)	$pK_a(QH_2)$	$E_7(Q/QH^-)$ (V)	k_{cat}/K_m (M^{-1} s^{-1}) ($k_{cat}[s^{-1}]$) NQO1	NR
1. 1,4-BQ	0.28	9.9	0.195	1.3×10^8 (769)	1.4×10^8 (1200)
2. 2-CH$_3$-1,4-BQ	0.21	10.0	0.12	1.3×10^8 (287)	5.0×10^7 (666)
3. 2,3-Cl$_2$-1,4-NQ	0.08				1.3×10^7 (250)
4. 2,5-(Az)$_2$-1,4-BQ (DZQ)	0.12	11.0	0.00	3.0×10^8 (1550)	1.5×10^7 (1020)
5. 2,5-(Az)$_2$-3,6-(NHCOOC$_2$H$_5$)$_2$-1,4-BQ (AZQ)	0.07	11.0	−0.046	1.0×10^5 (≥50)	1.5×10^5 (9.0)
6. 2,6-(Az)$_2$-3,6-(NHCOOCH$_3$)$_2$-1,4-BQ			−0.046^b	1.0×10^5 (≥50)	2.0×10^5 (6.0)
7. 2-CH$_3$-5-Az-1,4-BQ			0.03^b	1.1×10^8 (1540)	5.4×10^6 (500)
8. 2,5-(CH$_3$)$_2$-1,4-BQ	0.16	10.4	0.058	1.4×10^8 (500)	
9. 2,6-(CH$_3$)$_2$-1,4-BQ	0.16	10.4	0.058	4.9×10^7 (625)	1.8×10^7 (704)
10. 5-OH-1,4-NQ	−0.02	8.5	−0.06	2.4×10^8 (242)	1.2×10^7 (77)
11. Methylene Blue	0.00			5.6×10^6 (≥50)	8.5×10^4 (5.0)
12. 5,8-(OH)$_2$-1,4-NQ	−0.06	7.8	−0.084	2.6×10^8 (71)	2.5×10^7 (111)
13. 9,10-PQ	0.02	8.8	−0.034	2.6×10^6 (52)	2.4×10^7 (333)
14. 1,4-NQ	0.04	9.3	−0.029	5.4×10^8 (431)	2.2×10^7 (200)
15. 2-CH$_3$-5-OH-1,4-NQ	−0.08			1.6×10^8 (500)	1.4×10^6 (77)
16. (CH$_3$)$_3$-1,4-BQ	0.11	10.8	0.00	1.3×10^8 (839)	1.0×10^7 (400)
17. 2-CH$_3$-1,4-NQ	−0.03	9.8	−0.114	8.4×10^7 (380)	3.5×10^6 (50)
18. 2-CH$_3$-3,6-(Az)$_2$-5-CH$_2$OH-1,4-BQ (RH1)				3.6×10^7 (2000)	
19. 2,5-(CH$_3$)$_2$-3,6-(Az)$_2$-1,4-BQ (MeDZQ)	0.01	11.4^c	−0.128^c	1.9×10^6 (111)	1.0×10^5 (8.0)
20. 9,10-AQ-2,6-(SO$_3^-$)	−0.20	7.4	−0.21	2.0×10^5 (2.3)	1.2×10^4 (0.8)
21. 2,3,5-(CH$_3$)$_3$-6-Az-1,4-BQ			−0.107^b	6.5×10^6 (625)	7.0×10^4 (3.0)
22. (CH$_3$)$_4$-1,4-BQ	0.04	11.2	−0.086	6.7×10^7 (1000)	3.8×10^4 (17)
23. Mitomycin C	−0.13	≥10.0	−0.22	1.6×10^4 (1.7)	2.3×10^4 (1.0)
24. Riboflavin	−0.21	6.7	−0.21	2.5×10^4 (1.4)	1.0×10^5 (5.3)

(continued)

TABLE II (continued)

No. Compound	E_7^0 (V)	$pK_a(QH_2)$	$E_7(Q/QH^-)$ (V)	k_{cat}/K_m (M^{-1} s^{-1}) ($k_{cat}[s^{-1}]$)	
				NQO1	NR
25. 1,8-$(OH)_2$-9,10-AQ	-0.30			1.1×10^5 (0.5)	5.0×10^4 (1.0)
26. Adriamycin	-0.165	8.1	-0.34	300 (<0.1)	$\leq 10^3$ (<0.1)
27. 2,5-$(Az)_2$-3,6-$(NHC_2H_4OH)_2$-1,4-BQ (BZQ)		12.4^c	-0.32^c	$\leq 3.0 \times 10^3$ (<0.1)	4.0×10^4 (0.4)
28. 2,5-$(Az)_2$-3,6-$(NHC_2H_5)_2$-1,4-BQ			-0.32^b	3.2×10^5 (40)	
29. 9,10-AQ-2-SO_3^-	-0.23	8.65	-0.25	1.6×10^5 (3.6)	2.5×10^4 (0.8)
30. 1-OH-9,10-AQ				8.6×10^4 (2.5)	1.2×10^4 (1.5)
31. 2-OH-1,4-NQ	-0.14	9.0	-0.20	5.9×10^6 (232)	1.2×10^7 (160)
32. 2-CH_3-3-OH-1,4-NQ	-0.18	8.9	-0.24	1.0×10^6 (67)	3.7×10^6 (111)

Standard (two-electron) reduction potentials of quinones and related compounds at pH 7.0 (E_7^0),[a] pK_a of their hydroquinones ($pK_a[QH_2]$),[a] hydride transfer potentials ($E_7[Q/QH^-]$), and their bimolecular rate constants of reduction (k_{cat}/K_m) by rat liver NAD(P)H:quinone oxidoreductase (NQO1), and *Enterobacter cloacae* NAD(P)H:nitroreductase (NR). The catalytic constants (k_{cat}) of reactions given in parentheses. Rate constants determined at pH 7.0 and 25°, in the absence of activators. Abbreviations: AQ, anthraquinone; Az, aziridine; BQ, benzoquinone; NQ, naphthoquinone; PQ, phenanthrene quinone.

[a] E. Bishop, "Indicators," Vol. 2. Pergamon Press, Oxford (1972); G. M. Rao, A. Begleiter, J. W. Lown, and J. A. Plambeck, *J. Electrochem. Soc.* 124, 199 (1977); P. R. Rich and D. S. Bendall, *Biochim. Biophys. Acta* 592, 506 (1980); S. I. Bailey and I. M. Ritchie, *Electrochim. Acta* 30, 3 (1985); T. Mukherjee, E. J. Land, A. J. Swallow, P. M. Guyan, and J. M. Bruce, *J. Chem. Soc. Perkin Trans. I* 84, 2855 (1988); P. He, R. M. Crooks, and L. R. Faulkner, *J. Phys. Chem.* 94, 1135 (1990); R. J. Driebergen, J. J. M. Holthuis, J. S. Blauw, S. J. Postma Kelder, W. Verboom, D. N. Reinhoudt, and W. E. van der Linden, *Anal. Chim. Acta* 234, 285 (1990); R. J. Driebergen, E. E. Moret, L. H. M. Janssen, J. S. Blauw, J. J. M. Holthuis, S. J. Postma Kelder, W. Verboom, D. N. Reinhoudt, and W. E. van der Linden, *Anal. Chim. Acta* 257, 257 (1992).

[b] Calculated according to the $E_7^1(Q/QH^-)$ of their closest structural analog.

[c] Calculated according to Eq. (11).

$$NADH + Q + H^+ \rightarrow NAD^+ + QH_2. \tag{8}$$

The further studies of this reaction resulted in the two most commonly used models, a single-step (H^-) and a three-step (e, H^+, e^-) hydride transfer mechanism. The kinetic isotope effects (*KIE*) when using 4-H-deuterated NADH analog[21,22] show that during two-electron reduction, quinones may accept the hydrogen atom. The possibility of the two-step ($H^\bullet, H^+$) hydride transfer has been considered as well; however, it is not supported by substantial experimental data.[21,22]

Single-Step Hydride Transfer

The use of E_7^0 as the correlation parameter did not explain the pH-independent character of the reaction (Eq. [8]), since, according to the Nernst equation, the ΔE^0 of redox couples Q/QH_2 and $NAD^+/NADH$ should increase at lower pH, thus increasing the reaction rate. It was proposed that the reaction involves the rate-limiting single-step H^- transfer and subsequent fast protonization:

$$NADH + Q \rightarrow NAD^+ + QH^- \xrightarrow{H^+} NAD^+ + QH_2 \tag{9}$$

In this case, the potential of quinone/anionic hydroquinone couple $E_7[Q/QH^-]$ (or hydride-transfer potential) are used as the correlation parameter.[20] $E_7(Q/QH)$ is equal to E_7^0 of quinone, if pK_a of hydroquinone ($pK_a[QH_2]$, Table II) is equal or below 7.0, or

$$E_7(Q/QH^-) = E_7^0 - 0.029 \text{ V}(pK_a[QH_2] - 7.0) \tag{10}$$

The $E_7(Q/QH^-)$ values of quinones are given in Table II. The unavailable $pK_a(QH_2)$ for several aziridinylbenzoquinones, and, subsequently, their $E_7(Q/QH^-)$ values, were calculated according to a linear dependence of $pK_a(QH_2)$ on E_7^0 of substituted 1,4-benzoquinones ($r^2 = 0.9167$):

$$pK_a(QH_2) = 11.4616 \pm 0.1299 - (5.9909 \pm 0.8075)E_7^0 \tag{11}$$

The E^0 and $pK_a(QH_2)$ of quinones may be easily obtained by using standard voltammetric techniques. For the reversible redox process, E^0 is equal to the midpoint between the reduction and oxidation peak potentials in cyclic voltammetry. The $pK_a(QH_2)$ may be obtained from the pH dependence of E^0. It corresponds to the point where the slope $\Delta E^0/\Delta pH \approx -60$ mV changes into $\Delta E^0/\Delta pH \approx -30$ mV.

[21] S. Fukuzumi, M. Ishikawa, and T. Tanaka, *J. Chem. Soc. Perkin Trans. II* **1811** (1989), and references therein.

[22] S. Fukuzumi, K. Ohkubo, Y. Tokuda, and T. Suenobu, *J. Am. Chem. Soc.* **122,** 4286 (2000), and references therein.

By analogy with Eq. (9), one may propose a single-step hydride transfer scheme in the two-electron reduction of quinones by flavoenzymes:

$$E - FADH^- + Q \rightarrow [E - FADH^- \ldots Q] \rightarrow E - FAD + QH^- \quad (12)$$

In this case, one may expect a linear or parabolic dependence of log k_{cat}/K_m or log k_{cat} on $E_7(Q/QH^-)$ of quinones.

Three-Step Hydride Transfer

The isolation of paramagnetic charge-transfer intermediates in the oxidation of NADH analog by quinones[21,22] resulted in an alternative three-step (e^-, H^+, e^-) oxidation mechanism that is currently more widely accepted. It involves partly rate-limiting first electron (k_1) and proton transfers (k_H):

$$NADH + Q \underset{k_{-1}}{\overset{k_1}{\rightleftharpoons}} NADH^{\bullet+} \ldots Q^{\bullet-} \overset{k_H}{\rightarrow} NAD^{\bullet} \ldots QH^{\bullet} \rightarrow NAD^+ + QH^-, \quad (13)$$

The first electron transfer in Eq. (13) is thermodynamically unfavorable, since $E_7^1(NADH^{\bullet+}/NADH)$ is 0.93 V – 1.05 V,[23] or 0.75 V.[24] The endothermicity is decreased by the energy of stabilization of the $NADH^{+\bullet} \ldots Q^{\bullet-}$ ion-radical pair, 0.3 V.[21,22] The reaction is also driven forward by the exothermic proton transfer ($pK_a[NADH^{\bullet+}] = -3.5$,[23] or -0.5[24]). The log of the reaction rate constant, which may be expressed as $k = k_1 \times k_H/(k_{-1} + k_H)$, increased with an increase in E^1 and was accompanied by a decrease in KIE.[21,22] The increase in k^1 and/or decrease in k_{-1} at high redox potential of quinone is consistent with an outer-sphere electron transfer model. At high E^1 values, $k \approx k_1$, which leads to the disappearance of KIE.

By analogy with Eq. (13), the scheme of three-step hydride transfer in the two-electron reduction of quinones by flavoenzymes may be proposed:

$$E - FADH^- + Q \rightleftharpoons [E - FADH^- \ldots Q] \rightleftharpoons [E - FADH^{\bullet} \ldots Q^{\bullet-}]$$

$$\rightleftharpoons [E - FAD^{\bullet-} \ldots QH^{\bullet}] \rightarrow E - FAD + QH^- \quad (14)$$

In this scheme, the first electron transfer will be thermodynamically favorable, since the E_7^1 of flavin semiquinone/hydroquinone couples are much lower than that of $NADH^{\bullet+}/NADH$ (Eq. [13]). The energy of stabilization of $E\text{-}FADH^{\bullet} \ldots Q^{\bullet-}$ may be low because of the absence of electrostatic attraction; however, an additional stabilization may be provided by the

[23] B. W. Carlson, L. L. Miller, P. Neta, and J. Grodkowski, *J. Am. Chem. Soc.* **106**, 7233 (1984).
[24] F. M. Martens and J. W. Verhoeven, *Recl. Trav. Chim. Pays-Bas* **100**, 228 (1981).

enzyme active center. The proton transfer within the ion-radical pair E-FADH$^\bullet$...QH$^{\bullet-}$ also may take place, since, as a rule, flavoenzymes performing two-electron quinone reduction possess unstable red (anionic) semiquinones.[25,26] Thus one may expect the linear or parabolic dependence of log k on E_7^1 of quinones and an increase in the reactivity with an increase in pK_a(QH$^\bullet$) (Table I).

Our previous studies pointed to a possibility of a three-step (Eq. [14]) transfer in two-electron reduction of quinones. In reactions of *Clostridium kluyveri* diaphorase, the log k_{cat}/K_m of quinones exhibited parabolic dependence on their E_7^1.[27] The use of 4-*S*-^{2}H-NADH as substrate resulted in the appearance of *KIE* in quinone reduction rate constant. It means that ^{2}H, transferred from C-4 of dihydronicotinamide to N-5 of flavin, may be further transferred to quinone if the N-5 proton exchange rate with solvent is sufficiently low. *KIE* decreased with an increase in E_7^1—for example, $KIE = 1.0$ for 1,4-benzoquinone, and $KIE = 3.8$ for 2-methyl-1,4-naphthoquinone.[28] Lipoamide dehydrogenase (EC 1.6.4.3) contains FAD and redox active disulfide in the active center. It performs mixed single- and two-electron reduction of quinones (24% single-electron flux) via FAD cofactor.[29] Again, the parabolic log k_{cat}/K_m vs. E_7^1 dependence was found in the quinone reduction, accompanied by the *KIE* in k_{cat}/K_m of quinone reduction when using 4-*S*-^{2}H-NADH.[29] The kinetic isotope effect (1.85–3.0) also decreased with an increase in E_7^1 of quinones. These data point to a possibility of a common mechanism of two- and mixed single- and two-electron reduction of quinones; an initial electron transfer (Eq. [14]) may be followed either by flavin-quinone ion-radical pair dissociation (single-electron transfer) or by a subsequent proton and second electron transfer (two-electron transfer). The preceding results, obtained with a limited number of compounds, did not reveal the specificity of enzymes for the particular structures of quinones. Extending our previous studies,[30,31] the next sections of this chapter will address the role of hydride transfer energetics and steric substrate effects in the reactions of rat liver NAD(P):quinone oxidoreductase (NQO1, DT-diaphorase, EC 1.6.99.2), and *Enterobacter cloacae* NAD(P)H: nitroreductase (NR, EC 1.6.99.7). It is important to

[25] G. Tedeschi, S. Chen, and V. Massey, *J. Biol. Chem.* **270**, 1198 (1995).

[26] C. A. Haynes, R. L. Koder, A.-F. Miller, and D. W. Rodgers, *J. Biol. Chem.* **277**, 11513 (2002), and references therein.

[27] N. K. Čėnas, J. V. Vienožinskis, and J. J. Kulys, *Biokhimiya (in Russian)* **52**, 66 (1987).

[28] N. K. Čėnas, J. V. Vienožinskis, and J. J. Kulys, *Biokhimiya (in Russian)* **53**, 33 (1989).

[29] J. Vienožinskis, A. Butkus, N. Čėnas, and J. Kulys, *Biochem. J.* **269**, 101 (1990).

[30] H. Nivinskas, S. Staškevičienė, J. Šarlauskas, R. L. Koder, A. F. Miller, and N. Čėnas, *Arch. Biochem. Biophys.* **403**, 249 (2002).

[31] Ž. Anusevičius, J. Šarlauskas, and N. Čėnas, *Arch. Biochem. Biophys.* **404**, 254 (2002).

note that the multiparameter regression analysis using E_7^1 and $E_7(Q/QH^-)$ as variables may not provide the straightforward evidence for the single- or three-step mechanism of hydride transfer, since there exists a linear although scattered relationship between E_7^1 and $E_7(Q/QH)$ with $r^2 = 0.8367$ (Tables I and II).

Structure-Activity Relationship in Reactions of Rat Liver NQO1

Quinone reduction by NQO1 follows a "ping-pong" scheme with a rate-limiting reoxidation step. Quinone reactivity does not depend on their reduction potential[8] and lipophilicity.[32] The E_7^0 of FAD in the active center is -0.159 V. The anionic FAD semiquinone is unstable, since the potentials of FAD/FAD and $FAD^{\bullet-}/FADH^-$ couples are -0.200 and -0.118 V, respectively.[26] The formation of $FAD^{\bullet-}$ intermediate has not been detected during the reoxidation of reduced NQO1 by quinones.[26] This is consistent with single-step hydride transfer (Eq. [12]); however, it also does not contradict a three-step model (Eq. [14]) with a faster second electron transfer.

The k_{cat}/K_m and k_{cat} of quinone reduction by NQO1 (Table II) were determined like those in single-electron reduction reactions of quinones.[31] Since hydroquinones autoxidize much slower than semiquinones, the reaction rates determined at 340 nm should be corrected for the decrease in quinone absorbance, if necessary. The reduction of naphthoquinones and benzoquinones with $E_7^1 < 0.0$ V may be monitored in coupled cytochrome c reduction assay ($\Delta\varepsilon_{550} = 20$ mM^{-1} cm^{-1}, concentration of added cytochrome c, 50 μM). Alternatively, subnanomolar concentrations of NQO1 may be used to ensure the low reaction rate commeasurable with hydroquinone autoxidation rate. High quinone concentrations may complicate the data analysis, since quinones are competitive with respect to NAD(P)H inhibitors of NQO1, forming the dead-end complexes with the oxidized enzyme. The data may be analyzed by using Eq. (15):

$$v/[E] = \{k_{cat} \times K_{iQ} \times [Q] \times [NADPH]\}/\{K_{iQ} \times [Q] \times [NADPH] + K_{m(Q)} \times [NADPH] + K_{m(NADPH)} \times [Q] + K_{m(NADPH)} \times [Q]\}^2 \tag{15}$$

where K_{iQ} is quinone substrate inhibition constant. Alternatively, K_{iq} are determined at several fixed NADPH concentrations as the intercepts of plots $[E]/v$ vs. $[Q]$ with the x-axis, with further extrapolation to zero NADPH concentration.

In contrast to the reactions of P-450R and FNR (Fig. 1B), quinone reactivity toward NQO1 decreases with an increase in their $VdWvol$,

[32] G. D. Buffington, K. Ollinger, A. Brunmark, and E. Cadenas, *Biochem. J.* **257**, 561 (1989).

irrespective of the number of aromatic rings in quinone molecule (Fig. 2). By using a single-step hydride transfer model (Eq. [12]), a parabolic log k_{cat}/K_m vs. $E_7(Q/QH^-)$ dependence and the reactivity decrease with an increase in $VdWvol$ of quinones is demonstrated:

$$\log k_{cat}/K_m = 10.5836 \pm 0.4216 + (0.2242 \pm 1.5926)E_7(Q/QH^-)$$

$$-(16.0872 \pm 5.6069)(E_7(Q/QH^-))^2$$

$$-(0.0149 \pm 0.0020)VdWvol \quad (r^2 = 0.8937) \qquad (16)$$

Using a three-step hydride transfer model (Eq. [14]), an increase in quinone reactivity with an increase in $pK_a(QH^\bullet)$ and decrease in $VdWvol$ is demonstrated as well:

$$\log k_{cat}/K_m = (10.0660 \pm 0.535) + (6.1351 \pm 2.7984)E_7^1$$

$$-(24.9081 \pm 6.7954)(E_7^1)^2 + (0.1845 \pm 0.1079)pK_a(QH^\bullet)$$

$$-(0.0169 \pm 0.0018)VdWvol \quad (r^2 = 0.9251) \qquad (17)$$

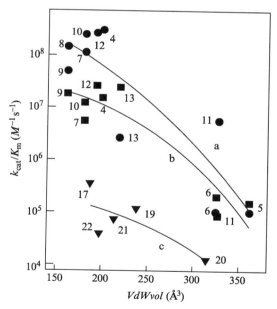

FIG. 2. Dependence of k_{cat}/K_m of quinones in rat liver NQO1 (a) and *Enterobacter cloacae* introreductase (NR) (b,c)-catalyzed reactions on their Van der Waals volume *(VdWvol)*. The E_7^1 of quinones are between -0.044 V and -0.12 V (a,b), and between -0.20 V and -0.260 V (c). The numbers of compounds, their E_7^1, $VdWvol$ and k_{cat}/K_m values are taken from Tables I and II.

The log k_{cat} values of NQO1 (Table II) showed the same trend when using both models. The introduction of $(VdWvol)^2$ as an additional parameter did not improve the correlations. The use of a number of aromatic rings (N) or $N + N^2$ instead of $VdWvol$ in the analysis showed the absence of correlation. The steric effects of quinone ring(s) and substituents may be assessed separately as well. The Van der Waals volume of substituents ($VdWvol_{subst}$) was defined as $VdWvol$ of compound, decreased by a $VdWvol$ of unsubstituted parent quinone (Table I). The use of N and $VdWvol_{subst}$ as variables results in less straightforward correlations as compared with Eq. (17, 18), but the use of N, N^2 and $VdWvol_{subst}$ gives equivalent and even better results:

$$\log k_{cat}/K_m = 7.1011 \pm 1.3429 + (1.6818 \pm 1.8929)\, E_7(Q/QH^-)$$
$$-(12.9391 \pm 5.8782)(E_7[Q/QH^-])^2$$
$$+(1.9751 \pm 1.5105)N - (0.6618 \pm 0.3713)N^2$$
$$-(0.0121 \pm 0.0025)VdWvol_{subst} \quad (r^2 = 0.9049) \quad (18)$$

and
$$\log k_{cat}/K_m = (5.8488 \pm 1.1763) + (3.8851 \pm 2.6747)E_7^1$$
$$-(17.7618 \pm 6.4765)(E_7^1)^2 + (0.1970 \pm 0.1116)$$
$$PK_a(QH^\bullet) + (2.7086 \pm 1.2298)N$$
$$-(0.9011 \pm 0.3094)N^2 - (0.0145 \pm 0.0022)$$
$$VdWvol_{subst} \quad (r^2 = 0.9485) \quad (19)$$

The Eq. (18, 19) point to a possibility that NQO1 may be equally specific toward benzo- and naphthoquinone structure ($N = 1$–2) but may discriminate quinones with a higher number of aromatic rings. The values of log k_{cat} show a similar trend when using both models. The use of the electrostatic charge of a molecule as an additional variable did not improve the correlations, pointing to the absence of an important role for electrostatic interactions.

NQO1 possesses a quinone/nicotinamide binding pocket close to 360 $\mathring{A}^3$, where the substrates interact with Tyr-128′, Phe-106, and Phe-232′ residues.[33] This is above the $VdWvol$ of most quinones investigated (Table I). In addition, NQO1 may accomodate large molecules such as Cibacron blue ($VdWvol \geq 750 \, \mathring{A}^3$), which occupies the ribose and adenosyl-binding sites of NADP(H).[34] Less-reactive bulky quinones possess less negative reaction

[33] M. Faig, M. A. Bianchet, S. Winski, R. Hargreaves, C. J. Moody, A. R. Hudnott, D. Ross, and L. M. Amzel, *Structure* **9**, 639 (2001).

activation enthropies than those with $VdWVol \leq 200 \text{ Å}^3$, which points to a weak electronic coupling between the reactants and/or a restricted ability to form a charge-transfer complex. Thus an increase in $VdWvol$ of quinones may cause their unfavorable orientation toward isoalloxazine ring of FAD. An inhibitor, competitive with NADPH, dicumarol, presumably binds at the nicotinamide-binding domain of NQO1. The studies of quinone reduction inhibition by Cibacron blue, and double inhibition experiments using dicumarol and quinones, provided some evidence that bulky quinones may bind distantly from the dicumarol-binding domain and partly occupy the Cibacron blue binding domain.[31,35]

Structure-Activity Relationship in Reactions of Enterobacter cloacae *Nitroreductase*

Enterobacter cloacae NR contains FMN ($E_7^0 = -0.19$ V) in the active center, with extremely unstable semiquinone state ($E_7[\text{FMN}^{\bullet-}/\text{FMNH}^-] \approx -0.01$ V, and $E_7[\text{FMN/FMN}^{\bullet-}] \approx -0.37$ V).[26] NR reduces nitroaromatic compounds and quinones in a two-electron way after a "ping-pong" scheme.[30,36] According to X-ray data, the pyrimidine ring of FMN of *E. cloacae* NR is close to the surface of the protein globule.[26] However, in homologous *E. coli* NR, the sterical hindrances, caused by conserved Phe-124 for the reduction of bulky oxidants, have been demonstrated.[37] The binding of aromatic compounds such as benzoate in the active center of *E. cloacae* NR is accompanied by the significant displacement of Phe-124 and Tyr-123 residues.[26] Our preliminary observations also point to the importance of the volume of isoalloxazine binding pocket on the rate of quinone reduction. Thus the substitution of Pro-162 residue in the active center of *E. cloacae* NR by less bulky amino acids enhanced the quinone reduction rate but did not affect the ferricyanide reduction rate. In P162A and P162G mutants, the k_{cat}/K_m of 2.6-dimethyl-1,4-benzoquinone, tetramethyl-1,4-benzoquinone, 2-methyl-5-hydroxy-1,4-naphthoquinone, 2-hydroxy-1,4-naphthoquinone and riboflavin were increased by 2 to 3.5 times, and the k_{cat} of the reaction was increased by 1.5 to 2 times, as compared with the wild-type NR.

[34] R. Li, M. A. Bianchet, P. Talalay, and L. M. Amzel, *Proc. Natl. Acad. Sci. USA* **92**, 8846 (1995).

[35] Ž. Anusevičius, J. Šarlauskas, L. Misevičienė, and N. Čėnas, *in* "Flavins and Flavoproteins 2002" (S. Chapman, R. Perham, and N. Scrutton, eds.), p. 113. Rudolf Weber, Berlin, 2002.

[36] H. Nivinskas, R. L. Koder, Ž. Anusevičius, J. Šarlauskas, A.-F. Miller, and N. Čėnas, *Arch. Biochem. Biophys.* **385**, 170 (2001).

[37] S. Zenno, H. Koike, M. Tanokura, and K. Saigo, *J. Bacteriol.* **178**, 4731 (1996).

The rate constants of quinone reduction by *E. cloacae* NR are given in Table II. Previously, we have found that the reactivities of 2-hydroxy-1, 4-naphthoquinones were much higher than expected from their reduction potential.[30] Most probably, their enhanced reactivity was determined by their binding at or close to the binding site of nicotinamide and dicumarol, whereas other quinones used the alternative, currently unidentified site. For this reason, the kinetic data of 2-hydroxy-1,4-naphthoquinones were not used in the analysis. It was found that the NR reactivity was equally well described by the first-order and second-order dependences on reduction potential and followed the same trend. Further, only the linear approximations are presented.

Using single-step hydride transfer model, one may demonstrate a decrease in quinone reactivity with an increase in *VdWvol*:

$$\log k_{cat}/K_m = 7.9306 \pm 0.5909 + (6.2064 \pm 1.8466)E_7(Q/QH^-)$$
$$-(0.0061 \pm 0.0028)VdWvol \quad (r^2 = 0.7747) \qquad (20)$$

One may note that the coefficient in Eq. (20) characterizing the influence of *VdWvol* is lower than that in Eq. (16), describing the reactivity of NQO1. This is in line with the data of Fig. 2, which show a less-pronounced *VdWvol* effect on reactivity of NR as compared with NQO1. The use of $(VdWvol)^2$ as an additional parameter did not improve the correlation. However, a better approximation has been obtained by using N, N^2 and $VdWvol_{subst}$ as additional variables:

$$\log k_{cat}/K_m = 3.6115 \pm 1.3970 + (8.4065 \pm 1.7200)E_7(Q/QH^-)$$
$$+(4.0479 \pm 1.5943)N - (0.9921 \pm 0.3939)N^2$$
$$-(0.0031 \pm 0.0027)VdWvol_{subst} \quad (r^2 = 0.8575) \qquad (21)$$

According to Eq. (21), *E. cloacae* NR may possess a preference for the naphthoquinone structure ($N = 2$). When using a three-step hydride transfer model, an increase in quinone reactivity with an increase in $pK_a(QH^\bullet)$ and decrease in *VdWvol* may be demonstrated as well, giving $r^2 = 0.8665$ in the correlation (data not shown). However, a better approximation was obtained by using N, N^2 and $VdWvol_{subst}$ as additional variables ($r^2 = 0.9243$):

$$\log k_{cat}/K_m = (4.2629 \pm 1.1.3167) + (7.8257 \pm 1.7302)E_7^1$$
$$+(0.2994 \pm 0.1152)pK_a(OH^\bullet) + (2.9465 \pm 1.3476)$$
$$N - (0.7448 \pm 0.3493)N^2 - (0.0087 \pm 0.0022)$$
$$VdWvol_{subst} \qquad (22)$$

The values of log k_{cat} were similar with both models. Summing up, the Eq. (21, 22) demonstrate the preference of NR for compounds with $N = 2$.

Conclusions

The multiparameter regression analysis that uses some sterical quinone parameters appeared to be a useful tool in the studies of their two-electron reduction by flavoenzymes such as NQO1, in which the simple log k vs. redox potential dependence is evidently absent. Although it is a significant simplification, it enabled us to explain the poor reactivity of anticancer compounds AZQ, BZQ, mitomycin C and adriamycin toward NQO1 (Table II). The extension of this approach for the larger amount of available data is problematic, since the studies were carried out under different conditions or even at a single concentration of quinone. However, the more thorough studies of bulky cyclopent[b]indole- and indole-based quinones, showing their modest or intermediate activity toward NQO1,[38] are in line with our findings. Our data provide the quantitative base for the selectivity of NQO1 and *E. cloacae* NR toward certain quinone structures. However, they do not enable us to conclude whether the existence of unstable anionic flavin semiquinone is a sufficient condition for a two-electron reduction of quinones by flavoenzymes or whether it may require some additional structure specific to their active centers.

Acknowledgments

We are grateful to Professor C. Gomez-Moreno and Dr. M. Martinez-Julvez (Zaragoza University, Spain) for their generous gift of ferredoxin:NADP$^+$ reductase and to Professor A.-F. Miller and Dr. R. L. Koder (University of Kentucky, Lexington, KY) for their generous gift of *E. cloacae* nitroreductase mutants.

[38] E. B. Skibo, C. Xing, and R. T. Dorr, *J. Med. Chem.* **44**, 3545 (2001).

[16]   p53-Dependent Apoptosis and NAD(P)H: Quinone Oxidoreductase 1

By GAD ASHER, JOSEPH LOTEM, LEO SACHS, and YOSEF SHAUL

Introduction

The Tumor Suppressor p53, Oxidoreductases and NAD

p53 is a tumor suppressor gene that is frequently mutated in various human malignancies.[1] The *p53* gene encodes a labile protein that is rapidly degraded in the proteasome. Proteasomal degradation of p53 in a ubiquitin-dependent pathway is induced through binding and ubiquitination of p53 by the E3 ubiquitin ligase Mdm2.[2,3] We have found another pathway for proteasomal degradation of p53 that is ubiquitin-independent and is regulated by NAD(P)H quinone oxidoreductase 1(NQO1).[4–6] Cells normally express low levels of p53, but in response to various types of stress including DNA damage, hypoxia and oxidative stress, p53 accumulates and induces either growth arrest[1,7,8] or apoptosis.[9–12] Many biological functions of p53 are attributed to its ability to function as a sequence-specific transcriptional activator of a variety of genes.[1,7,8] Several p53 target genes are oxidoreductases involved in reactive oxygen species (ROS) metabolism. The p53-inducible gene 3 *[PIG3]*,[13] *Ferredoxin reductase*

[1] B. Vogelstein, D. Lane, and A. J. Levine, *Nature* **408,** 307 (2000).

[2] Y. Haupt, R. Maya, A. Kazaz, and M. Oren, *Nature* **387,** 296 (1997).

[3] M. H. Kubbutat, S. N. Jones, and K. H. Vousden, *Nature* **387,** 299 (1997).

[4] G. Asher, J. Lotem, B. Cohen, L. Sachs, and Y. Shaul, *Proc. Natl. Acad. Sci. USA* **98,** 1188 (2001).

[5] G. Asher, J. Lotem, R. Kama, L. Sachs, and Y. Shaul, *Proc. Natl. Acad. Sci. USA* **99,** 3099 (2002).

[6] G. Asher, J. Lotem, L. Sachs, C. Kahana, and Y. Shaul, *Proc. Natl. Acad. Sci. USA* **99,** 13125 (2002).

[7] M. Oren, *FASEB J.* **6,** 3169 (1992).

[8] A. J. Levine, *Cell* **88,** 323 (1997).

[9] E. Yonish-Rouach, D. Resnitzky, J. Lotem, L. Sachs, A. Kimchi, and M. Oren, *Nature* **352,** 345 (1991).

[10] J. Lotem and L. Sachs, *Blood* **82,** 1092 (1993).

[11] S. W. Lowe, E. M. Schmitt, S. W. Smith, B. A. Osborne, and T. Jacks, *Nature* **362,** 847 (1993).

[12] A. R. Clarke, C. A. Purdie, D. J. Harrison, R. G. Morris, C. C. Bird, M. L. Hooper, and A. H. Wyllie, *Nature* **362,** 849 (1993).

[13] K. Polyak, Y. Xia, J. L. Zweier, K. W. Kinzler, and B. Vogelstein, *Nature* **389,** 300 (1997).

(FDXR)[14] and the p53 responsive gene 3 *(PRG3)*[15] are all transcriptionally activated by p53.

PIG3 was first identified in a screen for genes induced by p53 before the onset of apoptosis.[13] In addition to the p53 consensus-binding site, the *PIG3* promoter has an additional p53-binding site with a pentanucleotide microsatellite sequence. Despite its limited similarity to a p53-binding consensus, this pentanucleotide microsatellite sequence is necessary and sufficient for transcriptional activation of the *PIG3* promoter by p53.[16] Induction of *PIG3* expression by p53 following DNA damage occurs late when compared with other known p53 targets such as *p21* and *Mdm2*.[17] PIG3 is a cytoplasmatic protein that shares significant homology with oxidoreductases from several species, and its activation correlates with the generation of ROS in cells.[13] The proline-rich domain of p53 is essential for ROS generation, *PIG3* trans-activation and subsequent apoptosis.[18] However, expression of PIG3 by itself is not sufficient to induce apoptosis in p53-deficient cells.[13]

FDXR was identified as a p53 transcriptional target gene in human colorectal cancer cells on induction of apoptosis by the chemotherapeutic agent 5-fluorouracil (5-FU).[14] FDXR is a mitochondrial cytochrome P-450 NADPH reductase located on the matrix side of the inner mitochondrial membrane.[19] It is responsible for electron transfer from NADPH via ferredoxin as a single electron shuttle to substrates such as cholesterol.[20,21] Expression of wild-type FDXR, but not mutant FDXR that lacks the NADPH and flavin adenine dinucleotide (FAD) binding domains suppresses growth of colon cancer cells, suggesting that the oxidoreductase activity of FDXR is essential for growth inhibition.[14] Partial disruption of the *FDXR* gene by targeted homologous recombination decreased cell

[14] P. M. Hwang, F. Bunz, J. Yu, C. Rago, T. A. Chan, M. P. Murphy, G. F. Kelso, R. A. Smith, K. W. Kinzler, and B. Vogelstein, *Nat. Med.* **7,** 1111 (2001).

[15] Y. Ohiro, I. Garkavtsev, S. Kobayashi, K. R. Sreekumar, R. Nantz, B. T. Higashikubo, S. L. Duffy, R. Higashikubo, A. Usheva, D. Gius, N. Kley, and N. Horikoshi, *FEBS Lett.* **524,** 163 (2002).

[16] A. Contente, A. Dittmer, M. C. Koch, J. Roth, and M. Dobbelstein, *Nat. Genet.* **30,** 315 (2002).

[17] P. M. Flatt, K. Polyak, L. J. Tang, C. D. Scatena, M. D. Westfall, L. A. Rubinstein, J. Yu, K. W. Kinzler, B. Vogelstein, D. E. Hill, and J. A. Pietenpol, *Cancer Lett.* **156,** 63 (2000).

[18] C. Venot, M. Maratrat, C. Dureuil, E. Conseiller, L. Bracco, and L. Debussche, *EMBO J.* **17,** 4668 (1998).

[19] J. D. Lambeth, D. W. Seybert, J. R. Lancaster, Jr., J. C. Salerno, and H. Kamin, *Mol. Cell. Biochem.* **45,** 13 (1982).

[20] D. Lin, Y. F. Shi, and W. L. Miller, *Proc. Natl. Acad. Sci. USA* **87,** 8516 (1990).

[21] G. A. Ziegler, C. Vonrhein, I. Hanukoglu, and G. E. Schulz, *J. Mol. Biol.* **289,** 981 (1999).

sensitivity to 5-FU-induced p53-dependent apoptosis.[14] Furthermore, results obtained with pharmacologic inhibitors of ROS generation suggest that FDXR contributes to 5-FU-induced p53-mediated apoptosis by generating oxidative stress in the mitochondria.[14]

PRG3 was first identified as a p53 transcriptional target gene after induction of p53-dependent apoptosis in EB1 human colon carcinoma cells.[22] Analysis of the *PRG3* gene identified a p53 response element in an intron.[15] As with *PIG3,* transcriptional activation of *PRG3* on initiation of p53-dependent apoptosis is delayed as compared with other known p53 downstream targets such as *p21.*[15] PRG3 is a cytoplasmatic protein that shares homology with several bacterial oxidoreductases and with human AIF, a flavoprotein and quinone oxidoreductase that is normally confined to mitochondria and translocates to the nucleus on induction of apoptosis.[23] Both wild-type PRG3 and a mutant PRG3 that lack its conserved ADP binding motif induce apoptosis, suggesting that PRG3 oxidoreductase activity is not required for its apoptotic function.[15] Similarly, AIF that lacks NAD oxidase activity because of removal of the FAD cofactor from its molecule retains full apoptotic activity.[23] Thus, in contrast to PIG3 and FDXR, the apoptosis-inducing activity of both PRG3 and AIF is independent of their oxidoreductase activity.

Many proteins have been reported to bind p53 and modulate its stability and transcriptional activity, including several proteins with redox activity such as the WW domain containing oxidoreductase (WOX1),[24] the bacterial redox protein azurin,[25] thioredoxin (TXN),[26] and nuclear redox factor 1 (REF-1).[27]

WOX1 is a mitochondrial protein that contains two N-terminal WW domains, a nuclear localization sequence, and a C-terminal alcohol dehydrogenase (ADH) domain.[24] An increase in mitochondrial membrane permeability on treatment with apoptosis-inducers such as TNF results in

[22] N. Horikoshi, J. Cong, N. Kley, and T. Shenk, *Biochem. Biophys. Res. Commun.* **261,** 864 (1999).

[23] S. A. Susin, H. K. Lorenzo, N. Zamzami, I. Marzo, B. E. Snow, G. M. Brothers, J. Mangion, E. Jacotot, P. Costantini, M. Loeffler, N. Larochette, D. R. Goodlett, R. Aebersold, D. P. Siderovski, J. M. Penninger, and G. Kroemer, *Nature* **397,** 441 (1999).

[24] N. S. Chang, N. Pratt, J. Heath, L. Schultz, D. Sleve, G. B. Carey, and N. Zevotek, *J. Biol. Chem.* **276,** 3361 (2001).

[25] T. Yamada, M. Goto, V. Punj, V. Zaborina, M. L. Chen, K. Mimbara, D. Majumdar, E. Cunningham, T. K. Das Gupta, and A. M. Chakrabarty, *Proc. Natl. Acad. Sci. USA* **99,** 14098 (2002).

[26] M. Ueno, H. Masutani, R. J. Arai, A. Yamauchi, K. Hirota, T. Sakai, T. Inamoto, Y. Yamaoka, J. Yodoi, and T. Nikaido, *J. Biol. Chem.* **274,** 35809 (1999).

[27] L. Jayaraman, K. G. Murthy, C. Zhu, T. Curran, S. Xanthoudakis, and C. Prives, *Genes Dev.* **11,** 558 (1997).

translocation of WOX1 to the nucleus followed by cell death.[24] WOX1 enhances TNF cytotoxicity, and its down-regulation by anti-sense RNA protects cells from TNF-mediated apoptosis. Enhancement of TNF cyto-toxicity by WOX1 is associated with up-regulation of p53 and down-regulation of the apoptosis inhibitors Bcl-2 and Bcl-x. In addition, anti-sense WOX1 and a dominant negative WOX1 inhibited p53-mediated apoptosis.[24] Structure-function analysis revealed that the ADH domain, but not the WW domain, of WOX1 is sufficient to modulate the expression of p53, Bcl-2 and Bcl-x. p53 and WOX1 co-localize in the cytosol and co-immunoprecipitate. Yeast two hybrid analysis showed that the proline-rich domain of p53 binds to the WW domain of WOX1.[24] The results indicate that the oxidoreductase activity, rather than its ability to bind p53, is required for p53 up-regulation by WOX1.

Azurin is a bacterial redox protein involved in electron transfer during denitrification in *P. aeruginosa*.[28] Treatment of human melanoma cells with azurin elevates p53 level and induces p53-dependent apoptosis. In addition, azurin induces regression of human melanoma tumors in xeno-transplanted nude mice.[25] Glycerol gradient fractionations suggest that wild-type azurin forms a complex with p53.[25] Several azurin mutants with low redox activity still form a complex with p53, generate ROS and induce apoptosis, suggesting that complex formation of azurin with p53, ROS generation and subsequent apoptosis are independent of azurin redox activity.[29] Thus, as with PRG3 and AIF, the redox activity of azurin is independent of its apoptosis-inducing activity.

Mammalian thioredoxin reductase (TXNRD) is a homo-dimer and FAD-containing protein with a penultimate C-terminal selenocysteine residue that is essential for its enzymatic activity.[30] TXNRD is the only known enzyme responsible for NADPH-dependent reduction of TXN.[30] Both TXNRD and its substrate TXN have important roles in the regulation of various biological functions including cell growth, apoptosis, redox activity, transcription and protein folding.[30] The physiological role of TXNRD in p53 regulation was first identified in yeast. Expression of wild-type p53 in yeast induces growth inhibition. A mutant yeast strain that is partially re-sistant to p53-induced growth inhibition was found to have a mutation in the *trr1* gene, which has strong homology to TXNRD.[31] Studies with yeast

[28] M. van de Kamp, M. C. Silvestrini, M. Brunori, J. Van Beeumen, F. C. Hali, and G. W. Canters, *Eur. J. Biochem.* **194,** 109 (1990).

[29] M. Goto, T. Yamada, K. Kimbara, J. Horner, M. Newcomb, T. K. Gupta, and A. M. Chakrabarty, *Mol. Microbiol.* **47,** 549 (2003).

[30] G. Powis and W. R. Montfort, *Annu. Rev. Pharmacol. Toxicol.* **41,** 261 (2001).

[31] D. Casso and D. Beach, *Mol. Gen. Genet.* **252,** 518 (1996).

have shown that deletion of *trr1* does not affect the level and localization of the p53 protein but inhibits the ability of p53 to induce reporter gene expression. In cultured mammalian cells, wild-type but not a redox-inactive mutant of TXN increases the sequence specific binding of p53 to DNA and enhances transcription of *p21*.[26] These results suggest that TXNRD is an important regulator of p53 transcriptional activity, probably via reduction of its substrate TXN. Following exposure to the DNA damaging agent cisplatin, TXN and p53 co-localize in the nucleus.[32] TXN and p53 were also reported to co-immunoprecipitate.[33] It is not yet certain whether the interaction of p53 and TXN is direct or involves additional proteins. TXN was also reported to associate with and reduce REF-1.[34] REF-1 increases the DNA binding and transcriptional activity of p53,[35] possibly through binding to a negative regulatory region at the C-terminus of p53.[27] In addition, expression of REF-1 also elevates p53 protein level.[27]

The p53 protein undergoes many post-translational modifications such as phosphorylation, acetylation, poly ADP ribosylation and ubiquitination. These modifications are highly regulated and known to modulate p53 stability and function.[36] Some of these modifications including deacetylation[37,38] and poly ADP ribosylation[39] are NAD-dependent.

The yeast silent information regulator 2α (Sir2α) protein is a NAD-dependent histone deacetylase whose activity is essential for gene silencing and life span extension in yeasts.[40] The human homolog of yeast SIR2α binds to and deacetylates p53. Deacetylation of p53 by Sir2α is NAD-dependent and can be inhibited by nicotinamide. In addition, mammalian Sir2α represses p53 transcriptional activity and blocks p53-dependent apoptosis in response to DNA damage and oxidative stress.[37,38] Sir2α thus is an important negative regulator of p53 activity that provides a link between cellular levels of NAD and p53 function.

Another NAD-dependent p53 modification is poly ADP ribosylation. The enzyme poly ADP-ribose polymerase (PARP) is activated following induction of DNA strand breaks and hydrolizes its substrate NAD^+ to

[32] M. Ueno, Y. Matsutani, H. Nakamura, H. Masutani, M. Yagi, H. Yamashiro, H. Kato, T. Inamoto, A. Yamauchi, R. Takahashi, Y. Yamaoka, and J. Yodoi, *Immunol. Lett.* **75,** 15 (2000).

[33] P. Hainaut and K. Mann, *Antioxid. Redox Signal* **3,** 611 (2001).

[34] C. Abate, L. Patel, F. J. Rauscher, 3rd, and T. Curran, *Science* **249,** 1157 (1990).

[35] C. Gaiddon, N. C. Moorthy, and C. Prives, *EMBO J.* **18,** 5609 (1999).

[36] E. Appella and C. W. Anderson, *Pathol. Biol. (Paris)* **48,** 227 (2000).

[37] H. Vaziri, S. K. Dessain, E. Ng Eaton, S. I. Imai, R. A. Frye, T. K. Pandita, L. Guarente, and R. A. Weinberg, *Cell* **107,** 149 (2001).

[38] J. Luo, A. Y. Nikolaev, S. Imai, D. Chen, F. Su, A. Shiloh, L. Guarente, and W. Gu, *Cell* **107,** 137 (2001).

ADP-ribose. PARP subsequently attaches the ADP-ribose to target pro-teins and catalyzes the formation of poly ADP-ribose chains.[41,42] PARP binds to poly ADP ribosylates p53 on several arginines in its DNA binding domain.[39] Inhibition of PARP activity suppresses *p21* and *Mdm2* induction by p53 in response to DNA damage and inhibits p53 dependent apop-tosis.[43] Cells with reduced NAD$^+$ level caused by the use of nicotin-amide-deficient medium, as well as cells deficient in PARP, exhibit decreased basal levels of p53 protein.[44,45] In addition, cells with defective NAD/PARP metabolism showed resistance to DNA damage-induced p53 accumulation and subsequent apoptosis.[44] There is also an inverse corre-lation between NAD cellular levels and the occurrence of malignant skin lesions in humans.[45] It appears that poly ADP ribosylation of p53 is an NAD-dependent modification that positively regulates both p53 stability and p53 function as a transcriptional activator and apoptosis inducer.

The studies described previously indicate that there is a wide and inter-esting interplay between the tumor suppressor p53 and various mammalian and bacterial oxidoreductases. p53 transcriptionally activates certain oxi-doreductases, and other oxidoreductases modulate p53 transcriptional ac-tivity or protein stability. In some cases this modulation of p53 requires the oxidoreductase enzymatic activity, while in other cases the enzymatic ac-tivity is dispensable, and physical interaction with p53 is required. This raises the question of whether another oxidoreductase, NQO1, is also involved in the regulation of p53 stability and function, and if so, by what mechanism.

NAD(P)H Quinone Oxidoreductase 1 (NQO1)

In 1958, Drs. Lars Ernster and Franco Navazio discovered a highly active enzyme in the soluble fraction of rat liver homogenates that cata-lyzes the oxidation of NADH and NADPH.[46] They further purified and characterized the enzyme and named it DT diaphorase.[46] DT diaphorase, also referred today as NQO1, is a ubiquitous flavo-enzyme that catalyzes two-electron reduction of various quinones by utilizing NAD(P)H as an

---

[39] M. Malanga, J. M. Pleschke, H. E. Kleczkowska, and F. R. Althaus, *J. Biol. Chem.* **273**, 11839 (1998).

[40] L. Guarente, *Genes Dev.* **14**, 1021 (2000).

[41] W. M. Tong, U. Cortes, and Z. Q. Wang, *Biochim. Biophys. Acta* **1552**, 27 (2001).

[42] A. Chiarugi, *Trends Pharmacol. Sci.* **23**, 122 (2002).

[43] H. Vaziri, M. D. West, R. C. Allsopp, T. S. Davison, Y. S. Wu, C. H. Arrowsmith, G. G. Poirier, and S. Benchimol, *EMBO J.* **16**, 6018 (1997).

[44] C. M. Whitacre, H. Hashimoto, M. L. Tsai, S. Chatterjee, S. J. Berger, and N. A. Berger, *Cancer Res.* **55**, 3697 (1995).

[45] E. L. Jacobson, W. M. Shieh, and A. C. Huang, *Mol. Cell. Biochem.* **193**, 69 (1999).

[46] C. Lind, E. Cadenas, P. Hochstein, and L. Ernster, *Methods Enzymol.* **186**, 287 (1990).

electron donor.[46] NQO1-mediated reduction of quinones into hydroqui-nones is an important cellular defense mechanism against oxidative stress.[47] Studies of NQO1 structure and function show that NQO1 is a homo-dimer that functions via a "ping pong" mechanism whereby the reduced NAD(P)H binds to NQO1, reduces the FAD co-factor and is re-leased prior to binding of the quinone substrate. The NAD(P)H and the quinone binding sites of NQO1 have a significant overlap, thus providing a molecular basis for this "ping pong" mechanism.[48] NQO1 activity can be inhibited by dicoumarol (3-methylene-bis [4-hydroxycoumarin]), a potent and specific competitive inhibitor that competes with NAD(P)H for binding to NQO1.[49]

NQO1 is widely distributed in animals and ubiquitously expressed in all tissues.[46] Expression of the *NQO1* gene is induced in response to various agents including oxidants, anti-oxidants, xenobiotics, heavy metals, UV light and ionizing radiation.[50] The bulk of NQO1 activity in cells is found in the cytoplasm (95%), and a small portion is detected in mitochondria. It was recently reported that the enzyme is also present in the nucleus.[51]

To examine the *in vivo* role of NQO1, NQO1 knockout mice have been generated.[52] These mice are born alive and reproduce normally but exhibit increased sensitivity to the toxicity of the quinone menadione.[52] In ad-dition, these mice suffer from several metabolic defects including altered glucose levels, insulin resistance and a lower amount of abdominal adipose tissue. Biochemical analysis of NQO1 knockout mice revealed lower levels of pyridine nucleotides and an increase in the reduced/oxidized NAD(P)H/NAD(P) ratio.[53] NQO1 knockout mice are more susceptible to carcinogen-induced skin cancer and develop myeloid hyperplasia.[54,55]

Several polymorphisms in the human NQO1 gene were identified. The most common polymorphism is a C- to-T substitution at position 609 of the cDNA that causes a proline for serine replacement at position 187 of the protein and results in a biologically inactive enzyme. In addition,

[47] P. Joseph, D. J. Long, 2nd, A. J. Klein-Szanto, and A. K. Jaiswal, *Biochem. Pharmacol.* **60,** 207 (2000).

[48] R. Li, M. A. Bianchet, P. Talalay, and L. M. Amzel, *Proc. Natl. Acad. Sci. USA* **92,** 8846 (1995).

[49] S. Hosoda, W. Nakamura, and K. Hayashi, *J. Biol. Chem.* **249,** 6416 (1974).

[50] A. T. Dinkova-Kostova, and P. Talalay, *Free Radic. Biol. Med.* **29,** 231 (2000).

[51] S. L. Winski, Y. Koutalos, D. L. Bentley, and D. Ross, *Cancer Res.* **62,** 1420 (2002).

[52] V. Radjendirane, P. Joseph, Y. H. Lee, S. Kimura, A. J. Klein-Szanto, F. J. Gonzalez, and A. K. Jaiswal, *J. Biol. Chem.* **273,** 7382 (1998).

[53] A. Gaikwad, D. J. Long, 2nd, J. L. Stringer, and A. K. Jaiswal, *J. Biol. Chem.* **276,** 22559 (2001).

[54] D. J. Long, 2nd, R. L. Waikel, X. J. Wang, D. R. Roop, and A. K. Jaiswal, *J. Natl. Cancer Inst.* **93,** 1166 (2001).

[55] D. J. Long, 2nd, A. Gaikwad, A. Multani, S. Pathak, C. A. Montgomery, F. J. Gonzalez, and A. K. Jaiswal, *Cancer Res.* **62,** 3030 (2002).

only very low levels of the polymorphic NQO1 protein are detected in samples from individuals and cell lines homozygous for the C609T polymorphic NQO1.[56,57] These very low levels of the polymorphic NQO1 protein are probably due to its rapid degradation by the ubiqutin proteasome pathway.[58] The frequency of C609T homozygous genotype in the population ranges from 4% in Caucasians up to 22% in Chinese populations.[59] Epidemiological studies indicate that humans carrying a polymorphic C609T NQO1 gene are more susceptible to developing various solid tumors and leukemia.[60–62] These epidemiological studies and the finding that NQO1 knockout mice are more susceptible to carcinogen-induced cancer suggest that NQO1 functions as a tumor suppressor.

Regulation of p53 Stability and p53-Dependent Apoptosis by NQO1

The interplay between the tumor suppressor p53 and various oxidoreductases, and the possible function of NQO1 as a tumor suppressor, prompted us to investigate the role of NQO1 in the p53 pathway. We found that the NQO1-inhibitor dicoumarol induces proteasomal degradation of both basal and γ-irradiation induced p53 in normal and cancer cells.[4] Proteasomal degradation of the tumor suppressor p73, a p53 family member, is also induced by dicoumarol.[4,6] We also showed that p53-dependent apoptosis in γ-irradiated normal thymocytes (Fig. 1) and in M1 myeloid leukemic cells (Fig. 2) is inhibited on induction of p53 degradation by dicoumarol.[4] Degradation of p53 and inhibition of p53-mediated apoptosis is also induced by warfarin, and other coumarin and flavone NQO1 inhibitors.[63] Dicoumarol-induced p53 degradation is prevented in the presence of the tumor suppressor p14^ARF and the viral oncogenes SV40 large T antigen and adenovirus E1A that are known to stabilize p53.[4–6] NQO1 over-expression elevates the level of endogenous p53 and protects

[56] R. D. Traver, D. Siegel, H. D. Beall, R. M. Phillips, N. W. Gibson, W. A. Franklin, and D. Ross, *Br. J. Cancer* **75,** 69 (1997).

[57] D. Siegel, S. M. McGuinness, S. L. Winski, and D. Ross, *Pharmacogenetics* **9,** 113 (1999).

[58] D. Siegel, A. Anwar, S. L. Winski, J. K. Kepa, K. L. Zolman, and D. Ross, *Mol. Pharmacol.* **59,** 263 (2001).

[59] K. T. Kelsey, D. Ross, R. D. Traver, D. C. Christiani, Z. F. Zuo, M. R. Spitz, M. Wang, X. Xu, B. K. Lee, B. S. Schwartz, and J. K. Wiencke, *Br. J. Cancer* **76,** 852 (1997).

[60] M. T. Smith, *Proc. Natl. Acad. Sci. USA* **96,** 7624 (1999).

[61] M. T. Smith, Y. Wang, E. Kane, S. Rollinson, J. L. Wiemels, E. Roman, P. Roddam, R. Cartwright, and G. Morgan, *Blood* **97,** 1422 (2001).

[62] M. T. Smith, Y. Wang, C. F. Skibola, D. J. Slater, L. L. Nigro, P. C. Nowell, B. J. Lange, and C. A. Felix, *Blood* **100,** 4590 (2002).

[63] G. Asher, J. Lotem, P. Tsvetkov, V. Reiss, L. Sachs, and Y. Shaul, *Proc. Natl. Acad. Sci. USA* **100,** 15065–15070 (2003).

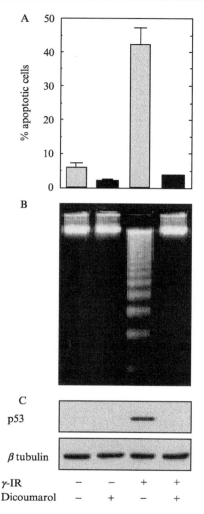

Fig. 1. Dicoumarol inhibits p53 accumulation and p53-dependent apoptosis in γ-irradiated thymocytes. Thymocytes that were not irradiated (−) or γ-irradiated at 4 Gy (+) were cultured for 5 h without (−) or with (+) 200 μM dicoumarol. (A) Analysis of the percent of apoptotic cells based on May-Grünwald Giemsa stained cytospin preparations. (B) DNA fragmentation at inter-nucleosomal sites. (C) Immunoblot analysis of p53 and β tubulin proteins.

p53 from dicoumarol-induced degradation.[4,5] NQO1 knockdown by specific small interfering RNA (siRNA) reduces the level of cellular p53.[6] In contrast to wild-type NQO1, the biologically inactive C609T polymorphic NQO1 fails to stabilize p53.[5] These findings indicate that NQO1, possibly

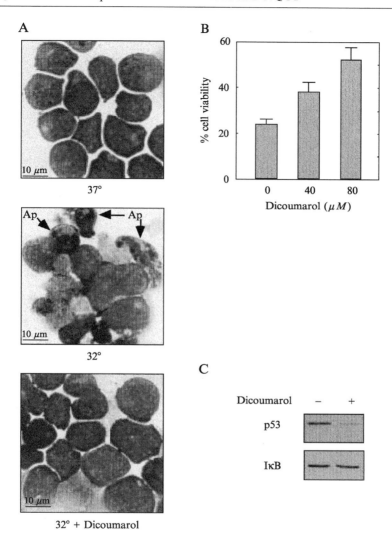

FIG. 2. Dicoumarol decreases p53 level and p53-dependent apoptosis in M1-t-p53 cells. (A) M1-t-p53 myeloid leukemic cells were cultured at $37°$ or $32°$ for 23 h without or with 100 μM dicoumarol and stained with May-Grünwald Giemsa. Ap, Apoptotic cells. (B) Analysis of the percent of viable M1-t-p53 cells at $32°$ treated with different concentrations of dicoumarol for 23 h. Incubation of M1-t-p53 cells for 23 h with concentrations above 125 μM dicoumarol are cytotoxic. (C) Immunoblot analysis of p53 and IkB proteins in M1-t-p53 cells cultured at $32°$ for 5 h without ($-$) or with ($+$) 200 μM dicoumarol. Similar results were obtained by culture of M1-t-p53 cells for 16 h with 100 μM dicoumarol. (See color insert.)

through its enzymatic activity, plays a direct role in p53 stabilization by blocking its proteasomal degradation.

We suggested that a physical interaction of p53 and NQO1 may also be required for p53 stabilization by NQO1.[5] Recently NQO1 and p53 were shown to co-immunoprecipitate both *in vitro* and *in vivo*.[63,64] This raises the possibility that both the enzymatic activity of NQO1 and its binding to p53 are required for p53 stabilization by NQO1. However, unlike dicoumarol or warfarin, a specific "suicide inhibitor" of NQO1 (ES936) was reported not to induce p53 degradation.[64] Further investigations should resolve this apparent difference between the effect of different inhibitors.

In further studies on the mechanism of NQO1-regulated p53 proteasomal degradation, we showed that NQO1 fails to inhibit p53 degradation mediated by the E3 ubiquitin ligase Mdm2.[5] A mutant p53 (p53[22,23]) that is resistant to Mdm2-mediated degradation is susceptible to dicoumarol-induced degradation.[6] These results show that the NQO1-regulated p53 proteasomal degradation is Mdm2-independent. Unlike Mdm2-mediated degradation, the NQO1-regulated p53 degradation pathway is not associated with accumulation of ubiquitin-conjugated p53.[6] *In vitro* degradation studies show that dicoumarol-induced p53 degradation is ubiquitin-independent and ATP-dependent.[6] In addition, inhibition of NQO1 by dicoumarol or by a specific NQO1 siRNA in cells with a temperature-sensitive E1 ubiquitin-activating enzyme induces p53 degradation and inhibits apoptosis at the restrictive temperature (39°) under conditions devoid of ubiquitination (Fig. 3).[6] Mdm2 fails to induce p53 degradation under these conditions.[6] These findings thus establish a novel Mdm2 and ubiquitin-independent mechanism for proteasomal degradation of p53 that is regulated by NQO1.

p53 transcriptionally activates several oxidoreductases including PIG3, FDXR and PRG3. This raises the question of whether NQO1 is also transcriptionally regulated by p53. Sequence analysis of the NQO1 promoter revealed a putative p53-binding element; however, NQO1 expression was not affected by p53 (unpublished data). NQO1 binds to[63,64] and stabilizes p53.[4–6] It will be interesting to determine whether PIG3, FDXR and PRG3 also bind to and stabilize p53, and if so, whether this depends on their oxidoreductase enzymatic activity.

NQO1 regulates the stability of the tumor suppressors p53 and p73.[4–6] This can explain the finding that NQO1 knockout mice have reduced levels

[64] A. Anwar, D. Dehm, D. Siegel, J. K. Keepa, L. J. Tang, J. A. Pietenpol, and D. Ross, *J. Biol. Chem.* **14,** 14 (2003).

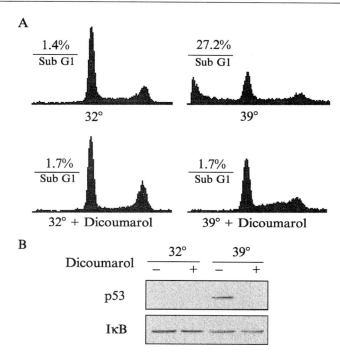

FIG. 3. Dicoumarol inhibits p53 accumulation and apoptosis in A31N-ts20. (A) A31N-ts20 cells were incubated at 32° or 39° for 48 h without or with 50 μM dicoumarol. Cell cycle analysis was carried out by flow cytometry and the percentage of sub-G1 (apoptotic) cells was determined (underlined). (B) Immunoblot analysis of p53 and IkB proteins in A31N-ts20 cells. Cells were preincubated for 24 h at 39° and then cultured for additional 5 h at 39° without (−) or with (+) 200 μM dicoumarol.

of p53 and p73[55] and that both NQO1 knockout mice and humans carrying a polymorphic inactive NQO1 are more susceptible to tumor development,[54,55,60–62] thus providing a molecular mechanism for NQO1 function as a tumor suppressor.

Principle of p53-Dependent Apoptosis Assays

Accumulation of wild-type p53 in many cell types either by exposure to various types of stress or by over-expression of p53 can lead to p53-dependent apoptosis.[9–12] To examine the regulation of p53 stability and p53-dependent apoptosis by NQO1, different cell types that undergo p53-dependent apoptosis are used: normal mouse thymocytes that undergo p53-dependent apoptosis after γ-irradiation,[10–12] M1-t-p53 myeloid leukemic cells that have a temperature-sensitive p53 [Val-135] and are viable and proliferate at 37° when the p53 is in a mutant conformation but undergo apoptosis at 32° when the p53 acquires wild-type conformation,[9] BALB/c mouse A31N-ts20 cells that have a temperature-sensitive E1 ubiquitin-activating enzyme.[65] These cells are viable and proliferate at 32° but undergo apoptosis at 39° because of accumulation of p53 and other short-lived proteins as a result of E1 enzyme inactivation and decreased protein ubiquitination and degradation.[65,66] These different cell types are treated with the NQO1 inhibitor dicoumarol, and the p53 protein levels and the percentage of apoptotic cells are determined.

Apoptotic cell death involves many changes in cell structure and morphology including cell shrinkage, condensation of cytoplasm and chromatin and nuclear fragmentation, which can be observed by using light microscopy.[67] Another method to identify apoptotic cell death is by a DNA fragmentation assay, a hallmark of apoptosis. DNA fragmentation in apoptotic cells specifically involves generation of large, 50 to 300 kb DNA fragments followed by the appearance of a "DNA ladder" composed of multiples of nucleosome size (180 bp) fragments that reflect internucleosomal DNA cleavage.[68] An additional method to determine apoptotic cell death is by measurement of cellular DNA content and cell cycle analysis by flow cytometry. Apoptotic cells have a diminished (sub-G1) DNA content caused by DNA degradation.[69]

Cells and Reagents

Cells and Cell Culture

To obtain thymocytes, two-month-old Balb/C mice are sacrificed and the thymiare removed and cut into pieces into cold phosphate-buffered saline solution (PBS). Thymocytes are teased out of the thymus by using tweezers and are centrifuged (800g, 5 min), and the cell pellet is washed twice with cold PBS. The cell pellet is then resuspended in Dulbecco's

[65] D. R. Chowdary, J. J. Dermody, K. K. Jha, and H. L. Ozer, *Mol. Cell. Biol.* **14,** 1997 (1994).
[66] L. Monney, I. Otter, R. Olivier, H. L. Ozer, A. L. Haas, S. Omura, and C. Borner, *J. Biol. Chem.* **273,** 6121–6131 (1998).
[67] J. F. Kerr, A. H. Wyllie, and A. R. Currie, *Br. J. Cancer* **26,** 239–257 (1972).
[68] S. H. Kaufmann, P. W. Mesner, Jr., K. Samejima, S. Tone, and W. C. Earnshaw, *Methods Enzymol.* **322,** 3–15 (2000).
[69] Z. Darzynkiewicz and E. Bedner, *Methods Enzymol.* **322,** 18–39 (2000).

modified Eagle's medium (DMEM) supplemented with 10% heat inactivated (56°, 30 min) horse serum and cultured at 37° in a humidified incubator with 10% CO_2. M1-t-p53 mouse myeloid leukemic cells that have a temperature-sensitive mutant p53 [Val-135] protein[9] are grown in DMEM supplemented with 10% heat inactivated (56°, 30 min) horse serum and cultured at 37° in an incubator with 10% CO_2. BALB/c mouse A31N-ts20 cells that have a temperature-sensitive E1 ubiquitin-activating enzyme[65] are grown in DMEM supplemented with 10% fetal bovine serum and cultured at 32° in a humidified incubator with 5.6% CO_2.

Compounds

Dicoumarol (Sigma) is dissolved in 0.13 N NaOH. May-Grünwald and Giemsa stains (Sigma) are dissolved in absolute methanol at 0.25% and 0.4%, respectively. Giemsa stain is freshly diluted 1:10 in water before staining.

Buffers

RIPA lysis buffer is as follows; 150 mM NaCl, 1% NP-40, 0.5% AB-deoxycholate, 0.1% SDS, 50 mM Tris-HCl (pH 8.0), 1 mM dithiothreitol (DTT) and 1 μg/ml each of leupeptin, aprotinin and pepstatin (Sigma cocktail). Laemmli protein sample buffer is as follows: 4% SDS, 20% glycerol, 10% 2-mercaptoethanol and 0.125 M Tris-Hcl (pH 8.0). DNA lysis buffer is as follows: 10 mM Tris-HCl (pH 8.0), 0.6% SDS, 10 mM EDTA (pH 8.0). DNA sample buffer is as follows: 10 mM Tris-HCl (pH 8.0), 10 mM EDTA (pH 8.0) and bromophenol blue.

Induction of Apoptotic Cell Death

Apoptosis in normal thymocytes is induced by γ-irradiation at 4 Gy (Co^{60} source 0.63 Gy/min) followed by culture at 37° for 5 h; in M1-t-p53 cells by culture at 32° for 23 h; and in A31N ts20 cells by culture at 39° for 48 h.

Methods

Detection of Apoptotic Cell Death by Light Microscopy

To determine cell morphology, glass slide cytospin preparations of thymocytes and M1-t-p53 cells are prepared by using a cytocentrifuge (Shandon-Elliot) (800g for 10 min). Slides are air dried for 1 min, fixed for 1 min in May-Grünwald solution, rinsed in 5% May-Grünwald in water, stained for 10 min in 10% Giemsa, rinsed again in water and air dried. The percent of apoptotic thymocytes is determined by light microscopy by counting 400 cells. Apoptotic cells are scored by their smaller size, condensed chromatin, and fragmented nuclei compared with non-apoptotic cells (Fig. 2A). Apoptotic M1-t-p53 cells undergo secondary "necrotic" changes including uptake of Trypan blue. The percentage of cells that uptake Trypan blue is determined by mixing 1×10^6 cells in a volume of 100 μl DMEM plus serum with 100 μl Trypan blue 1% in PBS and counting 400 cells in a hemacytometer. The percent of viable cells includes non-apoptotic cells that are not stained with Trypan blue.

Detection of Apoptotic Cell Death by DNA Fragmentation Assay

Thymocytes (2×10^6) are centrifuged (800g, 4° for 5 min) and the pellet is washed with PBS. The pellet is resuspended in 0.5 ml DNA lysis buffer, and RNAse A (Ambion) is added to a concentration of 15 $\mu g/ml$ and incubated for 20 min at 37°. NaCl is then added to a final concentration of 1 *M*, and samples are mixed by inversion and incubated for 2 h at 4° and centrifuged (13,000g, 4° for 30 min). The supernatant is collected, and 0.5 ml of phenol/chloroform is added, mixed thoroughly and centrifuged (13,000g, 4° for 5 min). The upper phase is collected, and 0.5 ml of chloroform is added, mixed and centrifuged (13,000g, 4° for 5 min). The upper phase is again collected, and 1 ml of ethanol ($-20°$) is added. The samples are incubated overnight at $-20°$ and centrifuged (13,000g, 4° for 10 min). The pellet is air dried and dissolved in 20 μl of DNA sample buffer. The samples undergo electrophores at 3 V/cm for 4 h in a 1.5% agarose gel with ethidium bromide (0.5 $\mu g/ml$) and are visualized under UV illumination (Fig. 1B).

Detection of Apoptotic Cell Death by FACS Analysis

Adherent A31N ts20 cells are removed by trypsinization, and both adherent and floating cells are harvested, washed with PBS and suspended in 0.5 ml PBS. Cells are fixed by dropwise addition of 5 ml 70% ethanol during vortexing. Samples are incubated at $-20°$ for at least 20 min and centrifuged (1000g for 3 min). The supernatant is discarded, and the cell pellet is washed with 5 ml of cold PBS, incubated for 30 min at 4° and centrifuged again (1000g for 3 min). The final cell pellet is resuspended in 1 ml PBS containing 50 $\mu g/ml$ of RNAse A and 25 $\mu g/ml$ of propiduim iodide (PI) (Sigma). Samples are incubated for 30 min at 37°, covered with aluminum foil and then analyzed by flow cytometry. The flow cytometer (FACSort Becton-Dickinson) is adjusted for detection of PI by excitation at 488 nm and emission at 585 nm and percent of cells with sub-G1 DNA content ($<$2N) is determined (Fig. 3A).

Immunoblot Analysis

Cells (5×10^6) are washed twice with cold PBS, centrifuged (800g, 4° for 5 min), resuspended in 100 μl of RIPA lysis buffer, incubated on ice for 20 min, and then centrifuged (13,000g, 4° for 15 min). The supernatant is collected, and the protein concentration is determined by using Bradford reagent (BioRad). Equal amounts of proteins are mixed with Laemmli sample buffer, heated at 95° for 5 min and loaded on an 8% polyacryl-amide-SDS gel. After electrophoresis at 20 V/cm for 2 h, proteins are transferred to cellulose nitrate 0.45 μm membranes (Schleicher & Schuell). The membranes are blocked with 5% low-fat milk in PBS containing 0.2% Tween 20 (Sigma) for 1 h at room temperature and then incubated over-night at 4° with mouse monoclonal antibody against mouse and human p53 (Pab240) (Santa Cruz). The blots are washed 5 times with PBS contain-ing 0.2% Tween 20 for 7 min, incubated with horseradish peroxidase (HRP) conjugated anti-mouse IgG antibody (Santa Cruz) for 1 h at room temperature, washed again, and developed by using Super Signal (Pierce) at 20° for 5 min. The membranes are exposed to X-ray film (Fuji) and de-veloped. To verify equal protein loading, membranes are stripped with 50 mM citric acid for 1 h, washed as stated previously, and re-probed with monoclonal mouse anti β tubulin (Sigma) or rabbit anti IkB (Santa Cruz), followed by incubation with the appropriate HRP-conjugated anti IgG antibody (Figs. 1C, 2C, and 3B).

[17]   The Role of Endogenous Catechol Quinones in the Initiation of Cancer and Neurodegenerative Diseases

By ERCOLE CAVALIERI, ELEANOR ROGAN, and DHRUBAJYOTI CHAKRAVARTI

Evolution of Fundamental Concepts And Principles of Chemical Carcinogenesis

Chemical carcinogenesis became a distinct branch of cancer research in the early 1960s when James and Elizabeth Miller hypothesized that chem-ical carcinogens bind covalently to cellular macromolecules.[1,2] This hy-pothesis, now considered a principle in chemical carcinogenesis, entails reaction of the ultimate electrophilic species of carcinogens with nucleo-philic groups of DNA, RNA and protein. The relevance of this principle is to unify the different structures of chemical carcinogens that directly or via metabolic activation produce electrophilic species reacting with cellular macromolecules. At that time, the discovery that chemical carcinogens react with DNA, RNA and protein did not allow identification of the cellu-lar macromolecule critical to initiation of cancer. Based on cellular func-tions, it is logical to hypothesize that DNA is the key macromolecule. This idea was supported by observing the development of malignant cells after transfection of mice with DNA from cells malignantly transformed by chemicals.[3] This result pointed indeed to the mutagenic basis of tumor initiation. Data obtained over the years make it clear that DNA is the critical macromolecule that can account for tumor initiation.

Determining the factors involved in the process of binding to DNA has led us to understand how various chemicals, such as polycyclic aromatic hydrocarbons (PAHs), become carcinogens. Furthermore, this knowledge has led us to understand how natural compounds, such as estrogens, can become endogenous carcinogens.

Mechanism of Tumor Initiation by PAHs

The main activation of PAHs by cytochrome P450 (CYP) occurs via one-electron oxidation with formation of PAH radical cations.[4–6] This pathway can lead to oxygenated metabolites[6–8] as well as covalent bonds with DNA.[4,5] For potent carcinogens such as benzo[a]pyrene (BP), 7,12-dimethylbenz[a]anthracene (DMBA) and dibenzo[a,l]pyrene (DB[a,l]P, Fig. 1), formation of radical cations constitutes the most important activat-ing mechanism for tumor initiation.[4,5] Among the various oxygenated metabolites of PAHs, the bay-region diol epoxide can also play a role in the mechanism of tumor initiation.[4,5]

[1] J. A. Miller, Cancer Res. 30, 559 (1970).

[2] E. C. Miller and J. A. Miller, Cancer 47, 2327 (1981).

[3] C. Shih, B.-Z. Shilo, M. P. Goldfarb, A. Dannenberg, and R. A. Weinberg, Proc. Natl. Acad. Sci. USA 76, 5714 (1979).

[4] E. L. Cavalieri and E. G. Rogan, Pharmacol. Ther. 55, 183 (1992).

[5] E. Cavalieri and E. Rogan, in "The Handbook of Environmental Chemistry" (A. H. Neilson, ed.), p. 81. Springer-Verlag, Berlin and Heidelberg, 1998.

[6] E. L. Cavalieri and E. G. Rogan, in "The Handbook of Environmental Chemistry" (A. H. Neilson, ed.), p. 277. Springer-Verlag, Berlin and Heidelberg, 2002.

[7] E. L. Cavalieri, E. G. Rogan, P. Cremonesi, and P. D. Devanesan, Biochem. Pharmacol. 37, 2173 (1988).

[8] P. P. J. Mulder, P. Devanesan, K. Van Alem, G. Lodder, E. G. Rogan, and E. L. Cavalieri, Free Radical Biol. and Med. 34, 734 (2003).

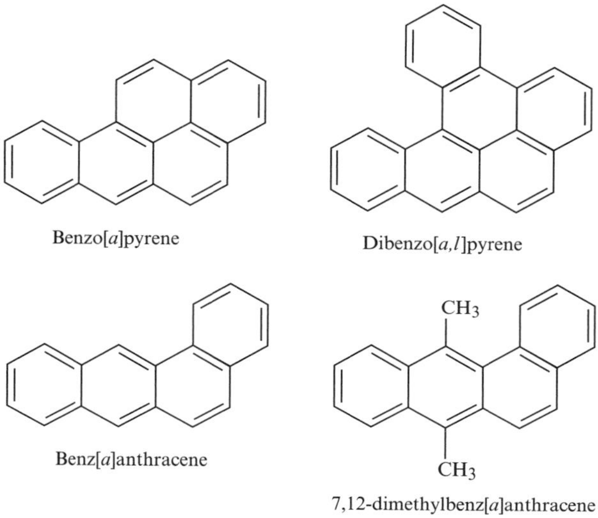

FIG. 1. Structures of the polycyclic aromatic hydrocarbons benzo[a]pyrene, dibenzo[a,l]-pyrene, 7,12-dimethylbenz[a]anthracene and benz[a]anthracene.

Cytochrome P450 in Cell Nuclei

Cytochrome P450 resides abundantly, but not exclusively, in the endoplasmic reticulum of the cell. In fact, isolated rat liver nuclei were found to contain cytochrome P450[9,10] and to catalyze the covalent binding of PAHs to DNA.[11,12] Thus the proximity between DNA, PAH and nuclear cytochrome P450 allows the metabolic activation of PAH by one-electron oxidation to yield covalent bonds between the PAH and DNA. Demonstration that cytochrome P450-catalyzed binding of PAH to DNA by one-electron oxidation was obtained by identifying the adduct formed by BP radical cation with the N-7 of guanine (Gua) after incubation of BP with isolated rat liver nuclei.[13]

[9] C. B. Kasper, J. Biol. Chem. **246,** 577 (1971).

[10] A. S. Khandwala and C. B. Kasper, Biochem. Biophys. Res. Commun. **54,** 1241 (1973).

[11] E. G. Rogan and E. Cavalieri, Biochem. Biophys. Res. Commun. **58,** 1119 (1974).

[12] E. G. Rogan, P. Mailander, and E. Cavalieri, Proc. Natl. Acad. Sci. USA **73,** 437 (1976).

[13] E. L. Cavalieri, E. G. Rogan, P. D. Devanesan, P. Cremonesi, R. L. Cerny, M. L. Gross, and W. J. Bodell, Biochemistry **29,** 4820 (1990).

Specificity in the Adduction of Chemical Carcinogens to DNA

An important factor for PAHs and other bulky carcinogens, such as estrogens, to trigger tumor initiation is related to their geometry. In fact, carcinogenic activity is observed only in PAHs containing three to seven condensed rings.[14] In the benz[a]anthracene series (Fig. 1), the presence of an angular ring is essential for eliciting carcinogenic activity, regardless of whether this ring is aliphatic or aromatic.[14,15] Optimal geometric configuration is necessary for appropriate physical complexes—for example, intercalation complexes—with DNA that are a prerequisite for the further formation of a covalent bond with DNA nucleophiles.[16,17] A strong piece of evidence that these complexes are formed derives from DNA adducts of PAH radical cations and catechol estrogen quinones (see the following), which occur predominantly with double-stranded DNA and very poorly, or not at all, with RNA or single-stranded DNA.[18–21]

The optimal geometry of the carcinogen is related not only to the formation of the physical complex with DNA, but also to the close proximity of the electrophilic site of the carcinogen and the nucleophilic site of the DNA base. Both of these characteristics can be provided only by specific sequences in DNA.

Sequence Specificity of Mutagenesis

PAHs induce specific transforming mutations that are found in a great majority of tumors induced by them. Evidence suggests that these mutations are initially found in the early preneoplastic period and then are clonally expanded to form tumors. It was therefore thought that PAHs induce specific mutations by reacting with specific DNA sequences. Studies support this idea. It is now understood that the mutational specificity of

[14] J. C. Arcos and M. F. Argus, "Chemical Induction of Cancer." Academic Press, New York. 1974.

[15] E. L. Cavalieri, E. G. Rogan, S. Higginbotham, P. Cremonesi, and S. Salmasi, *Polycyclic Aromat. Compd.* **1,** 59 (1990).

[16] S. A. Lesko, Jr., A. Smith, P. O. P. Ts'o, and R. S. Umans, *Biochemistry* **7,** 434 (1968).

[17] N. E. Geacintov, M. Shahbaz, V. Ibanez, K. Moussaoui, and R. G. Harvey, *Biochemistry* **27,** 8380 (1988).

[18] P. D. Devanesan, N. V. S. RamaKrishna, N. S. Padmavathi, S. Higginbotham, E. G. Rogan, E. L. Cavalieri, G. A. Marsch, R. Jankowiak, and G. J. Small, *Chem. Res. Toxicol.* **6,** 364 (1993).

[19] K.-M. Li, R. Todorovic, E. G. Rogan, E. L. Cavalieri, F. Ariese, M. Suh, R. Jankowiak, and G. J. Small, *Biochemistry* **34,** 8043 (1995).

[20] E. G. Rogan, P. D. Devanesan, N. V. S. RamaKrishna, S. Higginbotham, N. S. Padmavathi, K. Chapman, E. L. Cavalieri, H. Jeong, R. Jankowiak, and G. J. Small, *Chem. Res. Toxicol.* **6,** 356 (1993).

[21] E. Cavalieri, M. Saeed, S. Gunselman, and E. Rogan, *Proc. Amer. Assoc. Cancer Res.* **44,** 1324 (2003).

carcinogens can be determined at two levels: (1) the DNA sequence specificity of the adduct-forming reaction and (2) the DNA sequence specificity
of mutagenesis.

In brief, PAHs form physical complexes with DNA in a sequence-
selective manner.[17] PAH molecules in these complexes are metabolized
in situ by CYP to form radical cations and diol epoxides, which then react
with nucleophilic groups in DNA purines.[4,5] A moderate degree of sequence specificity is observed in PAH-DNA adducts.[22,23] DNA sequences
containing these adducts include the mutated sequences found in
early preneoplastic tissue, as well as in tumors (Chakravarti, unpublished
results).[22,24] In theory, preneoplastic mutations are induced from DNA
adducts and adduct-induced lesions such as apurinic sites by two mechanisms: (1) error-prone repair and (2) erroneous replication over any
unrepaired lesions. Studies suggest that apurinic sites generated by depurinating DNA adducts are mutated by error-prone repair,[25] whereas the
repair of stable adducts is error-free.[26,27] Replication over unrepaired
adducted bases, however, can sometimes cause mutations.[28] Studies suggest that error-prone repair at specific DNA sequences induces particular
mutations. For example, DB[*a,l*]P was found frequently to induce A to
G mutations in A residues 3′ to GN-doublet motifs,[25] and E_2-3,4-Q frequently induced the same mutations at A residues 5′ to G residues.[29] On
the other hand, mutagenesis by erroneous replication over stable adducts is dependent on the sequence context[30] and conformation of
these lesions.[31,32] Therefore the induction of specific mutations by these
carcinogens is determined by (1) mechanisms involved in adduct formation at specific DNA sequences and (2) mechanisms that determine

[22] D. Chakravarti, E. L. Cavalieri, and E. G. Rogan, *DNA Cell Biol.* **17**, 529 (1998).

[23] D. Chakravarti, *in* "Encyclopedia of Analytical Chemistry: Instrumentation and Application" (R. A. Meyers, ed.), p. 5144. John Wiley & Sons, Ltd. England 2000.

[24] D. Chakravarti, J. C. Pelling, E. L. Cavalieri, and E. G. Rogan, *Proc. Natl. Acad. Sci. USA*
92, 10422 (1995).

[25] D. Chakravarti, P. Mailander, E. L. Cavalieri, and E. G. Rogan, *Mutat. Res.* **456**, 17 (2000).

[26] M. Watanabe, V. M. Maher, and J. J. McCormick, *Mutat. Res.* **146**, 285 (1985).

[27] D. J. Choi, R. B. Roth, T. Liu, N. E. Geacintov, and D. A. Scicchitano, *J. Mol. Biol.* **264**,
213 (1996).

[28] M. Moriya, S. Spiegel, A. Fernandes, S. Amin, T. Liu, N. Geacintov, and A. P. Grollman,
Biochemistry **35**, 16646 (1996).

[29] D. Chakravarti, P. Mailander, K.-M. Li, S. Higginbotham, H. Zhang, M. L. Gross,
E. L. Cavalieri, and E. Rogan, *Oncogene* **20**, 7945 (2001).

[30] R. Shukla, T. Liu, N. E. Geacintov, and E. L. Loechler, *Biochemistry* **36**, 10256 (1997).

[31] R. Shukla, S. Jelinsky, T. Liu, N. E. Geacintov, and E. L. Loechler, *Biochemistry* **36**,
13263 (1997).

[32] R. A. Perlow and S. Broyde, *J. Mol. Biol.* **309**, 519 (2001).

the incorporation of mispaired bases (either by error-prone repair or by erroneous replication) at adduct-induced lesions in specific DNA sequence contexts.

Stable and Depurinating DNA Adducts

Carcinogens react with DNA to form two types of adducts: stable adducts and depurinating adducts. Investigators in chemical carcinogenesis have always dealt with stable adducts, which remain in DNA unless removed by repair. These adducts are usually detected by ^{32}P-postlabeling techniques, but their identification has rarely been accomplished.

In general, PAHs and catechol estrogens (see the following) predominantly form adducts at nucleophilic groups of adenine (Ade) and Gua, with destabilization of the glycosyl bond and subsequent depurination. As shown in Fig. 2, which represents a Watson-Crick DNA, the backbone is constituted by deoxyribose and phosphate, and the Gua is hydrogen bonded to cytosine and Ade to thymine. The Gua has an exocyclic NH_2 that reacts with electrophiles to form stable adducts (Fig. 2, hollow arrow). However, if reaction occurs at the N-7 (and sometimes, C-8), formation of depurinating adducts occurs (Fig. 2, solid arrows). When Ade reacts with an electrophile, it can form a stable adduct at the exocyclic NH_2 group and depurinating adducts at the N-3 and N-7 sites (Fig. 2, solid arrows). Following reaction at the N-3 of Ade (Fig. 2), destabilization of the glycosyl bond is assisted by the *anti*-elimination of the hydrogen atom at the C-2' of the deoxyribose moiety, with subsequent depurination. The resulting apurinic site in the DNA contains a C-2'-C-3' double bond, which by addition of water forms an apurinic site containing deoxyribose.[21]

The Role of Depurinating Adducts in Forming Tumorigenic Mutations

The first evidence that depurinating PAH-DNA adducts could be important for inducing the transforming mutations that lead to tumors was obtained when PAH-induced adducts were correlated with transforming Harvey (H)-*ras* mutations in SENCAR mouse skin tumors.[24] These mutations are detected by PCR amplification of the H-*ras* gene and directly sequencing the PCR products. This approach was useful in identifying mutations that were present in 14% to 47% of the tumor DNA.[33] These analyses indicated that the proportion of transforming H-*ras* mutations in tumors strongly correlates with the relative abundance of depurinating adducts. For example, both DMBA and DB[a,l]P formed primarily Ade-specific depurinating adducts (99% and 81% respectively)[5] in mouse skin

---

[33] D. Chakravarti, P. Mailander, J. Franzen, S. Higginbotham, E. L. Cavalieri, and E. G. Rogan, *Oncogene* **16,** 3203 (1998).

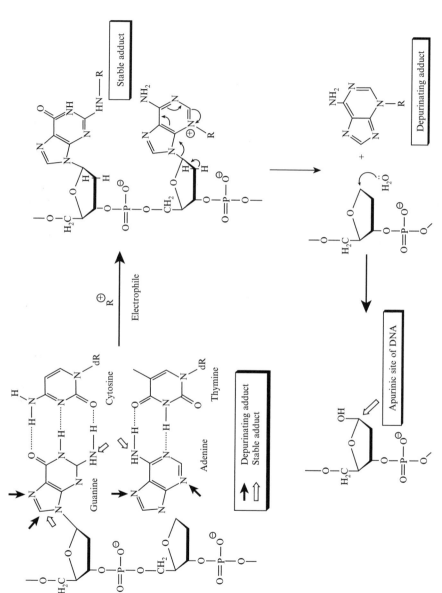

FIG. 2. Formation of stable and depurinating DNA adducts and generation of apurinic sites.

and induced codon 61 (CAA → CTA) mutations in the H-*ras* gene. On the other hand, BP formed 46% Gua-specific depurinating adducts and 25% Ade-specific depurinating adducts. Papillomas induced by BP contained 48% codon 13 (GGC → GTC) and 24% codon 61 (CAA → CTA) mutations in the H-*ras* gene. In a recent study, we have examined this relationship with BP-7,8-dihydrodiol. The dihydrodiol forms depurinating adducts in reverse proportions (12% Gua-specific and 25% Ade-specific depurinating adducts) as compared with its parent, BP. Papillomas induced by BP-7,8-dihydrodiol contained corresponding proportions (20% codon 13 and 30% codon 61) of specific H-*ras* mutations (Chakravarti *et al.*, unpublished results). More significantly, the relationship between depurinating adducts and mutations is also observed in preneoplastic tissue. Preneoplastic mutations were detected by PCR amplification of the H-*ras* gene from PAH-treated tissues For example, DB[*a,l*]P forms 81% Ade-depurinating adducts that are converted rapidly (by 12 h) into preneoplastic mutations, which consist primarily (90%) of A.T to G.C transitions.[25] The preneoplastic mutations were detected by PCR amplification of the H-*ras* gene at various times after PAH treatment, cloning PCR products into plasmids, isolating individual subclones and sequencing their H-*ras* inserts. More recently, the relationship between the relative abundance of depurinating adducts and preneoplastic mutations has been confirmed with other PAHs, including BP and DMBA (Chakravarti *et al.*, unpublished results).

Error-Prone Repair as a Mechanism of Mutagenesis

The rapidity (12 h) of induction of preneoplastic mutations suggested that these mutations are induced during a period when, as a consequence of DNA damage, replication is repressed and excision repair is induced.[34–36] The results suggest that these mutations are induced by error-prone repair. Further proof that these preneoplastic mutations are induced by error-prone repair was obtained when mismatched heteroduplexes corresponding to the mutations were observed. In brief, since excision repair removes a small section of the damaged strand and conducts fill-in DNA synthesis, if it is error-prone, misincorporation of nucleotides would generate mismatched heteroduplexes (Fig. 3). Thus A.T to G.C mutations are induced as G.T heteroduplexes by error-prone repair. These heteroduplexes are detected by treating preneoplastic DNA with a glycosylase enzyme (T.G-DNA glycosylase, which removes T residues from G.T heteroduplexes,

---

[34] T. J. Slaga, G. T. Bowden, B. G. Shapas, and R. K. Boutwell, *Cancer Res.* **34**, 771 (1974).
[35] T. W. Sawyer, R. D. Gill, T. Smith-Oliver, B. E. Butterworth, and J. DiGiovanni, *Carcinogenesis* **9**, 1197 (1988).
[36] R. D. Gill, B. E. Butterworth, A. N. Nettikumara, and J. DiGiovanni, *Environ. Mol. Mutagen.* **18**, 200 (1991).

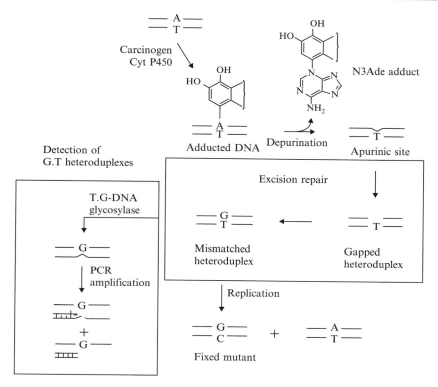

FIG. 3. Proposed pathway of formation of A to G mutations by error-prone repair of carcinogen-induced apurinic sites. A to G mutations by error-prone repair initially generate G.T heteroduplexes, which, after one round of replication, form fixed (G.C) mutations. *Left,* Detection of G.T heteroduplexes by the T.G-DNA glycosylase-PCR assay. Treatment of chromosomal DNA with T.G-DNA glycosylase converts G.T heteroduplexes into G. apyrimidinic sites, rendering the DNA refractory to PCR amplification. Compared with untreated DNA, T.G-DNA glycosylase-treated DNA shows a drastic reduction of A to G mutations.

forming abasic sites across G residues), followed by PCR amplification of the H-*ras* gene, cloning and sequence analysis to identify mutations. Since apurinic sites are refractory to PCR,[25,37] when G.T heteroduplexes are present, the glycosylase treatment drastically reduces the frequency of A.T to G.C transitions in the mutation spectra. However, when the A.T to G.C transitions are present as fixed mutations (after one round of replication following repair), glycosylase treatment does not alter the frequency of A.T to G.C mutations in the spectra. By using this technique, evidence

[37] B. Fromenty, C. Demeilliers, A. Mansouri, and D. Pessayre, *Nucleic Acids Res.* **28,** 50 (2000).

was obtained that error-prone repair of apurinic sites induced by PAHs—such as BP, DMBA and DB[a,l]P—generates preneoplastic mutations that correspond to the relative abundance of the depurinating adducts (Chakravarti, unpublished results).[25]

All depurinating adducts that have been studied dissociate from DNA instantaneously, with the exception of BP-6-C8Gua[38] and 4-OHE$_2$-1-N7Gua,[39] which dissociate slowly over a few hours. The rate of depurination appears to be of real consequence for the induction of mutations by error-prone repair. Results indicate that only those depurinating adducts that dissociate rapidly from DNA induce mutations by error-prone repair. This conclusion was reached from mutational studies with both BP and E$_2$-3,4-Q (see the following). These two carcinogens form significant amounts of slowly depurinating adducts (BP-6-C8Gua and 4-OHE$_2$-1-N7Gua), in addition to other depurinating adducts that dissociate rapidly. Preneoplastic mutations formed by these carcinogens correlate only with the rapidly depurinating adducts. This suggests that, in truth, mutations are related to the relative abundance of the rapidly depurinating adducts, which cause a burst of apurinic sites. Based on these studies, we hypothesize that the repair machinery becomes error-prone when cells are faced with a sudden abnormal burst of depurination, such as those induced by various carcinogens.

The critical role of depurinating PAH-DNA adducts in the initiation of cancer by these compounds has provided the impetus for discovering the specific estrogen metabolites, mostly estrogen-3,4-quinones (see the following), that form depurinating adducts and can become endogenous initiators of cancer.[40–42]

Catechol Quinones as Mutagens Initiating Cancer And Other Diseases

Experiments on estrogen metabolism,[43–46] formation of DNA adducts,[40–42] carcinogenicity,[47–49] and mutagenicity[29,50,51] have led to the hypothesis that reaction of certain estrogen metabolites, predominantly

[38] N. V. S. RamaKrishna, F. Gao, N. S. Padmavathi, E. L. Cavalieri, E. G. Rogan, R. L. Cerny, and M. L. Gross, *Chem. Res. Toxicol.* **5**, 293 (1992).

[39] K.-M. Li, W. Liang, P. Devanesan, E. Rogan, and E. Cavalieri, *Proc. Amer. Assoc. Cancer Res.* **40**, 46 (1999).

[40] E. L. Cavalieri, D. E. Stack, P. D. Devanesan, R. Todorovic, I. Dwivedy, S. Higginbotham, S. L. Johansson, K. D. Patil, M. L. Gross, J. K. Gooden, R. Ramanathan, R. L. Cerny, and E. G. Rogan, *Proc. Natl. Acad. Sci. USA* **99**, 10937 (1997).

[41] E. Cavalieri, K. Frenkel, J. G. Liehr, E. Rogan, and D. Roy in "INCI Monograph: Estrogens as Endogenous Carcinogens in the Breast and Prostate" (E. Cavalieri and E. Rogan, eds.), p. 75. Oxford Press, 2000.

[42] E. L. Cavalieri, E. G. Rogan, and D. Chakravarti, *Cell and Mol. Life Sci.* **59**, 665 (2002).

catechol estrogen-3,4-quinones, with DNA can generate the critical mutations initiating breast, prostate and other cancers.[42]

Formation of Estrogen Metabolites, Conjugates and DNA Adducts

Estrone (E_1) and E_2 are obtained by aromatization of androstenedione and testosterone, respectively, catalyzed by CYP19, aromatase (Fig. 4). E_1 and E_2 are biochemically interconvertible by the enzyme 17β-estradiol dehydrogenase. They are metabolized via two major pathways: formation of catechol estrogens and, to a lesser extent, 16α-hydroxylation (not shown in Fig. 4). The catechol estrogens formed are the major 2-hydroxyestrone (estradiol) ($OHE_1[E_2]$) and the minor $4\text{-}OHE_1(E_2)$.[52–54] In general, these two catechol estrogens are inactivated in the liver by conjugative reactions such as glucuronidation, sulfation and O-methylation. The most common pathway of conjugation in extrahepatic tissues occurs, however, by O-methylation catalyzed by the ubiquitous catechol-O-methyltransferase (COMT).[55] The level and/or induction of CYP1B1[56–58] and other 4-hydroxylases could render the $4\text{-}OHE_1(E_2)$, rather than the usual $2\text{-}OHE_1(E_2)$, as the major metabolites. In this case, conjugation of $4\text{-}OHE_1(E_2)$ by methylation in extrahepatic tissues might become insufficient, and competitive

[43] E. L. Cavalieri, S. Kumar, R. Todorovic, S. Higginbotham, A. F. Badawi, and E. G. Rogan, *Chem. Res. Toxicol.* **14**, 1041 (2001).

[44] P. Devanesan, R. J. Santen, W. P. Bocchinfuso, K. S. Korach, E. G. Rogan, and E. Cavalieri, *Carcinogenesis* **22**, 1573–1576. (2001).

[45] E. L. Cavalieri, P. Devanesan, M. C. Bosland, A. F. Badawi, and E. G. Rogan, *Carcinogenesis* **23**, 329 (2002).

[46] E. G. Rogan, A. F. Badawi, P. D. Devanesan, J. L. Meza, J. A. Edney, W. W. West, S. M. Higginbotham, and E. L. Cavalieri, *Carcinogenesis* **24**, 697 (2003).

[47] J. G. Liehr, W. F. Fang, D. A. Sirbasku, and A. Ari-Ulubelen, *J. Steroid Biochem.* **24**, 353 (1986).

[48] J. J. Li and S. A. Li, *Fed. Proc.* **46**, 1858 (1987).

[49] R. R. Newbold and J. G. Liehr, *Cancer Res.* **60**, 235 (2000).

[50] T. T. Rajah and J. T. Pento, *Res. Comm. Molecul. Pathol. and Pharmacol.* **89**, 85 (1995).

[51] L.-Y. Kong, P. Szaniszlo, T. Albrecht, and J. G. Liehr, *Intl. J. Oncol.* **17**, 1141 (2002).

[52] F. P. Guengerich, *Rev. Pharmacol. Toxicol.* **29**, 241 (1989).

[53] C. Martucci and J. Fishman, *Pharmacol. Ther.* **57**, 237 (1993).

[54] B. T. Zhu and A. H. Conney, *Carcinogenesis* **19**, 1 (1998).

[55] P. T. Männistö and S. Kaakola, *Pharmacol. Rev.* **51**, 593 (1999).

[56] U. Savas, K. K. Bhattacharya, M. Christou, D. L. Alexander, and C. R. Jefcoate, *J. Biol. Chem.* **269**, 14905 (1994).

[57] C. L. Hayes, D. C. Spink, B. C. Spink, J. Q. Cao, N. J. Walker, and T. R. Sutter, *Proc. Natl. Acad. Sci. USA* **93**, 9776 (1996).

[58] D. C. Spink, B. C. Spink, J. Q. Cao, J. A. DePasquale, B. T. Pentecost, M. J. Fasco, and T. R. Sutter, *Carcinogenesis* **19**, 291 (1998).

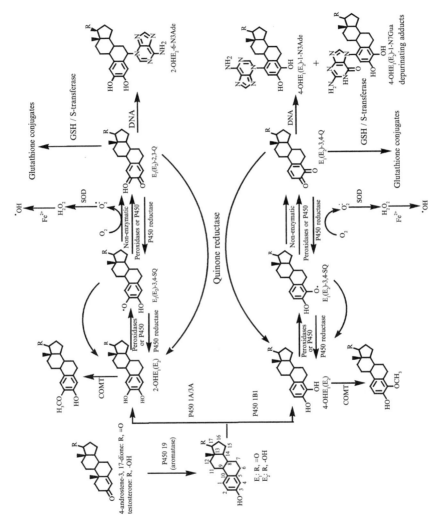

FIG. 4. Formation of estrogen metabolites, estrogen conjugates and depurinating estrogen–DNA adducts.

catalytic oxidation of catechol estrogens to $E_1(E_2)$-quinones $(E_1[E_2]$-Q) could occur (Fig. 4).

Redox cycling generated by reduction of $E_1(E_2)$-Q to semiquinones $(E_1 [E_2]$-SQ), catalyzed by CYP reductase, and subsequent oxidation back to $E_1(E_2)$-Q by oxygen forms superoxide anion radicals and, then, H_2O_2, catalyzed by superoxide dismutase. The presence of Fe^{++} generates formation of highly reactive hydroxyl radicals (Fig. 4). One of the possible types of damage by hydroxyl radicals is their reaction with lipids to form lipid hydroperoxides.[59,60] These compounds are unregulated cofactors of cytochrome P450, which lead to an abnormal increase in the oxidation of catechol estrogens to $E_1(E_2)$-Q.

$E_1(E_2)$-Q can be neutralized by conjugation with glutathione (GSH). A second inactivating pathway for $E_1(E_2)$-Q is their reduction to catechol estrogens by quinone reductase and/or CYP reductase.[61,62] If these two inactivating processes are insufficient, $E_1(E_2)$-Q may react with DNA to form stable and depurinating adducts (Fig. 4).[41,42] The carcinogenic 4-$OHE_1(E_2)$[47–49] are oxidized to form predominantly the depurinating adducts 4-$OHE_1(E_2)$-1-N3Ade and 4-$OHE_1(E_2)$-1-N7Gua,[40–42,63–65] whereas the borderline carcinogenic 2-$OHE_1(E_2)$[47–49] are oxidized to form much lower levels of the depurinating adducts, 2-$OHE_1(E_2)$-6-N3Ade, and higher levels of stable adducts than 4-$OHE_1(E_2)$.[41,42,63,65,66]

Imbalance in Estrogen Homeostasis

Estrogen metabolism involves a balance between activating and deactivating (protective) pathways. There are several factors that can unbalance estrogen homeostasis, that is, the equilibrium between activating and deactivating pathways, to limit the reaction of the endogenous carcinogenic $E_1(E_2)$-Q with DNA. The first critical factor is excessive synthesis of E_2

[59] H. Kappus, in "Oxidative Stress" (H. Sies, ed.), p. 273. Academic Press, New York, 1985.
[60] J. G. Liehr, *Environ. Health Perspect.* **105**(Suppl. 3), 565 (1997).
[61] DT diaphorase: A quinone reductase with special functions in cell metabolism and detoxication (L. Ernester, R. W. Estabrook, P. Hochstein, and S. Orrenius, eds.), *Chemica Scripta* **27A** (1987).
[62] D. Roy and J. G. Liehr, *J. Biol. Chem.* **263**, 3646 (1988).
[63] D. E. Stack, J. Byun, M. L. Gross, E. G. Rogan, and E. L. Cavalieri, *Chem. Res. Toxicol.* **9**, 851 (1996).
[64] K.-M. Li, R. Todorovic, P. Devanesan, S. Higginbotham, H. Köfeler, R. Ramanathan, M. L. Gross, E. G. Rogan, and E. L. Cavalieri, *Carcinogenesis* **24**, in press (2003).
[65] E. Cavalieri, E. Kohli, M. Zahid, and E. Rogan, *Proc. Am. Assoc. Cancer Res.* **44**, (2nd Ed.) 180 (2003).
[66] I. Dwivedy, P. Devanesan, P. Cremonesi, E. Rogan, and E. Cavalieri, E. *Chem. Res. Toxicol.* **5**, 828 (1992).

by over-expression of aromatase (CYP19) in target tissues[67–69] and/or the presence of sulfatase that excessively converts stored E_1 sulfate to E_1.[70,71] Synthesis of E_2 by breast tissue *in situ* suggests that much more E_2 is present in some locations of target tissues than predicted from plasma concentrations. A striking result of *in situ* production of E_2 in human breast tissue is the similar levels of E_2 in breast tissue in pre-menopausal and post-menopausal women, even though plasma levels are 10 to 50-fold higher in pre-menopausal than in post-menopausal women.[72] Both aromatase and sulfatase contribute to *in situ* estrogen production.[70,71]

A second critical factor leading to imbalances in estrogen homeostasis might be high levels of 4-$OHE_1(E_2)$ that are due to over-expression of CYP1B1, which metabolizes E_2 predominantly to form 4-OHE_2.[56–58] This could result in relatively large amounts of 4-$OHE_1(E_2)$ that, in turn, can lead to more extensive oxidation to the carcinogenic $E_1(E_2)$-3,4-Q. A third factor could be a lack or a low level of activity of the protective COMT enzyme. If this enzyme is insufficient, 4-$OHE_1(E_2)$ will not be effectively methylated in extrahepatic tissues but will be oxidized to the ultimate carcinogenic metabolites $E_1(E_2)$-3,4-Q. A fourth factor could be a low level of GSH and/or low levels of quinone oxidoreductase and/or CYP reductase, which could leave available higher levels of $E_1(E_2)$-3,4-Q that may react with DNA.

Imbalance in Estrogen Homeostasis in the Kidney of Syrian Golden Hamsters

The hamster is an excellent model for studying estrogen homeostasis, because implantation of E_1 or E_2 in male Syrian golden hamsters induces 100% of renal carcinomas but does not induce liver tumors.[73] Therefore comparison of the profile of estrogen metabolites, conjugates and DNA adducts in the two organs, after treatment of the hamster with E_2, should

[67] W. R. Miller and J. O'Neill, *Steroids* **50**, 537 (1987).

[68] E. R. Simpson, M. S. Mahendroo, G. D. Means, M. W. Kilgore, M. M. Hinshelwood, S. Graham-Lorence, B. Amarneh, Y. Ito, C. R. Fisher, M. D. Michael, C. R. Mendelson, and S. E. Bulun, *Endocrine Rev.* **15**, 342 (1994).

[69] C. R. Jefcoate, J. G. Liehr, R. J. Santen, T. R. Sutter, J. D. Yager, W. Yue, S. J. Santner, R. Tekmal, L. Demers, R. Pauley, F. Naftolin, G. Mor, and L. Berstein, *in* "INCI Monograph: Estrogens as Endogenous Carcinogens in the Breast and Prostate" (E. Cavalieri and E. Rogan, eds.), p. 95. Oxford Press, 2000.

[70] S. J. Santner, P. D. Feil, and R. J. Santen, *J. Clin. Endocrinol. Metab.* **59**, 29 (1984).

[71] J. R. Pasqualini, G. Chetrite, C. Blacker, M. C. Feinstein, L. Delalonde, M. Talbi, and C. Maloche, *J. Clin. Endocrinol. Metab.* **81**, 1460 (1996).

[72] A. A. J. Van Landeghem, J. Poortman, M. Nabuurs, and J. H. Thijssen, *Cancer Res.* **45**, 2900 (1985).

[73] J. J. Li, S. A. Li, J. L. Klicka, J. A. Parsons, and L. K. T. Lam, *Cancer Res.* **43**, 5200 (1983).

provide information on the relative imbalance in estrogen homeostasis in the two tissues.[43] In the liver, more O-methylation of 2-OHE$_1$(E$_2$) was observed, whereas more formation of E$_1$(E$_2$)-Q was detected in the kidney. These results suggest greater oxidation of catechol estrogens to E$_1$(E$_2$)-Q and less protective methylation of 2-OHE$_1$(E$_2$) in the kidney. When normal levels of GSH were depleted before the hamsters were treated with E$_2$, very low levels of catechol estrogens and methoxy catechol estrogens were observed in the kidney as compared with the liver, suggesting little protective reduction of E$_1$(E$_2$)-Q to catechol estrogens in the kidney. More importantly, the 4-OHE$_1$(E$_2$)-1-N7Gua depurinating adduct arising from reaction of E$_1$(E$_2$)-3,4-Q with DNA was detected in the kidney, but not in the liver.[43]

These results suggest that tumor initiation in the kidney occurs because of poor methylation of catechol estrogens, rendering more likely competitive oxidation of catechol estrogens to E$_1$(E$_2$)-Q, as well as poor reductase activity to remove the E$_1$(E$_2$)-Q. These two effects produce a large amount of E$_1$(E$_2$)-Q, which can react with the nucleophilic groups of DNA.

Imbalance in Estrogen Homeostasis in the Mammary Gland of ERKO/Wnt-1 Mice

Mammary tumors develop in female estrogen receptor-α knock-out (ERKO)/Wnt-1 mice despite their lack of functional estrogen receptor-α.[74] Extracts of hyperplastic mammary tissue and mammary tumors from these mice were analyzed by HPLC interfaced with an electrochemical detector.[44] Picomole amounts of the 4-catechol estrogens were detected, but their methoxy conjugates were not. Neither the 2-catechol estrogens nor 2-methoxy catechol estrogens were detected. 4-OHE$_1$(E$_2$)-GSH conjugates or their hydrolytic products (conjugates of cysteine and N-acetylcysteine) were detected in picomole amounts in both tumors and hyperplastic mammary tissue, demonstrating the formation of E$_1$(E$_2$)-3,4-Q. These preliminary findings indicate that estrogen homeostasis is unbalanced in the mammary tissue, in that the normally minor 4-catechol estrogen metabolites were detected in the mammary tissue, but not the normally predominant 2-catechol estrogens. Furthermore, methylation of catechol estrogens was not detected, whereas formation of 4-OHE$_1$(E$_2$)-GSH conjugates was. These results are consistent with the hypothesis that mammary tumor development is primarily initiated by metabolism of estrogens to E$_1$(E$_2$)-3,4-Q, which may react with DNA to induce oncogenic mutations.

[74] W. P. Bocchinfuso, W. P. Hively, J. F. Couse, H. E. Varmus, and K. S. Korach, *Cancer Res.* **59,** 1869 (1999).

Imbalances in Estrogen Homeostasis in the Prostate of Noble Rats

Estrogen metabolites and conjugates were analyzed in the ventral and anterior lobes of the rat prostate, which are not susceptible to estrogen-induced carcinogenesis, and in the susceptible dorsolateral and periurethral prostate of rats treated with 4-OHE$_2$ or E$_2$-3,4-Q.[45] The analyses revealed that the areas of the prostate susceptible to induction of carcinomas have less protection by COMT, quinone reductase and GSH, thereby favoring reaction of E$_1$(E$_2$)-3,4-Q with DNA.

Imbalances in Estrogen Homeostasis in the Breast of Women with Breast Carcinoma

A recent study of breast tissue from women with and without breast cancer provides key evidence in support of the concept of estrogen homeostasis.[46] In fact, relative imbalances in estrogen homeostasis were observed in analysis of women with breast cancer (Fig. 5). Levels of E$_1$ and E$_2$ in women with carcinoma were higher than those in controls. In women without cancer, a larger amount of 2-OHE$_1$(E$_2$) than 4-OHE$_1$(E$_2$) was observed. In women with carcinoma, the 4-OHE$_1$(E$_2$) were three times more abundant than the 2-OHE$_1$(E$_2$). The 4-OHE$_1$(E$_2$) were also four times higher than in women without cancer. Furthermore, a lower level of methylation was observed for the catechol estrogens in cancer cases *vs* the controls. Levels of E$_1$(E$_2$)-Q conjugates in women with cancer were three times those in controls, suggesting a larger probability for the

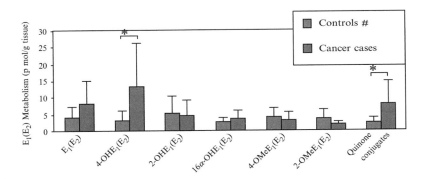

FIG. 5. Estrogen metabolites and estrogen conjugates in human breast tissue from women with and without breast carcinoma. Quinone conjugates are 4-OHE$_1$(E$_2$)-2-Cys, 4-OHE$_1$(E$_2$)-2-NAcCys, 2-OHE$_1$(E$_2$)-(1 + 4)-Cys and 2-OHE$_1$(E$_2$)-(1 + 4)-NAcCys. #Controls are benign fatty breast tissue and benign fibrocystic changes. *Statistically significant differences were determined by using the Wilcoxon rank sum test, $p < 0.01$. (See color insert.)

$E_1(E_2)$-Q to react with DNA in the breast tissue of women with carcinoma. Levels of 4-OHE$_1$(E$_2$) ($p < 0.01$) and quinone conjugates ($p < 0.003$) appear to be highly significant predictors of breast cancer.[46]

In summary, it is apparent from these animal and human studies that the oxidative stress leading to the formation of semiquinones and quinones from catechol estrogens is the result of an imbalance of one or more enzymes involved in the maintenance of estrogen homeostasis.

Stable and Depurinating Catechol Estrogen-DNA Adducts

Synthesis of Adduct Standards

To determine the possible DNA adducts of $E_1(E_2)$-2,3-Q and $E_1(E_2)$-3,4-Q, standard adducts were synthesized by reaction of these quinones with dG, dA and the nucleobase Ade.[63–65] By 1,4-Michael addition of $E_1(E_2)$-3,4-Q to dG (Fig. 6), the adduct 4-OHE$_1$(E$_2$)-1-N7Gua was obtained.[63] Reaction of $E_1(E_2)$-3,4-Q with dA produced no adducts (Fig. 6); however, reaction of $E_1(E_2)$-3,4-Q with Ade by 1,4-Michael addition resulted in the formation of 4-OHE$_1$(E$_2$)-1-N3Ade.[64] This adduct was obtained only with Ade, because in dA, the adjacent deoxyribose moiety at N-9 hinders the approach of the electrophile $E_1(E_2)$-3,4-Q to the N-3 of Ade.[75,76] PAH-N3Ade and 4-OHE$_1$(E$_2$)-1-N3Ade adducts, however, are formed in the reaction with DNA and are rapidly lost from the DNA by depurination.[5,64,65]

Reaction of $E_1(E_2)$-2,3-Q with dG or dA afforded predominantly the stable adducts 2-OHE$_1$(E$_2$)-6-N^{2}dG and 2-OHE$_1$(E$_2$)-6-N^{6}dA.[63] In this case, the $E_1(E_2)$-2,3-Q did not react as a quinone, but as the tautomeric $E_1(E_2)$-2,3-Q methide (Fig. 7).[77] These adducts retain the deoxyribose moiety and are referred to as stable adducts, because they remain in DNA unless removed by repair. By mass spectrometric techniques, other investigators have found the depurinating adducts, 2-OHE$_1$(E$_2$)-6-N7Gua and 2-OHE$_1$(E$_2$)-6-N7Ade, as minor products when $E_1(E_2)$-2,3-Q was reacted with dG and dA, respectively.[78] Reaction of E_2-2,3-Q with Ade afforded the depurinating adduct 2-OHE$_2$-6-N3Ade.[65]

[75] P. P. J. Mulder, L. Chen, B. C. Sekhar, M. George, M. L. Gross, E. G. Rogan, and E. L. Cavalieri, *Chem. Res. Toxicol.* **9**, 1264 (1996).

[76] A. A. Hanson, E. G. Rogan, and E. L. Cavalieri, *Chem. Res. Toxicol.* **11**, 1201 (1998).

[77] S. L. Iverson, L. Shen, N. Anlar, and J. L. Bolton, *Chem. Res. Toxicol.* **9**, 492 (1996).

[78] O. Convert, C. Van Aerden, L. Debrauwer, E. Rathahao, H. Molines, F. Fournier, J.-C. Tabet, and A. Paris, *Chem. Res. Toxicol.* **15**, 754 (2002).

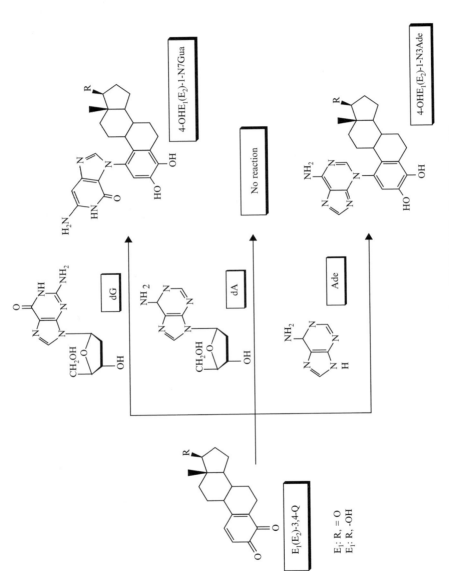

FIG. 6. Formation of adducts by reaction of $E_1(E_2)$-3,4-Q with dG, dA or Ade.

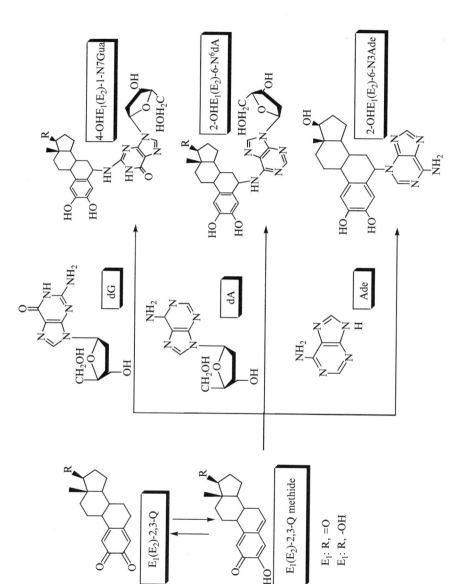

FIG. 7. Formation of adducts by reaction of $E_1(E_2)$-2,3-Q with dG, dA or Ade.

Rates of Depurination of Catechol Estrogen-DNA Adducts

The N3Ade adducts obtained from $E_1(E_2)$-3,4-Q and $E_1(E_2)$-2,3-Q after reaction with DNA depurinate instantaneously.[64,65] The reaction of $E_1(E_2)$-3,4-Q with dG or DNA,[39,65] however, produces adducts that depurinate slowly. The half time of depurination from dG is about 5 h at room temperature,[39] and the half time of depurination from DNA is 2–3 h at 37°.[65] The rate of depurination has a major impact on the level of mutations induced in animal tissues treated with $E_1(E_2)$-3,4-Q. Slow depurination of adducts at G residues does not induce mutations, whereas the burst of depurination at A residues generates extensive mutations (see the following).[29]

Competition Between E_2-3,4-Q and E_2-2,3-Q in the Reaction with DNA to Form Depurinating Adducts

The carcinogenicity of the catechol estrogens may depend in part on the relative reactivity of their quinones. To examine this idea the relative reactivity of E_2-3,4-Q and E_2-2,3-Q with DNA was determined.[65] The quinones were reacted individually with DNA, as well as in mixtures of 75% E_2-3,4-Q and 25% E_2-2,3 -Q to 5% E_2-3,4-Q and 95% E_2-2,3-Q (Fig. 8). After 10 h of incubation of E_2-3,4-Q with DNA, approximately 130 μmole

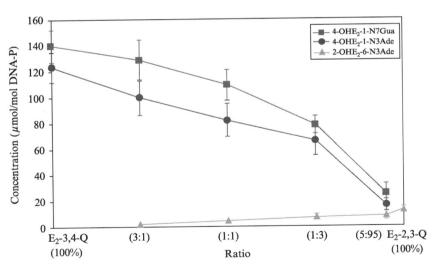

FIG. 8. Depurinating adducts formed after 10 h by mixtures of $E_1(E_2)$-3,4-Q and $E_1(E_2)$-2,3-Q reacted with DNA. The levels of stable adducts formed in the mixtures ranged from 0.1% to 1% of the total adducts. (See color insert.)

of both the N3Ade and N7Gua adducts per mole DNA-P were observed, whereas reaction of E_2-2,3-Q formed only 10 μmole of 2-OHE$_2$-6-N3Ade/mole DNA-P. Even with 5% E_2-3,4-Q and 95% E_2-2,3-Q, approximately 20 μmole of 4-OHE$_2$-1-N3Ade and 4-OHE$_2$-1-N7Gua and 10 μmole of 2-OHE$_2$-6-N3Ade per mole of DNA-P were observed. The levels of stable adducts ranged from 0.1% to 1% of the total adducts. Similar results were obtained when mixtures of 4-OHE$_2$ and 2-OHE$_2$ were incubated with lactoperoxidase, tyrosinase or prostaglandin H synthase in the presence of DNA. In every case, the binding of the activated 4-OHE$_2$ predominated over that of activated 2-OHE$_2$. These results suggest that the presence of even small amounts of E_2-3,4-Q can result in significant binding to DNA, generating depurinating adducts and apurinic sites.

Error-Prone Repair as an Initiating Mechanism of Breast Cancer

In a manner similar to the PAHs, studies with E_2-3,4-Q in SENCAR mouse skin and ACI rat mammary gland indicate that estrogens induce mutations by error-prone repair of apurinic sites generated by fast -depurinating adducts (Chakravarti, unpublished results).[29] Specifically, E_2-3,4-Q was found to induce primarily A.T to G.C mutations by error-prone repair that correlated with N3Ade depurinating adducts. These results suggest that error-prone repair could be a mechanism for initiation of breast cancer. If so, some individuals with particular base excision repair (BER) gene polymorphisms would be more susceptible to breast cancer. Recent studies suggest that the Arg 399 Gln polymorphic mutation in XRCC1 significantly accounts for increased risk of breast cancer. For example, among Korean women, the Gln/Gln mutant allele occurred in 7% of controls and 27% of cases.[79] Another study reported that among African-American women, the Arg/Gln allele occurred in 24% of controls and 32% of cases.[80] XRCC1 binds other BER proteins through its BRCT domains to participate in the initial and late stages of repair.[81] One study correlating aflatoxin B_1 (AFB$_1$) exposure, XRCC1 399 polymorphism and mutagenesis suggests that this polymorphic mutation in the XRCC1 gene could make BER more error-prone.[82] AFB$_1$ generates abundant amounts of apurinic sites in

[79] S. U. Kim, S. K. Park, K. Y. Yoo, K. S. Yoon, J. Y. Choi, J. S. Seo, W. Y. Park, J. H. Kim, D. Y. Noh, S. H. Ahn, K. J. Choe, P. T. Strickland, A. Hirvonen, and D. Kang, *Pharmacogenetics* **12**, 335 (2002).

[80] E. J. Duell, R. C. Millikan, G. S. Pittman, S. Winkel, R. M. Lunn, C. K. Tse, A. Eaton, H. W. Mohrenweiser, B. Newman, and D. A. Bell, *Prev.* **10**, 217 (2001).

[81] A. E. Vidal, S. Boiteux, I. D. Hickson, and J. P. Radicella, *EMBO J.* **20**, 6530 (2001).

[82] R. M. Lunn, R. G. Langlois, L. L. Hsieh, C. L. Thompson, and D. A. Bell, *Cancer Res.* **59**, 2557 (1999).

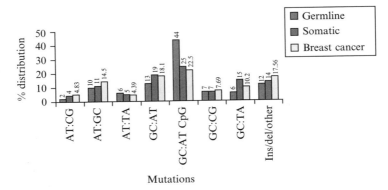

FIG. 9. Distribution of p53 gene mutations. The data are taken from the IARC database, which contains 17,914 p53 mutations in all cancers, consisting of 225 germline and 17,689 somatic mutations. The latter include 1523 breast cancer mutations. (See color insert.)

DNA.[83] AFB_1-exposed individuals who carried the XRCC1 399 Gln/Gln allele showed a 2-fold higher level of mutations in the glycophorin A gene compared to those who carried the wild type (Arg/Arg) gene.[82] Thus XRCC1 could participate in the error-prone BER involved in breast cancer.

Data in the IARC-p53 mutation database indicate that sporadic breast cancer patients carry a distinctive spectrum of mutations in their p53 gene (Fig. 9). These patients contain significantly higher levels of A.T to G.C mutations and significantly lower levels of G.C to T.A mutations. Therefore the induction of A.T to G.C mutations is significant for breast cancer induction. Our studies indicate that E_2-3,4-Q induces a remarkably similar spectrum of mutations (Chakravarti, unpublished results).[29] These results are consistent with our hypothesis that estrogen-induced mutations are involved in breast cancer, presumably by inducing specific A.T to G.C mutations.

Using the IARC database (http://www.iarc.fr/p53/), Greenblatt *et al.* have conducted an interesting analysis of p53 mutations among women who are hereditarily predisposed to breast cancer by carrying deleterious mutations in their BRCA1 and BRCA2 genes.[84,85] They reported that women with breast cancer who carried BRCA1 and BRCA2 mutations

[83] A. A. Stark, L. Malca-Mor, Y. Herman, and D. F. Liberman, *Cancer Res.* **48,** 3070 (1988).
[84] M. S. Greenblatt, P. O. Chappuis, J. P. Bond, N. Hamel, and W. D. Foulkes, *Cancer Res.* **61,** 4092 (2001).
[85] A. R. Hartman and J. M. Ford, *Nat. Genet.* **32,** 180, (2002).

also showed a greater abundance of Ade-specific mutations. Current data in the IARC database indicate that the incidence of A.T to G.C mutations among sporadic breast cancer patients (12%) is significantly less than those among BRCA1-mutated breast cancer patients (21%). This suggests that the lack of BRCA1 could be a factor in the induction of A.T to G.C mutations. It is very likely that BRCA1 acts in BER. Although not compared under isogenic conditions, BRCA1-deficient HCC1937 cells appear to be more sensitive to methylmethane sulfonate than MCF-7 cells,[86] which supports the idea. BRCA1 interacts with p53.[87] Because p53 participates in BER,[88–90] it is possible that BRCA1 acts in BER through a p53-mediated pathway. Although this is speculative, it appears possible that BRCA1 could be another gene that acts in breast cancer as a BER gene and modulates the formation of specific A.T to G.C mutations.

Unifying Mechanism of Initiation of Cancer by Endogenous and Synthetic Estrogens

The oxidation of catechols to semiquinones and quinones is the major pathway to initiate cancer by natural estrogens, as well as synthetic estrogens, such as the human carcinogen diethylstilbestrol[91] and its hydrogenated derivative hexestrol. These two compounds, analogously to the natural estrogens, are carcinogenic in the kidney of Syrian golden hamsters.[73,92] Their major metabolites are catechols,[92–95] which can be oxidized to catechol quinones. The chemical properties of the catechol quinone of hexestrol resemble those of $E_1(E_2)$-3,4-Q, in that, it specifically forms N7Gua and N3Ade adducts by 1,4-Michael addition to dG or Ade, respectively, as well as DNA (Fig. 10) (Saeed et al., unpublished results).[96]

[86] Q. Zhong, C. F. Chen, S. Li, Y. Chen, C. C. Wang, J. Xiao, P. L. Chen, Z. D. Sharp, and W. H. Lee, *Science* **285**, 747 (1999).

[87] H. Zhang, K. Somasundaram, Y. Peng, H. Tian, D. Bi, B. L. Weber, and W. S. El-Deiry, *Oncogene* **16**, 1713 (1998).

[88] H. Offer, M. Milyavsky, N. Erez, D. Matas, I. Zurer, C. C. Harris, and V. Rotter, *Oncogene* **20**, 581 (2001).

[89] H. Offer, N. Erez, I. Zurer, X. Tang, M. Milyavsky, N. Goldfinger, and V. Rotter, *Carcinogenesis* **23**, 1025 (2002).

[90] Y. R. Seo, M. L. Fishel, S. Amundson, M. R. Kelley, and M. L. Smith, *Oncogene* **21**, 731 (2002).

[91] A. L. Herbst, H. Ulfelder, and D. C. Poskanzer, *N. Engl. J. Med.* **284**, 878 (1971).

[92] J. G. Liehr, A. M. Ballatore, B. B. Dague, and A. A. Ulubelen, *Chem.-Biol. Interactions* **55**, 157 (1985).

[93] H. Haaf and M. Metzler, *Biochem. Pharmacol.* **34**, 3107 (1985).

[94] G. Blaich, M. Gåttlicher, P. Cikryt, and M. Metzler, *J. Steroid Biochem.* **35**, 201 (1996).

[95] M. Metzler and J. A. McLachlan, *Adv. Exp. Med. Biol.* **136A**, 829 (1981).

[96] S.-T. Jan, P. Devanesan, D. Stack, R. Ramanathan, J. Byun, M. L. Gross, E. Rogan, and E. Cavalieri, *Chem. Res. Toxicol.* **11**, 412 (1998).

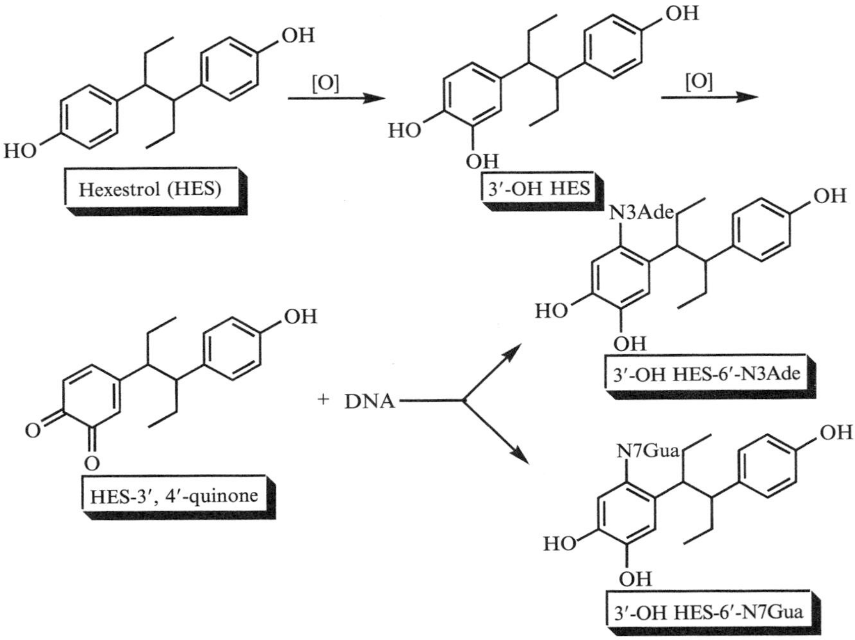

FIG. 10. Metabolic activation of hexestrol (HES) and formation of depurinating DNA adducts by reaction of HES-3′,4′-Q with DNA.

These results suggest that the hexestrol catechol quinone is involved in tumor initiation by this synthetic estrogen. Furthermore, they support the hypothesis that $E_1(E_2)$-3,4-Q may be the major endogenous tumor initiators because they react with DNA to form the N7Gua and N3Ade depurinating adducts by 1,4-Michael addition.

Unifying Mechanism of Initiation of Cancer and Other Diseases by Catechol Quinones

Oxidation of catechols to semiquinones and quinones is not only the major mechanism of tumor initiation by natural and synthetic estrogens but it could also be the major mechanism of tumor initiation for the leukemogen benezene. It has long been known that benzene causes acute myelogenous leukemia in humans.[97–100] Benzene metabolites include phenol,

---

[97] R. A. Rinsky, A. B. Smith, R. Hornung, T. G. Filloon, R. J. Young, A. H. Okun, and P. J. Landrigan, *N. Engl. J. Med.* **316,** 1044 (1987).

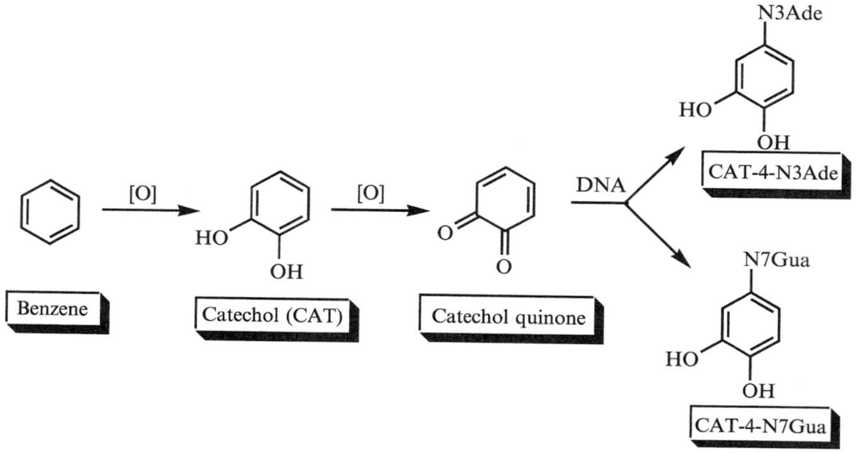

FIG. 11. Metabolic activation of benzene to form depurinating DNA adducts by reaction of benzene-1,2-quinone with DNA.

catechol (1,2-dihydroxybenzene, Fig. 11) and hydroquinone (1,4 -dihydrox-ybenzene).[101,102] Catechol and hydroquinone accumulate in the bone marrow,[103,104] where they can be oxidized by peroxidases, including myelo-peroxidase and prostaglandin H synthase.[105–108] The resulting quinones can produce DNA adducts.[108,109] In fact, catechol, one of the metabolites of benezene, when oxidized to its quinone reacts by 1,4-Michael addition with dG and Ade to form catechol-4-N7Gua and catechol-4-N3Ade, re-spectively (Fig. 11).[109] Oxidation of catechol by horseradish peroxidase,

[98] R. A. Rinsky, *Environ. Health Perspect.* **82,** 189 (1989).

[99] M. B. Paxton, *Environ. Health Perspect.* **104** (Suppl. 6), 1431 (1996).

[100] R. A. Rinsky, R. W. Hornung, S. R. Silver, and C. Y. Tseng, *Am. J. Ind. Med.* **42,** 474 (2002).

[101] P. J. Sabourin, W. E. Bechtold, L. S. Birnbaum, G. W. Lucier, and R. F. Henderson, *Toxicol. Appl. Pharmacol.* **99,** 421 (1989).

[102] P. M. Schlosser, J. A. Bond, and M. A. Medinsky, *Carcinogenesis* **14,** 2477 (1993).

[103] D. E. Rickert, T. S. Edgar, C. S. Barrow, J. Bus, and R. D. Irons, *Toxicol. Appl. Pharmacol.* **49,** 417 (1979).

[104] W. F. Greenlee, E. A. Gross, and R. D. Irons, *Chem.-Biol. Interact.* **33,** 285 (1981).

[105] D. A. Eastmond, M. T. Smith, O. Ruzo, and D. Ross, *Mol. Pharmacol.* **30,** 674 (1986).

[106] A. Sadler, V. V. Subrahmanyam, and D. Ross, *Toxicol. Appl. Pharmacol.* **93,** 62 (1988).

[107] M. J. Schlosser, R. D. Shurina, and G. F. Kalf, *Chem. Res. Toxicol.* **3,** 333 (1990).

[108] G. Levay, D. Ross, and W. J. Bodell, *Carcinogenesis* **14,** 2329 (1993).

[109] E. L. Cavalieri, K.-M. Li, N. Balu, M. Saeed, P. Devanesan, S. Higginbotham, J. Zhao, M. L. Gross, and E. Rogan, *Carcinogenesis* **23,** 1071, (2002).

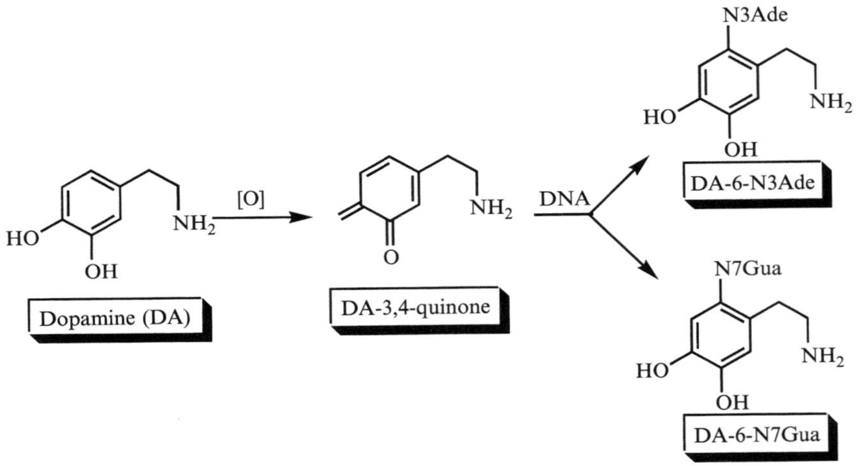

FIG. 12. Oxidative metabolism of dopamine to its quinone, which reacts with DNA to form depurinating adducts.

tyrosinase or phenobarbital -induced rat liver microsomes yields the catechol-4-N7Gua adduct, whereas catechol-4-N3Ade is obtained only with tyrosinase.[109] Formation of depurinating adducts specifically at the N-7 of Gua and N-3 of Ade by the 1,4-Michael addition of benzene-1,2-quinone to DNA suggests that oxidation of the metabolite catechol may play the major role in tumor initiation by benzene.

The catecholamine neurotransmitter dopamine produces semiquinones and quinones by autoxidation, metal ion oxidation and peroxidative enzyme or cytochrome P450 oxidation.[110–112] This oxidative process is similar to the one previously described for the benzene metabolite catechol and the catechols of natural and synthetic estrogens (see previous). In fact, reaction of dopamine-3,4-Q with DNA by 1,4-Michael addition forms specific N7Gua and N3Ade adducts (Fig. 12).[109] The same adducts are obtained by enzymatic oxidation of dopamine in the presence of DNA. The mutations generated by this DNA damage may be at the origin of Parkinson's and other neurodegenerative diseases.

[110] M. B. Mattammal, R. Strong, V. M. Lakshmi, H. D. Chung, and A. H. Stephenson, *J. Neurochem.* **64,** 1845 (1995).
[111] B. Kalyanaraman, C. C. Felix, and R. C. Sealy, *Environ. Health Perspect.* **64,** 185 (1985).
[112] B. Kalyanaraman, C. C. Felix, and R. C. Sealy, *J. Biol. Chem.* **259,** 7584 (1984).

Conclusions

Several lines of evidence point to the major role of endogenous catechol quinones in the etiology of cancer and neurodegenerative diseases. This picture is most developed for the role of catechol estrogen-3,4-quinones in the initiation of breast cancer. Imbalances of estrogen homeostasis in breast cells can result in the availability of $E_1(E_2)$-3,4-Q to react with DNA. Formation of depurinating 4-$OHE_1(E_2)$-1-N7Gua and 4-$OHE_1(E_2)$-1-N3Ade adducts[41,42] has specificity dependent on the physicochemical properties of the quinones and the sequence specificity of the DNA. The rapidly-depurinating N3Ade adducts result in a burst of apurinic sites that overwhelm the repair machinery of the cell and generate tumorigenic mutations by error-prone repair.[29] In contrast, repair of the apurinic sites that are formed by the slow depurination of N7Gua adducts can proceed correctly and leads to very few mutations. Mutations at Ade bases accumulate in breast cancer cells, presumably through this mechanism.

This pathway of activation for endogenous estrogens is mirrored by the activation of the synthetic estrogens hexestrol and DES. These human carcinogens can also be metabolized to their catechol and, then, their quinone derivatives, to react with DNA and form analogous N7Gua and N3Ade adducts. These adducts have properties similar to those of the endogenous estrogens, with the N3Ade adducts depurinating instantaneously and the N7Gua adducts depurinating over a period of a few hours.

A unifying mechanism of activation through catechol quinones has been hypothesized through the discovery of similar results with the metabolites of benzene, a leukemogen, and dopamine, a neurotransmitter.[109] Both of these compounds are metabolized to catechol quinones, which react with DNA to form N7Gua and N3Ade adducts. These results suggest that the metabolism of endogenous compounds to catechol quinones plays a major role in the initiation of cancer and other diseases. Abnormal proliferation of mutated cells may lead to the development of tumors via hormone-dependent and hormone-independent processes. A unifying mechanism of initiation of cancer and neurodegenerative diseases suggests that prevention can be adopted by inhibiting formation of the electrophilic catechol quinones and/or their reaction with DNA.

Acknowledgments

Writing of this chapter was supported by U.S. Public Health Service grants P01 CA49210 and R01 CA49917 from the National Cancer Institute. Core support in the Eppley Institute is provided by grant P30 CA36726 from the National Cancer Institute.

[18] Induction of NQO1 in Cancer Cells

By ASHER BEGLEITER and JEANNE FOURIE

Introduction

NQO1 (NAD[P]H:[quinone acceptor]oxidoreductase) (EC 1.6.99.2) (also known as DT-diaphorase) is a flavoenzyme that catalyzes obligatory two-electron reduction of quinones, quinone imines and nitrogen oxides and requires NADH or NADPH as an electron donor for enzymatic activity.[1] A major function of the enzyme may be to decrease formation of reactive oxygen species by decreasing one electron reductions and the associated redox cycling,[2] and it may have an additional role as an antioxidant enzyme through reduction of the oxidized form of vitamin E to a product with antioxidant properties.[3] However, NQO1 is also a phase II detoxifying enzyme[4] that may be involved in cancer prevention.[1,5,6] In addition, it is an activating enzyme for some anticancer drugs[1,5,7] that plays an important role in regulating the activity of these agents, and is a target for enzyme-directed tumor targeting.[1,8] Recent findings suggest that NQO1 may help to regulate the stability of p53 and apoptosis in human[9] and mouse cells.[10] Thus this enzyme may be an important regulator of a wide variety of biological functions and may play a particularly significant role in cancer.

Structure and Activity of NQO1

Structure of NQO1

The NQO1 gene, located on chromosome 16q22, is 20 kb in length and has 6 exons and 5 introns.[11] The sixth exon in the human NQO1 gene contains four polyadenylation sites and a single copy of human Alu

[1] R. J. Riley and P. Workman, *Biochem. Pharmacol.* **43**, 1657 (1992).

[2] L. Ernster, *Chemica Scripta* **27A**, 1 (1987).

[3] D. Siegel, E. M. Bolton, J. A. Burr, D. C. Liebler, and D. Ross, *Mol. Pharmacol.* **52**, 3005 (1997).

[4] H. J. Prochaska, P. Talalay, and H. Sies, *J. Biol. Chem.* **262**, 1931 (1987).

[5] D. Ross, D. Siegel, H. Beall, A. S. Prakash, R. T. Mulcahy, and N. W. Gibson, *Cancer Metast. Rev.* **12**, 83 (1993).

[6] A. Begleiter, M. K. Leith, T. J. Curphey, and G. P Doherty, *Oncol. Res.* **9**, 371(1997).

[7] A. Begleiter, E. Robotham, G. Lacey, and M. K. Leith, *Cancer Lett.* **45**, 173 (1989).

[8] P. Workman, *Oncol. Res.* **6**, 461 (1994).

[9] G. Asher, J. Lotem, B. Cohen, L. Sachs, and Y. Shaul, *Proc. Natl. Acad. Sci. USA* **98**, 1188 (2001).

repetitive sequence between the second and third polyadenylation signal site. The three transcripts originating from the human NQO1 gene are thought to result from the use of three of the four polyadenylation sites in exon six.[11]

Mutagenesis experiments have identified several essential cis-elements within the human NQO1 gene promoter region that are required for the expression and induction of this gene.[11,12] These elements include an anti-oxidant response element (ARE), a DNA fragment essential for 2,3,7,8-tet-rachlorodibenzo-p-dioxin (TCDD)-mediated induction of NQO1 gene expression, and an AP-2 element required for the cAMP-induced expression of the gene. The ARE element is necessary for basal expression as well as xenobiotic,[13] antioxidant[14] and oxidant[15]-induced NQO1 gene induction. This element contains one perfect and one imperfect AP-1 (TPA response element) element, which are arranged as inverse repeats of each other and are separated by three base pairs followed by a "GC" box.[16,17] TPA response elements normally require TPA to activate their expression. The human NQO1 gene and other detoxifying enzyme gene AREs are, however, unique in that these cis-elements do not require TPA but rather respond to xenobiotics and antioxidants.[16,17] The reason for the distinct function of the NQO1 ARE from those genes that respond to TPA is thought to be in part due to the other nucleotide sequences. These include the CTGACNNNCG core sequence present in the ARE, as well as additional cis-element and nucleotide sequences flanking the core sequence of the ARE. Mutational analysis has shown that these sequences contribute to the ARE-mediated expression and induction of NQO1.[12,16,18,19]

NQO1 is a 61-kDa protein consisting of two interlocked 30.7-kDa homo-monomeric subunits. Each subunit is comprised of two separate domains: a catalytic domain and a small C-terminal domain. The catalytic domain consists of a site that binds flavin adenine dinucleotide (FAD),

---

[10] D. J. Long, II, A. Gaikwad, A. Multani, S. Pathak, C. A. Montgomery, F. J. Gonzalez, and A. K. Jaiswal, *Cancer Res.* **62**, 3030 (2002).

[11] A. K. Jaiswal, *Biochemistry* **30**, 10647 (1991).

[12] T. Xie, M. Belinsky, Y. Xu, and A. K. Jaiswal, *J. Biol. Chem.* **270**, 6894 (1995).

[13] Y. Li and A. K. Jaiswal, *Eur. J. Biochem.* **226**, 31 (1994).

[14] S. Dhakshinamoorthy and A. K. Jaiswal, *Oncogene* **20**, 3906 (2001).

[15] P. Joseph, T. Xie, Y. Xu, and A. K. Jaiswal, *Oncol. Res.* **6**, 525 (1994).

[16] V. Radjenkirane, P. Joseph, and A. Jaiswal, *in* "Oxidative Stress and Signal Transduction" (H. Forman and E. Cadenas, eds.), p. 441. Chapman and Hall, New York, 1997.

[17] A. K. Jaiswal, *Biochem. Pharmacol.* **48**, 439 (1994).

[18] T. Prestera, W. D. Holtzclaw, Y. Zhang, and P. Talalay, *Proc. Natl. Acad. Sci. USA* **90**, 2965 (1993).

[19] W. W. Wasserman and W. E. Fahl, *Proc. Natl. Acad. Sci. USA* **94**, 5361 (1997).

a site that binds the adenine ribose portion of nicotinamide adenine dinucleotide (phosphate) (NAD[P]H), and the substrate donor/acceptor binding site that binds either the hydride acceptor (substrate) or the hydride donor (the nicotinamide moiety of NAD[P]H).

At the FAD binding site, the interaction of FAD with NQO1 requires primary amino acid residues from one of the two monomers of the dimer. Furthermore, the isoalloxazine moiety of FAD is planar when bound to human NQO1 and is required for making binding contacts with the substrate when it occupies the catalytic substrate binding site. The substrate donor/acceptor binding site of human NQO1 is composed of a catalytic substrate binding pocket that alternatively binds the nicotinamide portion of NAD(P)H and the substrate. The pocket consists of residues Phe-178, Tyr-126 and Tyr-128 from the first monomer, which form the roof of the pocket; the isoalloxazine ring of FAD, which forms the floor; and His-161, Gly-149, Gly-150 (amino acid residues from the second monomer) and the two conserved water molecules that form the side of the pocket.

Mechanism of Action of NQO1

NQO1-mediated reduction follows a ping-pong mechanism of reduction kinetics. NAD(P)H binds NQO1 to yield the complex NQO1-NAD(P)H. Subsequently, the FAD molecule is reduced by NAD(P)H to form $FADH_2$, which is followed by the release of $NAD(P)^+$ from the catalytic site. This allows for binding of the substrate to this site, transfer of the hydride from $FADH_2$ to the substrate and release of the reduced substrate product from the enzyme catalytic binding pocket.[20]

The anticoagulant dicoumarol is a competitive inhibitor that competes with NAD(P)H for binding to the oxidized form of the enzyme at the NAD(P)H binding site.[21,22] Because of the competitive nature of this inhibition by dicoumarol, as well as its ability to inhibit other cellular reductases, there has been interest in the development of a more selective inhibitor for NQO1. Recently, Winski *et al.*[23] reported the development of the NQO1 inhibitor 5-methoxy-1,2-dimethyl-3-[(4-nitrophenoxy)methyl]indole-4,7-dione (ES936), which inhibits NQO1 by formation of an NQO1-ES936 adduct.

[20] M. Faig, M. A. Bianchet, P. Talalay, S. Chen, S. Winski, D. Ross, and L. M. Amzel, *Proc. Natl. Acad. Sci. USA* **97,** 3177 (2000).

[21] P. M. Hollander and L. Ernster, *Arch. Biochem. Biophys.* **169,** 560 (1975).

[22] S. Hosoda, W. Nakamura, and K. Hayashi, *J. Biol. Chem.* **249,** 6416 (1974).

[23] S. L. Winski, M. Faig, M. A. Bianchet, D. Siegel, E. Swann, K. Fung, M. W. Duncan, C. J. Moody, L. M. Amzel, and D. Ross, *Biochemistry* **40,** 15135 (2001).

NQO1 Distribution

NQO1 is a mainly cytosolic flavoprotein, with this compartment containing 84% or more of the total amount of NQO1 as determined by studies using rat liver.[24,25] Similar analysis revealed that the remaining fractions of this enzyme were located in the mitochondria (13%), golgi apparatus (1%) and microsomes (2%).[25] This distribution is thought to hold in the case of the human NQO1 enzyme, as the protein shares 85% amino acid homology with that of the rat; however, some discrepancies have been identified. In a recent study that used confocal immunoelectron microscopy in NCI-H661 human non-small-cell lung carcinoma cells and HT-29 human colon carcinoma cells, NQO1 could not be identified in the mitochondrial fraction. This finding was attributed to possible methodological constraints or the possibility that mitochondrial localization only occurred under cellular stress.[26] This study was also the first to indicate, through confocal immunoelectron microscopy and cell fractionation methodology, that a significant proportion of NQO1 is found in the nuclei of NCI-H661 and HT-29 cancer cell lines. The significance of nuclear NQO1 is unknown at present; however, it may have implications in the possible nuclear activation or inactivation of quinone based bioreductive alkylating agents and other quinone substrates.

NQO1 is ubiquitous in eukaryotes, but levels vary in different tissues.[5,27,28] NQO1 activity is high in normal stomach and kidney tissue, moderately high in normal breast and colon tissue and low in normal lung, liver and hematopoetic tissues.[27–29]

NQO1 Polymorphisms

A polymorphism involving a base change from cytosine to thymine at base 609 of the NQO1 gene occurs at an incidence of ~50% in the human population,[30–32] with 10% of the population being homozygous for the [609]T form of the gene.[31,32] However, these percentages vary in different ethnic

[24] T. Conover and L. Ernster, *Biochim. Biophys. Acta* **58,** 189 (1962).

[25] C. Edlund, A. Elhammer, and G. Dallner, *Biosci. Rep.* **2,** 861 (1982).

[26] S. L. Winski, Y. Koutalos, D. L. Bentley, and D. Ross, *Cancer Res.* **62,** 1420 (2002).

[27] M. Belinsky and A. K. Jaiswal, *Cancer Metastasis Rev.* **12,** 103 (1993).

[28] J. J. Schlager and G. Powis, *Int. J. Cancer* **45,** 403 (1990).

[29] G. B. Gordon, H. J. Prochaska, and L. Y. Yang, *Carcinogenesis* **12,** 2393 (1991).

[30] J. K. Wiencke, M. R. Spitz, A. McMillan, and K. T. Kelsey, *Cancer Epidemiol. Biomarkers Prev.* **6,** 87 (1997).

[31] R. D. Traver, D. Siegel, H. D. Beall, R. M. Phillips, N. W. Gibson, W. A. Franklin, and D. Ross, *Br. J. Cancer* **75,** 69 (1997).

[32] K. T. Kelsey, D. Ross, R. D. Traver, D. C. Christiani, Z.-F. Zuo, M. R. Spitz, M. Wang, X. Xu, B-K. Lee, B. S. Schwartz, and J. K. Wiencke, *Br. J. Cancer* **76,** 852 (1997).

groups, with an incidence rate of 2 to 5% for the ^{609}T:^{609}T genotype in Caucasians and African Americans and 20% in Asians.[30,32] The ^{609}T form of the NQO1 protein has little or no activity,[31,33] and cells with two ^{609}T alleles have no enzyme activity, while cells with one ^{609}T allele have reduced activity.[31,34,35] Thus the 10% of humans who are homozygous for the ^{609}T form of NQO1 have no enzyme activity, and the 40% of humans with one ^{609}T allele have reduced enzyme activity. Clinical studies have shown a link between the ^{609}T form of NQO1 and the incidence of renal, urothelial and lung cancers and some leukemias.[30,36,37] Several other variations have been found in the NQO1 gene including a C to T substitution at base 465[38] and an alternatively spliced mRNA lacking exon 4.[38,39] The frequency of these changes seems to be limited and their significance is unknown.

Expression and Induction of NQO1

Expression of NQO1

Expression of NQO1 appears to be transcriptionally regulated.[1] Several nuclear transcription factors have been identified to be part of the human ARE-nuclear protein complex, which binds to the ARE, and so regulates transcription of the human NQO1 gene. Transcription factors that bind to the human NQO1 ARE include c-Jun, Jun-B, Jun-D, c-Fos, Fra1, Nrf1, and Nrf2.[40,41] Nrf1 and Nrf2, which are b-zip (leucine zipper) proteins that do not heterodimerize with each other, but rather require a second leucine zipper protein for activation, are positive regulators of ARE-induced expression and induction of the NQO1 gene in response to oxidants and xenobiotics.[40] Nrf2 has been implicated in the *in vivo* regulation of NQO1 gene expression by knockout studies in mice. Specifically, Nrf2$^{-/-}$ mice lacking Nrf2 expression showed significantly decreased expression and induction of the NQO1 gene.[42] Furthermore, a cytosolic protein,

[33] V. Misra, H. J. Klamut, and A. M. Rauth, *Br. J. Cancer* **77**, 1236 (1998).

[34] R. D. Traver, T. Horikoshi, K. D. Danenberg, T. H. Stadlbauer, P. V. Danenberg, D. Ross, and N. W. Gibson, *Cancer Res.* **52**, 797 (1992).

[35] A. Begleiter, M. K. Leith, and S. Pan, *Biochem. Pharmacol.* **61**, 955 (2001).

[36] W. A. Schulz, A. Krummeck, I. Rosinger, I. P. Eickelmann, C. Neuhaus, T. Ebert, B. J. Schmitz-Drager, and H. Sies, *Pharmacogenetics* **7**, 235 (1997).

[37] R. A. Larson, Y. Wang, M. Banerjee, J. Wiemels, C. Hartford, M. M. Le Beau, and M. T. Smith, *Blood* **94**, 803 (1999).

[38] L. T. Hu, J. Stamberg, and S. S. Pan, *Cancer Res.* **56**, 5253 (1996).

[39] P. Y. Gasdaska, H. Fisher, and G. Powis, *Cancer Res.* **55**, 2542 (1995).

[40] R. Venugopal and A. K. Jaiswal, *Proc. Natl. Acad. Sci. USA* **93**, 14960 (1996).

[41] R. Venugopal and A. K. Jaiswal, *Oncogene* **17**, 3145 (1998).

Kcap1, is suggested to act as an oxidative stress sensor protein, as it was shown to retain Nrf2 in the cytoplasm under physiological conditions, but on exposure to xenobiotics and antioxidants, the association was terminated, allowing for nuclear translocation of Nrf2.[43] In addition, Venugopal and Jaiswal[41] have shown that nuclear transcription factors Nrf2 and Nrf1 heterodimerize with c-Jun, Jun-B and Jun-D proteins. They provided evidence that these Nrf-Jun complexes (in addition to unknown cytosolic factors), bind to the ARE in response to antioxidants and xenobiotics. This is followed by subsequent expression and synchronized induction of other detoxifying enzymes.

ARE-mediated expression of the NQO1 gene is thought to be controlled by a balance between positive regulatory factors such as Nrf2 and proteins that repress NQO1 gene expression. For instance, c-Fos is thought to act in the repression of ARE-mediated gene expression as shown with c-Fos$^{-/-}$ knockout mice, which demonstrated increased expression of NQO1 as compared to wild-type c-Fos$^{+/+}$ mice.[44] This repression of detoxifying enzymes such as NQO1 is hypothesized to be required to modulate the significant and potent effects of decreasing free radicals below base line physiological concentrations by genes such as NQO1.

Induction of NQO1

NQO1 is induced in many tissues by a wide variety of structurally dissimilar chemicals, categorized by Prochaska and Talalay[45,46] as bifunctional and monofunctional inducers. Bifunctional inducers, which include polycyclic aromatic hydrocarbons, dioxins and azo dyes, induce both phase I and phase II detoxifying enzymes. They bind to the aromatic hydrocarbon receptor, are translocated to the nucleus where the inducer-receptor complex binds to the xenobiotic response elements of the phase I enzyme genes, and activate transcription of the genes. Phase I enzymes, which include the cytochrome P450 1A enzymes, functionalize the inducers to highly reactive, electrophilic metabolites, which can increase transcription of phase II enzymes.[45] In contrast, monofunctional inducers, which can

[42] K. Itoh, T. Chiba, S. Takahashi, T. Ishii, K. Igarashi, Y. Katoh, T. Oyake, N. Hayashi, K. Satoh, I. Hatayama, M. Yamamoto, and Y. Nabeshima, *Biochem. Biophys. Res. Commun.* **236,** 313 (1997).

[43] K. Itoh, N. Wakabayashi, Y. Katoh, T. Ishii, K. Igarashi, J. D. Engel, and M. Yamamoto, *Genes Dev.* **13,** 76 (1999).

[44] J. T. Wilkinson, V. Radjendirane, G. R. Pfeiffer, A. K. Jaiswal, and M. L. Clapper, *Biochem. Biophys. Res. Commun.* **253,** 855 (1998).

[45] H. J. Prochaska and P. Talalay, *Cancer Res.* **48,** 4776 (1988).

[46] T. Prestera, Y. Zhang, S. R. Spencer, C. A. Wilczak, and P. Talalay, *Adv. Enzyme Regul.* **33,** 281 (1993).

have diverse chemical structures, are highly electrophilic, or are converted to electrophilic species, and induce only phase II enzymes.[45] Monofunctional inducers include diphenols, phenylene diamines, quinones, coumarin analogues, unsaturated carboxylic acids and esters, isothiocyanates, 1,2-dithiole-3-thiones, thiocarbamates, allyl sulfides and other Michael reaction acceptors.[47] Phase II enzymes include NQO1, glutathione S-transferases (GST), epoxide hydrolase and UDP-glucuronosyltransferases, and assist in conjugating electrophilic metabolites and carcinogens with endogenous ligands facilitating their removal from the cell.[45]

Although many positive and negative regulators of ARE-mediated NQO1 expression have been identified, the precise mechanisms involved in basal expression and coordinated induction and repression of NQO1 expression still need to be fully elucidated. Recently, Jaiswal has described a hypothesized model by which ARE-mediated expression might occur.[48] It is known that xenobiotics and antioxidants undergo metabolism to produce superoxide and related reactive oxygen species.[49] The superoxide may serve as a trigger that activates a battery of defensive genes such as NQO1, which protect cells against oxidative stress. This is supported by the finding that hydrogen peroxide leads to induction of the ARE-mediated expression of rat and human NQO1 genes.[13,50] It is thought that the superoxide signal involves first, interactions with intermediary proteins leading ultimately to interactions with cytosolic factors. These factors may be important in the modification of the interaction between Nrf2 and Kcap1, which sequesters Nrf2 in the cytoplasm. Nrf2 may then translocate to the nucleus, and form Nrf2/c-Jun heterodimers, which may on binding to the ARE, induce NQO1 expression. Further research is needed to identify all the factors involved in these pathways, including the identification of unknown cytosolic factors.

Induction of NQO1 in Normal and Cancer Cells

NQO1 has been shown to play a role in the detoxification of xenobiotics,[51] in cancer prevention[5,6] and in activation of cancer chemotherapeutic agents.[1,5,6] This has led to many studies of induction of NQO1 in cells and tissues as a way of enhancing these activities. Studies to identify agents that might be useful in detoxification or chemoprevention strategies have

---

[47] P. Talalay, M. J. De Long, and H. J. Prochaska, *Proc. Natl. Acad. Sci. USA* **85,** 8261 (1988).
[48] A. K. Jaiswal, *Free Radic. Biol. Med.* **29,** 254 (2000).
[49] M. J. De Long, A. B. Santamaria, and P. Talalay, *Carcinogenesis* **8,** 1549 (1987).
[50] L. V. Favreau and C. B. Pickett, *J. Biol. Chem.* **266,** 4556 (1991).
[51] L. E. Twerdok, S. J. Rembish, and M. A. Trush, *Toxicol. Appl. Pharmacol.* **112,** 273 (1992).

generally used induction of NQO1 in model cancer cells or normal tissues. Hepa 1c1c7 murine hepatoma cells have been extensively used to identify dithiolethione,[52] isothiocyanate,[53] diphenol[54] and other[45,47,55] analogs that induce NQO1 and other phase II detoxifying enzymes for prevention strategies. Alternatively, other tumor cell lines[56,57] or mouse,[58,59] rat[60] or human[29] tissues have been used for similar studies. The role of NQO1 in cancer prevention and chemotherapy, and the potential clinical importance of induction of this enzyme to achieve these goals are outlined later.

NQO1 and Cancer Chemoprevention

Cancer chemoprevention is an attempt to use natural or synthetic compounds to interfere with the carcinogenesis process at an early stage in order to prevent the development of cancer. Epidemiological studies in humans and cancer prevention studies in animals have identified a large number of compounds that have chemopreventive ability. These compounds can be generally classified as compounds that inhibit the formation of carcinogens, blocking agents that prevent carcinogens from reaching or reacting with critical target sites, and suppressing agents that block the neoplastic process when administered after the initial actions of the carcinogen.[61] A wide variety of chemical compounds including dithiolethiones, isothiocyanates, diphenols, quinones, Michael reaction acceptors, and heavy metals[18] can inhibit the carcinogenesis process by acting as blocking agents.[61] Thus these blocking agents, which are inducers of NQO1, may exert their cancer chemopreventive effects by inducing phase II enzymes that detoxify carcinogens either by destroying their reactive centers or by conjugating them with endogenous ligands, facilitating their excretion from the cell.[18]

Inducers of phase II detoxifying enzymes are currently under active investigation as cancer chemopreventive agents. Dithiolethiones occur

---

[52] P. A. Egner, T. W. Kensler, T. Prestera, P. Talalay, A. H. Libby, H. H. Joyner, and T. J. Curphey, *Carcinogenesis* **15**, 177 (1994).

[53] G. H. Posner, C. Cho, J. V. Green, Y. Zhang, and P. Talalay, *J. Med. Chem.* **37**, 170 (1994).

[54] H. J. Prochaska, M. J. De Long, and P. Talalay, *Proc. Natl. Acad. Sci. USA* **82**, 8232 (1985).

[55] S. C. Sisk and W. R. Pearson, *Pharmacogenetics* **3**, 167 (1993).

[56] P. J. O'Dwyer, M. Clayton, T. Halbherr, C. B. Myers, and K. Yao, *Clin. Cancer Res.* **3**, 783 (1997).

[57] W. Wang, L. Q. Liu, C. M. Higuchi, and H. Chen, *Biochem. Pharmacol.* **56**, 189 (1998).

[58] X. Hu, P. J. Benson, S. K. Srivastava, L. M. Mack, H. Xia, V. Gupta, H. A. Zaren, and S. V. Singh, *Arch. Biochem. Biophys.* **336**, 199 (1996).

[59] G. Wagner, U. Pott, M. Bruckschen, and H. Sies, *Biochem. J.* **252**, 825 (1988).

[60] T. M. Buetler, E. P. Gallagher, C. Wang, D. L. Stahl, J. D. Hayes, and D. L. Eaton, *Toxicol. Appl. Pharmacol.* **135**, 45 (1995).

[61] L. W. Wattenberg, *Cancer Res.* **45**, 1 (1985).

naturally in cruciferous vegetables such as broccoli, cauliflower and cabbage. A large number of dithiolethione analogs have been shown to be effective inducers of NQO1 and other phase II enzymes in cells *in vitro*[52] and *in vivo*.[62] Dithiolethiones, as a whole, appear to function as chemoprotective agents against a variety of chemical carcinogens in animal model systems,[63] with the parent compound being among the most active of those tested.[64] An especially noteworthy member of this class of compounds is the 4-methyl-5-pyrazinyl derivative, oltipraz. Originally developed and used as an antischistosomal agent in humans,[65] oltipraz is in clinical trials as a cancer chemoprotective agent, including a large-scale phase II trial in China in an area having a high incidence of aflatoxin B_1-induced liver cancer.[66,67] These human trials were prompted by animal studies in which oltipraz was shown to protect against the development of pulmonary and forestomach cancers induced by benzo[a]pyrene in ICR/Ha mice[68] and against hepatocellular carcinomas induced by aflatoxin B_1 in F344 rats.[69] Additionally, recent bioassays demonstrate that oltipraz also protects against chemically induced carcinogenesis in the trachea, colon, breast, skin and urinary bladder in a variety of rodent models.[70] The low toxicity exhibited by the dithiolethiones in animal studies, the use of oltipraz in the clinic as an antischistosomal agent, and the clinical use of related dithiolethiones as choleretic agents suggest that these compounds would be suitable for use as chemopreventive agents in high-risk individuals.[70]

Phase II detoxifying enzymes are also induced by isothiocyanates.[18] Zhang *et al.*[71] isolated and identified the isothiocyanate derivative,

[62] T. W. Kensler, P. A. Egner, P. M. Dolan, J. D. Groopman, and B. D. Roebuck, *Cancer Res.* **47**, 4271 (1987).

[63] T. W. Kensler, J. D. Groopman, B. D. Roebuck, and T. J. Curphey, *in* "Food Phytochemicals for Cancer Chemoprevention" (C.-T. Ho, M.-T. Huang, T. Osawa, and R. T. Rosen, eds.), p. 154. American Chemical Society, Washington, D.C., 1994.

[64] T. W. Kensler, J. D. Groopman, D. L. Eaton, T. J. Curphey, and B. D. Roebuck, *Carcinogenesis* **13**, 95 (1992).

[65] S. Archer, *Ann. Rev. Pharmacol. Toxicol.* **25**, 485 (1985).

[66] P. J. O'Dwyer, C. E. Szarka, K. S. Yao, T. C. Halbherr, G. R. Pfeiffer, F. Green, J. M. Gallo, J. Brennan, H. Frucht, E. B. Goosenberg, T. C. Hamilton, S. Litwin, A. M. Balshem, P. F. Engstrom, and M. L. Clapper, *J. Clin. Invest.* **98**, 1210 (1996).

[67] J. S. Wang, X. Shen, X. He, Y. R. Zhu, B. C. Zhang, J. B. Wang, G. S. Qian, S. Y. Kuang, A. Zarba, P. A. Egner, L. P. Jacobson, A. Munoz, K. J. Helzlsouer, J. D. Groopman, and T. W. Kensler, *J. Natl. Cancer Inst.* **91**, 347 (1999).

[68] L. W. Wattenberg and E. Bueding, *Carcinogenesis* **7**, 1379 (1986).

[69] B. D. Roebuck, Y. Liu, A. E. Rogers, J. D. Groopman, and T. W. Kensler, *Cancer Res.* **51**, 5501 (1991).

[70] T. W. Kensler, J. D. Groopman, and B. D. Roebuck, *in* "Cancer Chemoprevention" (L. Wattenberg, M. Lipkin, C. Boone, and G. Kelloff, eds.), p. 205. CRC Press, Boca Raton, 1992.

sulforaphane, as a major and very potent phase II enzyme inducer found in broccoli. This compound induced NQO1 and GST activity in mouse cells and tissues *in vitro* and *in vivo,*[71] and blocked the formation of 9,10-dimethyl-1,2-benzanthracene-induced mammary tumors in rats.[72] Based on these findings, a phase I clinical trial examined the effect of broccoli supplements on the level of GST activity in peripheral blood lymphocytes and colon mucosa.[73]

These studies demonstrate the potential for using natural or synthetic non-toxic inducers of phase II detoxifying enzymes in cancer chemoprevention programs; however, the exact role of NQO1 remains unclear. NQO1 may block the carcinogenesis process by reducing quinone or nitrogen oxide groups on carcinogens by direct transfer of two-electrons. This may prevent one-electron redox cycling of these groups and inhibit the formation of DNA-damaging reactive oxygen species. Reduction of the quinone and nitrogen oxide groups may also make them available for conjugation with ligands such as UDP-glucuronic acid, facilitating their excretion from the cell.[18] However, induction of NQO1 is often accompanied by induction of other phase II enzymes, which also have detoxifying functions. Thus it has not been possible to isolate the specific contribution of NQO1 induction to the overall chemopreventive effect of phase II enzyme inducers such as oltipraz or sulforaphane.

In a recent study, we examined the effect of NQO1 inducers on colon carcinogenesis using an aberrant crypt foci (ACF) rat model.[74] ACF are early preneoplastic lesions that predict for colon tumor formation in this model. Sprague-Dawley rats were fed control diet or diet containing oltipraz and then were treated with the carcinogens, azoxymethane or methyl nitrosourea. Oltipraz selectively increased NQO1 activity in the rat colon and liver, with little or no increase in the activities of other phase II detoxifying enzymes. Both carcinogens produced ACF in all the rat colons, but rats fed the oltipraz diet had significantly fewer ACF than those fed control diet. This protective effect was reversed in rats treated with the NQO1 inhibitor dicoumarol. This study provides the first direct evidence that induction of NQO1 can play a significant role in inhibiting initiation of colon carcinogenesis and that this enzyme may play a role in preventing colon

[71] Y. Zhang, P. Talalay, C. G. Cho, and G. H. Posner, *Proc. Natl. Acad. Sci. USA* **89,** 2399 (1992).

[72] Y. Zhang, T. W. Kensler, C. G. Cho, G. H. Posner, and P. Talalay, *Proc. Natl. Acad. Sci. USA* **91,** 3147 (1994).

[73] M. L. Clapper, C. E. Szarka, G. R. Pfeiffer, T. A. Graham, A. M. Balshem, S. Litwin, E. B. Goosenberg, H. Frucht, and P. F. Engstrom, *Clin. Cancer Res.* **3,** 25 (1997).

[74] A. Begleiter, K. Sivananthan, T. J. Curphey, and R. P. Bird, *Cancer Epidemiol. Biomark. Prev.* **12,** 566 (2003).

cancer. In addition, this model will be useful to further investigate the role of NQO1 in the prevention of colon cancer.

NQO1 and Cancer Chemotherapy

Bioreductive antitumor agents are an important class of anticancer drugs.[75] The prototype drug in this class, mitomycin C (MMC),[76] is widely used for treating bladder, lung, colorectal, head and neck, breast and other solid tumors.[77] Interest in this class of anticancer drugs has been raised by clinical and pre-clinical trials with new bioreductive agents such as tirapazamine,[78] EO9[79] and RH1.[80] These drugs have varied chemical structures including quinones, nitroimidazoles, benzotriazine-di-N-oxides and dinitroaromatics, but they are characterized by a common requirement for reductive activation.[75] They produce their antitumor activity by a variety of mechanisms but are generally preferentially toxic to hypoxic cells.[75] Thus there has been a major interest in using bioreductive agents in combination with radiation and drugs that kill oxygenated cells. As these agents are activated intracellularly, they may also be effective in treating solid tumors because they are able to diffuse to cells in the interior of the tumor without being inactivated.

The mechanisms involved in the activation of bioreductive antitumor agents have been extensively studied.[1,5,76] These agents can be activated by one-electron reducing enzymes such as NADPH:cytochrome P450 reductase (EC 1.6.2.4),[76,81] NADH:cytochrome b_5 reductase (EC 1.6.2.2)[82] and xanthine oxidase (EC 1.1.3.22)[81] and by two-electron reducing enzymes such as NQO1[1,5,7] and xanthine dehydrogenase (EC 1.1.1.204).[83] NQO1 is a significant contributor to activation of bioreductive antitumor agents in many systems.[1,5,7,76] Many studies have shown, in both human and animal cell lines, that cells with elevated NQO1 levels are more sensitive to MMC and that drug activity is inhibited by the NQO1 inhibitor

[75] P. Workman and I. J. Stratford, *Cancer Metastasis Rev.* **12,** 73 (1993).

[76] S. Rockwell, A. C. Sartorelli, M. Tomasz, and K. A. Kennedy, *Cancer Metastasis Rev.* **12,** 165 (1993).

[77] A. Begleiter, *Front. Biosci.* **5,** 153 (2000).

[78] N. Doherty, S. L. Hancock, S. Kaye, C. N. Coleman, L. Shulman, C. Marquez, C. Mariscal, R. Rampling, S. Senan, and R. V. Roemeling, *Int. J. Radiat. Oncol. Biol. Phys.* **29,** 379 (1994).

[79] J. H. M. Schellens, A. S. T. Planting, B. A. C. Van Acker, W. J. Loos, M. De Boer-Dennert, M. E. L. Van der Burg, I. Koier, R. T. Krediet, G. Stoter, and J. Verweij, *J. Natl. Cancer Inst.* **86,** 906 (1994).

[80] S. L. Winski, R. H. Hargreaves, J. Butler, and D. Ross, *Clin. Cancer Res.* **4,** 3083 (1998).

[81] S. S. Pan, P. A. Andrews, C. J. Glover, and N. R. Bachur, *J. Biol. Chem.* **259,** 959 (1984).

[82] W. F. Hodnick and A. C. Sartorelli, *Cancer Res.* **53,** 4907 (1993).

[83] D. L. Gustafson and C. A. Pritsos, *J. Natl. Cancer Inst.* **84,** 1180 (1992).

dicoumarol.[7,84] NQO1 has also been shown to be an important activating enzyme for the aziridinyl quinone, diaziquone,[1,85] the indoloquinone, EO9[1,86] and other bioreductive agents.[75] The cytotoxic activities of EO9[86] and RH1[80] appear to be particularly sensitive to the level of NQO1 activity. In contrast, this enzyme may protect cells from toxicity by the benzotriazine di-N-oxide, tirapazamine[87] and the quinone agents menadione,[88] HBM, BM and BDM.[89] However, the contribution of NQO1 to activation of these agents varies in different cells, with oxygen tension[75,76,90] and with pH.[5,75,76,84,91] While human and other mammalian NQO1 enzymes display similar activities, the ability of NQO1 from different species to activate a particular bioreductive agent can vary greatly.[1]

Various approaches have been suggested to exploit the activation of bioreductive agents by NQO1 in the clinical setting. Tumors with high levels of NQO1 activity can be identified and treated with agents activated by NQO1, such as EO9 and MMC.[1] New bioreductive agents that are good substrates for activation by NQO1 are being designed and synthesized.[1,80,92] The antitumor efficacy of bioreductive agents could also be enhanced by selectively increasing NQO1 activity in cancer cells as compared with normal cells either by selective induction of the enzyme in cancer cells[6,93] or by selective transfer of the NQO1 gene into these cells.[94]

NQO1 levels vary widely in different tissues, with activity generally being relatively high in liver, stomach, bladder, intestine, colon and kidney, possibly because of the exposure of these tissues to dietary or environmental inducers of the enzyme. In contrast, NQO1 levels are usually low in hematopoetic cells.[27–29] NQO1 activity is also generally higher in tumor cells as compared with normal cells of the same origin,[1,27,28] but the levels of enzyme activity vary markedly in different tumor cells and may be higher

[84] D. Siegel, N. W. Gibson, P. C. Preusch, and D. Ross, *Cancer Res.* **50,** 7483 (1990).

[85] D. Siegel, N. W. Gibson, P. C. Preusch, and D. Ross, *Cancer Res.* **50,** 7293 (1990).

[86] J. A. Plumb, M. Gerritsen, R. Milroy, P. Thomson, and P. Workman, *Int. J. Radiat. Oncol. Biol. Phys.* **29,** 295 (1994).

[87] A. V. Patterson, N. Robertson, S. Houlbrook, M. A. Stephens, G. E. Adams, A. L. Harris, I. J. Stratford, and J. Carmichael, *Int. J. Radiat. Oncol. Biol. Phys.* **29,** 369 (1994).

[88] A. S. Atallah, J. R. Landolph, L. Ernster, and P. Hochstein, *Biochem. Pharmacol.* **37,** 2451 (1988).

[89] A. Begleiter and M. K. Leith, *Cancer Res.* **50,** 2872 (1990).

[90] A. Begleiter, E. Robotham, and M. K. Leith, *Mol. Pharmacol.* **41,** 677 (1992).

[91] A. Begleiter and M. K. Leith, *Mol. Pharmacol.* **44,** 210 (1993).

[92] J. Fourie, C. Oleschuk, F. Guziec, Jr., L. Guziec, D. J. Fitterman, C. Monterrosa, and A. Begleiter, *Cancer Chemother. Pharmacol.* **49,** 101 (2002).

[93] A. Begleiter and M. K. Leith, *Biochem. Pharmacol.* **50,** 1281 (1995).

[94] M. F. Belcourt, W. F. Hodnick, S. Rockwell, and A. C. Sartorelli, *Biochem. Pharmacol.* **51,** 1669 (1996).

in tumor cells *in vitro* than in similar cells or primary tumors *in vivo*.[95] Since hematopoetic cells are a major target for the toxic side effects of many anticancer drugs, including bioreductive agents, we suggested that the antitumor efficacy of bioreductive agents can be enhanced by the selective induction of NQO1 in tumor cells as compared with hematopoetic cells.[6,93] Studies to evaluate the feasibility of this strategy are described later.

Selective Induction of NQO1 in Cancer Cells to Enhance Antitumor Activity

Methods

In general, for studies of induction of NQO1 in cancer or normal cells *in vitro,* cells were incubated with or without the NQO1 inducers at 37° in 5% CO_2 for 48 hr. Concentrations of inducers that were not toxic to the cells during the incubation time were used. Following incubation, cells were washed with PBS, suspended in 200 μl of 0.25 M sucrose, sonicated and stored at −80°. Protein concentration was measured by using the Bio-Rad DC Kit with gamma globulin as the standard, then NQO1 activity was measured spectrophotometrically by a modification of the procedure of Prochaska and Santamaria[96,97] with menadione as the electron acceptor. NQO1 activity is reported as dicoumarol-inhibitable activity and expressed as $nmole.min.^{-1}$ mg $prot^{-1}$. A dicoumarol concentration of 10 μM was used.

For cytotoxicity studies with tumor or normal cells *in vitro,* cells were incubated with or without the NQO1 inducer for 48 hr and then were treated with various concentrations of bioreductive agent for 1 hr at 37°. For experiments where the NQO1 inhibitor dicoumarol was used, dicoumarol was added to the cells at 100 μM for 15 min prior to addition of the bioreductive agent. For tumor cells, the surviving cell fraction was determined by MTT assay[98] after incubation at 37° for 4 to 9 days; this length of time was sufficient to allow at least three cell doublings. In some cases the surviving cell fraction was determined by clonogenic assay.[7,97] For normal marrow cells the surviving cell fraction was determined by methylcellulose clonogenic assay.[99] The D_0 (concentration of drug required to

[95] S. A. Fitzsimmons, P. Workman, M. Grever, K. Paull, R. Camalier, and A. D. Lewis, *J. Natl. Cancer Inst.* **88,** 259 (1996).

[96] H. J. Prochaska and A. B. Santamaria, *Anal. Biochem.* **169,** 328 (1988).

[97] G. P. Doherty, M. K. Leith, X. Wang, T. J. Curphey, and A. Begleiter, *Br. J. Cancer* **77,** 1241 (1998).

[98] J. B. Johnston, L. Verburg, T. Shore, M. Williams, L. G. Israels, and A. Begleiter, *Leukemia* **8** (Suppl. S140), (1994).

[99] A. Begleiter, L. Verburg, A. Ashique, K. Lee, L. G. Israels, M. R. A. Mowat, and J. B. Johnston, *Leukemia* **9,** 1875 (1995).

reduce the surviving cell fraction to 0.37) was calculated from the linear regression line of the surviving cell fraction versus drug concentration curve.

For studies of induction of NQO1 in tumor and normal cells *in vivo,* CD-1 nude mice were implanted subcutaneously in each flank with 5×10^6 tumor cells and then were fed either a control diet or a diet containing 0.3% or 0.4% DMF, 0.05% sulforaphane or 0.05% 1,2-dithiole-3-thione (D3T) for 3 to 14 days. The mice were euthanized; tumors and normal tissues were excised, and NQO1 activity in the tissues was measured.[96,97] For *in vivo* studies of enhancement of MMC antitumor activity by induction of NQO1, CD-1 nude mice were implanted subcutaneously in the flank with 5×10^6 HCT-116 human colon cancer cells. The mice were fed control diet or diet containing 0.3% DMF for 7 to 10 days and then were treated with a single tail-vein injection of 2.0 mg/kg MMC. Tumor volumes were measured for 15 days. The mice were euthanized and organs were excised and examined histologically for evidence of toxicity.

Induction of NQO1 in Mouse Cancer Cells and Effect on Antitumor Activity

When EMT6 murine mammary tumor cells were treated with the anthraquinone antitumor agent, doxorubicin (DOX), *in vitro,* we observed a 40% increase in the level of NQO1 activity in these cells. In contrast, similar treatment of murine bone marrow cells had no effect on the activity of this enzyme in the marrow cells. These findings demonstrated that DOX can selectively induce NQO1 in tumor cells.

As elevated levels of NQO1 activity can increase the cytotoxic activity of MMC,[7,75,76] we investigated whether pretreatment of EMT6 cells with DOX to increase the level of NQO1 could potentiate the antitumor activity of MMC. We did observe increased tumor kill of EMT6 cells pretreated with DOX as compared with cells treated with MMC alone, and the combination therapy resulted in a 1.4-fold greater cell kill than was expected for an additive cytotoxic effect with these two agents. However, we observed a similar level of synergy for combination therapy with DOX and MMC in EMT6 cells when the cells were treated simultaneously with the two agents. Furthermore, we also obtained synergistic cell kill in normal murine bone marrow cells pretreated with DOX and then treated with MMC. These results demonstrated that combination therapy with DOX and MMC can produce synergistic cell kill; however, this effect may not be related to an increase in the level of NQO1 activity. In addition, this enhanced cell kill did not appear to be selective for tumor cells as compared with normal marrow cells. The mechanism responsible for the synergistic

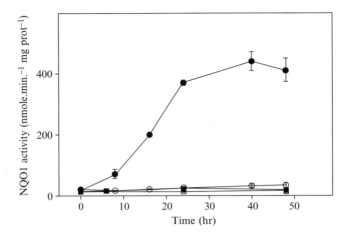

FIG. 1. Induction of NQO1 activity by D3T in L5178Y and normal murine marrow cells. L5178Y (○, ●) or normal marrow (□, ■) cells were incubated at 37° with (●, ■) or without (○, □) 75 μM D3T and at various times the level of NQO1 activity was measured. The points represent the mean ± SE of three experiments. (Adapted with permission from A. Begleiter, M. K. Leith, and T. J. Curphey, *Br. J. Cancer* **74** (Suppl. XXVII), S9 [1996].)

cell kill observed with DOX and MMC is unknown; however, both agents produce DNA damage.[100–102] The combined accumulation of DNA damage in the cells may result in greater than expected cell kill. Alternatively, both agents can induce apoptosis in cells. It is possible that the induction of the cell death process by each agent via different mechanisms may result in enhanced cell kill.

In another study with mouse cells we investigated whether the NQO1 inducer 1,2-dithiole-3-thione, D3T, could selectively induce NQO1 in tumor cells as compared with normal marrow cells. *In vitro* treatment with a non-toxic concentration of D3T increased the level of NQO1 activity in L5178Y murine lymphoma cells by 22-fold without affecting the enzyme level in normal murine marrow cells (Fig. 1). Previous studies had shown that D3T produced a 4-fold increase in NQO1 activity in murine hepatoma cells[52] and a 10-fold increase in enzyme activity in human myeloid leukemia cells.[103] Twerdok *et al.*[51] demonstrated a 2- to 3-fold increase in NQO1 activity in primary bone marrow stromal cells from DBA/2 and

[100] C. Myers, *Cancer Chemother.* **8,** 52 (1986).
[101] J. W. Lown, A. Begleiter, D. Johnson, and A. R. Morgan, *Can. J. Biochem.* **54,** 110 (1976).
[102] M. Tomasz, R. Lipman, D. Chowdary, L. Pawlak, G. L. Verdine, and K. Nakanishi, *Science (Washington D.C.)* **235,** 1204 (1987).
[103] Y. Li, A. Lafuente, and M. A. Trush, *Life Sciences* **54,** 901 (1994).

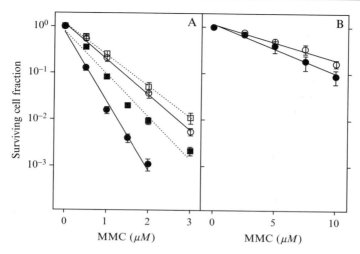

Fig. 2. Combination therapy with D3T and MMC in L5178Y (A) and normal murine marrow (B) cells. Cells were incubated at 37° with (●, ■) or without (○, □) 75 μM D3T for 40 hr. Cells were then treated with various concentrations of MMC for 1 hr (○, ●), or with 100 μM dicoumarol for 15 min and then with various concentrations of MMC for 1 hr (□, ■). Surviving cell fraction was determined by clonogenic assay for the L5178Y cells and by methylcellulose clonogenic assay for the marrow cells. The points represent the mean ± SE of three to seven experiments. The lines are linear regression lines. (Adapted with permission from A. Begleiter, M. K. Leith, and T. J. Curphey, *Br. J. Cancer* **74** (Suppl. XXVII), S9 [1996].)

C57B1/6 mice treated *in vitro* with D3T; however, *in vivo* feeding of DBA/2 mice with 0.1% D3T in their diet produced only a 30% and 50% increase in NQO1 in whole bone marrow and primary stromal cells, respectively.[104] It is not clear why we observed selectivity in the induction of NQO1 in the tumor cells as compared with the normal marrow cells, but this may relate to differences in the activation of transcription factors between the tumor and normal cells.

We investigated whether pretreatment of L5178Y cells with D3T to increase the level of NQO1 activity could potentiate the antitumor activity of the bioreductive agents, MMC and EO9. Combination therapy with D3T resulted in a 2- and 7-fold enhancement of MMC and EO9 cytotoxicity, respectively, in the tumor cells (Fig. 2). In contrast, D3T treatment produced only a small increase in the toxicity of these agents in normal murine marrow cells.

---

[104] L. E. Twerdok, S. J. Rembish, and M. A. Trush, *Environ. Health Perspect.* **101,** 172 (1993).

The results of this study were consistent with our hypothesis that the antitumor efficacy of bioreductive agents that are activated by NQO1 can be enhanced by selectively increasing the level of this enzyme in tumor cells. The role of NQO1 in the increased antitumor activity of MMC and EO9 was confirmed by the finding that the NQO1 inhibitor dicoumarol significantly reduced the effect of D3T on the cytotoxicity of these agents (Fig. 2). Dicoumarol also inhibited the activity of the bioreductive agents in the absence of D3T, but this effect was smaller than that observed in the presence of D3T and likely represents the effect of inhibiting activation of MMC and EO9 by the base level of NQO1 in the L5178Y cells.

Thus, this study demonstrated that NQO1 could be selectively induced in murine L5178Y lymphoma cells compared with normal murine bone marrow cells. Combination therapy with D3T and MMC or EO9 produced significantly increased tumor cell kill that was due to the elevated levels of NQO1, with only a small increase in cytotoxicity to normal marrow cells. This suggests that D3T, or other inducers of NQO1, might be used to enhance the antitumor efficacy of bioreductive antitumor agents.

Induction of NQO1 in Human Cancer Cells and Effect on Antitumor Activity

To determine whether NQO1 could be induced in human tumors, we investigated whether D3T could increase the level of NQO1 activity in human tumor cell lines *in vitro*. We found that D3T increased the level of NQO1 activity in most human tumor cell types. Enzyme induction with D3T in NCI-H661 human lung tumor cells reached a maximum at 48 hr. Induction of NQO1 activity in HL-60 human leukemia cells increased with increasing D3T concentrations, reaching a maximum at 100 μM of D3T. Overall, 28 of 37 human tumor cell lines showed significant increases in enzyme activity following treatment with D3T that ranged from 1.3- to 7.0-fold (Table I). Leukemia, lung, colon, stomach, breast, ovary, prostate and melanoma tumor cell lines were all inducible. In contrast, three head and neck tumors and one liver tumor were not induced by D3T. In addition, some tumor types appeared to be more readily induced. For example, all the leukemia cell lines were induced, as were all the colon tumor, gastric tumor, ovarian tumor and malignant melanoma cell lines. However, the differences in inducibility of NQO1 activity in the different cell types may simply reflect the particular cell lines of the tumor types examined in this study. The mechanisms responsible for the different levels of induction are unknown.

There was no obvious relationship between the base level of NQO1 activity in the tumor cells and the ability of D3T to increase enzyme activity.

TABLE I
INDUCTION OF NQO1 ACTIVITY IN HUMAN TUMOR CELL LINES BY D3T[a]

Tumor Type	Cell Line	NQO1 Activity[b] (nmole.min.$^{-1}$ mg prot^{-1})		P[c]
		Control	D3T Induced	
Leukemia	HL-60	4.5 ± 0.5	31.6 ± 3.6	< 0.001
	THP-1	26.1 ± 2.3	102.8 ± 5.4	< 0.001
	WIL-2	58.8 ± 4.4	139.2 ± 8.9	< 0.001
Head and Neck	SSC-25	39.7 ± 7.8	66.1 ± 16.7	NS
	Detroit 562	167.7 ± 27.9	199.7 ± 39.3	NS
	FaDu	463.2 ± 29.9	444.1 ± 45.2	NS
Lung	NCI-H596	1.0 ± 0.2	3.5 ± 0.6	< 0.005
	NCI-H209	8.3 ± 0.8	40.2 ± 3.1	< 0.001
	NCI-H661	112.7 ± 12.4	284.8 ± 27.3	< 0.001
	NCI-H520	230.1 ± 31.3	267.0 ± 36.5	NS
	NCI-H125	689.5 ± 39.0	1136.0 ± 52.1	< 0.001
Colon	HCT116	91.5 ± 15.2	224.4 ± 38.1	< 0.02
	LS174T	314.6 ± 39.0	776.0 ± 97.4	< 0.005
	Colo205	671.2 ± 64.5	1117.3 ± 111.5	< 0.005
	HT29	713.1 ± 47.4	944.6 ± 69.1	< 0.02
Stomach	RF-48	6.9 ± 0.3	22.7 ± 1.0	< 0.002
	RF-1	7.7 ± 0.5	22.7 ± 0.4	< 0.001
	AGS	113.0 ± 1.6	266.0 ± 15.3	< 0.001
	Kato III	167.0 ± 6.7	289.0 ± 11.0	< 0.001
Breast	MDA-MB-231	0.9 ± 0.3	1.9 ± 0.5	NS
	T47D	25.8 ± 1.0	92.2 ± 5.5	< 0.001
	MDA-MB-468	90.5 ± 4.8	87.1 ± 4.9	NS
	BT474	191.5 ± 10.0	430.8 ± 36.1	< 0.001
	SK-Br-3	213.0 ± 1.5	331.0 ± 13.7	< 0.002
	MDA-MB-435	232.2 ± 16.2	291.7 ± 16.0	< 0.02
	HS578T	237.9 ± 13.9	420.8 ± 19.0	< 0.001
	ZR-75-1	355.5 ± 43.0	592.5 ± 71.7	< 0.02
	MCF-7	939.1 ± 88.3	981.2 ± 96.5	NS
Ovary	OVCAR-3	47.1 ± 5.9	129.3 ± 10.9	< 0.001
	SK-OV-3	177.8 ± 8.6	259.0 ± 15.9	< 0.002
Prostate	PC-3	149.3 ± 5.0	183.0 ± 1.7	< 0.005
	LnCAP	166.0 ± 15.1	266.2 ± 34.5	< 0.02
	DU145	652.3 ± 29.9	632.0 ± 20.1	NS
Skin	SK-MEL-28	586.7 ± 19.6	828.7 ± 38.4	< 0.01
	SK-MEL-2	666.5 ± 30.6	796.3 ± 36.5	< 0.05
	SK-MEL-5	2120.0 ± 51.3	2710.0 ± 60.8	< 0.005
Liver	HepG2	1292.5 ± 162.0	1356.3 ± 141.3	NS

[a] Cells were incubated without, or with, 100 μM D3T at 37° for 48 hr. Cells were washed, pelleted, suspended in 100 to 200 μl of 0.25 M sucrose, sonicated and stored at −80°. NQO1 activity was measured spectroscopically using menadione as the electron acceptor.
[b] The data represent the mean ± SE of 3 to 15 determinations.
[c] Statistical significance was determined by a 2-tailed t-test comparing the significance of the difference of the mean NQO1 activity in control and D3T treated cells. NS, not significant. (Adapted with permission from G. P. Doherty, M. K. Leith, X. Wang, T. J. Curphey, and A. Begleiter, *Br. J. Cancer* **77,** 1241 [1998].)

Although 4 of the 9 tumor cell lines that were not induced by D3T had high base levels of NQO1 activity, all the melanoma tumors, and other cell lines, that had high base levels of enzyme activity were induced. While the increase in enzyme activity relative to the base level was low in these cells, the absolute increase in enzyme activity was very high.

The base levels of NQO1 activity were very low in the normal marrow and kidney cells we examined. While D3T produced statistically significant increases in enzyme activity in these cells, the actual increases were only 10.8 and 4.9 nmole.min.$^{-1}$ mg prot^{-1}, respectively—less than half the increase observed in HL-60 leukemia cells. In contrast, D3T increased the level of NQO1 activity in normal human lung cells by 2.4-fold.

Combination treatment with D3T and EO9 increased the antitumor activity of EO9 in HL-60, human promyelocytic leukemia cells (Fig. 3). D3T increased the level of NQO1 activity in these cells by 5-fold ($p < 0.05$) and also increased the cytotoxic activity of EO9 by 2-fold in these

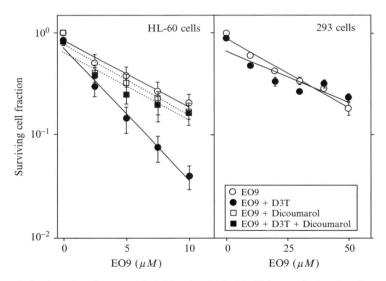

FIG. 3. Combination therapy with D3T and EO9 in HL-60 human leukemia cells and 293 human normal kidney cells. Cells were incubated at 37° with (●, ■) or without (○, □) 50 μM D3T for 48 hr. Cells were then treated with various concentrations of EO9 for 1 hr (○, ●), or with 50 μM dicoumarol for 15 min and then with various concentrations of EO9 for 1 hr (□, ■). Surviving cell fraction was determined by clonogenic assay for the HL-60 leukemia cells and by MTT assay for the 293 kidney cells. The points represent the mean ± SE of five or seven determinations. The lines are linear regression lines. (Adapted with permission from G. P. Doherty, M. K. Leith, X. Wang, T. J. Curphey, and A. Begleiter, *Br. J. Cancer* **77,** 1241 [1998].)

cells ($p < 0.001$). The increased cytotoxic activity was due to the elevated NQO1 activity, as dicoumarol reversed the effect of D3T ($p < 0.005$). In addition, D3T did not increase the level of NADPH:cytochrome P450 reductase activity, another activating enzyme for some bioreductive agents, in these cells. Combination treatment with D3T and MMC increased the antitumor activity of MMC in T47D, human breast carcinoma cells. D3T increased the level of NQO1 activity in these cells by 4-fold ($p < 0.001$) and also increased the cytotoxic activity of MMC by 3-fold in these cells ($p < 0.001$).

D3T produced statistically significant increases in the levels of NQO1 activity in normal human marrow and kidney cells. The major toxicities observed with MMC and EO9 are bone marrow and kidney toxicity, respectively.[79,105] However, the increases in NQO1 activities were very small in these cells, making it unlikely that the toxicity to these cells would be significantly increased. Indeed, D3T did not effect the toxicity of EO9 in the human normal kidney cells (Fig. 3), and did not increase the toxicity of MMC to mouse marrow cells.[106] In contrast, the increase in enzyme activity in the normal human lung cells was considerably larger. MMC has been shown to produce pulmonary fibrosis in approximately 5% of patients[107] by a mechanism that likely involves redox cycling and the formation of reactive oxygen species. Thus it is possible that the increase in NQO1 activity in normal lung cells with D3T treatment may serve to decrease this lung toxicity, because two-electron reduction of MMC to its hydroquinone by NQO1 would decrease reactive oxygen species formation.

GSTs are phase II detoxifying enzymes that can be induced by inducers of NQO1.[18,45,46] These enzymes have been shown to play an important role in resistance to a variety of antitumor agents, including MMC, by aiding in the removal of the anticancer drugs from cells.[108] Thus, if D3T were to increase the levels of both NQO1 and GST activities in tumor cells, this might not result in a net increase in antitumor activity. This did not occur in our studies, as we did observe an increase in antitumor activity with both EO9 and MMC in two tumor cell lines. Furthermore, we did not find any increase in GST activity in HL-60 cells treated with D3T, as has been observed previously.[103] Thus it appears that it is possible to increase NQO1 activity in tumor cells without increasing GST.

These results demonstrated that it is possible to enhance the antitumor activity of bioreductive agents in human tumor cells with inducers of

[105] G. N. Hortobagyi, *Oncology (Basel)* **50** (Suppl. 1), 1 (1993).
[106] A. Begleiter, M. K. Leith, and T. J. Curphey, *Br. J. Cancer* **74** (Suppl. XXVII), S9 (1996).
[107] D. S. Klein and P. R. Wilds, *Can. Anaesth. Soc. J.* **30**, 399 (1983).
[108] D. J. Waxman, *Cancer Res.* **50**, 6449 (1990).

NQO1. This approach appeared to be applicable to different bioreductive agents and in different tumor cells.

In a further study, we extended these investigations to examine combination therapy with MMC and D3T in six human tumor cell lines. D3T increased NQO1 activity in all six cell lines, and pretreatment with the enzyme inducer significantly enhanced the cytotoxicity of MMC in four of the cell lines. Combination treatment with D3T and MMC increased the cytotoxic activity of MMC by 2.3-fold in H661 non-small cell lung cancer cells, by 2.4-fold in T47D breast cancer cells, by 1.4-fold in HS578T breast cancer cells and by 2-fold in HCT116 human colon cancer cells (Table II). These results demonstrated that this combination treatment is effective in enhancing the cytotoxicity of MMC in a wide variety of different tumor types. In contrast, D3T did not increase MMC cytotoxic activity in SK-MEL-28 melanoma cells or AGS stomach cancer cells. Both of these cell lines have relatively high base levels of NQO1 activity. This suggests that the combination therapy approach may be restricted by the level of NQO1 in the cells. If the base or induced level of NQO1 is above

TABLE II

EFFECT OF INDUCERS OF NQO1 ON THE CYTOTOXIC ACTIVITY OF MMC IN HUMAN CELLS[a]

Tumor Type	Cell Line	Inducer	D_0 $(\mu M)^b$		P^c
			Control	With Inducer	
Colon	HCT116	D3T	3.31 ± 0.25	1.67 ± 0.11	< 0.001
Lung	H661	D3T	3.24 ± 0.19	1.43 ± 0.15	< 0.001
Skin	MEL-28	D3T	4.40 ± 0.41	4.06 ± 0.34	NS
Stomach	AGS	D3T	0.78 ± 0.08	0.66 ± 0.05	NS
Breast	HS578T	D3T	3.72 ± 0.30	2.63 ± 0.10	< 0.01
Breast	T47D	D3T	3.27 ± 0.20	1.34 ± 0.13	< 0.001
Breast	T47D	DMM	3.27 ± 0.20	1.13 ± 0.09	< 0.001
Breast	T47D	PG	3.27 ± 0.20	1.85 ± 0.14	< 0.001
Normal Human Bone Marrow		D3T	4.66 ± 0.41	2.95 ± 0.23	< 0.05

[a] Cells were treated with, or without, inducers at $37°$ for 48 hrs, then were incubated with various concentration of MMC for 1 hr. Surviving cell fractions were measured by MTT assay for tumor cells, or by methylcellulose clonogenic assay for normal marrow cells. The cytotoxic activity of MMC is presented as the D_0, which was obtained from the linear regression line of the surviving cell fraction versus drug concentration curve.

[b] The data represent the mean $\pm$ SE of three to eight determinations.

[c] A t-test comparing the significance of the difference of the slopes of the linear regression lines was used to compare the D_0 for cells treated with, or without, inducers. NS, not significant.

(Adapted with permission from X. Wang, G. P. Doherty, M. K. Leith, T. J. Curphey, and A. Begleiter, *Br. J. Cancer* **80,** 1223 [1999].)

an upper threshold, further induction may not lead to an increase in MMC cytotoxicity. Our results, and a previous study by Beall et al.[109] suggest that this upper threshold may be close to 300 nmole.min.$^{-1}$ mg prot^{-1} for MMC. While this may limit the use of this combination therapy approach in some situations, it should not significantly impair the use of NQO1 inducers to increase the antitumor activity of bioreductive agents in the clinic, since primary tumors generally have base levels of NQO1 that are <100 nmole.min.$^{-1}$ mg prot^{-1}.[28,110]

To investigate whether the level of enhancement of MMC activity is dependent on the level of NQO1 induction, we carried out combination therapy studies in T47D cells with MMC and enzyme inducers that increased the level of NQO1 to different extents. Pretreatment of cells with propyl gallate (PG), D3T or dimethyl maleate (DMM) increased MMC activity in the order PG < D3T < DMM, and this paralleled the increase in NQO1 activity (Fig. 4). This suggests that there is a relationship between the level of induction in enzyme activity and the enhancement of MMC cytotoxicity, provided the NQO1 activity does not exceed an upper threshold level. Since primary tumors usually have lower levels of NQO1 activity than tumor cells grown *in vitro*, it may be possible to achieve greater enhancement of antitumor activity of bioreductive agents in the clinic by using more potent inducers of NQO1.

Since many enzymes can activate bioreductive antitumor agents, we examined the effect of DMM on other enzymes that have been reported to be involved in MMC activation. When T47D cells were treated with 50 μM DMM, there were no changes in NADPH:cytochrome P450 reductase, or NADH:cytochrome b_5 reductase activities. Xanthine dehydrogenase activity was too low to be detected either before or after DMM treatment. Although we did not see an increase in GST activity in HL-60 human leukemia cells following treatment with 100 μM D3T,[97] in this study we did observe an increase in GST activity when T47D cells were treated with DMM. However, the effect was small and there was still a large enhancement of MMC cytotoxic activity in these cells.

The major toxicity associated with the use of MMC is bone marrow toxicity. We previously found that D3T did not increase the toxicity of MMC to mouse bone marrow.[106] In contrast, pretreatment with D3T produced a 1.5-fold increase in toxicity to human marrow cells (Fig. 5); however, this

[109] H. D. Beall, A. M. Murphey, D. Siegel, R. H. Hargreaves, J. Butler, and D. Ross, *Mol. Pharmacol.* **48**, 499 (1995).

[110] D. Ross, H. Beall, R. D. Traver, D. Siegel, R. M. Phillips, and N. W. Gibson, *Oncol. Res.* **6**, 493 (1994).

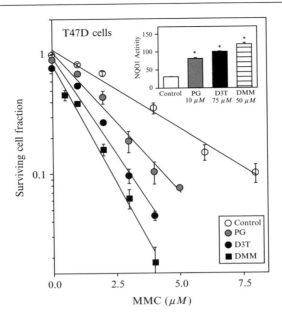

Fig. 4. Combination therapy with PG, D3T, or DMM, and MMC in T47D, human breast cancer cells. Cells were incubated with or without 10 μM PG, 75 μM D3T or 50 μM DMM for 48 hrs at 37°, then were treated with various concentrations of MMC for 1 hr. Surviving cell fraction was determined by MTT assay. NQO1 activity, shown in the inserted figure, was measured spectrophotometrically by using menadione as the electron acceptor and is reported as dicoumarol-inhibitable activity and expressed as nmole.min.$^{-1}$ mg prot^{-1}. Data represent the mean $\pm$ SE of three to eight determinations. The lines are linear regression lines. ANOVA analysis was used to determine the significance of the difference of the NQO1 activities. *$p < 0.05$. (Adapted with permission from X. Wang, G. P. Doherty, M. K. Leith, T. J. Curphey, and A. Begleiter, *Br. J. Cancer* **80**, 1223 [1999].)

was small compared with the enhancement of MMC cytotoxic activity in T47D, H661 and HCT116 cells. This result suggests that combination therapy with NQO1 inducers and MMC may increase the therapeutic index for MMC for appropriate tumors. In addition, NQO1 does not appear to be the major enzyme involved in activating MMC. Thus we would expect a greater increase in therapeutic index if this approach was used with bioreductive agents that are selectively activated by NQO1 like EO9[86] or RH1.[80] Indeed, we did see a greater enhancement of EO9 cytotoxic activity as compared with MMC in mouse lymphoma cells pretreated with D3T.[106]

These studies provided further support for the hypothesis that inducers of NQO1 could be used to increase the effectiveness of bioreductive agents

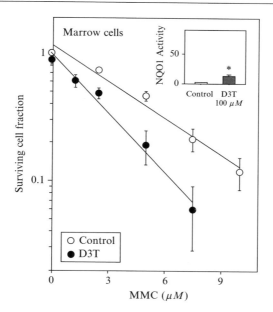

Fig. 5. Combination therapy with MMC and D3T in normal human marrow cells. Cells were incubated at 37° with or without 100 μM D3T for 48 hrs, then were treated with various concentrations of MMC for 1 hr. Surviving cell fraction was measured by methylcellulose clonogenic assay. NQO1 activity, shown in the inserted figure, was measured spectro-photometrically by using menadione as the electron acceptor, and is reported as dicoumarol-inhibitable activity and expressed as nmole.min.$^{-1}$ mg prot^{-1}. Data represent the mean ± SE of four to five determinations. The lines are linear regression lines. Two-tailed t-tests were used to determine the significance of the difference of the NQO1 activities. *$p < 0.05$. (Adapted with permission from X. Wang, G. P. Doherty, M. K. Leith, T. J. Curphey, and A. Begleiter, *Br. J. Cancer* **80**, 1223 [1999].)

in the clinic; however, Nishiyama et al.[111] found a negative correlation between NQO1 activity and MMC antitumor activity *in vivo*. In contrast, Malkinson et al.[112] saw greater MMC activity in human tumor xenografts with higher NQO1 activity compared with xenografts with low activity. Thus the ability of NQO1 inducers to enhance the antitumor activity of bioreductive agents must still be demonstrated *in vivo*. In addition, our findings suggest that this strategy may not be effective with all tumors, and that for clinical application it will be necessary to measure the level

[111] M. Nishiyama, S. Saeki, K. Aogi, N. Hirabayashi, and T. Toge, *Int. J. Cancer* **53**, 1013 (1993).
[112] A. M. Malkinson, D. Siegel, G. L. Forrest, A. F. Gazdar, H. K. Oie, D. C. Chan, P. A. Bunn, M. Mabry, D. J. Dykes, S. D. Harrison, and D. Ross, *Cancer Res.* **52**, 4752 (1992).

of NQO1 and the inducibility of the enzyme *in vitro* in tumor biopsy samples prior to the start of therapy.

Natural and Synthetic Inducers of NQO1 in Cancer Cells

To further improve the clinical potential of this new treatment strategy for enhancing the antitumor efficacy of bioreductive agents, we have investigated additional potential inducers of NQO1. We evaluated D3T, oltipraz and 11 other dithiolethiones for their ability to induce NQO1 activity in eight human tumor cell lines from five different tissues. Some of the D3T analogs were better inducers of NQO1 than others in the human tumor cell lines examined, indicating some useful structure-activity relationships. The parent compound, D3T, was the most consistent inducer and was also one of the best overall inducers of NQO1 activity in terms of the magnitude of enzyme induction; however, other D3T analogs also consistently produced large increases in NQO1 activity.

To identify other potential non-toxic enhancers of bioreductive agents, we investigated the ability of dietary components and pharmaceutical agents to induce NQO1 in human tumor cell lines. We found a number of compounds that were as good as or better than D3T as inducers of NQO1 activity in human tumors (Table III). The antioxidant PG and metabolites of fumaric and maleic acid, which are found in fruits and vegetables, produced 2- to 5-fold increases in NQO1 activity in T47D human breast cancer cells. These dietary components and a metabolite of the antioxidant BHA also increased NQO1 activity in HL-60 cells. These studies provide evidence that dietary components can be good inducers of NQO1 in human tumor cells. Thus, combining non-toxic dietary components with bioreductive drugs may provide a safe and effective way to increase the antitumor efficacy of these agents. As described previously, combination therapy studies with PG, D3T and DMM together with MMC in T47D cells increased MMC activity in the order PG < D3T < DMM, and this paralleled the increase in NQO1 activity (Fig. 4). This result demonstrated that dietary inducers of NQO1 could be used to enhance the antitumor activity of bioreductive agents.

Induction of NQO1 in Cancer Cells and Enhancement of Antitumor Activity In Vivo

A study by Shao *et al.*[113] provides support for using dietary inducers of NQO1 to enhance the antitumor activity of bioreductive agents *in vivo*. They observed that feeding high levels of the dietary fish oil menhaden

[113] Y. Shao, L. Pardini, and R. S. Pardini, *Lipids* **30**, 1035 (1995).

TABLE III
INDUCTION OF NQO1 ACTIVITY IN T47D HUMAN BREAST CANCER CELLS BY DIETARY AND
PHARMACEUTICAL INDUCERS[a]

Inducers	Concentration	NQO1 Activity[b] $(nmole.min.^{-1} mg prot^{-1})$	P[c]
Control		27.8 ± 1.2	
D3T	$100\ \mu M$	101.0 ± 2.8	< 0.001
13-cis-Retinoic Acid	$10\ \mu M$	61.6 ± 4.9	< 0.001
Genistein	$5\ \mu M$	23.8 ± 1.4	NS
Ursolic Acid	$15\ \mu M$	29.6 ± 4.4	NS
Chalcone	$20\ \mu M$	41.4 ± 3.4	< 0.001
Aspirin	$1\ mM$	40.8 ± 1.2	< 0.001
Ibuprofen	$200\ \mu M$	16.1 ± 1.1	NS
Caffeic Acid	$500\ \mu M$	33.4 ± 2.3	NS
Folic Acid	$100\ \mu M$	17.3 ± 2.5	NS
Vitamin K	$10\ \mu M$	17.3 ± 1.3	NS
Dimethyl Maleate (DMM)	$50\ \mu M$	121.0 ± 3.9	< 0.001
Propyl Gallate (PG)	$10\ \mu M$	80.5 ± 5.3	< 0.001
Dimethyl Fumarate (DMF)	$50\ \mu M$	128.5 ± 5.6	< 0.001
Sulforaphane	$10\ \mu M$	102.3 ± 0.7	< 0.001

[a] Cells were incubated with, or without, inducers at the concentrations shown at 37° for
48 hrs. Cells were washed, pelleted, suspended in 200 μl of 0.25 M sucrose and sonicated.
NQO1 activity was measured spectroscopically using menadione as the electron acceptor.
[b] Data represent the mean ± SE of three to five determinations.
[c] Two-tailed t-tests were used to compare the significance of the difference of NQO1
activity in control and inducer treated cells. NS, not significant.
(Adapted with permission from X. Wang, G. P. Doherty, M. K. Leith, T. J. Curphey, and
A. Begleiter, *Br. J. Cancer* **80,** 1223 [1999].)

oil to athymic mice decreased the growth rate of an MX-1, human breast
carcinoma tumor xenograft in these mice and also increased the response
of the tumor to MMC. In addition, the authors found that the fish oil
increased NQO1 activity in the tumor. Although this study used only a
small number of mice, it provides support for the hypothesis that it is possible to induce NQO1 in a human tumor *in vivo* and that this can produce
enhancement of the antitumor activity of the bioreductive agent MMC.

To examine whether we could observe a similar effect *in vivo*, CD-1
nude mice were implanted with human HCT116 colon cancer cells and
then were fed control diet or diet containing the NQO1 inducer dimethyl
fumarate (DMF).[114] DMF is a metabolite of fumaric acid, which occurs
naturally in fruits and vegetables. Mice were sacrificed to measure NQO1

[114] M. K. Leith, T. Digby, and A. Begleiter, *Proc. Amer. Assoc. Cancer Res.* **43,** 585 (2002).

activity in tumors and normal tissues. The DMF diet increased NQO1 activity in the tumors by 2.5-fold, but also increased enzyme activity in kidney and forestomach. Enzyme activity changes in other tissues, including bone marrow, where MMC produces dose limiting toxicity, were small.

To optimize the DMF schedule, tumor-bearing mice were fed DMF diet for 3, 7 or 14 days and NQO1 activities in tumors and normal tissues were measured. NQO1 activity reached a maximum in tumor at 7 days, while enzyme activity in various organs increased to 14 days. To investigate whether other NQO1 inducers could produce greater enhancement of antitumor activity than DMF, mice bearing HCT116 tumors were fed control, sulforaphane-, D3T- or DMF-containing diet for 7 days and then were treated with MMC. All of the inducers increased NQO1 activity in the tumor and enhanced MMC antitumor activity, but the effects with sulforaphane and D3T were small. Preliminary results using optimized doses and schedules for MMC and DMF treatment showed enhancement of MMC antitumor activity in mice bearing human HCT116 colon tumors, without any obvious increase in toxicity. However, the number of mice used was small, and additional studies with larger numbers of mice are required to confirm these findings. With this caution, these results suggest that it is possible to induce NQO1 activity selectively in tumors *in vivo* and that this can increase the antitumor activity of bioreductive agents such as MMC.

Difficulties Related to Induction of NQO1 as a Method for Enhancing Antitumor Activity

Our studies demonstrated that it was possible to induce NQO1 activity selectively in cancer cells as compared with normal cells and that this could enhance the antitumor activity of bioreductive antitumor agents. However, we encountered a number of difficulties that may limit the utility of this approach. In some cases we were unable to produce significant induction of NQO1 in cancer cells; in other cases we were unable to obtain enhanced antitumor activity even though NQO1 activity was increased in the cancer cells.

We showed that the dithiolethione D3T could selectively increase NQO1 activity in a wide variety of human[6,97,115] and murine tumors[106] and that this could enhance the antitumor activity of MMC and EO9. However, NQO1 was not induced in approximately 25% of the human tumor cell lines investigated.[97] Therefore we examined mechanisms that might

[115] X. Wang, G. P. Doherty, M. K. Leith, T. J. Curphey, and A. Begleiter, *Br. J. Cancer* **80,** 1223 (1999).

be responsible for the lack of induction of NQO1 in the human tumor cell lines.

We used fourteen human tumor cell lines representing various tissue types that were either induced or not induced by D3T. Two cell lines that were induced by D3T were homozygous for the active ^{609}C form of the NQO1 gene, while 4 cell lines that were not induced were also homozygous for the active NQO1 gene. Two of the cell lines that were induced by D3T were heterozygous for the base 609 C to T polymorphism, while two of the cell lines that were not induced were also heterozygous for the polymorphism. All 4 cell lines that were homozygous for the inactive ^{609}T form of NQO1 had little NQO1 activity which was not increased by D3T. Thus, cells required at least one ^{609}C allele for the enzyme activity to be induced; however, having one or even two ^{609}T alleles did not assure that NQO1 activity would be increased. The lack of induction of NQO1 activity in cells with at least one ^{609}C allele was due to a lack of increased transcription of this allele. In cells that were homozygous for the inactive ^{609}T form of the gene, the level of mutant protein increased but there was not an increase in enzyme activity. These results provided evidence that induction of NQO1 in human tumor cells is likely regulated at, or prior to, the transcription level. Since many of the cells that were induced by D3T had moderate to high base levels of NQO1 activity, this further suggested that basal and induced expression of NQO1 are regulated independently.

NQO1 activity is dependent on formation of a dimer of the NQO1 protein.[1] Thus cells that are heterozygous for the NQO1 polymorphism may form a heterodimer. If this heterodimer has low NQO1 activity, the mutant NQO1 protein might act as a dominant-negative to decrease enzyme activity. Since our studies suggested that mutant NQO1 protein may be more easily induced than wild-type protein, this might make it more difficult to induce NQO1 activity in cells that are heterozygous for the mutation. To test whether the mutant NQO1 protein acts as a dominant-negative, we transfected a plasmid containing a ^{609}T NQO1 cDNA into HCT116 human colon cancer cells and PC-3 human prostate cancer cells, which have two ^{609}C NQO1 alleles. NQO1 activity in the mutant transfected cells was not significantly different from control cells, suggesting that the mutant NQO1 protein does not act as a dominant-negative to decrease NQO1 activity.

The lack of NQO1 induction in some cells may be due to the absence of, or mutations in, transcriptional elements that control enzyme induction in the regulatory region of the NQO1 gene. Alternatively, there may be defects that prevent induction of necessary transcription factors, or changes in the cellular target for the inducing agent. To determine whether the lack of increased transcription in some of the cell lines was due to a mutation in the promoter region of the NQO1 gene in these cells, we sequenced approximately 2000 bases of the promoter region of human tumor cell lines that showed significant increases in NQO1 transcription after treatment with D3T, that showed small increases in NQO1 transcription, or that showed no increases in NQO1 transcription after NQO1 inducer treatment. These were compared with the base sequence for this same region of the NQO1 gene reported in Genbank.[116] The sequence analysis showed no changes in the sequences of major transcriptional elements like the XRE, CAT box, ARE, AP1 site or AP2 site in the tumor cells as compared with the published sequences. Although there were some random changes in single bases, these did not appear to correlate with differences in induced transcription of the NQO1 gene. These results suggested that the differences in transcription of the NQO1 gene after treatment with NQO1 inducers were not due to alterations in the promoter region of the gene. The observed differences in induced transcription are likely due to upstream factors such as alterations in induction of transcription factors, changes in the signal transduction pathway or differences in the cellular target(s) for the inducers.

Overall, these studies suggest that while it will likely be possible to induce NQO1 in most tumors, it will not be possible to induce this enzyme in all tumors for a variety of different reasons. This will somewhat limit the use of NQO1 induction to enhance the antitumor activity of bioreductive agents to those tumors in which this enzyme can be induced.

Another potential difficulty limiting the use of the NQO1 induction strategy for enhancing the antitumor activity of bioreductive agents is a possible upper threshold of NQO1 for increased activity. As described previously, the upper threshold appears to be close to 300 $nmole.min.^{-1}$ mg $prot^{-1}$ for MMC, which should not significantly impair the use of NQO1 inducers to increase the antitumor activity of MMC in the clinic. However, we observed a similar effect with the new bioreductive agent RH1 (2,5-diaziridinyl-3-3[hydroxymethyl]-6-methyl-1,4-benzoquinone). This agent is a water-soluble analog of the bioreductive alkylating agent MeDZQ that is activated selectively by NQO1. We investigated whether induction of NQO1 would produce a greater enhancement of the antitumor activity of RH1 compared with MMC. DMF pretreatment increased NQO1 activity 2.1-fold, from 94.0 to 194.0 $nmole.min.^{-1}$ mg $prot^{-1}$ ($p < 0.001$) in HCT116 human colon cancer cells, and 2.6-fold, from 13.8 to 36.4 ($p < 0.05$) in T47D human breast cancer cells. Similarly, sulforaphane pretreatment increased NQO1 activity 2.2-fold ($p < 0.001$) in HCT116 cells and 2.8-fold ($p < 0.001$) in T47D cells. The cells were then treated with RH1 and cytotoxic activity was determined. Induction of NQO1 did not enhance

---

[116] Genbank. Accession number, AH005427.

the antitumor activity of RH1 in HCT116 or T47D cells. Since RH1 is an excellent substrate for NQO1, the base levels of enzyme activity found in these cells have sufficient NQO1 to fully activate the low concentrations of RH1 used in these studies. Thus the upper threshold of NQO1 for increased antitumor activity for RH1 may be relatively low. This suggestion is supported by a study using a series of cell lines derived from cells that had no base level of NQO1 activity but were transfected with an NQO1 cDNA to express different levels of NQO1.[117] RH1 cytotoxic activity increased in the transfected cells compared with the parent cells up to an NQO1 activity of approximately 30 nmole.min.$^{-1}$ mg prot^{-1}, but did not increase in cells having higher enzyme activity. While this result might limit the use of the NQO1 induction strategy with RH1, there are many primary tumors that have very low levels of NQO1 activity in which this approach could still be used.

Future Perspectives

NQO1 is an important enzyme with a wide range of biological functions, including the maintenance of appropriate redox levels in cells, detoxification of xenobiotics and carcinogens, and activation of bioreductive cancer chemotherapeutic agents such as MMC and RH1. We have shown that this enzyme can be selectively induced in cancer cells *in vitro* and *in vivo*.[97,106,114,115] We also demonstrated that combining an inducer of NQO1 with an antitumor agent that is activated by this enzyme can enhance antitumor activity *in vitro*,[97,106,115] and we have obtained preliminary evidence that this also occurs *in vivo* when using a human tumor mouse xenograft model.[114] However, additional studies are required to optimize this strategy and to assess its applicability to the clinical situation.

Studies with normal cells *in vitro* and some preliminary studies in mice suggest that, in general, induction of NQO1 is lower in normal cells than in tumor cells and that combining an NQO1 inducer with MMC does not significantly enhance toxicity to normal bone marrow cells or produce any obvious organ toxicity. However, a more detailed assessment of possible increased toxicity is required to ensure that the induction strategy does not unduly increase the side effects of the antitumor agent to patients. This would require special attention to normal tissues that are known to be sensitive to each specific bioreductive agent. For example, the dose-limiting toxicity of MMC is bone marrow toxicity. Studies to date did not show any significant increase in toxicity to these cells, but MMC can also have

---

[117] S. L. Winski, E. Swann, R. H. J. Hargreaves, D. L. Dehn, J. Butler, C. J. Moody, and D. Ross, *Biochem. Pharmacol.* **61,** 1509 (2001).

toxic effects on other organs.[77] Thus additional studies that focus on possible toxicities to these other tissues are required.

We have demonstrated that NQO1 can be induced in a wide variety of human tumor types and have shown that induction of NQO1 can enhance the antitumor activity of bioreductive agents in a number of different types of human cancers *in vitro*. However, it is not known which tumor types can be induced *in vivo* and whether the NQO1 induction strategy will be useful in different types of tumors. Studies with a variety of different tumors types are required to identify which tumors would be suitable for this therapeutic approach.

Additional studies are also required with other bioreductive agents to determine whether their antitumor activity can also be enhanced by induction of NQO1. We have shown that this approach can be used with MMC and EO9; however, we observed some difficulty in obtaining enhanced cytotoxic activity with RH1 because of its potency and highly efficient activation by NQO1. The NQO1 induction strategy to enhance antitumor activity would be most effective with a bioreductive agent that was specifically activated by NQO1 (unlike MMC, which is activated by both NQO1 and NADPH:cytochrome P450 reductase) and an agent that required a relatively high level of NQO1 for full activation. Such an agent should have greater selectivity for killing tumor cells as compared with normal cells because of the higher base level of NQO1 already present in the tumor cells and the selective induction of NQO1 in these cells. Several structure-activity studies showed that various functional groups can influence the ability of NQO1 to reduce bioreductive agents.[92,118,119] This information would be useful for designing new bioreductive agents that were best suited for use with the NQO1 induction strategy.

Another important area of study would be the identification of better inducers of NQO1. While a wide variety of compounds can induce NQO1,[18,46,52] the ideal enzyme inducer for clinical enhancement of antitumor activity would produce a high level of induction, would be selective for induction in tumor cells, and would have no toxicity. In addition, from a practical perspective, an inducer that was already approved for human use would facilitate the application of the NQO1 induction strategy in the clinic. We have shown that a number of pharmaceutical agents, including oltipraz[97] and aspirin,[115] can induce NQO1 activity in tumor cells. We also found that dietary components such as sulforaphane, DMM and DMF

---

[118] H. D. Beall, A. R. Hudnott, S. Winski, D. Siegel, E. Swann, D. Ross, and C. J. Moody, *Bioorg. Med. Chem. Lett.* **8,** 545 (1998).

[119] R. M. Phillips, M. A. Naylor, M. Jaffar, S. W. Doughty, S. A. Everett, A. G. Breen, G. A. Choudry, and I. J. Stratford, *J. Med. Chem.* **42,** 4071 (1999).

were very good enzyme inducers in tumor cells.[115] We are currently evaluating other dietary components as possible enzyme inducers that could be suitable for use in the NQO1 induction strategy.

Acknowledgments

This work was supported by grants from the Canadian Institutes of Health Research, the American Institute for Cancer Research and the National Cancer Institute of Canada with funds from the Canadian Cancer Society.

Section III

Quinone Reductases: Chemoprevention and Nutrition

[19]   Role of Nicotinamide Quinone Oxidoreductase 1 (NQO1) in Protection against Toxicity of Electrophiles and Reactive Oxygen Intermediates

By PAUL TALALAY and ALBENA T. DINKOVA-KOSTOVA

Introduction

NQO1 is a broadly distributed FAD-dependent flavoprotein that catalyzes the reduction of a wide variety of quinones, quinone imines, nitroaromatics, and azo dyes. Its notable characteristics include: (1) equal efficiency of NADH and NADPH as electron donors; (2) an obligatory two-electron reduction mechanism without semiquinone intermediates; (3) potent inhibition by dicumarol and similar anticoagulants which has been a valuable tool for assessing the functional role of this enzyme; and (4) induction by a plethora of chemically diverse inducers that activate the regulatory antioxidant response element (ARE) of the NQO1 gene.

The question of the physiological function of NQO1 has been of continuing interest since the discovery of the enzyme by Lars Ernster[1] and the demonstration of the identity of the enzyme with the dicumarol-inhibited vitamin K reductase described by Martius.[2] Our two earlier reviews[3,4] summarized the growing evidence that NQO1 is a phase 2 detoxication enzyme and plays a major role in protecting cells against the toxicities of quinones. This chapter describes more recent evidence for the important protective functions of NQO1 and places them in their historical perspective.

Quinones are highly electrophilic compounds that are dietary plant components and arise also from the metabolism of benzene, phenols, and other aromatics, including polycyclic aromatics of environmental origin. The damaging effects of quinones result from two properties: (1) generation of reactive oxygen species by oxidative cycling via semiquinone intermediates and (2) depletion of cellular thiol groups by virtue of their Michael reaction acceptor functions (olefinic groups conjugated to electron-withdrawing carbonyl functions).

---

[1] L. Ernster and F. Navazio, *Acta. Chem. Scand.* **12**, 595 (1958).

[2] F. Märki and C. Martius, *Biochem. J.* **333**, 111 (1960).

[3] H. J. Prochaska and P. Talalay, in "Oxidative Stress: Oxidants and Antioxidants" (H. Sies, ed.), Academic Press, London, 1991.

[4] A. T. Dinkova-Kostova and P. Talalay, *Free Rad. Biol. Med.* **29**, 231 (2000).

Elucidation of the Physiological Functions of NQO1

More than 40 years ago, Williams-Ashman and Huggins[5] reported the chance observation that NQO1 (which these authors referred to as menadione reductase) was potently induced in cytosols of mammary gland and adipose tissue of rats by administration of 7,12-dimethylbenz[a]anthracene (DMBA), a carcinogenic polycyclic aromatic hydrocarbon. The induction of NQO1 (designated DT diaphorase by Ernster) by polycyclics was subsequently rediscovered independently by Ernster and colleagues,[6] who suggested that it might promote the detoxication of the quinone metabolites of polycyclic hydrocarbons by catalyzing their reduction to hydroquinones and thereby facilitating their conversion to more easily excreted glucuronide conjugates.

The observation that NQO1 was induced by polycyclics[5] was part of extensive efforts by Huggins[7] to develop methods to prevent DMBA-induced mammary tumors in rats. Huggins and colleagues demonstrated that mammary cancer and adrenal hemorrhage evoked by DMBA could be prevented by administration of azo dyes and aromatic olefins that likewise were inducers of NQO1.[8,9] Moreover, there was a remarkable quantitative relationship between the potencies of azo dyes in protecting rats against DMBA tumor formation and adrenal hemorrhage and the induction of NQO1 in liver (Fig. 1). Huggins further established that the degree of this protection increased with dose of protector, required time, lasted for several days, and depended on protein synthesis. On the basis of these findings, Huggins[7] made the imaginative suggestion that the capacity of compounds to induce NQO1 be used as a rapid screen to identify protectors against adrenal hemorrhage and tumor production by polycyclics. This prescient recommendation was ignored for many years, presumably because of the then-prevailing skepticism that cancer could be prevented and the absence of any insight into how NQO1 levels and tumor formation might be mechanistically related.

Further insight into the aforementioned relationship between NQO1 induction and protection against carcinogen-dependent neoplasia arose from the pursuit of unrelated experiments. Beginning in the early 1970s, Wattenberg[10] made a series of landmark observations demonstrating that

[5] H. G. Williams-Ashman and C. Huggins, Med. Exper. 4, 223 (1961).

[6] C. Lind, H. Vadi, and L. Ernster, Arch. Biochem. Biophys. 190, 97 (1978).

[7] C. B. Huggins, in "Experimental leukemia and mammary cancer: induction, prevention, cure," The University of Chicago Press, Chicago, 1979.

[8] C. Huggins and R. Fukunishi, J. Exp. Med. 119, 923 (1964).

[9] C. Huggins and J. Pataki, Proc. Natl. Acad. Sci. USA 53, 791 (1965).

[10] L. W. Wattenberg, Adv. Cancer Res. 26, 197 (1979).

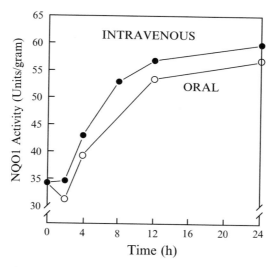

FIG. 1. Kinetics of induction of NQO1 in liver cytosols of rats after administration of 5 mg of DMBA by feeding or intravenous injection. NQO1 specific activity is expressed as units/gram—that is, μmoles of NADPH oxidized per minute by cytosol obtained from 1 g of fresh liver. [Modified from C. Huggins and R. Fukunishi, *J. Exp. Med.* **119**, 923 (1964).]

phenolic antioxidants such as BHA (2(3)-*tert*-butyl-4-hydroxyanisole) and BHT (3,5-di-*tert*-butyl-4-hydroxytoluene) protected against a variety of carcinogen-evoked tumors in rodent models. Unlike the potentially toxic protectors (aromatic azo dyes and polycyclic aromatics) used by Huggins, the phenolic antioxidants were already present as preservatives in the human diet and were therefore assumed to be relatively safe. Moreover, phenolic antioxidants protected against a variety of carcinogens with different target organ specificities. In the late 1970s, in collaboration with the late Ernest Bueding, we initiated efforts to understand the mechanism of the Wattenberg phenomenon. Treatment of mice with phenolic antioxidants markedly decreased the mutagenic urinary metabolites originating from administration of benzo[*a*]pyrene,[11] suggesting that major changes in the metabolism of this carcinogenic polycyclic aromatic had taken place. Whereas only negligible changes in cytochromes P450 were detected by Wattenberg,[10] we observed that the phenolic antioxidants that blocked neoplasia also raised the specific activities of glutathione transferases, epoxide hydrolase, and glutathione levels in the livers and other organs.[12–15]

[11] R. P. Batzinger, S.-Y. L. Ou, and E. Bueding, *Cancer Res.* **38**, 4478 (1978).
[12] A. M. Benson, R. P. Batzinger, S.-Y. L. Ou, E. Bueding, Y.-N. Cha, and P. Talalay, *Cancer Res.* **38**, 4486 (1978).

The marked elevations of glutathione transferases were shown to be due to enhanced transcription.[16] Consequently we made the explicit suggestion that these changes in phase 2 enzyme levels were largely responsible for the protective effects. Since we were familiar with the aforementioned observations of Huggins, we extended the repertoire of induced enzymes by demonstrating that BHA administration to mice raised the specific activities of NQO1 in nearly all tissues that were examined.[17] In a paper titled "Increase of NAD(P)H:quinone reductase by dietary antioxidants: Possible role in protection against carcinogenesis and toxicity," we proposed that NQO1 might play an important role in protecting cells against electrophilic carcinogens and reactive oxygen intermediates. Although prior views that NQO1 was involved in electron transport and blood coagulation reactions were not well documented, this radically new view of the functional role of this enzyme was not greeted with significant interest, although those familiar with NQO1 soon added evidence for its protective role.[18,19]

Evidence That NQO1 Protects Against Electrophile Toxicity, Oxidative Stress, and Neoplasia

There is now a large body of persuasive testimony that NQO1 in concert with other inducible phase 2 enzymes provides major cellular protection against oxidative and electrophile stress. The main lines of evidence supporting this view are extensively developed in several chapters of this volume, and the following account outlines and references these observations.

Direct Demonstration That NQO1 Quenches Quinone-Dependent Generation of Reactive Oxygen Species

Exposure of supernatant fractions of mouse liver to the synthetic vitamin K analog menadione (2-methyl-1,4-naphthoquinone) in the presence of NADPH results in an oxygen-dependent low-level emission of red

[13] A. M. Benson, Y.-N. Cha, E. Bueding, H. S. Heine, and P. Talalay, *Cancer Res.* **39**, 2971 (1979).
[14] P. Talalay, J. W. Fahey, W. D. Holtzclaw, T. Prestera, and Y. Zhang, *Toxicol. Letts.* **82/83**, 173 (1995).
[15] P. Talalay, *BioFactors* **12**, 5 (2000).
[16] W. R. Pearson, J. J. Windle, J. F. Morrow, A. M. Benson, and P. Talalay, *J. Biol. Chem.* **258**, 2052 (1983).
[17] A. M. Benson, M. J. Hunkeler, and P. Talalay, *Proc. Natl. Acad. Sci. USA* **77**, 5216 (1980).
[18] C. Lind, P. Hochstein, and L. Ernster, *Arch. Biochem. Biophys.* **216**, 178 (1982).
[19] P. L. Chesis, D. E. Levin, M. T. Smith, L. Ernster, and B. N. Ames, *Proc. Natl. Acad. Sci. USA* **81**, 1696 (1984).

light.[20] This chemiluminescence is believed to arise largely from the one-electron reduction of menadione to its semiquinone by the NADPH-cytochrome P-450 of the microsomal membranes present in these preparations, and the cyclic reoxidation of the semiquinone generating superoxide and singlet oxygen. In contrast, similar supernatant fractions from BHA-treated mice, which have 13-fold higher NQO1 activities, gave rise to only a very small light signal. The addition of dicumarol to the supernatant liver fractions of BHA-treated mice restored the light emission to levels comparable to those observed in fractions from untreated animals. Furthermore, exposure of control fractions to dicumarol increased the chemiluminescence by 34%. These results were interpreted as showing that the increased NQO1 levels, resulting from BHA feeding, converted the menadione to its hydroquinone by the obligatory two-electron reduction mechanism and thereby diverted menadione from oxidative one-electron cycling.

When pure and crystalline mouse liver NQO1 became available,[21,22] these experiments were extended in collaboration with Helmut Sies and colleagues.[23] We showed that the menadione-dependent chemiluminescence of mouse liver supernatant fractions could be abolished by the direct addition of pure NQO1.[23] Titration of the chemiluminescence of these preparations with increasing amounts of pure NQO1 resulted in progressive suppression of menadione-dependent chemiluminescence (Fig. 2). Moreover this suppression declined to identical low levels observed with supernatant preparations from BHA-treated animals when the NQO1 levels became equal. These experiments provided conclusive evidence that the menadione-dependent chemiluminescence of postmitochondrial supernatant fractions of mouse liver is regulated by their NQO1 levels, and that the protective actions of BHA treatment can be accounted for quantitatively and completely on the basis of their elevated NQO1 activities.

Effects of Induction, Overexpression, and Inhibition of NQO1 on the Tolerance to Electrophile and Oxidative Toxicity

The role of NQO1 in preserving the antioxidant status of cell membranes, especially with respect to maintaining coenzyme Q and α-tocopherol in the reduced state, have been reviewed by us elsewhere.[4]

---

[20] H. Wefers, T. Komai, P. Talalay, and H. Sies, *FEBS Lett.* **169,** 63 (1984).
[21] H. J. Prochaska and P. Talalay, *J. Biol. Chem.* **261,** 1372 (1986).
[22] L. M. Amzel, S. H. Bryant, H. J. Prochaska, and P. Talalay, *J. Biol. Chem.* **261,** 1370 (1986).
[23] H. J. Prochaska, P. Talalay, and H. Sies, *J. Biol. Chem.* **262,** 1931 (1987).

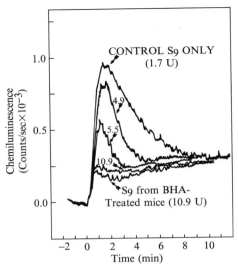

Fig. 2. Effect of additions of pure NQO1 on the menadione-mediated low-level chemiluminescence of postmitochondrial supernatant fractions (S$_9$) of mouse liver supplemented with an NADPH generating system. The uppermost and lowest tracings are from untreated (1.7 units NQO1) and BHA-fed (10.9 units NQO1) mice, respectively. The intermediate tracings show the chemiluminescence of liver fractions of untreated mice that were supplemented with crystalline mouse NQO1 to contain 4.9, 5.5, and 10.9 units of NQO1, respectively. Note that the chemiluminescence of control preparation was quenched to almost the same level as that from BHA-treated mice when the NQO1 activities were equalized. (Reproduced with permission from H. J. Prochaska, P. Talalay, and H. Sies, *J. Biol. Chem.* **262**, 1931 [1987].)

Induction of NQO1 as Guide to Isolation, Identification, and Structure-Activity Analysis of Anticarcinogens

Our conviction that tissue NQO1 levels reflect the degree of protection against quinone toxicities, and that these could be modulated by induction or inhibition of this enzyme, motivated the development of a simple and rapid cell culture assay system for detecting inducer activity, isolating inducers from natural sources, and determining their potencies quantitatively. The 96-well microtiter plate assay system that uses Hepa 1c1c7 murine hepatoma cells was developed by the late Hans Prochaska[24,25] based on earlier work by De Long and colleagues.[26] A full description of this system and its subsequent technical improvements are provided in

[24] H. J. Prochaska and A. B. Santamaria, *Anal. Biochem.* **169**, 328 (1988).
[25] H. J. Prochaska, A. B. Santamaria, and P. Talalay, *Proc. Natl. Acad. Sci. USA* **89**, 2394 (1992).
[26] M. J. De Long, H. J. Prochaska, and P. Talalay, *Proc. Natl. Acad. Sci. USA* **83**, 787 (1986).

this volume.[27] This assay has become exceptionally useful and has been widely adopted by many investigators for the isolation of inducers from plant sources. It has made possible the isolation of the potent isothocyanate inducer sulforaphane and its glucosinolate precursor from broccoli, broccoli sprouts, and seeds.[28,29] The assay has been extensively used by Pezzuto and his colleagues[30] at the University of Illinois for the isolation of a wide variety of anticarcinogenic compounds. The induction of NQO1 has also been useful in carrying out structure-activity correlations in the synthesis of analogs of isothiocyanates, 1,2-dithiole-3-thiones, and double Michael reaction acceptors such as curcuminoids.[31–34] The finding that many of the isolated inducers also blocked chemical carcinogenesis in animal models is consistent with the protective function of NQO1. But since many other phase 2 proteins are also induced, this evidence alone cannot be considered persuasive.

Regulation of NQO1 by Antioxidant Response Elements (ARE) and Consequences of Disruption of the NQO1 Gene

Both human and rodent NQO1 are regulated by antioxidant response elements located in the upstream regulatory regions of their genes.[35,36] Disruption of the NQO1 gene in mice (NQO1−/−) resulted in markedly increased susceptibility to the toxicity of menadione[37] and enhanced skin tumor formation in response to administration of benzo[a]pyrene and DMBA.[38,39] Although quinone metabolites may play a major role in the neoplastic effects of benzo[a]pyrene, they may be less important in the neoplastic action of DMBA. Therefore the observation that tumor formation

[27] J. W. Fahey, A. T. Dinkova-Kostova, K. K. Stephenson, and P. Talalay, *Methods Enzymol.* **382**, 243 (2004).

[28] Y. Zhang, P. Talalay, C.-G. Cho, and G. H. Posner, *Proc. Natl. Acad. Sci. USA* **89**, 2399 (1992).

[29] J. W. Fahey, Y. Zhang, and P. Talalay, *Proc. Natl. Acad. Sci. USA* **94**, 10367 (1997).

[30] J. Pezzuto and Y.-H. Kang, *Methods Enzymol.* **382**, 380 (2004).

[31] G. H. Posner, C.-G. Cho, J. V. Green, Y. Zhang, and P. Talalay, *J. Med. Chem.* **37**, 170 (1994).

[32] C. Gerhäuser, M. You, J. Liu, R. M. Moriarty, M. Hawthorne, R. J. Mehta, R. C. Moon, and J. M. Pezzuto, *Cancer Res.* **57**, 272 (1997).

[33] P. A. Egner, T. W. Kensler, T. Prestera, P. Talalay, A. H. Libby, H. H. Joyner, and T. J. Curphey, *Carcinogenesis* **15**, 177 (1994).

[34] A. T. Dinkova-Kostova and P. Talalay, *Carcinogenesis* **20**, 911 (1999).

[35] L. V. Favreau and C. B. Pickett, *J. Biol. Chem.* **268**, 19875 (1993).

[36] T. Nguyen, P. J. Sherratt, and C. B. Pickett, *Annu. Rev. Pharmacol. Toxicol.* **43**, 233 (2003).

[37] V. Radjendirane, P. Joseph, Y. H. Lee, S. Kimura, A. J. Klein-Szanto, F. J. Gonzalez, and A. K. Jaiswal, *J. Biol. Chem.* **273**, 7382 (1998).

[38] D. J. Long 2nd, R. L. Waikel, X.-J. Wang, L. Perlaky, D. R. Roop, and A. K. Jaiswal, *Cancer Res.* **60**, 5913 (2000).

[39] D. J. Long 2nd, R. L. Waikel, X.-J. Wang, D. R. Roop, and A. K. Jaiswal, *J. Natl. Cancer Inst.* **93**, 1166 (2001).

by both hydrocarbons is enhanced in the NQO1-deficient mouse suggests that the protective effects of the enzyme may extend beyond the reduction of quinones, and may involve other oxidative processes.

Epidemiological Links of NQO1 Polymorphism to the Risk of Developing Cancer

Edwards et al.[40] first noted the absence of a diaphorase (with properties identical to those of NQO1) from various tissues obtained at autopsy in 4% of 628 unrelated British subjects. Further evidence that NQO1 is the product of a polymorphic human gene residing on chromosome 16q2.2 was subsequently obtained.[41,42] The frequencies of the wild-type, homozygous, and heterozygous mutants vary markedly among ethnic groups. The homozygous null genotype is carried by 4% of Caucasians, 5% of African Americans, 16% of Mexican Americans, and 22% of Asians.[41,43] The heterozygous genotype is more common, and is found in 40% of Caucasians, 34% of African Americans, 52% of Mexican Americans, and 57% of Asians. The enzymatic activity is absent from homozygous mutants, and is variable among heterozygotes. This interesting polymorphism has provided the opportunity for examining the question of whether polymorphic individuals have a higher susceptibility to developing malignancy. This issue is reviewed elsewhere in this volume.[42] In relation to the question of the protective functions of NQO1, it appears that NQO1 deficiency is a risk factor for urothelial carcinoma,[44] for cutaneous basal cell carcinoma, and for esophageal squamous cell carcinoma.[45,46] The null NQO1 genotype was twice as prevalent in colorectal cancer patients as in controls and six times more prevalent in patients with tumors with K-ras mutants in codon 12.[47] Association with lung cancer appears complex. Some studies find an increased lung risk with the wild-type genotype,[48] but Lewis et al.[49] have

[40] Y. H. Edwards, J. Potter, and D. A. Hopkinson, Biochem. J. **187**, 429 (1980).
[41] D. Ross, J. K. Kepa, S. L. Winski, H. D. Beall, A. Anwar, and D. Siegel, Chem.-Biol. Interactions **129**, 77 (2000).
[42] D. Ross and D. Siegel, Methods Enzymol. **382**, 115 (2004).
[43] D. W. Nebert, A. L. Roe, S. E. Vandale, E. Bingham, and G. G. Oakley, Genet. Med. **4**, 62 (2002).
[44] W. A. Schulz, A. Krummeck, I. Rosinger, P. Eickelman, C. Neuhaus, T. Ebert, B. J. Schmitz-Drager, and H. Sies, Pharmacogenetics **7**, 235 (1997).
[45] A. Clairmont, H. Sies, S. Ramachandran, J. T. Lear, A. G. Smith, B. Bowers, P. W. Jones, A. A. Fryer, and R. C. Strange, Carcinogenesis **20**, 1235 (1999).
[46] J. Zhang, W. A. Schulz, Y. Li, R. Wang, R. Zotz, D. Wen, D. Siegel, D. Ross, H. E. Gabbert, and M. Sarbia, Carcinogenesis **24**, 905 (2003).
[47] M. J. Lafuente, X. Casterad, M. Trias, C. Ascaso, R. Molina, A. Ballesta, S. Zheng, J. K. Wiencke, and A. Lafuente, Carcinogenesis **21**, 1813 (2000).
[48] H. Chen, A. Lum, A. Seifried, L. R. Wilkens, and L. Le Marchand, Cancer Res. **59**, 3045 (1999).
[49] S. J. Lewis, N. M. Cherry, R. M. Niven, P. V. Barber, and A. C. Povey, Lung Cancer **34**, 177 (2001).

shown that patients carrying at least one variant allele have a 4-fold increased risk for developing small cell lung cancer.

There is a correlation between the NQO1 null genotype and hemato-poietic malignancies. This genotype is a significant risk factor for benzene-induced hematotoxicity,[50,51] for chemotherapy-related acute myeloid leukemia,[52] and in children for mixed-lineage leukemia.[53–55]

There is thus abundant epidemiological evidence that human popula-tions that are genetically deficient in NQO1 are more susceptible to the development of malignancies.

Stabilization of Tumor Suppressor Gene p53 by Binding to NQO1

p53 is very commonly mutated in human cancers and is activated by oxidative stress and DNA damage resulting in growth arrest and apoptosis. Recent experiments have demonstrated that NQO1 binds and stabilizes p53 against proteasomal degradation.[56–58] This stabilization does not appear to depend on catalytic activity and provides a novel mechanism of protection by NQO1 against carcinogenesis.

Conclusions

We have reviewed the main lines of evidence that NQO1 plays a major role in protecting animals and their cells against oxidative and electrophile stress. This evidence includes: (1) suppression of the chemiluminescence resulting from oxidative cycling of quinones by direct addition of pure NQO1; (2) demonstration that induction and overexpression of NQO1 results in decreased toxicity of oxidants and electrophiles, whereas in-hibition of NQO1 by dicumarol enhances these toxicities; (3) the finding

[50] N. Rothman, M. T. Smith, R. B. Hayes, R. D. Traver, B. Hoener, S. Campleman, G. L. Li, M. Dosemeci, M. Linet, L. Zhang, L. Xi, S. Wacholder, W. Lu, K. B. Meyer, N. Titenko-Holland, J. T. Stewart, S. Yin, and D. Ross, *Cancer Res.* **57,** 2839 (1997).

[51] M. T. Smith, *Proc. Natl. Acad. Sci. USA* **96,** 7624 (1999).

[52] R. A. Larson, Y. Wang, M. Banerjee, J. Wiemels, C. Hartford, M. M. Le Beau, and M. T. Smith, *Blood* **94,** 803 (1999).

[53] J. L. Wiemels, A. Pagnamenta, G. M. Taylor, O. B. Eden, F. E. Alexander, and M. F. Greaves, *Cancer Res.* **59,** 4095 (1999).

[54] M. Krajinovic, D. Labuda, G. Mathonnet, M. Labuda, A. Moghrabi, J. Champagne, and D. Sinnett, *Clin. Cancer Res.* **8,** 802 (2002).

[55] M. T. Smith, Y. Wang, C. F. Skibola, D. J. Slater, L. Lo Nigro, P. C. Nowell, B. J. Lange, and C. A. Felix, *Blood* **100,** 4590 (2002).

[56] G. Asher, J. Lotem, R. Kama, L. Sachs, and Y. Shaul, *Proc. Natl. Acad. Sci. USA* **99,** 3099 (2002).

[57] G. Asher, J. Lotem, L. Sachs, C. Kahana, and Y. Shaul, *Proc. Natl. Acad. Sci. USA* **99,** 13125 (2002).

[58] A. Andwar, D. Dehn, D. Siegel, J. K. Kepa, L. J. Tang, J. A. Pietenpol, and D. Ross, *J. Biol. Chem.* **278,** 10368 (2003).

that measurement of induction of NQO1 in cells in culture is an effective tool for identifying and isolating anticarcinogenic inducers from plant sources and is a useful guide for elucidating the relation of structure to inducer activity in various chemical families of inducers; (4) the finding that disruption of the NQO1 gene in mice increases their susceptibility to the toxicity of quinones and carcinogens; and (5) the finding that polymorphism of the NQO1 gene in human populations increases risk of developing some human malignancies.

[20] Activation and Detoxification of Naphthoquinones by NAD(P)H: Quinone Oxidoreductase

By REX MUNDAY

It is known that 1,4-naphthoquinones (Fig. 1) are widely distributed in nature, and many derivatives have been isolated from plants, bacteria and fungi.[1] Naturally occurring naphthoquinones have long been used in folk medicine,[2,3] and one such compound, 2-hydroxy-1,4-naphthoquinone, the active ingredient of henna, has been in continuous use as a cosmetic for more than 3,000 years.[4] More recently, certain naphthoquinone derivatives have been shown to be effective as parasiticides[5] and as anti-cancer agents.[6–10] Naphthoquinones are also toxic, both *in vitro* and *in vivo*.

Two major mechanisms of naphthoquinone toxicity have been identified. First, some compounds of this type are powerful electrophiles that arylate tissue components, thereby severely disrupting cellular metabolism and causing cell death.[11] Since nucleophilic attack occurs only in the quinone ring, arylation is possible only when at least one of the two positions

---

[1] R. H. Thomson, "Naturally Occurring Quinones III. Recent Advances." Chapman & Hall, London, 1986.

[2] J. A. Duke, "CRC Handbook of Medicinal Herbs." CRC Press, Boca Raton, 1985.

[3] J. M. Watt and M. G. Breyer-Brandwijk, "The Medicinal and Poisonous Plants of Southern and Eastern Africa," Second Edition. Livingstone, Edinburgh, 1963.

[4] G. R. Hughes, *J. Soc. Cosmet. Chem.* **10,** 159 (1959).

[5] M. E. Hanke and S. M. Talaat, *Trans. Roy. Soc. Trop. Med. Hyg.* **55,** 56 (1961).

[6] M. Krishnaswamy and K. K. Purushothaman, *Ind. J. Exp. Biol.* **18,** 876 (1980).

[7] R. V. Rao, T. J. McBride, and J. J. Oleson, *Cancer Res.* **28,** 1952 (1968).

[8] V. K. Tandon, M. Vaish, J. M. Khanna, and N. Anand, *Arch. Pharm. (Weinhiem)* **323,** 383 (1990).

[9] A. B. Pardee, Y. Z. Lee, and C. J. Lee, *Curr. Cancer Drug Targets* **2,** 227 (2002).

[10] P. Buc Calderon, J. Cadrobbi, C. Marques, N. Hong-Ngoe, J. M. Jamison, J. Gilloteaux, J. L. Summers, and H. S. Taper, *Curr. Med. Chem.* **9,** 2271 (2002).

[11] A. Brunmark and E. Cadenas, *Free Rad. Biol. Med.* **7,** 435 (1989).

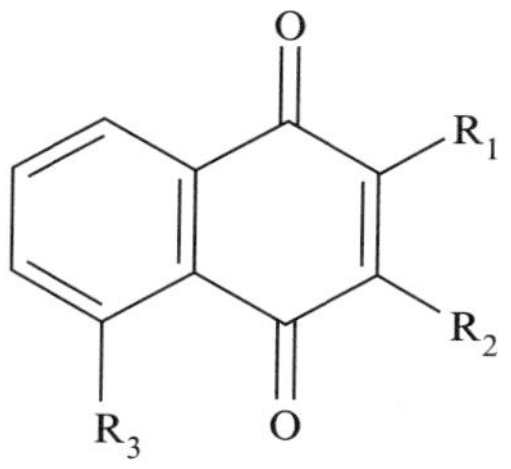

FIG. 1. Structure of 1,4-naphthoquinones.

(carbons 2 and 3) in this ring is unsubstituted. The presence of electron donating groups in the quinone ring decreases electrophilicity but has comparatively little effect in the aromatic ring. Thus 1,4-naphthoquinone itself (Fig. 1, $R_1=R_2=R_3=H$), 2-methyl-1,4-naphthoquinone (menadione, Fig. 1, $R_1=CH_3$; $R_2=R_3=H$) and 5-hydroxy-1,4-naphthoquinone (Fig. 1, $R_1=R_2=H$; $R_3=OH$) readily react with nucleophiles, whereas 2-hydroxy-1,4-naphthoquinone (Fig. 1, $R_1=OH$; $R_2=R_3=H$) does not.[12] Arylation by doubly substituted compounds such as 2,3-dimethyl-1,4-naphthoquinone (Fig. 1, $R_1=R_2=CH_3$; $R_3=H$) and 2,3-dimethoxy-1,4-naphthoquinone (Fig. 1, $R_1=R_2=OCH_3$; $R_3=H$) is not possible.

The second mechanism of naphthoquinone toxicity involves redox cycling, with generation of "active oxygen" species. The latter are powerful oxidants that are known to be harmful in a variety of biological systems. Redox cycling is a general property of naphthoquinones, which may be initiated by either one- or two-electron reduction. One-electron reduction of a naphthoquinone—which is mediated by enzymes such as NADH:ubiquinone reductase, NADPH-cytochrome P450 reductase and NADH:cytochrome b_5 reductase[13]—forms the corresponding semiquinone. Semiquinones are very unstable substances and readily react with molecular oxygen at physiological pH to form "active oxygen" species, and one-electron reduction of quinones is generally regarded as an activation reaction, leading inevitably to toxic change.[13] Two-electron reduction of the naphthoquinone, which is mediated by NAD(P)H: quinone oxidoreductase (DT-diaphorase, QR), leads to hydroquinone formation. As first suggested by Ernster,[14,15] reduction of quinones by QR may

---

[12] K. Öllinger and A. Brunmark, *J. Biol. Chem.* **266**, 21496 (1991).
[13] G. Powis, *Pharmacol. Ther.* **35**, 57 (1987).
[14] L. Ernster, *Meth. Enzymol.* **10**, 309 (1967).
[15] C. Lind, P. Hochstein, and L. Ernster, *Arch. Biochem. Biophys.* **216**, 178 (1982).

constitute a detoxification process, since rapid reduction of the quinone to the hydroquinone would circumvent one-electron processes. Furthermore, hydroquinones may be conjugated with sulfate or glucuronic acid to form unreactive substances that are readily eliminated from the body. It must be recognized, however, that under certain circumstances, hydroquinones readily undergo autoxidation, and if this process is rapid, two-electron reduction of a naphthoquinone may, like the one-electron process, lead to redox cycling and initiate toxic change.

The involvement of redox cycling in the mechanism of naphthoquinone toxicity and the involvement of QR in this process has attracted considerable attention. From the previous considerations, the rate of reduction of a naphthoquinone, and the rate of autoxidation of the corresponding naphthohydroquinone, will be crucial in determining whether a naphthoquinone is activated or detoxified by QR. Data on these parameters are now available for a number of naphthoquinones and naphthohydroquinones, as described later.

Rates of Naphthoquinone Reduction by QR and Rates of Naphthohydroquinone Autoxidation

Purification of QR

Several methods are available for purification of QR. The method of Lind et al.[16] has proved very satisfactory for purifying the enzyme from rat liver.

Synthesis of Naphthohydroquinones

Naphthoquinones may be reduced to the corresponding hydroquinones by sodium dithionite, stannous chloride or zinc and sulfuric acid, as detailed previously.[17] Naphthohydroquinones are unstable even in the solid state and should be kept desiccated at $-20°$ in an inert atmosphere.

Rates of Reduction of Naphthoquinones by QR

Rates of reduction of naphthoquinones by purified QR are measured as initial rates of reduction of cytochrome c at 550 nm.[18,19]

[16] C. Lind, E. Cadenas, P. Hochstein, and L. Ernster, *Meth. Enzymol.* **186,** 287 (1990).
[17] R. Munday, *Free Rad. Res.* **32,** 245 (2000).
[18] R. Munday, *Redox Rep.* **3,** 189 (1997).
[19] R. Munday, *Free Rad. Res.* **35,** 145 (2001).

TABLE I

RATES OF REDUCTION OF VARIOUS NAPHTHOQUINONES BY PURIFIED QR AND RATE OF
AUTOXIDATION OF THE CORRESPONDING HYDROQUINONES

1,4-Naphthoquinone	Rate of reduction by purified QR[*]	Rate of autoxidation of hydroquinone[**]
Unsubstituted	4.85 ± 0.10	12.6 ± 0.3
2-Methyl (menadione)	10.17 ± 0.14	23.8 ± 0.8
2,3-Dimethyl	6.76 ± 0.09	28.6 ± 1.1
2,3-Dimethoxy	3.24 ± 0.08	67.3 ± 5.2
2-Hydroxy	0.12 ± 0.01	175 ± 4
2-Amino	1.20 ± 0.08	184 ± 7
5-Hydroxy	6.77 ± 0.28	313 ± 6

[*] Initial rate of cytochrome c reduction (μM/min) in the presence of 10 μM quinone, 200 μM NADH and 0.003 U/ml QR at pH 7.4 and 25°. Data from R. Munday, *Free Rad. Res.* **35,** 145 (2001) and unpublished.

[**] Initial rate of oxygen uptake (μM/min) by 50 μM hydroquinone at pH 7.4 and 25°. Data from R. Munday, *Free Rad. Res.* **32,** 245 (2000).

Rates of Autoxidation of Hydroquinones

Solutions of hydroquinones are highly unstable. They should be made immediately before use, in acidified solvent, under an atmosphere of nitrogen. Rates of oxidation of hydroquinones are measured by initial rates of oxygen uptake.[18,19] The use of purified buffers or addition of a chelating agent, such as DTPA, is essential, since trace amounts of copper that are present in unpurified buffers inhibit the autoxidation of some naphthohydroquinones.[20,21]

The rates of reduction of several naphthoquinone derivatives by purified QR and the rates of autoxidation of the corresponding chemically-pure hydroquinones are shown in Table I.

Menadione was found to be rapidly reduced by QR, while the 2-hydroxy and 2-amino derivatives reacted only slowly. 2,3-Dimethyl-, 2,3-dimethoxy- and 5-hydroxy-1,4-naphthoquinone and unsubstituted 1,4-naphthoquinone were reduced at an intermediate rate. Amino- and hydroxy-naphthohydroquinones underwent very rapid autoxidation under the conditions of these experiments. 2,3-Dimethoxy-1,4-naphthohydroquinone oxidized more slowly, and the alkyl and unsubstituted derivatives slower still. It should be noted, however, that none of the hydroquinones

[20] R. Munday, *Free Rad. Biol. Med.* **26,** 1475 (1999).
[21] R. Munday, *Free Rad. Biol. Med.* **22,** 689 (1997).

could be classified as "stable" under the conditions of these experiments, since even the slowest compound, 1,4-naphthohydroquinone, was completely oxidized to the quinone in a matter of minutes.

The data on the autoxidation rates of the chemically pure naphthohydroquinones are not in accord with rates determined with hydroquinones prepared *in situ* by reduction with QR.[22] In the case of menadione, for example, reduction by QR was associated with rapid oxidation of NAD(P)H in an amount corresponding to that required for total reduction of the quinone to the hydroquinone. Following this, NAD(P)H oxidation was very slow, indicating that the hydroquinone was quite stable under these conditions. Similar results were obtained with unsubstituted naphthoquinone, and these compounds were defined as "redox stable."[23] The disparity between these sets of data may be explained by the effect of QR itself on the rate of autoxidation of hydroquinones.

Inhibition of Hydroquinone Autoxidation by QR

The rate of autoxidation of pure 2-methyl-1,4-naphthohydroquinone was not affected by NADH alone or by QR alone. In the presence of both the enzyme and its cofactor, however, oxidation of the hydroquinone was inhibited. The degree of inhibition increased with increasing enzyme concentration up to a maximum of ~80%, beyond which no further decrease in reaction rate was recorded. Inhibition of the enzyme with dicumarol restored oxidation rates to control values.[18]

Comparative data for a number of naphthohydroquinones are shown in Table II. The autoxidation of 1,4-naphthohydroquinone and its 2-methyl, 2,3-dimethyl, and 2,3-dimethoxy derivatives was strongly inhibited by QR. In contrast, the enzyme had no effect on the rate of autoxidation of the 2-hydroxy- and 2-amino derivatives.

In the previous studies with naphthoquinones,[22] the concentration of the quinone was 20 μM and the concentration of QR was 0.1 U/ml. Under these conditions, the rate of oxidation of both 1,4-naphthohydroquinone and 2-methyl-1,4-naphthohydroquinone is decreased by more than 75%.[24] The concept of "redox stable" naphthohydroquinones is therefore not correct. The hydroquinones from compounds such as menadione are not intrinsically stable, but they are stabilized in the presence of QR.

Since the degree of inhibition of hydroquinone autoxidation is dependent on the concentration of QR, it is possible that redox cycling could

[22] G. D. Buffinton, K. Öllinger, A. Brunmark, and E. Cadenas, *Biochem. J.* **257,** 561 (1989).
[23] E. Cadenas, *Biochem. Pharmacol.* **49,** 127 (1995).
[24] R. Munday, unpublished observation.

TABLE II

INHIBITION OF NAPHTHOHYDROQUINONE AUTOXIDATION BY QR

Naphthohydroquinone	Percent inhibition of autoxidation by QR
Unsubstituted	75 ± 3
2-Methyl	80 ± 2
2,3-Dimethyl	50 ± 4
2,3-Dimethoxy	66 ± 5
2-Hydroxy	0
2-Amino	0

Rates of oxidation were measured as initial rates of oxygen uptake by 75 μM hydroquinone plus NADH (200 μM) in the presence or absence of 1.0 U/ml QR.

occur with the alkyl, dialkyl and methoxy derivatives at low levels of the enzyme. Furthermore, not all hydroquinones are stabilized in this way, and redox cycling by compounds such as 2-hydroxy- and 2-amino-1,4-naphthoquinone could be facilitated by QR at all enzyme concentrations.

Redox Cycling of Naphthoquinones in the Presence of QR

Rates of redox cycling by naphthoquinones in the presence of various concentrations of QR are measured as rates of oxygen uptake or rates of NADH oxidation at pH 7.4.[18,19] Sucrose, at 0.25 M in the buffer (0.05 M phosphate containing 50 μM DTPA), should be used to activate the enzyme, since other activators, such as BSA and detergents, markedly influence the rate of naphthohydroquinone autoxidation.[18]

With menadione, redox cycling was recorded at low concentrations of QR, but the rate decreased at higher levels of enzyme (Fig. 2). With 2-hydroxy-1,4-naphthoquinone, however, the rate of redox cycling increased with increasing enzyme concentration over the whole range studied (Fig. 2).

2,3-Dimethyl-1,4-naphthoquinone also underwent redox cycling at low concentrations of QR, with inhibition at high concentrations. A similar response was seen with 2,3-dimethoxy-1,4-naphthoquinone at a concentration of 20 μM,[18] but less inhibition was recorded at a quinone concentration of 150 μM.[19] The behavior of 2-amino- and 5-hydroxy-1,4-naphthoquinone was similar to that of the 2-hydroxy derivative.[19]

Superoxide dismutase (SOD) greatly decreased the rate of redox cycling of menadione and 2,3-dimethyl- and 2,3-dimethoxy-1,4-naphthoquinone in

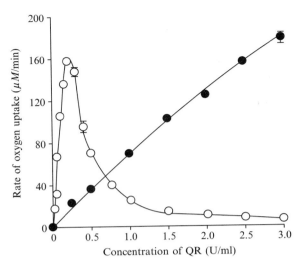

Fig. 2. Redox cycling by menadione and 2-hydroxy-1,4-naphthoquinone in the presence of various concentrations of QR. The rate of redox cycling was measured as rate of oxygen uptake in the presence of 1 mM NADH, 50 μM DTPA and 0.25 M sucrose in 0.05 M phosphate buffer at pH 7.4 and 25°. The concentration of quinone was 150 μM.

the presence of QR,[19] indicating a complementary effect of these enzymes in preventing redox cycling of such naphthoquinones, as shown previously.[25] SOD also decreased the rate of QR-induced redox cycling of 2-hydroxy-1,4-naphthoquinone but had little or no effect on that of 2-amino- and 5-hydroxy-1,4-naphthoquinone.[19]

The rates of autoxidation of different naphthohydroquinones and the effects of QR and SOD on redox cycling of naphthoquinones can be explained by reference to the mechanism of autoxidation of naphthohydroquinones.

Mechanism of Naphthohydroquinone Autoxidation

Ionization of the hydroquinone (QH_2) (reaction 1) is a prerequisite for autoxidation.[26–28] Hydroquinones with relatively low acid dissociation constants, such as the hydroxy derivatives, are therefore oxidized more rapidly

[25] E. Cadenas, D. Mira, A. Brunmark, C. Lind, J. Segura-Aguilar, and L. Ernster, *Free Rad. Biol. Med.* **5,** 71 (1988).
[26] T. Ishii and I. Fridovich, *Free Rad. Biol. Med.* **8,** 21 (1990).
[27] P. Eyer, *Chem.-Biol. Interact.* **80,** 159 (1991).
[28] B. Bandy, J. Moon, and A. J. Davison, *Free Rad. Biol. Med.* **9,** 143 (1990).

than less acidic hydroquinones.[17] Oxidation is initiated by reaction of the anion with molecular oxygen (reaction 2), forming superoxide and the semiquinone ($Q^{\bullet-}$). The rate of this reaction is related to the half-wave potential of the hydroquinone,[17] and compounds with negative potentials (such as 2-amino and 2-hydroxy-1,4-naphthohydroquinone) oxidize faster than those with more positive potentials (such as 2-methyl and 2,3-dimethoxy-1,4-naphthohydroquinone). Reaction of the semiquinone with molecular oxygen via reaction 3 yields the quinone (Q). More semiquinone is formed through the comproportionation reaction between the quinone and hydroquinone (reaction 4), while superoxide generated in reactions 2 and 3 may initiate a radical chain reaction by oxidizing the hydroquinone anion, with formation of the semiquinone and hydrogen peroxide (reaction 5).

$$QH_2 \rightleftharpoons QH^- + H^+ \tag{1}$$

$$QH^- + O_2 \rightarrow Q^{\bullet-} + O_2^{\bullet-} + H^+ \tag{2}$$

$$Q^{\bullet-} + O_2 \rightleftharpoons Q + O_2^{\bullet-} \tag{3}$$

$$Q + QH^- \rightleftharpoons Q^{\bullet-} + H^+ \tag{4}$$

$$QH^- + O_2^{\bullet-} + H^+ \rightarrow Q^{\bullet-} + H_2O_2 \tag{5}$$

With pure hydroquinones, oxygen uptake accelerates after a brief lag phase.[17] The initial rate of reaction reflects the rate of the initiation reaction (reaction 1), and the acceleration after some hydroquinone is oxidized is due to the contribution of superoxide (via reaction 5) and/or quinone (via reaction 4) to the oxidative process. If, however, the quinone is re-reduced to the hydroquinone by QR, reaction 4 will be eliminated and the overall rate of oxidation decreased. This appears to be the case only with naphthoquinones that are reduced comparatively rapidly, forming hydroquinones that are oxidized comparatively slowly, such as 2-methyl-, 2,3-dimethyl- and 2,3-dimethoxy-1,4-naphthoquinone. With the 2-hydroxy and 2-amino derivatives, which are reduced slowly and oxidized rapidly, the quinone cannot be reduced fast enough to eliminate reaction 4 and hence QR does not inhibit hydroquinone oxidation. Furthermore, if reaction 5 is important in the oxidation reaction, SOD will decrease the rate, as seen with alkyl-, amino- and 2-hydroxy-naphthohydroquinones.[17] If this reaction is not significant in promoting hydroquinone oxidation, SOD will not inhibit, and this is the case with 5-hydroxy-1,4-naphthohydroquinone and unsubstituted naphthohydroquinone.[17]

With QR in the presence of the naphthoquinone rather than the naphthohydroquinone, the situation is rather different. At low levels of enzyme, the quinone will be only partially reduced, and both quinone and hydroquinone will be present in solution. Oxidation of the latter will be facilitated by reaction 4, and redox cycling will occur. As the concentration of QR is increased, however, the concentration of quinone will decrease, and with the rapidly reduced quinones, its concentration may be reduced to very low levels. In this situation, reaction 4 will be effectively eliminated and redox cycling inhibited, as seen in Fig. 2 for menadione. Furthermore, with hydroquinones for which reaction 5 is important, the rate of redox cycling will be further decreased in the presence of both QR and SOD, since both reaction 4 and reaction 5 will be eliminated. With slowly reduced naphthoquinones, such as the 2-hydroxy derivative, increasing levels of QR will increase the concentration of hydroquinone in solution, but quinone will never be eliminated completely. The rate of redox cycling will therefore increase with increasing enzyme concentration, as shown in Fig. 2 for 2-hydroxy-1,4-naphthoquinone. Furthermore, SOD will have comparatively little effect on the rate of redox cycling by these compounds, because although reaction 5 will be eliminated, reaction 4 will continue unabated.

Two Classes of Naphthoquinone

From the foregoing experiments, two classes of naphthoquinone can be distinguished, as detailed later.

Naphthoquinones That Form Hydroquinones That Can Be Stabilized by QR

With these compounds, the rate of redox cycling decreases with increasing QR activity. Such compounds, which include menadione, 2,3-dimethyl-1,4-naphthoquinone and 2,3-dimethoxy-1,4-naphthoquinone, would be expected to be detoxified by QR. If this is the case, increasing cellular or tissue levels of QR would decrease toxicity, while inhibition of the enzyme would lead to an increase in the severity of the harmful effects.

Naphthoquinones That Form Hydroquinones That Cannot Be Stabilized by QR

With these compounds, which include 2-hydroxy-, 5-hydroxy- and 2-amino-1,4-naphthoquinone, the rate of redox cycling increases with increasing QR activity. Such compounds would be expected to be activated by QR through facilitation of redox cycling by the enzyme. In this situation, naphthoquinone toxicity would be increased by increasing cellular or tissue levels of QR, while toxicity would be decreased by inhibiting the enzyme.

The utility of these classifications in rationalizing the results of experiments on naphthoquinone toxicity *in vitro* and *in vivo* is discussed later.

Cytotoxicity of Naphthoquinones *In Vitro*

Naphthoquinones are toxic to a variety of cell types *in vitro*. The role of QR in cytotoxicity has been explored by comparing cells with different levels of QR, by selecting clones of one cell type with disparate levels of QR and by transfection of cells with QR cDNA.[29-32] The effect of decreasing cellular levels of QR has been investigated by inhibiting the enzyme. In most experiments, dicumarol has been employed as inhibitor, although xanthenone and flavone derivatives have also been used.[33]

The effects of differences in cellular QR activities on the cytotoxicity of various naphthoquinone derivatives are summarized later.

Menadione

This is by far the most intensively studied naphthoquinone, whose toxicity has been evaluated in a wide range of cells *in vitro,* with toxicity being assessed in terms of cell death, membrane damage, mutagenicity and DNA strand breaks. In 33 of the 34 studies reported with this material, cells high in QR were shown to be resistant to the cytotoxic effects of menadione. Furthermore, in all but one experiment, QR inhibitors increased toxicity. The only exceptions to these findings were a study in MCF-7 cells, in which menadione toxicity was decreased by dicumarol,[34] and another in NIH 3T3 cells, in which transfection with QR did not decrease toxicity.[35]

The results on the cytotoxicity studies are thus largely in accord with expectation, with menadione being detoxified by QR. Whether such detoxification simply results from inhibition of redox cycling through maintenance of the hydroquinone in the reduced form (thereby preventing both redox

[29] D. Ross, J. K. Kepa, S. L. Winski, H. D. Beall, A. Anwar, and D. Siegel, *Chem.-Biol. Interact.* **129,** 77 (2000).

[30] P. L. Gutierrez, *Free Rad. Biol. Med.* **29,** 263 (2000).

[31] T.-J. Chiou and W.-F. Tzeng, *Toxicology* **154,** 75 (2000).

[32] P. Joseph, D. J. Long, A. J. P. Klein-Szanto, and A. K. Jaiswal, *Biochem. Pharmacol.* **60,** 207 (2000).

[33] R. M. Phillips, *Biochem. Pharmacol.* **58,** 303 (1999).

[34] L. M. Nutter, E. O. Ngo, G. R. Fisher, and P. L. Gutierrez, *J. Biol. Chem.* **267,** 2474 (1992).

[35] G. Powis, P. Y. Gasdaska, A. Gallegos, K. Sherrill, and D. Goodman, *Anticancer Res.* **15,** 1141 (1995).

cycling and arylation), or whether the hydroquinone is further inactivated by conjugation is not presently known. Similarly, whether the increased toxicity of menadione through inhibition of QR is due to redox cycling promoted by enzymes catalyzing the one-electron reduction of the quinone or is due to arylation is not known. Since dicumarol does not totally inhibit QR in cells,[36] it is also possible that, in the presence of this substance, residual levels of QR are within the window of activity in which redox cycling is promoted by this enzyme.

Unsubstituted 1,4-Naphthoquinone

All 6 cytotoxicity studies with this substance showed decreased toxicity in cells with increased cellular levels of QR and increased toxicity in the presence of dicumarol. Although no data on the effect of different levels of QR on the rate of redox cycling are available for this compound, it falls in the group for which detoxification is to be expected, since 1,4-naphthoquinone is rapidly reduced by QR, yielding a hydroquinone that is stabilized by the enzyme.

2,3-Dimethyl-1,4-Naphthoquinone

Two studies on the toxicity of this substance to hepatocytes have been reported.[37,38] Both showed an increase in toxicity after addition of dicumarol, suggesting that QR is detoxifying the quinone. Again, detoxification is to be expected from the rates of naphthoquinone reduction and hydroquinone oxidation and the ability of QR to inhibit redox cycling.

2,3-Dimethoxy-1,4-Naphthoquinone

Again, the biochemical properties of this substance would predict detoxification by QR. Experiments in lymphoblasts[39] and in Caco-2 and HT-29 cells[36] are in accord with this prediction.

2-Hydroxy- and 5-Hydroxy-1,4-Naphthoquinone

It would be expected that these compounds would be activated by QR. No data on the effect of increased cellular QR on the toxicity of the hydroxy compounds are available, but experiments with dicumarol are not in

---

[36] J. M. Karczewski, J. G. P. Peters, and J. Noordhoek, *Biochem. Pharmacol.* **57,** 27 (1999).

[37] D. Ross, H. Thor, M. D. Threadgill, M. S. Sandy, M. T. Smith, P. Moldéus, and S. Orrenius, *Arch. Biochem. Biophys.* **248,** 460 (1986).

[38] M. G. Miller, A. Rodgers, and G. M. Cohen, *Biochem. Pharmacol.* **35,** 1177 (1986).

[39] A. Halinska, T. Belej, and P. J. O'Brien, *Br. J. Cancer* **74** (Suppl. 27), S23 (1996).

accord with expectation. In an experiment with hepatocytes, toxicity was increased in the presence of dicumarol,[12] while no effect was seen in HepG2 or 3T3 cells,[40] rather than the expected decrease.

5,8-Dihydroxy-1,4-Naphthoquinone

This compound is slowly reduced by QR,[24] and although no data on the rate of oxidation of the corresponding hydroquinone are available, by analogy with the 2-hydroxy and 5-hydroxy compounds, rapid oxidation would be expected. Activation by QR would therefore be predicted. However, results on the effects of modulating cellular levels of QR on the cytotoxicity of 5,8-dihydroxy-1,4-naphthoquinone are conflicting. In hepatocytes, toxicity was increased by dicumarol,[12] while inhibition of QR had no effect in HepG2 and 3T3 cells.[40] In lymphoblasts, however, toxicity was increased in cells containing high levels of QR and was decreased by dicumarol.[39]

3,4-Dihydro-2,2-Dimethyl-2H-Naphtho[1,2-b]Pyran-5,6-Dione (β-Lapachone)

No data on the rate of reduction of this 1,2-naphthoquinone by QR or on the rate of autoxidation of the hydroquinone so formed are available. However, the rate of redox cycling of this substance increased with increasing QR activity, with no inhibition at high levels of the enzyme (Fig. 3), suggesting that this compound would be activated by QR. In accord with expectation, the toxicity of β-lapachone was increased in cells containing high levels of QR and decreased in the presence of dicumarol.[9]

5-Amino-6-(7-Amino-5,8-Dihydro-6-Methoxy-5,8-Dioxo-2-Quinolinyl)- 4-(2-Hydroxy-3,4-Dimethoxyphenyl)-3-Methyl-2-Pyridine Carboxylic Acid (Streptonigrin)

Streptonigrin is structurally related to the naphthoquinones. It is a quinolinequinone, with carbon-5 of the aromatic ring being replaced by nitrogen. It is fairly rapidly reduced by QR,[41] but no information on the rate of autoxidation of the hydroquinone is available. As in the case of β-lapachone, the rate of redox cycling of this substance increased with increasing activity of QR (Fig. 4), suggesting that streptonigrin would be activated by QR. This is in accord with the results of cytotoxicity studies, which consistently show increased toxicity in cells high in QR and decreased toxicity after inhibition of the enzyme.[42,43]

---

[40] H. Babich and A. Stern, *J. Appl. Toxicol.* **13**, 353 (1993).
[41] H. D. Beall, R. T. Mulcahy, D. Siegel, R. D. Traver, N. W. Gibson, and D. Ross, *Cancer Res.* **54**, 3196 (1994).

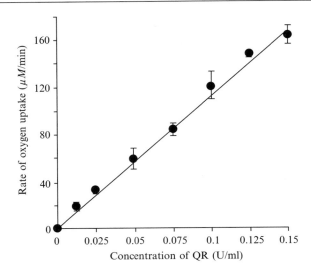

FIG. 3. Redox cycling by β-lapachone in the presence of various concentrations of QR. The rate of redox cycling was measured as rate of oxygen uptake in the presence of 1 mM NADH, 50 μM DTPA and 0.25 M sucrose in 0.05 M phosphate buffer at pH 7.4 and 25°. The concentration of quinone was 150 μM.

In general, the results of the cytotoxicity experiments support the concept that naphthoquinones forming hydroquinones that can be stabilized by QR are inactivated by the enzyme, while those giving hydroquinones that are not stabilized by QR are activated. Some anomalies obviously exist, particularly with regard to the hydroxynaphthoquinones, and more work with these substances is required.

Toxicity of Naphthoquinones In Vivo

In vivo, two major pathological changes have been recorded in animals and humans dosed with naphthoquinone derivatives. First, these substances are toxic to red blood cells, causing hemolytic anemia. Hemolysis appears to be a general property of naphthoquinones, having been observed with alkylnaphthoquinones,[44] dialkynaphthoquinones,[45]

[42] H. D. Beall, Y. Liu, D. Siegel, E. M. Bolton, N. W. Gibson, and D. Ross, *Biochem. Pharmacol.* **51,** 645 (1996).

[43] S. Y. Sharp, L. R. Kelland, M. R. Valenti, L. A. Brunton, S. Hobbs, and P. Workman, *Mol. Pharmacol.* **58,** 1146 (2000).

[44] R. Munday, E. A. Fowke, B. L. Smith, and C. M. Munday, *Free Rad. Biol. Med.* **16,** 725 (1994).

[45] R. Munday, B. L. Smith, and C. M. Munday, *Free Rad. Biol. Med.* **19,** 759 (1995).

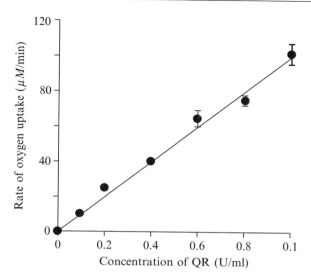

Fig. 4. Redox cycling by streptonigrin in the presence of various concentrations of QR. The rate of redox cycling was measured as rate of oxygen uptake in the presence of 1 mM NADH, 50 μM DTPA and 0.25 M sucrose in 0.05 M phosphate buffer at pH 7.4 and 25°. The concentration of quinone was 150 μM.

hydroxynaphthoquinones[46,47] and aminonaphthoquinones.[48] The hemolysis is of the oxidative type, characterized by the presence of Heinz bodies within the erythrocytes, and there is evidence for the involvement of "active oxygen" species in initiating red cell damage. Second, some naphthoquinones are nephrotoxic, causing necrosis of the renal tubular epithelium. This effect has not been recorded with alkylnaphthoquinones, even at high dose-levels,[49] but slight renal damage was observed with high doses of 2,3-dimethyl-1,4-naphthoquinone.[50] In contrast, the hemolytic action of 2-hydroxy-1,4-naphthoquinone,[46] 2-alkyl-3-hydroxy-1,4-naphthoquinones[47] and 2-amino-1,4-naphthoquinone[48] is invariably accompanied by renal tubular necrosis. The mechanism of the renal toxicity of naphthoquinones is unknown.

Comparatively few studies on the role of QR in naphthoquinone toxicity *in vivo* have been reported, with detailed studies only on menadione,

[46] R. Munday, B. L. Smith, and E. A. Fowke, *J. Appl. Toxicol.* **11,** 85 (1991).
[47] R. Munday, B. L. Smith, and C. M. Munday, *Chem.-Biol. Interact.* **98,** 185 (1995).
[48] R. Munday, B. L. Smith, and C. M. Munday, manuscript in preparation.
[49] R. Munday, B. L. Smith, and C. M. Munday, *Chem.-Biol. Interact.* **108,** 155 (1998).
[50] R. Munday, B. L. Smith, and C. M. Munday, *Chem.-Biol. Interact.* **134,** 87 (2001).

2,3-dimethyl-1,4-naphthoquinone and 2-hydroxy- and 2-amino-1,4-naphtho-quinone. QR-knockout mice were employed in two studies, while the effect of modulating QR activity *in vivo* on the toxicity of naphthoquinones has been explored by increasing tissue levels of the enzyme by use of inducers (butylated hydroxyanisole [BHA], butylated hydroxytoluene [BHT], ethoxyquin [EQ], dimethyl fumarate [DMF] and disulfiram [DIS]) and by inhibiting the enzyme by administration of dicumarol. It was shown[51] that none of these compounds *per se* influenced parameters associated with hemolysis or nephrotoxicity.

Menadione

The acute toxicity of menadione was higher in QR-knockout mice than in wild-type controls,[52] as was the degree of hepatic lipid peroxidation in animals dosed with this quinone.[32] The severity of menadione-induced hemolytic anemia was decreased in animals with high tissue levels of QR and increased in those pre-treated with dicumarol.[49,51]

2,3-Dimethyl-1,4-Naphthoquinone

The hemolytic action of this substance was decreased in animals with high levels of QR, and the slight nephrotoxicity of 2,3-dimethyl-1, 4-naphthoquinone was eliminated in animals treated with QR inducers. Di-cumarol caused only a marginal increase in the severity of hemolysis but greatly increased the severity of renal tubular necrosis.[50]

2-Hydroxy-1,4-Naphthoquinone

Hemolysis was increased in rats treated with inducers of QR and then challenged with 2-hydroxy-1,4-naphthoquinone. In contrast, there was no significant change in the severity of the renal lesions in rats receiving BHA, BHT or EQ, while the nephrotoxicity of 2-hydroxy-1,4-naphthoquinone was decreased in rats pre-dosed with DMF and DIS.[51,53]

2-Amino-1,4-Naphthoquinone

Pronounced increases in the severity of hemolysis were recorded in rats with high tissue activities of QR. However, the extent of renal tubular necrosis was significantly decreased in animals pre-treated with all the QR inducers.[48]

[51] R. Munday, B. L. Smith, and C. M. Munday, *Chem.-Biol. Interact.* **123**, 219 (1999).
[52] V. Radjendirane, P. Joseph, Y.-H. Lee, S. Kimura, A. J. P. Klein-Szanto, F. J. Gonzalez, and A. K. Jaiswal, *J. Biol. Chem.* **273**, 7382 (1998).
[53] R. Munday, B. L. Smith, and C. M. Munday, *Chem.-Biol. Interact.* **117**, 241 (1999).

In the case of menadione, the results of the *in vivo* studies on acute toxicity, lipid peroxidation and hemolysis are again in accord with a detoxification role for QR. The hemolytic anemia induced by this substance, and by other alkylnaphthoquinones, is initiated by one-electron reduction of the quinone in the erythrocyte by oxyhemoglobin.[44] The "active oxygen" species so formed cause oxidative damage to the red cells, leading to their destruction. Reduction of menadione to the hydroquinone, and the maintenance of the latter in the reduced form, would favor conjugation reactions and thus the elimination of the toxic material from the body. Conversely, inhibition of QR would increase the residence time of the quinone, increasing the extent of interaction with erythrocytes and thus increasing the severity of hemolysis.

2,3-Dimethyl-1,4-naphthoquinone is also detoxified by QR, as shown by decreased hemolysis and nephrotoxicity in rats with elevated tissue levels of QR. Inhibition of the enzyme increased toxicity, but the target site specificity was also altered, with a far greater increase in renal damage than in erythrocyte destruction.

The situation is more complex with the hydroxy- and amino-naphthoquinones. The presence of Heinz bodies in the erythrocytes of animals treated with these substances is again consistent with "active oxygen" generation, but unlike menadione, these substances do not generate such species in erythrocytes *in vitro,* indicating that metabolic activation is required.[47] The increase in hemolysis seen with these substances in animals with high QR activities is consistent with activation by this enzyme. However, while hemolysis was increased, nephrotoxicity was decreased, indicating that QR in some way protects the kidney from the harmful effects of the quinones. It has been suggested that activation of these quinones takes place in the intestine, with "active oxygen" species crossing the gut wall into the circulation, where they would initiate hemolysis.[51] It is possible that this redox cycling in the gastrointestinal tract slows the uptake of the quinone and thus decreases the amount reaching the kidney. Studies on the effect of inducers of QR on the metabolism and distribution of the hydroxy- and amino-naphthoquinones are required.

Conclusion

QR, at appropriate concentrations, inhibits the redox cycling of certain naphthoquinones, and the results of both *in vitro* and *in vivo* experiments indicate that such naphthoquinones are detoxified by QR. In contrast, redox cycling of other naphthoquinones is promoted by QR, and there is evidence that the toxicity of these compounds is increased in cells and in animals with high levels of this enzyme. It should be noted, however, that modulation of QR activities *in vivo* may modify not only the severity of the toxic effects of the naphthoquinone but also the target site specificity.

It is interesting to note that promotion of redox cycling by QR is a common feature of certain anti-cancer drugs such as β-lapachone and streptonigrin. On the basis of structural similarities, this behavior is also to be expected with 2-hydroxy-3-methyl-1,4-naphthohydroquinone and 5,8-dihydroxy-1,4-naphthoquinone, which also show anti-cancer activity.[6,8] It is possible that *in vitro* experiments on QR-induced redox cycling could provide a useful initial screen for identification of potential anti-tumor quinones, particularly compounds that may show selectivity toward tumors with high activities of QR.[29,54]

[54] R. J. Riley and P. Workman, *Biochem. Pharmacol.* **43**, 1657 (1992).

[21] Induction of Quinone Reductase as a Primary Screen for Natural Product Anticarcinogens

By YOUNG-HWA KANG and JOHN M. PEZZUTO

Introduction

Cancer is a complicated group of diseases characterized by the uncontrolled growth and spread of abnormal cells.[1] It is anticipated that about 1.3 million new cases of cancer will be diagnosed in the United States this year, and about 555,500 persons are expected to die of the disease.[2] Furthermore, despite small decreases in overall cancer incidence and mortality rates in the United States since the early 1990s, the total number of recorded cancer deaths continues to increase because of an aging and expanding population. Importantly, cancer can be considered as the end stage of a chronic disease process characterized by abnormal cell and tissue differentiation.[3] This chronic process, carcinogenesis, eventually leads to

[1] W. B. Pratt, R. W. Ruddon, W. D. Ensminger, and J. Maybaum, "The Anticancer Drugs." Oxford University Press: Oxford, UK, 1994.

[2] A. Jemal, A. Thomas, T. Murray, and M. Thun, *Cancer J. Clin.* **52**, 23 (2002).

[3] D. S. Alberts, G. M. Colvin, A. H. Conney, V. L. Ernster, J. E. Garber, P. Greenwald, L. J. Gudas, W. K. Hong, G. J. Kelloff, R. A. Kramer, C. E. Lerman, D. J. Mangelsdorf, A. Matter, J. D. Minna, W. G. Nelson, J. M. Pezzuto, F. Prendergast, V. W. Rusch, M. B. Sporn, L. W. Wattenberg, and I. B. Weinstein, *Cancer Res.* **59**, 4743 (1999).

the final outcome of invasive and metastatic cancer.[4,5] Recent advances that have defined the cellular and molecular events associated with carcinogenesis, along with a growing body of experimental, epidemiological and clinical trial data, provide a foundation for relatively new strategies of cancer prevention.[6–10] One such strategy involves suppression of carcinogen metabolic activation or blocking the formation of ultimate carcinogens.[11] In particular, the induction of phase II enzymes can offer protection against toxic and reactive chemical species.[12–15] Many recent studies have shown that elevation of phase II enzymes correlates with protection against chemical-induced carcinogenesis in animal models.[16–18]

NAD(P)H:quinone reductase (NQO1), one of the phase II drug-metabolizing enzymes, plays an important role in the mechanism of cancer chemoprevention, presumably at the initiation stage of carcinogenesis.[19,20] Since induction of quinone reductase can be detected readily by using murine hepatoma Hepa 1c1c7 cells, this cell line has been used for the discovery of novel natural product anticarcinogens.[12] A large number of natural product quinone reductase inducers are known,[16,21–23] and several

[4] G. J. Kelloff, C. W. Boone, V. E. Steele, J. R. Fay, and C. C. Sigman, in "Chemical Induction of Cancer" (J. C. Arcos, ed.), p. 401. Birkauser: Basel, 1995.

[5] C. C. Harris, Cancer Res. **51** (Suppl.), 5023S (1991).

[6] G. J. Kelloff, C. C. Sigman, and P. Greenwald, Eur. J. Cancer **35**, 1755 (1999).

[7] T. W. Kensler, W. Thomas, Environ. Health Perspect. **105**, 965 (1997).

[8] P. Talalay, J. W. Fahey, W. D. Holtzclaw, T. Prestera, and Y. Zhang, Toxicol. Lett. **82/83**, 173 (1995).

[9] G. J. Kelloff, C. C. Sigman, and P. Greenwald, Eur. J. Cancer **35**, 1755 (1999).

[10] M. A. Morse and G. D. Stoner, Carcinogenesis **4**, 1737 (1993).

[11] L. Wattenberg, Cancer Res. **45**, 1 (1985).

[12] P. Talalay, Biofactors **12**, 5 (2000).

[13] P. L. Gutierrez, Free Radic. Biol. Med. **29**, 263 (2000).

[14] A. Begleiter, M. K. Leith, T. J. Curphey, and G. P. Doherty, Oncol. Res. **9**, 371 (1997).

[15] H. J. Prochaska, A. B. Santamaria, and P. Talalay, Proc. Natl. Acad. Sci. USA **89**, 2394 (1992).

[16] J. M. Pezzuto, L. L. Song, S. K. Lee, L. A. Shamon, E. Mata-Greenwood, M. Jang, H. J. Jeong, E. Pisha, R. G. Mehta, and A. D. Kinghorn, in "Chemistry, Biological and Pharmacological Properties of Medicinal Plants from the Americas" (K. Hostettmann, M. P. Gupta, and A. Marston, eds.), p. 81. Chur, Switzerland: Harwood Academic Publishers, 1998.

[17] P. Talalay, Adv. Enzyme Regul. **28**, 237 (1989).

[18] C. W. Boone, V. E. Steele, and G. J. Kelloff, Mutat. Res. **267**, 251 (1992).

[19] J. M. Pezzuto, J. W. Kosmeder, E. J. Park, S. K. Lee, M. Cuendet, J. Gills, K. Bhat, S. Grubjesic, H. S. Park, E. Mata-Greenwood, Y. M. Tan, R. Yu, D. D. Lantvit, and A. D. Kinghorn, in "Strategies for Cancer Chemoprevention" (G. J. Kelloff, E. T. Hawk, and C. C. Sigman, eds.), Totowa, New Jersey: The Humana Press, In Press.

[20] P. Talalay, Acad. Med. **76**, 238 (2001).

[21] K. Hashimoto, S. Kawamata, N. Usui, A. Tanaka, and Y. Uda, Cancer Lett. **180**, 1 (2002).

[22] P. Talalay, in "Cellular Targets for Chemoprevention" (V. Steele, ed.), p. 193. Boca Raton: CRC Press, 1992.

laboratories throughout the world have been involved in this area of drug discovery.[24–27] We currently describe phytochemical inducers that include flavonoids, steroids, isothiocyanates, and dithiolethiones, with special emphasis on compounds that have been reported by our group. In addition, details of using the quinone reductase induction assay as a monitor for drug discovery are provided, as well as a strategy for the discovery of active compounds from plant sources.

Role of Phase II Enzymes in Cancer

It is generally accepted that carcinogenesis is a multistage process that may be caused by carcinogen-induced genetic and epigenetic damage in susceptible cells.[28] The first stage of the carcinogenic process, tumor initiation, involves exposure of normal cells to electrophilic carcinogen metabolites or reactive oxygen species.[29,30] One of the major mechanisms to protect against the toxic electrophilic metabolites of carcinogens and reactive oxygen is induction of phase II detoxification enzymes such as glutathione S-transferases, UDP-glucuronosyltransferases, and NAD(P)H: quinone reductase (NQO1).[13,31–33] Enzyme inducers are of two types: monofunctional and bifunctional.[34] Bifunctional inducers increase phase II enzymes as well as phase I enzymes such as aryl hydrocarbon hydroxylase, and bind with high affinity to the aryl hydrocarbon receptor.[35,36] Monofunctional inducers induce phase II enzymes selectively and are independent of the aryl hydrocarbon receptor. Since phase I enzymes can activate procarcinogens to their ultimate reactive species, monofunctional

[23] S. K. Lee, L. Song, E. Mata-Greenwood, G. J. Kelloff, V. E. Steele, and J. M. Pezzuto, *Anticancer Res.* **19**, 35 (1999).

[24] A. D. Kinghorn, B. N. Su, D. Lee, J. Q. Gu, and J. M. Pezzuto, *Curr. Org. Chem.* **7**, 213 (2003).

[25] A. D. Kinghorn, H. H. S. Fong, N. R. Farnsworth, R. G. Mehta, R. C. Moon, R. M. Moriarty, and J. M. Pezzuto, *Curr. Org. Chem.* **2**, 597 (1998).

[26] Y. Zhang, P. Talalay, C. G. Cho, and G. H. Posner, *Proc. Natl. Acad. Sci. USA* **89**, 2399 (1992).

[27] Y. H. Heo, S. Kim, J. E. Park, L. S. Jeong, and S. K. Lee, *Arch. Pharm. Res.* **24**, 597 (2001).

[28] M. B. Sporn, *Lancet* **347**, 1377 (1996).

[29] H. J. Prochaska and P. Talalay, *in* "Oxidative Stress. Oxidants and Antioxidants" (H. Sies, ed.), p. 195. London: Academic Press, 1991.

[30] R. Li, M. A. Bianchet, P. Talalay, and L. M. Amzel, *Proc. Natl. Acad. Sci. USA* **12**, 8846 (1995).

[31] D. Ross, J. K. Kepa, S. L. Winski, H. D. Beall, A. Anwar, and D. Siegel, *Chem.-Biol. Interact.* **129**, 77 (2000).

[32] P. Talalay, M. J. Long, and H. J. Prochaska, *Proc. Natl. Acad. Sci. USA* **85**, 8261 (1988).

[33] G. Cantelli-Forti, P. Hrelia, and P. Paolini, *Mut. Res.* **402**, 179 (1998).

[34] H. J. Prochaska and P. Talalay, *Cancer Res.* **48**, 4776 (1988).

[35] K. Sogawa and Y. Fujii-Kuriyama, *J. Biochem. (Tokyo)* **122**, 1075 (1997).

[36] C. S. Yang, T. J. Smith, and J. Y. Hong, *Cancer Res.* **54**, 1982s (1994).

agents that induce phase II enzymes selectively would theoretically appear to be more desirable candidates for cancer chemoprotection.[12] In addition, selective phase II enzyme inducers would be anticipated to serve as anticarcinogens early in the process of carcinogenesis.

Quinone Reductase Induction and Cancer Chemoprevention

Cancer chemoprevention is an approach to reducing cancer risk by using natural or synthetic compounds to suppress, prevent, or reverse the process of carcinogenesis.[3,9,37–40] The term *cancer chemoprevention* was coined by Dr. Michael B. Sporn to distinguish between preventing the occurrence of cancer and cancer chemotherapy.[41] Relative to chemotherapy, the prevention of cancer is a preferable option. Chemopreventive agents have been further classified into "blocking" and "suppressing" agents by Wattenberg.[7] Suppressing agents are chemopreventive chemicals that inhibit the progression of the overall neoplastic process and prevent cells from becoming malignant. Blocking agents can prevent the activation of carcinogens, enhance detoxification, or trap reactive carcinogens prior to the initiation of target organ damage.

Bearing in mind the importance of phase II enzyme induction in cancer chemoprevention, methods to determine phase II enzyme inducer potencies of pure compounds and extracts of natural products are necessary. Quinone reductase is a flavoprotein that catalyzes the two-electron reduction of electrophilic quinones into stable hydroquinones and reduces oxidative cycling.[29] It is a representative phase II detoxifying enzyme based on wide distribution in mammalian tissues. The enzyme shows a strong induction response and is easily measured by a coupled tetrazolium dye reduction assay.[12] In addition, with *in vitro* and *in vivo* systems, induction has been shown to correlate with elevation of other protective phase II enzymes,[42] and induction provides a reasonable biomarker for the

[37] G. J. Kelloff, J. A. Crowell, E. T. Hawk, V. E. Steele, R. A. Lubet, C. W. Boone, J. M. Covey, L. A. Doody, G. S. Omenn, P. Greenwald, W. K. Hong, D. R. Parkinson, D. Bagheri, G. T. Baxter, M. Blunden, M. K. Doeltz, K. M. Eisenhauer, K. Johnson, G. G. Knapp, D. G. Longfellow, W. F. Malone, S. G. Nayfield, H. E. Seifried, L. M. Swall, and C. C. Sigman, *J. Cell. Biochem.* **26** (Suppl.), 54 (1996).

[38] M. B. Sporn and N. Suh, *Carcinogenesis* **21,** 525 (2000).

[39] P. Greenwald, G. J. Kelloff, C. Burch-Whitman, and B. S. Kramer, *CA Cancer J. Clin.* **45,** 31 (1995).

[40] R. I. Sanchez, S. Mesia-Vela, and F. C. Kauffman, *Curr. Cancer Drug Targets* **1,** 1 (2001).

[41] M. B. Sporn, N. M. Dunlop, D. L. Newton, and J. M. Smith, *Fed. Proc.* **35,** 1331 (1976).

[42] J. M. Pezzuto, *in* "Recent Advances in Phytochemistry, Phytochemistry of Medicinal Plants" (J. T. Arnason, R. Mata, and J. T. Romeo, eds.), Vol. 29, p. 19. New York: Plenum Press, 1995.

potential chemoprotective effect of test agents against the initiation of cancer.

The murine hepatoma cell line Hepa 1c1c7 contains inducible quinone reductase that is easily measurable and provides a reliable, high-throughput system for the detection of induction.[15] This assay can also be used to determine if an agent induces phase II enzymes only (monofunctional) or both phase I and II enzymes (bifunctional). This is accomplished by comparing the induction capability of a compound in wild-type Hepa 1c1c7 cells with that observed in two mutant cell lines, TAOc1 and BP^{r}c1, which are defective in a functional aryl hydrocarbon receptor or unable to translocate the receptor-ligand complex to the nucleus, respectively.[43] Compounds that have similar inducing ability in the wild-type and mutant Hepa cell lines are considered monofunctional inducers.

In Vitro Quinone Reductase Assay

The assay we have used for detecting the induction of cellular quinone reductase is based on the method described by Prochaska et al.[44] with some modifications. This system provides a highly quantitative and reproducible method for determining inducer potencies of pure compounds and plant extracts and can be used for evaluating a large number of natural products as a high-throughput system that uses 96-well plates. Quinone reductase–specific activity is determined by measuring NADPH-dependent menadiol-mediated reduction of 3-(4,5-dimethylthiazo-2-yl)-2,5-diphenyl-tetrazolium bromide (MTT) to a blue fromazan. Protein is determined by crystal violet staining with an identical set of test plates. Details of the procedure follow.

Materials

Chemicals were purchased from Sigma Chemical Co. (St. Louis, MO), and cell culture media and supplements were obtained from Life Technologies (Grand Island, NY).

Digitonin: D5628
Glucose 6-phosphate: G7772
Menadione: M5625
NADP$^+$: N0505
FAD$^+$: F6625

---

[43] C. Gerhäuser, S. K. Lee, J. W. Kosmeder, R. M. Moriarty, E. Hamel, R. G. Mehta, R. C. Moon, and J. M. Pezzuto, Cancer Res. **57,** 3429 (1997).

[44] H. J. Prochaska and A. B. Santamaria, Anal. Biochem. **169,** 328 (1988).

BSA: A7906
Glucose 6-phosphate dehydrogenase: G6378
3-(4,5-Dimethylthiazo-2-yl)-2,5-diphenyltetrazolium bromide (MTT): M2128
Tween 20: P1379
Tris-base: T3253
Crystal violet: C3886
SDS: L4509

Methods

Cell Cultures

Hepa 1c1c7 murine hepatoma cells and mutants were maintained in α-MEM (without ribonucleosides or deoxyribonucleosides) supplemented with 10% fetal calf serum (Atlanta Biologicals), 100 units of penicillin G sodium/ml and 100 μg of streptomycin/ml, at 37 °C in a humidified atmosphere at 5% CO_2 in air.

Assay Procedure

1. Hepa lclc7 cells are seeded in 96-well plates at a density of 10,000 cells/ml in 200 μl of α-minimal essential medium supplemented with 10% fetal calf serum. The cells are grown for 24 h in a humidified incubator in 5% CO_2 at 37 °C.

2. The medium is decanted after a 24 h incubation, 190 μl fresh medium and 10 μl of 10% DMSO containing test samples are added to each well. The cells are incubated for an additional 48 h.

3. After the cells are treated with test samples for 48 h, the medium is decanted and the cells are incubated at 37° for 10 min with 50 μl of 0.8% digitonin and 2 mM EDTA solution (pH 7.8).

4. The plates are then agitated on an orbital shaker (100 rpm) for 10 min at room temperature and 200 μl of reaction mixture (see later) are added to each well. Menadione solution (1 μl of 50 mM menadione dissolved in acetonitrile per ml of reaction mixture) is added just before the mixture is dispensed into the microtiter plates.

5. The reaction generates a blue color, and this is arrested after 5 min by the addition of 50 μl of a solution containing 0.3 mM dicoumarol in 0.5% DMSO and 5 mM potassium phosphate, pH 7.4.

6. The plates are then scanned at 595 nm. The first column of wells in each plate normally contains the reaction mixture only and serves as the nonenyzmatic blank. The second column of wells are the control cells treated with medium containing 0.5% DMSO. β-Naphthoflavone

(bifunctional inducer) and BHA (monofunctional inducer) are used as positive control groups.

Reaction Mixture

The following stock solution (150 ml) is prepared for each set of assays:

7.5 ml of 0.5 M Tris-HCl (pH 7.4)
100 mg of bovine serum albumin
1 ml of 1.5% Tween-20
0.1 ml of 7.5 mM FAD$^+$
1 ml of 150 mM glucose 6-phosphate
90 μl of 50 mM NADP$^+$
300 U of yeast glucose 6-phosphate dehydrogenase
45 mg of MTT
150 μl of 50 mM menadione
Distilled water to volume

Protein Determination by Crystal Violet Staining

Since some substances being tested for inducer activity depress the rate of cell growth, it is desirable to relate the observed quinone reductase activity to the number of cells or to the amount of protein in each microtiter well. This normalization can be accomplished by staining a set of microtiter plates treated identically to those used for the MTT assay with crystal violet. After the cells are treated with test samples for 48 h, the medium is decanted and the plates are submerged with 200 μl of 0.2% crystal violet solution in 2% ethanol for 10 min. The plates are rinsed for 2 min with tap water. The bound dye is solublized by incubation at 37°C for 1 h with 200 μl of 0.5% SDS in 50% ethanol. The plates are then scanned at 595 nm.

Data Analysis

The specific activity of QR is defined as nmol MTT blue formazan formed per mg protein and per min. A plot of the ratio of quinone reductase specific activities of agent-treated cells to solvent-treated control cells as a function of inducer concentration permits the determination of the CD (concentration required to double specific activity) or CQ (concentration required to quadruple the specific activity) values. A chemopreventive index (CI) is obtained by dividing the IC$_{50}$ (concentration for 50% inhibition of cell viability) values by the respective CD values.

$$\text{Specific activity} = \frac{\text{absorbance change of MTT/min} \times 3247 \text{ nmol/mg}}{\text{absorbance of crystal violet}}$$

$$\text{Fold of induction (ratio)} = \frac{\text{specific activities of treated group}}{\text{specific activity of DMSO control group}}$$

Screening of Medicinal Plants

In this article, plants will be classified as edible or medicinal plants. We presently define a medicinal plant as a plant that has pharmacological activity to treat disease, as compared with an edible plant that is used in daily life as a food. We have procured plant materials from throughout the world for investigation of natural inhibitors of carcinogenesis. The overall experimental approach for obtaining natural product anticarcinogens from these plants has been described in detail.[16,42,45–48] Priority in plant selection, based in part on information contained in the NAPRALERT computer database,[49] has been accorded to edible plants as well as to species with reported biological activity relating to cancer chemoprevention, plants with no history of toxicity, and finally, plants that have not been investigated phytochemically. Initially, a small amount (300 g to 1 kg) of each dried plant sample is collected for preliminary investigation. More than 3000 plants have been obtained from throughout the world and they have been taxonomically identified.

Crude non-polar and polar extracts are prepared from each plant collected and are then evaluated for their activity using a battery of *in vitro* bioassays.[16] Since the water extract often does not contain useful new agents, most effort is directed toward organic solvent extracts. In general, "active" plant extracts are tested in a secondary model of greater physiological complexity. One example is a test to assess potential to inhibit 7,12-dimethylbenz(*a*)anthracene (DMBA)-induced preneoplastic lesion formation in the mouse mammary organ culture (MMOC) model.[50–52] In

[45] R. G. Mehta and J. M. Pezzuto, *Curr. Oncol. Rep.* **4**, 478 (2002).
[46] J. M. Pezzuto, *Biochem. Pharmacol.* **53**, 121 (1997).
[47] J. M. Pezzuto, C. K. Angerhofer, and H. Mehdi, *in* "Studies in Natural Product Chemistry" (Atta-ur-Rahman, ed.), Vol. 20, p. 507. Amsterdam: Elsevier Science, 1998.
[48] J. M. Pezzuto, *in* "Human Medicinal Agents from Plants" (Symposium Series No. 534) (A. D. Kinghorn and M. F. Balandrin, eds.), p. 205. Washington, DC: American Chemical Society Books, 205, 1993.
[49] W. D. Loub, N. R. Farnsworth, D. D. Soejarto, and M. L. Quinn, *J. Chem. Inf. Computer Sci.* **25**, 99 (1985).
[50] R. G. Mehta, J. Liu, A. Constantinou, C. F. Thomas, M. Hawthorne, M. You, C. Gerhäuser, J. M. Pezzuto, R. C. Moon, and R. M. Moriarty, *Carcinogenesis* **16**, 399 (1995).
[51] R. G. Mehta, K. P. L. Bhat, M. E. Hawthorne, L. Kopelovich, R. R. Mehta, K. Christov, G. J. Kelloff, V. E. Steele, and J. M. Pezzuto, *J. Natl. Cancer Inst.* **93**, 1103 (2001).
[52] R. G. Mehta and R. C. Moon, *Anticancer Res.* **11**, 593 (1991).

the next stage, plant extracts showing potency and/or selectivity in the *in vitro* bioassay models are selected for bioassay-guided fractionation to uncover the active principles. Active initial extracts are subjected to solvent partitioning, and chromatography over standard chromatographic materials, with final purification typically effected by semi-preparative HPLC. Active isolates in the previously-mentioned bioassays are characterized by the usual physical, spectral, and chromatographic measurements, and every effort is made to obtain unambiguous NMR assignments for each compound of interest through the use of combinations of conventional one- and two-dimensional NMR methods. Increasingly, Mosher ester methodology has been used in our recent work for the determination of absolute stereochemistry of secondary alcohols,[53] and recently a simplified Mosher ester method has been developed to prepare Mosher esters directly in NMR tubes.[54] Finally, the structure of pure compounds may be identified by small-molecule x-ray crystallography. Pure active compounds are then evaluated in the mouse mammary organ culture model.[55,56] Finally, the *in vivo* cancer chemopreventive activity of highly promising pure plant constituents is evaluated in full-term tumorigenesis models, including the two-stage mouse skin model that uses DMBA as an initiator and 12-*O*-tetradecanoylphorbol 13-acetate (TPA) as a promoter and the rat mammary carcinogenesis model with *N*-methyl-*N*-nitrosourea (MNU) or DMBA as a carcinogen.[57,58]

To date, we have evaluated 2475 plant extracts in the quinone reductase induction assay system by using cultured Hepa 1c1c7 cells. Of these, 136 plant extracts (5.4%) showed quinone reductase induction activity. Table I gives a partial listing of the plant materials selected for bioassay-guided fractionation. When CD values of plant extracts are below 10 μg/ml, plants are considered as active leads. About 100 active compounds have been identified from plant materials, some of which are listed in Table II.

The edible fruit of *Physalis philadelphica* Lam. (Solanaceae), commonly known as tomatillo, is used as an ingredient in foods such as enchiladas and salsas in Latin America.[59] The whole plants of *Physalis*

[53] B. N. Su, E. J. Park, Z. H. Mbwambo, B. D. Santarsiero, A. D. Mesecar, H. H. S. Fong, J. M. Pezzuto, and A. D. Kinghorn, *J. Nat. Prod.* **65,** 1278 (2002).
[54] B. N. Su, L. C. Chang, E. J. Park, M. Cuendet, B. D. Santarsiero, A. D. Mesecar, R. G. Mehta, H. H. S. Fong, J. M. Pezzuto, and A. D. Kinghorn, *Planta Med.* **68,** 730 (2002).
[55] R. G. Mehta, J. Liu, A. Constantinou, M. Hawthorne, J. M. Pezzuto, R. C. Moon, and R. M. Moriarty, *Anticancer Res.* **14,** 1209 (1994).
[56] E. Mata-Greenwood, J. F. Daeuble, P. A. Grieco, J. Dou, J. D. McChesney, R. G. Mehta, A. D. Kinghorn, and J. M. Pezzuto, *J. Nat. Prod.* **64,** 1509 (2001).
[57] G. O. Udeani, C. Gerhäuser, C. F. Thomas, R. C. Moon, J. W. Kosmeder, A. D. Kinghorn, R. M. Moriarty, and J. M. Pezzuto, *Cancer Res.* **57,** 3424 (1997).
[58] M. Jang and J. M. Pezzuto, *Meth. Cell Sci.* **19,** 25 (1997).

TABLE I
PLANT MATERIALS IDENTIFIED AS ACTIVE INDUCERS OF QUINONE REDUCTASE

Family	Plant Name	Plant Part
Apcynaceae	*Tabernaemontana angulata*	ST
Asteraceae	*Artemisia ludoviciana*	PX
Asteraceae	*Carpesium abrotanoides*	ST
Boraginaceae	*Heliotropium indicum*	PX
Brassicaceae	*Raphanus sativus*	RT
Burseraceae	*Santiria apiculata*	LF
Chrysobalanaceae	*Cyphomandra oblongifolia*	LF
Elaeocarpaceae	*Muntingia calabura*	LF
Euphorbiaceae	*Amanoa oblongifolia*	SB
Fabaceae	*Tephrosia purpurea* Pers.	AC
	Tephrosia toxicarina	LF
	Tephrosia toxicarina	ST
	Phaseolus vulgaris	FR
	Muntingia calabura	LF
	Dipteryx sp.	SD
Malvaceae	*Sida acuta*	PL
Musaceae	*Musa* sp.	FR
Rubiaceae	*Capirona decorticans*	BK
	Coussarea brevicaulis	WS
Solanaceae	*Physalis philadelphica*	FR
	Solanum altissimum	PL
Zingiberaceae	*Renealmia nicolaioides*	NA

The above plants mediated induction of quinone reductase with CD values <10 μg/ml. ST, stem; PX, aerial part; RT, root; LF, leaf; SB, stembark; AC, flower, fruit, leaf, root, stem; FR, fruit; SD, seed; PL, entire plant; BK, bark; WS, stemwood; NA, bud, flower.

philadelphica have been used for the treatment of gastric intestinal disorders and for treating leprosy in traditional medicine.[60,61] Ethyl acetate extracts of this plant induced quinone reductase potently, and equally well in the wild-type Hepa 1c1c7 and mutant cell lines, suggesting the presence of monofunctional inducers. Over 30 withanolides were isolated and characterized as inducers (Fig. 1, Table III).[62–65] The withanolides are a group of natural C-28 steroids that occur mainly in plants of certain genera

[59] A. Cáceres, M. F. Torres, S. Ortiz, F. Cano, and E. Jauregui, *J. Ethnopharmacol.* **39,** 73 (1993).
[60] R. E. Dimayuga, M. Virgen, and N. Ochoa, *Pharm. Biol.* **36,** 33 (1998).
[61] M. A. Bock, J. Sanchez-Pilcher, L. J. McKee, and M. Ortiz, *Plant Foods Hum. Nutr.* **48,** 127 (1995).
[62] B. N. Su, R. Misico, E. J. Park, B. D. Santarsiero, A. D. Mesecar, H. H. S. Fong, J. M. Pezzuto, and A. D. Kinghorn, *Tetrahedron* **58,** 3453 (2002).
[63] R. I. Misico, L. L. Song, A. S. Veleiro, A. M. Cirigliano, M. C. Tettamanzi, G. Burton, G. M. Bonetto, V. E. Nicotra, G. L. Silva, R. R. Gil, J. C. Oberti, A. D. Kinghorn, and J. M. Pezzuto, *J. Nat. Prod.* **65,** 677 (2002).

TABLE II
PHYTOCHEMICAL ISOLATES THAT INDUCE QUINONE REDUCTASE

Plant name	Compound
Acnistus arborescens	7β-Acetoxywithanolide D
	7β,16α-Diacetoxywithanolide D
	4-Deoxy-7β,16α-Diacetoxywithanolide D
Tephrosia purpurea	7,4'-Dihydroxy-3',5'-dimethoxyisoflavone
	(−)-3-Hydroxy-4-methoxy-8,9-methylenedioxypterocarpan
	Lanceolatin B
	(−)-Maackiain
	(−)-Medicarpin
	Pongamol
	(+)-Purpurin
	(+)-Tephropurpurin
Physalis philadelphica	2,3-Dihydro-3-methoxywithanolide D
	23,25-Dihydrowithanolide
	Withaphysacarpin
	4-Acetyllixocarpalactone B
	18-Hydroxywithanolide D
	Ixocarpalactone A
	Ixocarpalactone B
	Philadelphicalactone A
	Philadelphicalactone B
	Withaphysacarpin
	2,3-Dihydroixocarpalactone B
	4β,7β,20R-trihydroxyl-1-oxowitha-2,5-dien-22,26-olide
Musa × paradisiacal	1,2-Dihydro-1,2,3-trihydroxy-9-(4-methoxyphenyl)phenalene
	Hydroxyanigorufone
	Rel-(3S,4aR,10bR)-8-hydroxy-3-(4-hydroxyphenyl)-9-methoxy-4a,5,6,10b-Tetrahydro-3H-naphtho[2,1-b]pyran
	1,7-*bis*(4-hydroxyphenyl)hepta-4(E),6(E)-dien-3-one

[64] J. Q. Gu, W. Li, Y. H. Kang, B. N. Su, H. H. S. Fong, R. B. Van Breemen, J. M. Pezzuto, and A. D. Kinghorn, *Chem. Pharm. Bull.* In Press.

[65] E. J. Kennelly, C. Gerhäuser, L. L. Song, J. G. Graham, C. W. W. Beecher, J. M. Pezzuto, and A. D. Kinghorn, *J. Agric. Food Chem.* **45,** 3771 (1997).

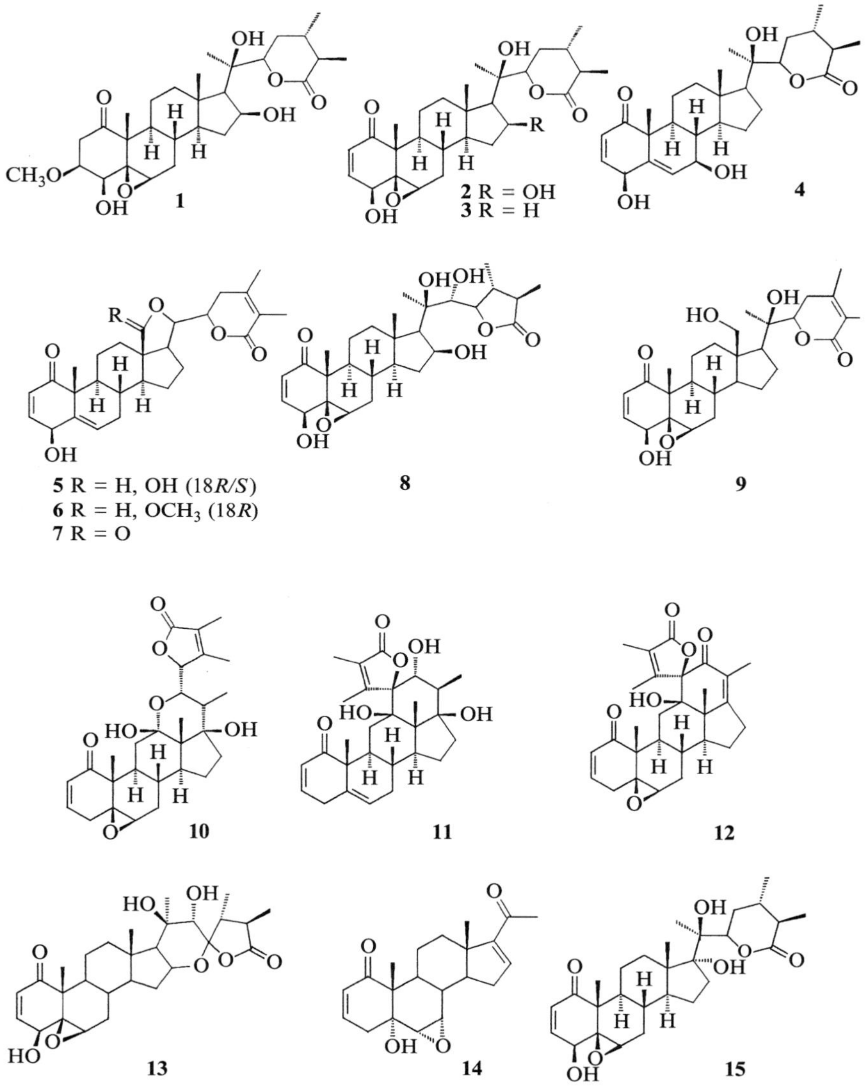

FIG. 1. Withanolides inducing quinone reductase activity.

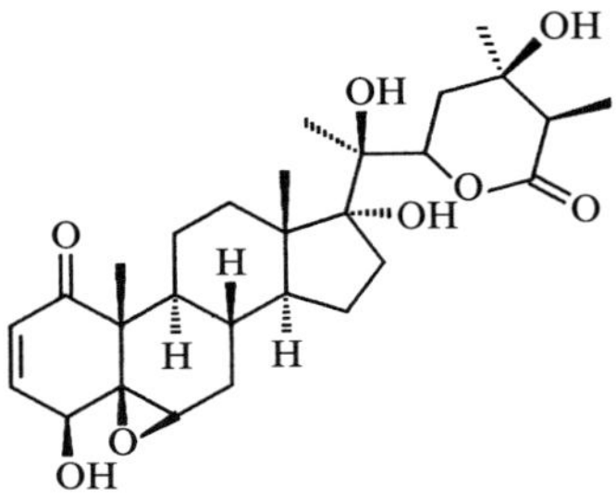

FIG. 1. *(continued)*

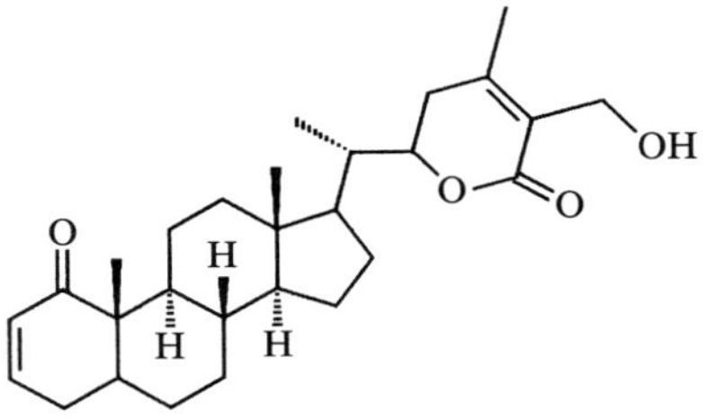

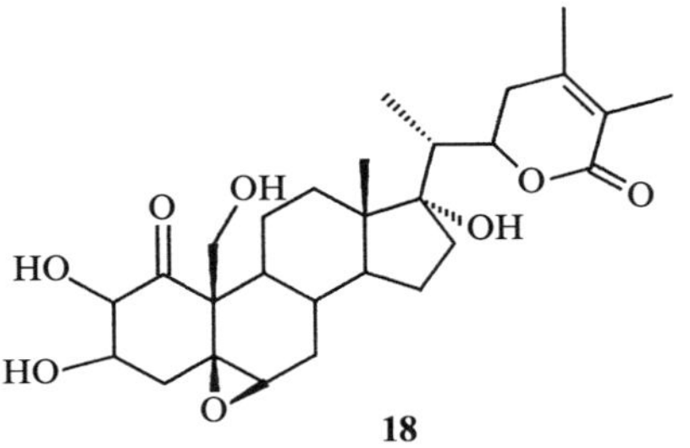

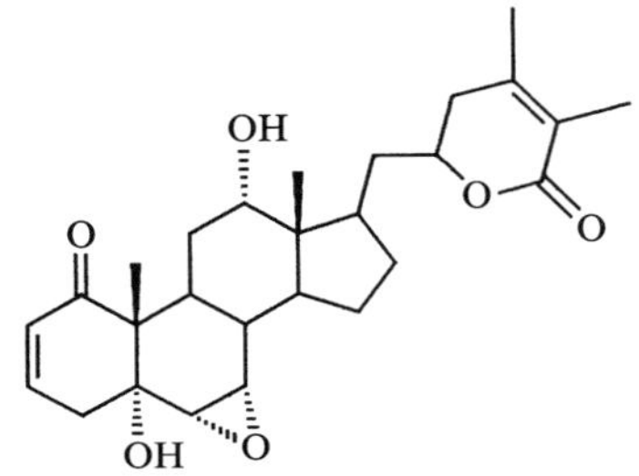

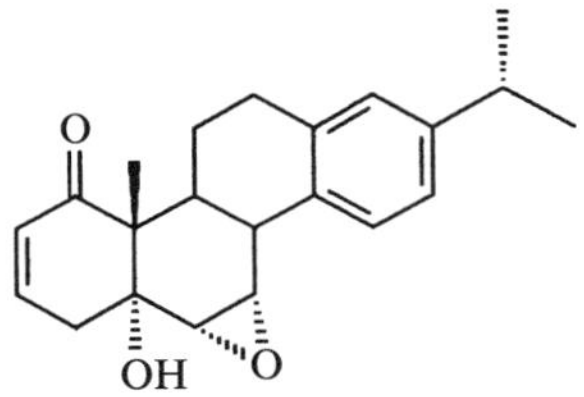

of the Solanaceae. In addition to quinone reductase induction, withanolides have been studied previously for their antifeedant, anti-inflammatory, antitumor, cytotoxic, and immunomodulating activities and for protection against CCl_4-induced hepatotoxicity.[66–68] These compounds are being further investigated in various animal models as lead compounds.

The whole flowering and fruiting parts of the medicinal plant *Tephrosia purpurea* were shown to have several active compounds. 7,4'-Dihydroxy-

[66] C. H. Chiang, S. M. Jaw, C. F. Chen, and W. S. Kan, *Anticancer Res.* **12**, 837 (1992).
[67] K. R. S. Ascher, N. E. Nemny, M. Eliyahu, I. Kirson, A. Abraham, and E. Glotter, *Experientia* **36**, 998 (1980).
[68] E. Glotter, *Nat. Prod. Rep.* **8**, 415 (1982).

TABLE III
QUINONE REDUCTASE INDUCTION EFFECTS OF WITHANOLIDES

Compound	Trivial Name[a]	CD (μM)	IC$_{50}$ (μM)	CI
1	2,3-Dihydro-3-methoxywithaphysacarpin	7.8	47	6
2	Withaphysacarpin	0.4	4.8	11
3	24,25-Dihydrowithanolide D	0.7	5.5	8
4	NA[a]	1.1	24	21
5	Withaphysalin G	0.5	10	19
6	Withaphysalin H	0.5	5	10
7	Withaphysalin J	0.4	11	28
8	Ixocarpalactone A	0.3	8	24
9	18-Hydroxywithanolide D	0.06	0.3	5
10	Trechonolide A	0.3	8	29
11	Jaborosalactone P	0.8	43	57
12	Jaborosalactone I	0.3	8	29
13	Ixocarpalactone B	0.2	1	7
14	NA[a]	35	>58	>2
15	Philadelphicalactones A	0.04	0.5	12
16	Philadelphicalactones B	0.3	0.6	2
17	Jaborosalactone A	0.3	1.8	6
18	Jaborosalactone O	1.5	>42	>28
19	Withaferoxolide	14.5	>42	>3
20	Nicandrenone	10.5	43	>4
	Sulforaphane[b]	0.2	10	42

CD: Concentration to double QR activity in Hepa 1c1c7 cells.
IC$_{50}$: Concentration to inhibit Hepa 1c1c7 cell growth by 50%.
CI: Chemoprevention Index (IC$_{50}$/CD).
[a] Trivial name not available.
[b] QR assay positive control.

3',5'-dimethoxyisoflavone, (+)-tephropurpurin, (+)-purpurin, pongamol, lanceolatin B, (−)-maackiain, (−)-3-hydroxy-4-methoxy-8,9-methylenedioxypterocarpan, and (−)-medicarpin were isolated as quinone reductase inducers from ethyl acetate extracts.[69]

The genus *Renealmia* comprises about 75 species, of which approximately one-third grow in tropical Africa, while others are found in the neotropics.[70–71] Various plant parts belong to genus *Renealmia*—such as *R. alpinia, R. asplundii,* and *R. cincinnata*—that have been used traditionally

[69] L. C. Chang, C. Gerhäuser, L. Song, N. R. Farnsworth, J. M. Pezzuto, and A. D. Kinghorn, *J. Nat. Prod.* **60,** 869 (1997).
[70] M. H. Chuendem, J. A. Mbah, A. Tsopmo, J. F. Ayafor, O. Sterner, C. C. Okunjic, M. M. Iwu, and B. M. Schuster, *Phytochemistry* **52,** 1095 (1999).
[71] W. Milliken and B. Albert, *Econ. Bot.* **50,** 10 (1996).

to treat fever, headache, and stomachache.[71,72] Species in this genus also have medicinal uses as antivenins, febrifuges, and tonics, and some are edible. *Renealmnia nicolaioides* Loes. (Zingiberaceae) is commonly known as "mishqui panga" in the Quechua dialect in Peru, which means "tasty leaf."[73] A methanolic extract of the roots of *R. nicolaioides* was found to significantly induce quinone reductase activity. Bioassay-guided fractionation of *n*-hexane and ethyl acetate extracts using cultured Hepa 1c1c7 mouse hepatoma cells led to two inducers, 2'-hydroxy-4',6'-dimethoxychalcone and (±)-5-hydroxy-7-methoxyflavanone.[74] Concentrations to double activity (CD) values are 1.7 and 0.9 μg/mL, respectively.

Musa × *paradisiaca* L. (Musaceae), with orange-yellow flowers and long, narrow, starchy, edible fruits, is the common name for the cultivar French plantain, which is of hybrid origin (*Musa acuminata* × *Musa balbisiana* Colla).[75] The fruits of *Musa* × *paradisiaca* cultivar were chosen for further investigation, since the ethyl acetate–soluble fraction of the methanol extract significantly induced quinone reductase in cultured Hepa 1c1c7 cells. Bioassay-guided fractionation of the ethyl acetate-soluble fraction of the methanol extract of *Musa* × *paradisiaca* cultivar using the induction assay led to two diarylheptanoids: 1,7-*bis*(4-hydroxyphenyl)-hepta-4*(E)*,-6*(E)*-dien-3-one and *rel*-(3*S*,4a*R*,10b*R*)-8-hydroxy-3-(4-hydroxyphenyl)-9-methoxy-4a,5,6,10b-tetrahydro-3*H*-naphtho[2,1-*b*]pyran, as well as hydroxyanigorufone and 1,2-dihydro-1,2,3-trihydroxy-9-(4-methoxyphenyl)phenalene.[76]

Quinone Reductase Inducers Present in Edible Plants

Numerous epidemiological studies together with data from *in vivo* and *in vitro* studies have shown that vegetables, especially cruciferous vegetables, have an important role in protection against various cancers.[77–83]

[72] E. W. Davis and J. A. Yost, *J. Ethnopharmacol.* **9**, 273 (1983).

[73] P. J. M. Mass, "Flora Neotropica: Monograph No. 18. Renealmia (Zingiberaceae-Zingiberoideae) and Costoideae (Additions)," p. 100. The New York Botanical Garden: Bronx, NY, 1997.

[74] J. Q. Gu, E. J. Park, J. S. Vigo, J. G. Graham, H. H. Fong, J. M. Pezzuto, and A. D. Kinghorn, *J. Nat. Prod.* **65**, 1616 (2002).

[75] G. C. G. Argent, *in* "The European Garden Flora. Volume II, Monocotyledons (Part II) Juncaceae to Orchidaceae" (S. M. Walters, A. Brady, C. D. Brickell, J. Cullen, P. S. Green, J. Lewis, V. A. Matthews, D. A. Webb, P. F. Yeo, and J. C. M. Alexander, eds.), p. 117. Cambridge University Press: Cambridge, UK, 1984.

[76] D. S. Jang, E. J. Park, M. E. Hawthorne, J. S. Vigo, J. G. Graham, F. Cabieses, B. D. Santarsiero, A. D. Mesecar, H. H. Fong, R. G. Mehta, J. M. Pezzuto, and A. D. Kinghorn, *J. Agric. Food Chem.* **50**, 2330 (2002).

[77] K. A. Steinmetz and J. D. Potter, *J. Am. Diet Assoc.* **96**, 1027 (1996).

Some of the most convincing evidence for the health benefits of fruit and vegetable consumption relates to the reduced risk of gastrointestinal cancers, such as those associated with the stomach, colon, mouth, pharynx, esophagus, and rectum.[81] Cohort studies involving prostate, breast, bladder,[84] lung cancer,[85] and non-Hodgkin's lymphoma[86] have shown that high crucifer diets reduce cancer risk significantly. The intake of *Brassica* vegetables was inversely associated with the risk of breast cancer. Notably, 1 to 2 servings per day reduced the risk of breast cancer by as much as 20% to 40%.[87]

Multiple mechanisms may be involved in the protective effects of fruits and vegetables.[77,88–90] Cruciferous vegetables contain a wide variety of phytochemicals such as organosulfides, isothiocyanates, indoles, and polyphenols that have the potential to modulate cancer development (Table IV).[91] Crucifers are especially rich in glucosinolates, which are converted by plant myrosinase and gastrointestinal microflora to isothiocyanates.[92,93] Many Cruciferae vegetables *(Brassica)* such as cabbage, broccoli, Brussels sprouts, watercress, and cauliflower induce phase II enzymes.[94] Glucosinolates found in Brussels sprouts induced phase II enzymes such as glutathione *S*-transferase and quinone reductase in animal models.[95] A number of isothiocyanates are known to effectively protect

[78] G. Block, B. Patterson, and A. Subar, *Nutr. Cancer* **18**, 1 (1992).

[79] V. L. W. Go, D. A. Wong, M. S. Resnick, and D. Heber, *J. Nutr.* **131**, 179S (2001).

[80] D. M. Eisenberg, R. B. Davis, S. L. Ettner, S. Appel, S. Wilkey, M. Von Rompay, and R. C. Kessler, *J. Am. Med. Assoc.* **280**, 569 (1998).

[81] P. Terry, E. Giovannucci, K. B. Michels, L. Bergkvist, H. Hansen, L. Holmberg, and A. Wolk, *J. Natl. Cancer Inst.* **93**, 525 (2001).

[82] E. J. Park and J. M. Pezzuto, *Cancer Metastasis Rev.* **21**, 231 (2002).

[83] M. A. S. van Duyn and E. Pivonka, *J. Am. Diet Assoc.* **100**, 1511 (2000).

[84] D. S. Michaud, D. Spiegelman, S. K. Clinton, E. B. Rimm, W. C. Willett, and E. L. Giovannucci, *J. Natl. Cancer Inst.* **91**, 605 (1999).

[85] M. R. Spitz, C. M. Duphorne, M. A. Detry, P. C. Pillow, C. I. Amos, L. Lei, M. de Andrade, X. Gu, W. K. Hong, and X. Wu, *Cancer Epidemiol. Biomarkers Prev.* **9**, 1017 (2000).

[86] S. M. Zhang, D. J. Hunter, B. A. Rosner, E. L. Giovannucci, G. A. Colditz, F. E. Speizer, and W. C. Willett, *Cancer Epidemiol. Biomarkers Prev.* **9**, 477 (2000).

[87] P. Terry, A. Wolk, I. Persson, and C. Magnusson, *J. Am. Med. Assoc.* **285**, 2975 (2001).

[88] K. A. Steinmetz and J. D. Potter, *Cancer Causes Control* **2**, 325 (1991).

[89] K. A. Steinmetz and J. D. Potter, *Cancer Causes Control* **2**, 427 (1991).

[90] P. Talalay and J. W. Fahey, *J. Nutr.* **131**, 3027 (2001).

[91] W. R. Bidlack, S. T. Omaye, M. S. Meskin, and D. K. W. Topham, eds. Technomic Publishing Company, Lancaster, PA, 2000.

[92] T. A. Shapiro, J. W. Fahey, K. L. Wade, K. K. Stephenson, and P. Talalay, *Cancer Epidemiol. Biomark. Prev.* **10**, 501 (2001).

[93] J. W. Fahey, A. T. Zalcmann, and P. Talalay, *Phytochemistry* **56**, 5 (2001).

TABLE IV
PHYTOCHEMICALS PRESENT IN EDIBLE PLANTS THAT INDUCE QUINONE REDUCTASE

Common Name	Plant Name	Phytochemicals
Broccoli	*Brassica oleracea*	Dithiolethiones
		Sulforaphane
		Phenylethylisothiocyanate
Brussels sprouts	*Brassica oleracea gemmifera*	Isothiocyanate
Cabbage	*Brassica oleracea*	Brassinin
Onion	*Allium cepa* L.	Allyl sulfide
		Quercetin-4'-glucoside
Garlic	*Allium sativum*	Allyl sulfide
Watercress	*Rorippana sturtium*	Isothiocyanate
Wasabi	*Eutrema wasabi*	6-(Methylsufinyl)hexylisothiocyanate
Turmeric	*Curcuma longa*	Curcumin

against chemical carcinogenesis in various animal models.[96,97] Sulforaphane is a representative isothiocyanate isolated from broccoli that has potent monofunctional phase II enzyme inducing activity.[26] Sulforaphane was shown to significantly inhibit chemical carcinogenesis in rats.[98] The anticarcinogenic effects of broccoli may be due to modification of phase II detoxification enzymes by isothiocyanates.

Indole-3-carbinol, phenethyl isothiocyanate, sinigrin, and crambene, found in cruciferous vegetables, also induced phase II enzymes.[99–102] Indole-3-carbinol and phenethyl isothiocyanate are potent bifunctional agents (inducing both phase I and II activities).[103] Indole-3-carbinol also shows anti-estrogenic activities and has been associated with the prevention of cervical and breast cancer.[104–106] It was reported that the nitrile

[94] P. Talalay and Y. Zhang, *Trans. Biochem. Soc.* **24,** 806 (1996).
[95] C. W. Nho and E. Jeffery, *Toxicol. Appl. Pharmacol.* **174,** 146 (2001).
[96] Y. Zhang and P. Talalay, *Cancer Res.* **54** (Suppl.), 1976s (1994).
[97] S. S. Hecht, *J. Cell. Biochem.* **22** (Suppl.), 195 (1995).
[98] Y. Zhang, T. W. Kensler, C. G. Cho, G. H. Posner, and P. Talalay, *Proc. Natl. Acad. Sci. USA* **91,** 3147 (1994).
[99] H. G. Shertzer and A. P. Senft, *Drug Metabol. Drug Interact.* **17,** 159 (2000).
[100] M. M. Manson, H. W. Ball, M. C. Barrett, H. L. Clark, D. J. Judah, G. Williamson, and G. E. Neal, *Carcinogenesis* **18,** 172 (1997).
[101] R. Staack, S. Kingston, M. A. Wallig, and E. H. Jeffery, *Toxicol. Appl. Pharmacol.* **149,** 17 (1998).
[102] A. S. Keck, P. Staack, and E. H. Jeffery, *Nutr. Cancer* **42,** 233 (2002).
[103] E. A. Hudson, L. Howells, H. W. Ball, A. M. Pfeifer, and M. M. Manson, *Biochem. Soc. Trans.* **26,** S370 (1998).
[104] M. S. Brignall, *Altern. Med. Rev.* **6,** 580 (2001).
[105] Q. Meng, I. D. Goldberg, E. M. Rosen, and S. Fan, *Breast Cancer Res. Treat.* **63,** 147 (2000).
[106] F. Yuan, D. Z. Chen, D. W. Sepkovic, H. L. Bradlow, and K. Auborn, *Anticancer Res.* **19,** 1673 (1999).

crambene (1-cyano-2-hydroxy-3-butene), present in most *Brassica* vegetables, induced hepatic quinone reductase activity when adminis-tered to rats.[107] Crambene showed hepatic quinone reductase activity comparable to that of sulforaphane in an animal model. Crambene (5 m*M*) induced quinone reductase activity and caused cell cycle arrest in the G2/M phase in mouse Hepa 1c1c7 cells, rat H4IIEC3 cells, and human HepG2, cells without cytotoxicity. Two components found in crucif-erous vegetables, crambene (1-cyano-2-hydroxy-3-butene) and indole-3-carbinol, were reported as monofunctional and bifunctional inducers, respectively.[108]

Brassinin, a phytoalexin, is found in Chinese cabbage, Brussels sprouts, and cauliflower. Brassinin, the indole dithiocarbamate, showed induction of quinone reductase with *in vitro* and *in vivo* experiments. Brassinin in-duced phase II detoxication enzymes such as quinone reductase and glu-tathione *S*-transferase in Hepa 1c1c7 cells, and elevated phase II mRNA levels in H4IIE rat hepatoma cells and in rat organs. This induction was found to be regulated at the transcriptional level by interaction with two transcription factors, the XRE (xenobiotic response element)[35,109] and ARE (antioxidant response element).[110,111] Brassinin and analogs did not induce quinone reductase protein in the mutant Hepa 1c1c7 cell lines. Therefore, brassinin is regarded as a bifunctional inducer, with similarities to compounds such as indole-3-carbinol.[50] Cyclobrassinin, a naturally oc-curring brassinin analog, showed more potent induction of quinone reduc-tase than brassinin. Brassinin inhibited DMBA-induced mammary lesions in organ culture.[55] Moreover, it reduced significantly the formation of the DMBA/phorbol ester-induced papillomas in the two-stage mouse skin car-cinogenesis model.[55] We synthesized several analogs of brassinin and evaluated their effectiveness in the mouse mammary gland organ culture model. Cyclobrassinin was more effective than brassinin. Spirobrassinin and *N*-ethyl-2,3-dihydrobrassinin also significantly inhibited mammary lesion formation. These effects of brassinin and analogs may, in part, be mediated by induction of phase II detoxifying enzymes (Table V). Interest-ingly, brassinin may be effective as a chemopreventive agent during the initiation and promotion phase of carcinogenesis.

[107] M. A. Wallig, S. Kingston, R. Staack, and E. H. Jefferey, *Food Chem. Toxicol.* **36,** 365 (1998).
[108] H. L. Bradlow, D. W. Sepkovic, N. T. Telang, and M. P. Osborne, *Ann. N.Y. Acad. Sci.* **889,** 204 (1999).
[109] J. Mimura and Y. Fujii-Kuriyama, *Biochim. Biophys. Acta* **17,** 1619 (2003).
[110] A. T. Dinkova-Kostova, W. D. Holtzclaw, R. N. Cole, K. Itoh, N. Wakabayashi, Y. Katoh, M. Yamamoto, and P. Talalay, *Proc. Natl. Acad. Sci. USA* **99,** 11908 (2002).
[111] T. H. Rushmore and A. N. Kong, *Curr. Drug Metab.* **3,** 481 (2002).

TABLE V
QUINONE REDUCTASE INDUCTION EFFECTS OF BRASSININ DERIVATIVES

Structure	Compound	CD (μM)
	Brassinin	4.0
	Cyclobrassinin	1.2
	Spirobrassinin	7.9
	N-Ethyl-2,3-dihydrobrassinin	0.13
	S-Selenomethylbrassinin	0.5

CD: Concentration to double QR activity in Hepa 1c1c7 cells.

Garlic and onion have received great attention as potential cancer chemopreventive agents.[112,113] Several studies have shown that the consumption of allium vegetables is associated with a reduction of cancer risk in humans. It has been suggested that this effect is due to the organosulfur

[112] B. S. Reddy, C. V. Rao, A. Rivenson, and G. Kelloff, *Cancer Res.* **53**, 3493 (1993).
[113] R. Munday and C. M. Munday, *Nutr. Cancer* **40**, 205 (2001).

compounds of these vegetables.[114] The organosulfur compounds (OSCs) of the *Allium* genus are major components in garlic and onion.[115,116] Induction of phase II detoxification enzymes by these agents seems to be responsible for chemoprotective action. Diallyl trisulfide (DATS) and diallyl disulfide (DADS) isolated from galic are potent inducers of the phase II enzymes, quinone reductase and glutathione transferase.[117] Dipropenyl sulfide, derived from onion, is also an active inducer. DADS induced quinone reductase activity in liver, intestine, kidney, and lung. A good correlation between the inhibitory effects of OSCs against benzo(*a*)pyrene (BP)-induced pulmonary tumorigenesis and their effects on quinone reductase expression in the lung was reported. DADS and DATS were much more potent inducers of forestomach quinone reductase activity. Inhibitory effects of OSCs against BP-induced tumorigenesis might be due to induction activity of OSCs, since quinone reductase is an enzyme implicated in the detoxification of activated quinone metabolites of benzo(a)pyrene (BP). Recently, it was also reported that antimutagenic activity of organosulfur compounds from *Allium* was associated with phase II enzyme induction.[118]

Curcumin (diferuloylmethane) is a yellow pigment of turmeric (*Curcuma longa* L., Zingiberaceae) that is widely used as a food flavoring and coloring agent (e.g., in curry).[119] Curcumin has a long history in traditional medicine of Southeast Asia for the treatment of several diseases.[120] Antioxidant and anti-inflammatory effects of this compound have been assessed in various *in vitro* models and in experimental animal systems.[121] Chemopreventive properties of curcumin have been extensively investigated and well documented.[122–127] One of the most plausible mechanisms underlying the

[114] S. V. Singh, S. S. Pan, S. K. Srivastava, H. Xia, X. Hu, H. A. Zaren, and J. L. Orchard, *Biochem. Biophys. Res. Commun.* **244,** 917 (1998).

[115] R. Munday and C. M. Munday, *Nutr. Cancer* **34,** 42 (1999).

[116] D. Guyonnet, M. H. Siess, A. M. Le Bon, and M. Suschetet, *Toxicol. Appl. Pharmacol.* **154,** 50 (1999).

[117] M. H. Siess, A. M. Le Bon, M. C. Canivenc-Lavier, and M. Suschetet, *Cancer Lett.* **120,** 195 (1997).

[118] D. Guyonnet, C. Belloir, M. Suschetet, M. H. Siess, and A. M. Le Bon, *Mutat. Res.* **495,** 135 (2001).

[119] H. P. T. Ammon and M. A. Wahl, *Planta Med.* **57,** 1 (1991).

[120] Y. J. Surh, *Mutat. Res.* **428,** 305 (1999).

[121] E. Kunchandy and M. N. A. Rao, *Int. J. Pharmacol.* **58,** 237 (1990).

[122] M. A. Mukundan, M. C. Chacko, V. V. Annapurna, and K. Krishnaswamy, *Carcinogenesis* **14,** 493 (1993).

[123] J. L. Arbiser, N. Klauber, R. Rohan, R. van Leeuwen, M. T. Huang, C. Fisher, E. Flynn, and H. R. Byers, *Mol. Med.* **4,** 376 (1998).

[124] M. Nagabhushan and S. V. Bhide, *J. Am. Coll. Nutr.* **11,** 192 (1992).

chemopreventive effects of curcumin involves suppression of tumor promo-tion. Curcumin inhibits TPA (12-*O*-tetradecanoylphorbol-13-acetate)--induced inflammation, proliferation, ROS (reactive oxygen species) generation, and ODC (ornithine decarboxylase), and shows anti-mutagenesis, anti-angiogenesis, and blocking effects for carcinogen-DNA adduct formation.[128] Curcumin also inhibits COX-2 and lipoxygenase activities in TPA-treated mouse epidermis.[129] In addition, it was reported that curcu-min induces phase II detoxification enzymes,[130] while inhibiting procarci-nogen activating phase I enzymes, such as cytochrome P4501A1.[131] Recently, Talalay and colleagues examined structural features of the curcu-min molecule that contribute to enzyme induction (Table VI).[132] The intro-duction of aromatic *o*-hydroxyl groups into the curcuminoid skeleton raises inducer potency >30-fold. The activities of these compounds are generally attributable to their Michael reaction acceptor centers. This study also shows that dibenzoylmethane, a β-diketone that is not a classical Michael reaction acceptor, is also a potent inducer. This suggests that the β-diketone moiety of curcuminoids may play a significant role in quinone reductase activities. A phase I clinical study of curcumin showed there was no toxicity up to 8 g/kg body weight, and there is significant interest in developing curcumin as a cancer chemoprevention agent.[133] Data obtained from the *in vitro* quinone reductase assay has consistently provided excellent leads, many of which have shown good activity in animal carcinogenesis models and are being considered for further devel-opment. Chemopreventive agents from some vegetables are summarized in Table VII; some of these agents may be worthy of further consideration.

A great deal of evidence from cruciferous vegetables supports the idea that induction of phase II detoxification enzymes is an effective strategy in

[125] A. A. Conney, Y. R. Lou, J. G. Xie, T. Osawa, H. L. Newmark, Y. Liu, R. L. Chang, and M. T. Huang, *Proc. Soc. Exp. Biol. Med.* **216,** 234 (1997).

[126] Y. Shukla, A. Arora, and P. Taneja, *Teratog. Carcinog. Mutagen.* **23,** 323 (2003).

[127] M. T. Huang, Y. R. Lou, W. Ma, H. L. Newmark, K. R. Reuhl, and A. H. Conney, *Cancer Res.* **54,** 5841 (1994).

[128] Y. J. Surh, K. S. Chun, H. H. Cha, S. S. Han, Y. S. Keum, K. K. Park, and S. S. Lee, *Mutat. Res.* **480–481,** 243 (2001).

[129] M. T. Huang, T. Lusz, T. Ferraro, T. F. Abidi, J. D. Laskin, and A. H. Conney, *Cancer Res.* **51,** 813 (1991).

[130] A. Khafif, S. P. Schantz, T. C. Chou, D. Edelstein, and P. G. Sacks, *Carcinogenesis* **19,** 419 (1998).

[131] H. P. Ciolino, P. J. Daschner, T. T. Y. Wang, and G. C. Yeh, *Biochem. Pharm.* **56,** 197 (1998).

[132] A. T. Dinkova-Kostova and P. Talalay, *Carcinogenesis* **20,** 911 (1999).

[133] A. L. Cheng, C. H. Hsu, J. K. Lin, M. M. Hsu, Y. F. Ho, T. S. Shen, J. Y. Ko, J. T. Lin, B. R. Lin, W. Ming-Shiang, H. S. Yu, S. H. Jee, G. S. Chen, T. M. Chen, S. A. Chen, M. K. Lai, Y. S. Pu, M. H. Pan, Y. J. Wang, C. C. Tsai, and C. Y. Hsieh, *Anticancer Res.* **21,** 2895 (2001).

TABLE VI
QUINONE REDUCTASE INDUCTION EFFECTS OF CURCUMIN AND ANALOGS

Structure	Compound	CD (μM)
	3,4-Dimethoxycin-namaldehyde	52.8
	Methyl *m*-coumarate	83
	Dibenzoylmethane	0.8
	Benzoylacetone	92
	Curcumin	7.3
	Tetrahydro-curcumin	35.7
	Salicylcurcuminoid	0.3
	1,7-*bis*(4-Hydroxyphenyl) hepta-4(E),6(E)-dien-3-one	6.4

Data from references 76 and 132.
CD: Concentration to double QR activity in Hepa 1c1c7 cells.

TABLE VII
QUINONE REDUCTASE INDUCTION EFFECTS OF VARIOUS PHYTOCHEMICALS

	Compound	CD (μM)	IC$_{50}$ (μM)	CI
1	Ceramide −1	1.8	3.9	2
2	Ceramide −2	2.5	> 7.3	> 3
3	(−)-Maackiain	8.8	> 70	> 8
4	(−)-3-Hydroxy-4-methoxy-8,9-methylene-dioxypterocarpan	14.7	> 64	> 4
5	(−)-Medicarpin	13.7	> 74	> 5
6	Hydroxyanigorufone	13.2	> 23	2
7	1,2-Dihydro-1,2,3-trihydroxy-9-(4-methoxyphenyl)phenalene	16.4	> 62	> 4
8	Rel-(3S,4aR,10bR)-8-Hydroxy-3-(4-hydroxyphenyl)-9-methoxy-4a,5,6,10b-Tetrahydro-3H-naphtho[2,1-b]pyran	28.7	> 62	> 2
9	26-Hydroxy-5α-lanosta-7,9(11),24-triene-3,22-dione	6.9		
10	Coussaric acid A	17.9		

CD: Concentration to double QR activity in Hepa 1c1c7 cells.
IC$_{50}$: Concentration to inhibit Hepa 1c1c7 cell growth by 50%.
CI: Chemoprevention Index (IC$_{50}$/CD).

cancer chemoprevention.[134] Recently, knock-out mouse experiments in which the specific transcription factor, nrf2, which is essential for induction of phase II proteins, was deleted, showed low basal levels of phase II enzymes. These mice were much more susceptible than their wild-type counterparts to BP forestomach carcinogenesis and were not protected by phase II inducers.[135] This provides further evidence that novel anticarcinogens can be identified on the basis of their ability to induce phase II enzymes.

Phytochemicals Inducing Quinone Reductase

Isothiocyanates

Sulforaphane, an aliphatic isothiocyanate isolated from broccoli, is a representative potent monofunctional inducer of quinone reductase. A novel sulforaphane derivative, sulforamate [(±)-4-methylsulfinyl-1-(S-methyldithiocarbamyl)-butane], was synthesized and studied in our

[134] D. T. H. Verhoeven, H. Verhagen, R. A. Goldbohm, and G. van den Brandt, *Chem. Biol. Interact.* **103**, 79 (1997).
[135] M. K. Kwak, K. Itoh, M. Yamamoto, T. R. Sutter, and T. W. Kensler, *Mol. Med.* **7**, 135 (2001).

TABLE VIII
QUINONE REDUCTASE INDUCTION EFFECTS OF ISOTHIOCYANATE COMPOUNDS

Structure	Compound	CD (μM)
	Sulforaphane	0.4–0.8
	β-Phenylethyl isothiocyanate	5
	7-Methylsulfinylheptyl isothiocyanate	0.2
	8-Methylsulfinyloctyl isothiocyanate	0.5
	Sulforamate	0.26
	Oxomate	0.96

CD: Concentration to double QR activity in Hepa 1c1c7 cells.

systems (Table VIII).[39] Induction potential was comparable to that observed with sulforaphane (CD range: 0.4 to 0.8 μM), but cytotoxicity was reduced by 3- to 5-fold (IC$_{50}$: 30 to 50 μM). This compound also increased glutathione S-transferase levels about 2-fold in cultured Hepa 1c1c7 mouse and H4IIE rat hepatoma cells. By using Western blotting techniques, expression of two major detoxification isoforms of GST (GST Ya: alpha type; and GST Yb: mu type) were observed in time- and dose-dependent fashions in 4HIIE cell culture. This compound was shown to be a monofunctional inducer by interaction with ARE (antioxidant response element). By using Northern blotting and RT-PCR, time- and dose-dependent induction of quinone reductase mRNA levels were demonstrated. Sulforaphane and the analogs were identified as potent inhibitors of preneoplastic lesion formation in carcinogen-treated MMOC (75% to 85% inhibition at 1 μM).

Sulforamate increased glutathione levels 2-fold in Hepa 1c1c7 and H4IIE rat hepatoma cells and was shown to interact with the ARE without involvement of XRE. However, the difficulty of synthesis of either sulforaphane or sulforamate has hindered their development for further study. A more synthetically accessible analog, oxomate, was devised. Oxomate is a keto-analog to sulforamate and is substantially easier to produce multi-kilogram quantities. Oxomate has weaker induction activity when compared to sulforaphane and sulforamate (CD = 0.96 μM), but toxicity is also substantially reduced (IC$_{50}$ = 67 μM). Oxomate, sulforamate and sulforaphane have shown similar dose-response patterns for the inhibition of DMBA-induced lesions in mouse mammary organ culture, and oxomate significantly reduced tumor multiplicity in DMBA-treated female Sprague-Dawley rats when fed as 3% in the diet.[16]

Dithiocarbamates

The indole dithiocarbamate, brassinin, is found in cruciferous vegetables such as Brussels sprouts, cauliflower and Chinese cabbage. Several natural and synthetic analogs induced phase II enzymes with *in vitro* and *in vivo* models. Of the 27 brassinin derivatives tested, cyclobrassinin (CD = 1.2 μM), spirobrassinin (CD = 7.9 μM), N-ethyl-2,3-dihydrobrassinin (CD = 0.13 μM), and S-selenomethyl brassinin (CD = 0.5 μM) were the more active (Table V).[52]

Steroids

As noted previously, high-throughput screening of quinone reductase activity in murine hepatoma cells has led to the isolates of withanolides (Table III). The withanolides are a group of natural C-28 steroids that occur mainly in plants of certain genera of the Solanaceae. We obtained 12 withanolides from *Physalis philadelphica* Lam. "tomatillos" (Solanaceae) by activity-guided fractionation.[62–65] They are built on an ergostane skeleton in which C-22 and C-26 are appropriately oxidized to form a δ-lactone or δ-lactol ring. Biogenetic transformations can produce highly modified compounds both in the steroid nucleus and the side chain, including the formation of additional rings. The withanolides may be classified according to their structural skeleton; compounds within each group usually differ in the nature and number of oxygenated substituents and the degree of unsaturation of the rings. The diversity of pharmacological effects exhibited by these molecules is obviously a function of the differences in their chemical structure. At first, three withanolides were isolated and shown to have potent inducing activity. CD values of two known compounds, withaphysacarpin and 24,25-dihydrowithanolide D, were 0.70 μM, and the CD value of a novel substance, 2,3-dihydro-3-methoxywithaphysa-carpin, was 7.8 μM. Nine additional withanolides were also isolated as quinone reductase inducers. Philadelphicalactones A showed the most potent activity (CD = 0.04 μM). Subsequent analysis of an additional 37 withanolides isolated from a variety of species from Solanaceae revealed 16 withanolides with CD values below 1 μM. Of those 16 withanolides, only six exhibited a CI above 10: withaphysalin G (CD = 0.70 μM), with-aphysalin H (CD = 0.52 μM), withaphysalin J (CD = 0.39 μM), jaborosa-lactone P (CD = 0.75 μM), jaborosalactone 1 (CD = 0.28 μM), and trechonolide A (CD = 0.27 μM).[63] Spiranoid and 18-functionalized witha-nolides were found to be potent inducers, while 5α-substituted derivatives exhibited weak activity. The stereochemistry at C-18 did not affect inducer potency, since both withaphysalin H (18R 3) and withaphysalin I (18S 4) gave similar CD values (0.52 and 1.04 μM, respectively). When the 2,5-dien-1,4-diketone system was present, no induction was found, although these compounds also presented a functionalized C-18. This finding indi-cates that these structural features have a decisive influence in promoting QR activity. Large-scale isolation of ixocarpalactone A from *Physalis philadelphica* has provided sufficient amounts to begin preliminary animal testing.

Flavonoids

Flavonoids are the most common and widely distributed polyphenolic group in the plant kingdom,[136] comprising several classes, including flavo-nols, flavonones, flavones, anthocyanins and isoflavones. Flavonoids are omnipresent in a human diet, including vegetables, fruits, tea and wine. Consumption has been estimated to reach approximately 1 g/day from food sources.[137] A number of flavonoids have been shown to suppress carcino-genesis in various animal models.[138,139] Multiple mechanisms have been identified for the anti-neoplastic effects of flavonoids, including antioxi-dant, anti-inflammatory and anti-proliferative activities, inhibition of bioactivating enzymes and induction of detoxifying enzymes.[140,141] The antioxidant activity of flavonoids was the first mechanism of action studied, with particular emphasis on protective effects against cardiovascular

[136] J. B. Harbone, "Flavonoids." Chapman & Hall, London, 1993.
[137] L. Bravo, *Nutr. Rev.* **11**, 317 (1998).
[138] L. Sampson, E. Rimm, P. C. Hollman, J. H. de Vries, and M. B. Katan, *J. Am. Diet Assoc.* **102**, 1414 (2002).
[139] A. K. Verma, J. A. Johnson, M. N. Gould, and M. A. Tanner, *Cancer Res.* **48**, 5754 (1998).
[140] K. E. Heim, A. R. Tagliaferro, and D. J. Bobilya, *J. Nutr. Biochem.* **13**, 572 (2002).
[141] L. Marchand, *Biomed. Pharmacother.* **56**, 296 (2002).

diseases.[142] Flavonoids have been shown to be effective scavengers of oxidizing molecules, including singlet oxygen and various free radicals, which are possibly involved in DNA damage and tumor promotion.

Flavonoids may also have a beneficial effect through their impact on the bioactivation of carcinogens by phase II enzymes.[143–146] The structural requirements of naturally occurring flavonoids have been examined with Hepa 1c1c7 cells. It was found that the flavonols were the most effective inducers of quinone reductase, and flavonols and flavans were ineffective.[147] Many flavonoids from a variety of plants have shown quinone reductase inducing activities (Table IX; Fig. 2). To help assess the structural characteristics responsible for this activity, we tested about 60 natural or synthetic flavonoids for inducing potential.[148] Flavone, 3'-bromoflavone, 4'-bromoflavone, 3'-chloroflavone, 4'-chloroflavone, 3'-fluoroflavone, 2'-methoxyflavone and 2'-methoxyflavanone were identified as potent inducers, with CD values of 2.5, 0.1, 0.01, 0.27, 0.02, 0.67, 0.63 and 0.63 μM, respectively. A variety of B-ring substituted 2'-hydroxychalcones and their related flavanones and flavones were examined for inducing activity in cultured Hepa 1c1c7 cells and over 10 compounds exhibited CD values under 1 μM with CI values above 50. The most potent compounds were para-substituted flavones, followed by several members of ortho-substituted chalcones.

Among the active compounds, 4'-bromoflavone (4'BF) was selected for further development because this compound showed not only potent inducing capability (CD = 0.01 μM) in Hepa 1c1c7 cell culture and mouse organs (5-fold induction in liver; 3-fold in mammary gland), but also showed no cytotoxicity in Hepa 1c1c7 and H4IIE cells (IC$_{50}$ >100 μM). The CI value of 4'-bromoflavone was above 10,000. 4'-Bromoflavone increases GSH levels in H4IIE cell culture (2-fold) and mouse organs (3-fold in liver; 2-fold in mammary gland), and was shown to reduce mammary tumor incidence from 89.5 to 30 and 20% in the 2 and 4 g 4'BF/kg diet groups of DMBA-induced Sprague-Dawley rats, respectively. The compound also reduced tumor multiplicity significantly.[148]

---

[142] M. G. L. Hertog, D. Kromhout, C. Aravanis, H. Blackburn, R. Buzina, F. Fidanza, S. Giampaoli, A. Jansen, A. Menotti, S. Nedeljkovic et al., Arch. Intern. Med. 155, 381 (1995).

[143] E. Wollenweber, J. F. Stevens, K. Klimo, J. Knauft, N. Frank, and C. Gerhäuser, Planta Med. 69, 15 (2003).

[144] R. K. Gandhi and K. L. Khanduja, J. Clin. Biochem. Nutr. 14, 107 (1993).

[145] S. Yannai, A. J. Day, G. Williamson, and M. J. Rhodes, Food Chem. Toxicol. 36, 623 (1998).

[146] K. Katiyar and H. Mukhtar, Int. J. Oncol. 8, 221 (1996).

[147] Y. Uda, K. R. Price, G. Williamson, and M. J. Rhodes, Cancer Lett. 120, 213 (1997).

[148] L. L. Song, J. W. Kosmeder 2nd, S. K. Lee, C. Gerhäuser, D. Lantvit, R. C. Moon, R. M. Moriarty, and J. M. Pezzuto, Cancer Res. 59, 578 (1999).

TABLE IX
QUINONE REDUCTASE INDUCTION EFFECTS OF FLAVONOIDS

	Compound	CD (μM)
1	Chrysin	25
2	Galangin	11
3	Kaempferol	7
4	Quercetin	3–20
5	Myricetin	58-NA[a]
6	Apigenin	NA
7	Lanceolatin B	22.9
8	7,4'-Dihydroxy-3',5'-dimethoxyisoflavone	17.2
9	(+)-Purpurin	5.6
10	(+)-Tephrorins A	4.0
11	2'-Methoxyflavanone	0.63
12	5-Hydroxy-7-methoxyflavanone	0.9
13	Catechin	NA
14	Pinocembrin	110
15	5-Methoxy-7-hydroxyflavanone	11
16	5,7-Dimethoxyflavanone	2.0
17	Pinostrobin	0.5
18	Taxifolin	NA
19	Querectin-4'-glucoside	17–30
20	4'-Bromoflavone	0.01
	Sulforaphane[b]	0.2

CD: Concentration to double QR activity in Hepa 1c1c7 cells.
[a] NA: not active.
[b] QR assay positive control.

Two flavonoids, 2'-hydroxy-4',6'-dimethoxychalcone and (±)-5-hydroxy-7-methoxyflavanone, were isolated from the roots of *Renealmia nicolaioides* and induced quinone reductase with observed CD values of 1.7 and 0.9 μg/ml, respectively.[74] The whole flowering and fruiting parts of *Tephrosia purpurea* were chosen for activity-guided fractionation, because petroleum ether- and ethyl acetate-soluble extracts were both found to significantly induce quinone reductase activity with cultured Hepa 1c1c7 mouse hepatoma cells. Flavonoid, lanceolatin B, 7,4'-dihydroxy-3',5-dimethoxyisoflavone, (+)-purpurin were active compounds. These compounds all significantly induced enzyme activity, with CD values of 22.9, 17.2 and 5.6 μM, respectively.[69,149,150] A novel flavonoid, (+)-tephrorins, A, was also isolated and

[149] L. C. Chang, L. L. Song, E. J. Park, L. Luyengi, K. J. Lee, N. R. Farnsworth, J. M. Pezzuto, and A. D. Kinghorn, *J. Nat. Prod.* **63**, 1235 (2000).
[150] L. C. Chang, D. Chávez, L. L. Song, N. R. Farnsworth, J. M. Pezzuto, and A. D. Kinghorn, *Org. Lett.* **2**, 515 (2000).

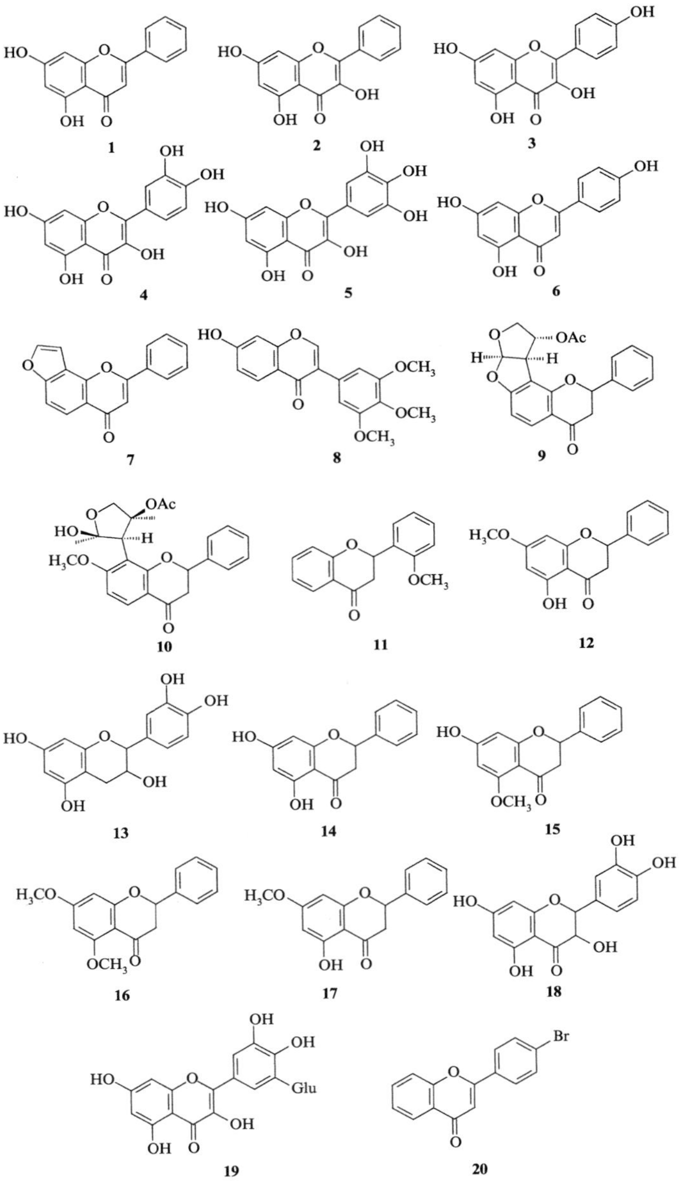

FIG. 2. Flavonoids inducing quinone reductase activity.

induced quinone reductase activity, with an observed CD value of 4.0 μM. The CI value of this compound was 11.8.[149] The induction of phase II detoxification enzymes may be important for the anticarcinogenic effects of these flavonoids.

Chalcones

Chalcones are C ring-open biosynthetic precursors of flavonoids. We have tested 30 naturally occurring chalcone and synthetic derivatives for their potential to induce quinone reductase activity in cultured Hepa 1c1c7 cells (Table X; Fig. 3).[71] 2'-Hydroxy-2-methoxychalcone, 2'-hydroxy-2-methylchalcone, 2'-hydroxy-2-nitrochalcone, 2-bromo-2'-hydroxychalcone and 2'-hydroxy-2,6-dimethoxychalcone were identified as inducers (CD: 0.31, 0.67, 0.30, 0.53, and 0.54 μM, respectively). Preliminary experiments have shown that these compounds induce quinone reductase and elevate glutathione levels in mouse liver and mammary glands. A methanolic extract of the roots of *Renealmia nicolaioides* was found to significantly induce quinone reductase activity with cultured Hepa lclc7 mouse hepatoma cells. Bioassay-guided fractionation of *n*-hexane and ethyl acetate extracts using the induction bioassay has led to 2'-hydroxy-4',6'-dimethoxychalcone as an active substance (CD value: 1.7 μM).[74]

Three quinone reductase inducers, pongamol, (+)-tephrosone, (+)-tephropurpurin were isolated from bioassay-guided fractionation of an EtOAc-soluble residue of *Tephrosia purpurea* Pers. (Leguminosae). Pongamol showed a significant inducing effect with CD and CI values of 3.1 μM and 6.1, respectively. The CD value of (+)-tephrosone was

TABLE X
QUINONE REDUCTASE INDUCTION EFFECTS OF CHALCONES

	Compound	CD (μM)
1	2'-Hydroxy-2,6-dimethoxychalcone	0.54
2	2'-Hydroxy-2-nitrochalcone	0.3
3	2'-Hydroxy-2-methylchalcone	0.67
4	2'-Hydroxy-2-methoxychalcone	0.31
5	2-Bromo-2'-hydroxychalcone	0.53
6	2'-Hydroxy-4,6-dimethoxychalcone	1.7
7	Pongamol	6.1
8	(+)-Tephropurpurin	0.15
9	(+)-Tephrosone	3.1
	Sulforaphane[a]	0.2

CD: Concentration to double QR activity in Hepa 1c1c7 cells.
[a] QR assay positive control.

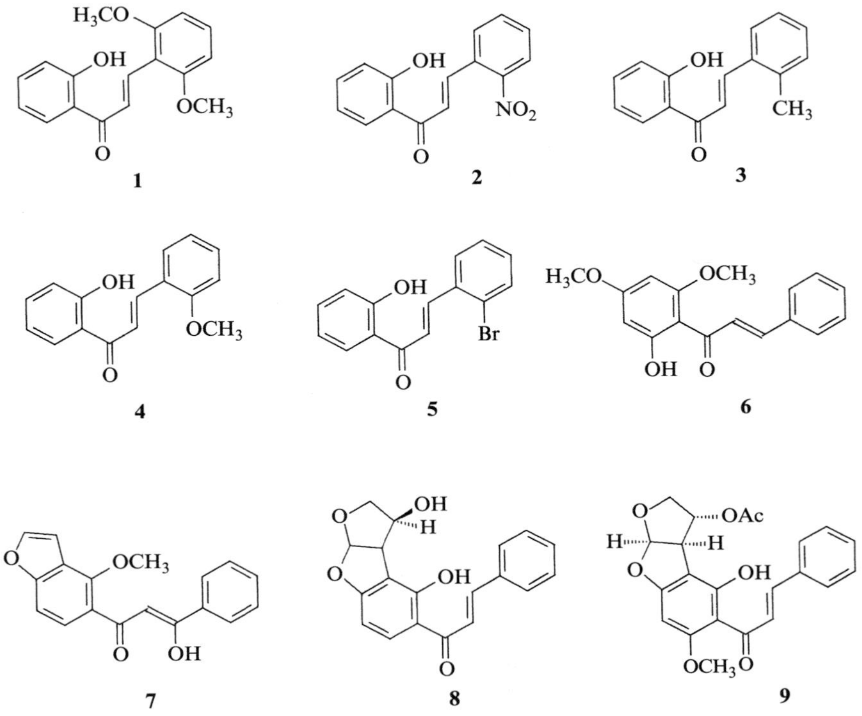

FIG. 3. Chalcones inducing quinone reductase activity.

4.1 μM and the CI value was 6.2. The presence of the bulky cinnamic acid group at C-4' in this compound may affect its biological activity. (+)-Tephropurpurin is the most active compound isolated from *Tephrosia purpurea*. The CD value of (+)-tephropurpurin was 0.15 μM and CI value was 89. (+)-Tephropurpurin was approximately 3-fold more active than sulforaphane, the positive control compound used for this assay, and a superior chemopreventive index (CI) value was obtained as a result of limited cytotoxicity. Based on induction pattern experiments, (+)-tephropurpurin was characterized as a monofunctional inducer by selectively inducing phase II enzymes.

Diarylheptanoid

Two diarylheptanoids, a new bicyclic diarylheptanoid, rel-(3S,4aR, 10bR)-8-hydroxy-3-(4-hydroxyphenyl)-9-methoxy-4a,5,6,10b-tetrahydro-3H-naphtho[2,1-b]pyran and a known 1,7-*bis*(4-hydroxyphenyl)-hepta-4(E),

6(E)-dien-3-one are quinone redutase inducers from *Musa* × *paradi-siaca* L. (Musaceae).[76] Their CD values are 28.7 and 6.4 μM, respectively (Table VI). A new bicyclic diarylheptanoid, rel-(3S,4aR,10bR)-8-hydroxy-3-(4-hydroxyphenyl)-9-methoxy-4a,5,6,10b-tetrahydro-3H-naphtho[2,1-b]-pyran was marginally active (55.6% inhibition at 10 $\mu g/ml$) in a mouse mammary organ culture assay used to evaluate the potential of inhibiting carcinogen-induced preneoplastic lesion formation.

Miscellaneous Compounds

Natural product quinone reductase inducers isolated in our studies showed significant molecular diversity (Fig. 4). Two ceramides—one new ceramide, (2S,3S,4R,9E)-1,3,4-trihydroxy-2-[(2′R)-2′-hydroxytetracosanoy-lamino]-9-octadecene, and one known ceramide, (2S,3S,4R)-2-[(2′R)-2′-hy-droxytetracosanoylamino]-1,3,4-octadecanetriol—were isolated from an ethyl acetate-soluble extract of the leaves and stems of *Physalis philadelphica*.[62] Their CD values were 1.76 and 2.45 μM, respectively. (−)-Maackiain, (−)-3-hydroxy-4-methoxy-8,9-methylenedioxypterocarpan, and (−)-medi-carpin were obtained as active compounds from *Tephrosia purpurea*.[165] These compounds demonstrated significant inducing potency with CD values in a range of 8 to 15 μM. Hydroxyanigorufone was isolated as an active substance from the ethyl acetate-soluble fraction of the methanol ex-tract of *Musa* × *paradisiaca* cultivar,[84] with a CD value of 3.8 $\mu g/ml$. This compound was chosen for evaluation in a mouse mammary organ culture assay to evaluate the potential of inhibiting carcinogen-induced preneoplas-tic lesion formation and was shown to mediate a significant response (63% inhibition at 10 $\mu g/ml$). 1,2-Dihydro-1,2,3-trihydroxy-9-(4-methoxyphenyl)-phenalene was also isolated as an active inducer from the methanol extract of *Musa* × *paradisiaca*. The CD value was 5.3 $\mu g/ml$. A new lanostanoid, 26-hydroxy-5α-lanosta-7,9(11),24-triene-3,22-dione, was isolated from the basi-diocarp of *Ganoderma lucidum*. This lanostan-type triterpenoid doubled the specific activity of quinone reductase at a concentration of 3.0 $\mu g/ml$.[151] In addition, this compound inhibited sheep vesicle cyclooxygenase-1 activity at a test concentration of 40 $\mu g/ml$. A new triterpenoid, coussaric acid A, isolated from *Coussarea brevicaulis*, showed evident induction ability with a CD value of 17.9 μM.[152] Some triterpenoids have been found to be signifi-cant inhibitors of mouse skin carcinogenesis in experimental animals and

[151] T. B. Ha, C. Gerhäuser, W. D. Zhang, N. Ho-Chong-Line, and I. Fouraste, *Planta Med.* **66**, 681 (2000).
[152] B. N. Su, Y. H. Kang, R. E. Pinos, B. D. Santarsiero, A. D. Mesecar, D. D. Soejarto, H. H. S. Fong, J. M. Pezzuto, and A. D. Kinghorn, *Phytochemistry* In Press.

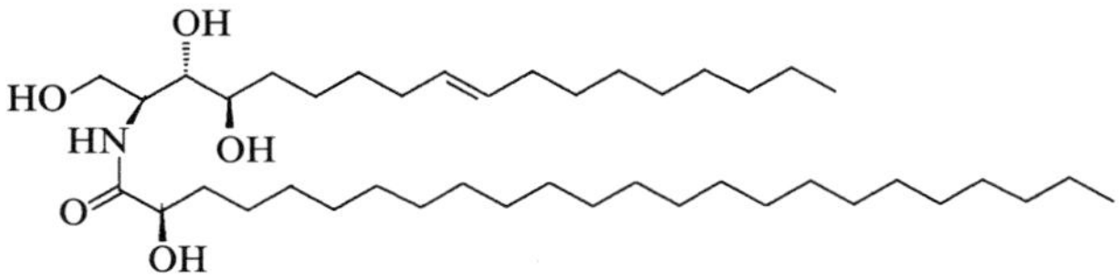

1

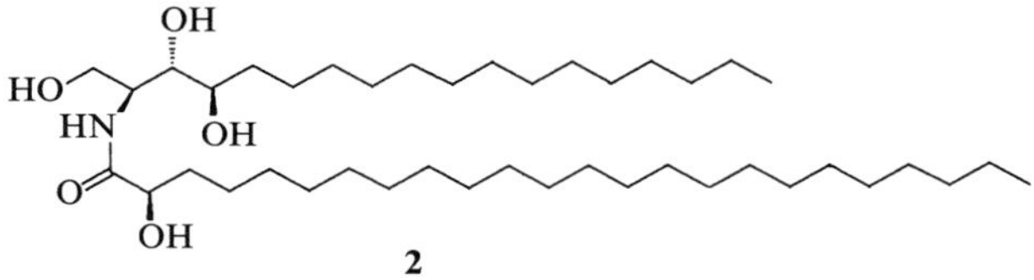

2

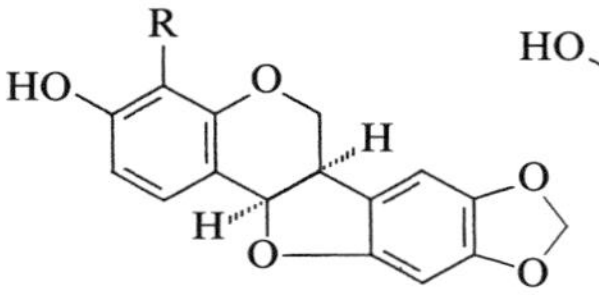

3 R = H
4 R = OCH_3

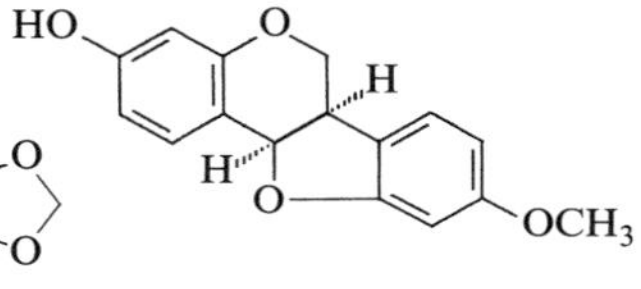

5

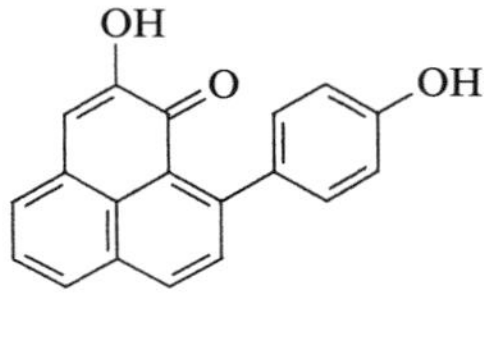

6

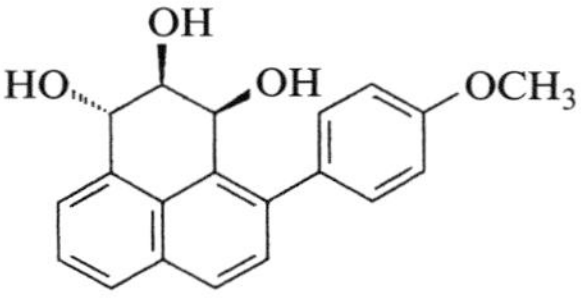

7

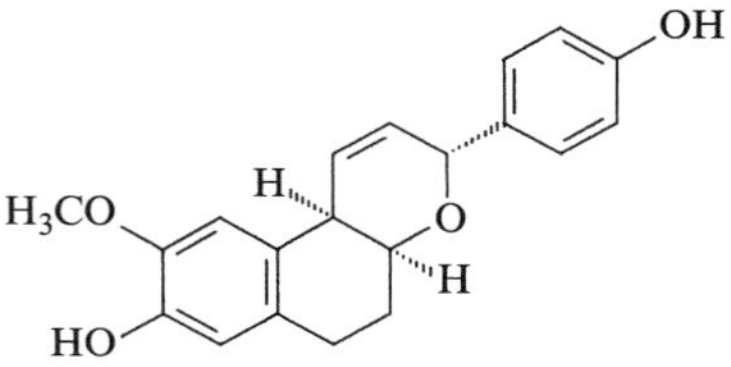

8

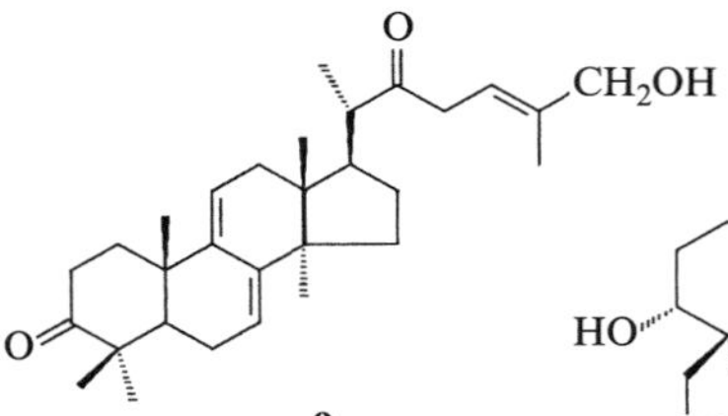

9

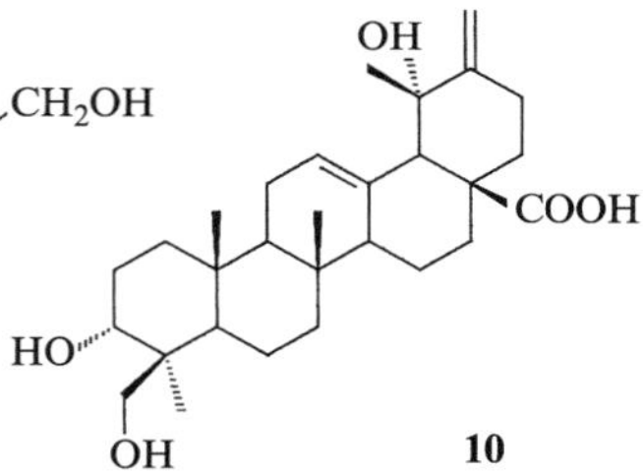

10

FIG. 4. Miscellaneous compounds inducing quinone reductase activity.

inhibitors against production of nitric oxide induced by interferon-γ in mouse macrophages.[153]

Conclusions

Cancer currently claims the lives of over 6 million people each year. While some progress has been made in the diagnosis and treatment of cancer, it is still a rare event when any metastatic malignant condition can be definitively managed, and it is estimated that the number of deaths attributable to cancer will exceed 10 million each year by 2010. Recently, it has been recognized that the logic of cancer prevention is overwhelming, relative to cancer treatment. Induction of the phase II detoxification enzymes such as quinone reductase is a useful strategy for cancer chemo-prevention. As described herein, many edible plants have been found to contain cancer chemopreventive agents capable of inducing phase II enzymes. Notably, since some of the lead compounds are found in vegetables, administration of cancer chemopreventive agents through the diet may be viewed as a convenient and effective strategy in cancer prevention.

As a part of a continuous effort to obtain novel potential cancer chemo-preventive agents from natural products, we have examined potential to induce phase II detoxification enzymes. Following evaluation of approximately 2500 plant extracts, the number characterized as "active" was found to be in the range of 5% of the total. Edible plants were selected with priority for activity-guided fractionation. Over 60 active compounds were isolated as quinone reductase inducers, and substantial molecular diversity was observed. These active compounds comprise ceramides, terpenoids, witha-nolides, flavonoids, chalcones, alkaloids, and diarylheptanoids. In addition to the active phytochemicals, semi-synthetic or synthetic derivatives of bioactive compounds can provide more promising lead compounds. For example, 4'-bromoflavone was found to be an extremely potent inducer of quinone reductase. A number of these agents, such as sulforaphane and sul-foramate, 4'-bromoflavone, brassinin, and withanolides, are considered promising lead anticarcinogens to further investigation in preclinical and clinical studies.

Finally, chemopreventive and chemotherapeutic agents have been traditionally considered as different classes of compounds. While this is true to a certain extent, there is no clear boundary where the function of a chemo-preventive agent ends and a therapeutic agent begins. Therefore, it may be

---

[153] T. Konoshima, M. Takasaki, H. Tokuda, K. Masuda, Y. Arai, K. Shiojima, and H Ageta, *Biol. Pharm. Bull.* **19**, 962 (1996).

useful to evaluate the effectiveness of non-toxic chemopreventive agents as possible chemotherapeutic agents.[45]

Acknowledgments

Work performed in the laboratory of the authors is supported by the National Cancer Institute under the auspices of P01 CA48112 "Natural Inhibitors of Carcinogenesis." The authors gratefully acknowledge Dr. Jerome W. Kosmeder II for critically reviewing the manuscript.

[22]   Chemoprevention by 1,2-Dithiole-3-Thiones Through Induction of NQO1 and Other Phase 2 Enzymes

By MI-KYOUNG KWAK, MINERVA RAMOS-GOMEZ, NOBUNAO WAKABAYASHI, and THOMAS W. KENSLER

Introduction

Cancer chemoprotection can be achieved by diverse classes of molecules with different biological effects including anti-mutagenic, antioxidative, and anti-inflammatory activities.[1,2] One effective strategy of chemoprevention is the modulation of the metabolism of carcinogenic chemicals. Altered biotransformation of carcinogens can result in decreased activation of procarcinogens to reactive intermediates and enhanced elimination of reactive intermediates from the body. Dithiolethiones such as $3H$-1,2-dithiole-3-thione (D3T), anethole dithiolethione (5-[p-methoxyphenyl]-1,2-dithiole-3-thione) and oltipraz (4-methyl-5-[2-pyrazinyl]-1,2-dithiole-3-thione) inhibit the toxicity and carcinogenicity of many chemical carcinogens in multiple target organs and are undergoing preclinical and clinical evaluations for use as cancer chemopreventive agents.[3-6] The broad-based

[1] L. W. Wattenberg, *Cancer Res.* **45,** 1 (1985).

[2] W. K. Hong and M. B. Sporn, *Science* **278,** 1073 (1997).

[3] L. W. Wattenberg and E. Bueding, *Carcinogenesis* **7,** 1379 (1986).

[4] C. V. Rao, A. Rivenson, M. Katiwalla, G. J. Kelloff, and B. S. Reddy, *Cancer Res.* **53,** 2502 (1993).

[5] T. W. Kensler, P. A. Egner, P. Dolan, J. D. Groopman, and B. D. Roebuck, *Cancer Res.* **47,** 4271 (1987).

[6] T. W. Kensler, J. D. Groopman, D. L. Eaton, T. J. Curphey, and B. D. Roebuck, *Carcinogenesis* **13,** 95 (1992).

protective effects of dithiolethiones have been largely associated with the induction of a battery of genes that encode phase 2 and antioxidant enzymes involved in the detoxification of exogenous and endogenous carcinogens.[7] Regulation of both basal and inducible expression of these protective genes is mediated in part by the antioxidant response element (ARE), a cis-acting sequence found in the 5'-flanking region of genes encoding many phase 2 enzymes including mouse glutathione S-transferase (GST) A1, human NAD(P)H quinone oxidoreductase (NQO1) and human γ-glutamylcysteine ligase.[8–10] Recently, Maf and CNC-bZIP ("cap" n collar family of basic region leucine-zipper proteins) transcription factors such as Nrf2 have been shown to be ARE binding proteins. Overexpression of Nrf2 in human hepatoma cells enhanced both basal and inducible activation of an ARE-reporter gene.[11] Studies with *nrf2*-disrupted mice indicated that Nrf2 was essential for the induction of GST and NQO1 activities *in vivo* by different classes of chemopreventive agents including dithiolethiones, isothiocyantes and phenolic antioxidants.[11–13] Moreover, the anticarcinogenic activities of these agents are lost in *nrf2*-deficient mice.[14] Itoh *et al.*[15] have identified and characterized Keap1, an actin-binding protein localized to the cytoplasm that sequesters Nrf2 by specific binding to its amino-terminal regulatory domain. Administration of sulfhydryl reactive compounds abolishes Keap1 repression of Nrf2 activity in cells and facilitates the nuclear accumulation of Nrf2.[15] Direct interaction of enzyme inducers with cysteine residues of Keap1 has been demonstrated by using the irreversible sulfhydryl reactant dexamethasone mesylate.[16] Therefore Keap1 has been proposed

[7] T. W. Kensler, J. D. Groopman, T. R. Sutter, T. J. Curphey, and B. D. Roebuck, *Chem. Res. Toxcol.* **12,** 113 (1999).

[8] T. H. Rushmore, R. G. King, K. E. Paulson, and C. B. Pickett, *Proc. Natl. Acad. Sci. USA* **87,** 3826 (1990).

[9] Y. Li and A. K. Jaiswal, *J. Biol. Chem.* **267,** 15097 (1992).

[10] H. R. Moinova and R. T. Mulcahy, *J. Biol. Chem.* **273,** 14683 (1998).

[11] K. Itoh, T. Chiba, S. Takahashi, T. Ishii, K. Igarashi, Y. Katoh, T. Oyake, N. Hayashi, K. Satoh, I. Iatayama, M. Yamamoto, and Y. Nabeshima, *Biochem. Biophys. Res. Commun.* **236,** 313 (1997).

[12] M.-K. Kwak, K. Itoh, M. Yamamoto, T. R. Sutter, and T. W. Kensler, *Mol. Med.* **7,** 135 (2001).

[13] M. MacMahon, K. Itoh, M. Yamamoto, S. A. Chanas, C. J. Henderson, L. I. McLellan, C. R. Wolf, C. Cavin, and J. D. Hayes, *Cancer Res.* **61,** 3299 (2001).

[14] M. Ramos-Gomez, M.-K. Kwak, P. M. Dolan, K. Itoh, M. Yamamoto, P. Talalay, and T. W. Kensler, *Proc. Natl. Acad. Sci. USA* **98,** 3410 (2001).

[15] K. Itoh, N. Wakabayashi, Y. Katoh, T. Ishii, K. Igarashi, J. D. Engel, and M. Yamamoto, *Genes Dev.* **13,** 76 (1999).

[16] A. T. Dinkova-Kostova, W. D. Holtzclaw, R. N. Cole, K. Itoh, N. Wakabayashi, Y. Katoh, M. Yamamoto, and P. Talalay, *Proc. Natl. Acad. Sci. USA* **99,** 11908 (2002).

as a putative cellular sensor of oxidative stress. The critical role of Keap1 in the regulation of Nrf2 function has been established in a study using *keap1*-knockout mice. *Keap1*-disrupted mice are not viable after 3 weeks of age; however, expression levels of phase 2 detoxifying genes such as NQO1 are much higher in neonatal *keap1*-disrupted animals than in wild-type mice.[17] These results collectively support that Keap1-Nrf2 signaling system is a main signaling pathway in the regulation of protective genes against cellular stress conditions such as oxidative stresses. This chapter highlights the essential role of Keap1-Nrf2 signaling system in the chemopreventive efficacy of dithiolethiones and focuses on the quinone-detoxifying enzyme NQO1 as a prototypical phase 2 enzyme.

Induction of NQO1 by Chemopreventive Dithiolethiones and a Role of NRF2 in the Induction of NQO1 in Mice

NQO1 Enzyme Activity Following Oltipraz Treatment In Wild-Type and nrf2-Knockout Mice

Female ICR mice (7 to 9 weeks old), which were either wild-type or *nrf2*-deficient, were fed AIN-76A diet and water *ad libitum* and treated by gavage with oltipraz (500 mg/kg, suspended in 1% cremophor and 25% glycerol) or vehicle. Mice were killed 48 h after treatment for measurement of enzyme activity. Liver, lung, kidney and forestomach were harvested from mice, and cytosolic fractions ($105,000 \times g$) were prepared. NQO1 activity was determined by using menadione ($50 \ \mu M$) as substrate in the presence of an NADPH-generating system and the obtained activity normalized by measuring protein content with the bicinchoninic acid assay.[18] Oltipraz treatment significantly increased NQO1 enzyme activity in these four tissues of wild-type mice. Livers showed 4-fold induction of enzyme activity by oltipraz but only marginal induction in lungs (Fig. 1). Basal activity of hepatic NQO1 in *nrf2*-deficient mice was 70% lower than in wild-type mice; similar reductions of constitutive activity were detected in forestomach and kidney. Moreover, the induction of NQO1 activity by oltipraz was largely attenuated in all tissues of the *nrf2*-deficient mice.

[17] N. Wakabayashi, K. Itoh, J. Wakabayashi, H. Motohashi, S. Noda, S. Takahashi, S. Imakado, T. Kotsuji, F. Otsuka, D. R. Roop, T. Harada, J. D. Engel, and M. Yamamoto, *Nature Genetics* **35**, 238 (2003).

[18] P. K. Smith, R. I. Krohn, G. T. Hermanson, A. K. Mallia, F. H. Gartner, M. D. Provenzano, E. K. Fujimoto, N. M. Goeke, B. J. Olson, and D. C. Klenk, *Anal. Biochem.* **150**, 76 (1985).

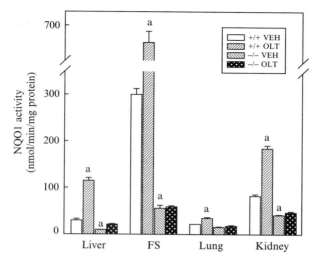

Fig. 1. Effect of *nrf2* genotype and oltipraz on NQO1 activity in mouse liver, forestomach (FS), lung and kidney. NQO1 activity was measured in cytosols prepared from wild-type (+/+) and *nrf2*-deficient mice (−/−) treated with vehicle (VEH) or oltipraz (OLT, 500 mg/kg) 48 h before sacrifice. Values are mean ± SE for three to four animals in each group. [a], $P < 0.05$ compared with wild-type vehicle-treated control.

NQO1 mRNA Levels Following Oltipraz Treatment in Wild-Type and nrf2-*Knockout Mice*

Mice were sacrificed 24 h after treatment with oltipraz (500 mg/kg) and liver, lung, kidney and forestomach were collected from each mouse for mRNA analysis (Fig. 2). Total RNA was isolated from these tissues by the procedure of Chomczynski and Sacchi. RNA samples were subjected to electrophoresis on 1% agarose gels containing 2.2 *M* formaldehyde and transferred to nylon membranes. Membranes were hybridized with [α-^{32}P] dCTP-labeled mouse NQO1 cDNA at 42° overnight, washed and exposed to x-ray film. Imaged densities of NQO1 mRNA were normalized by using GAPDH mRNA levels (Table I). In wild-type mice, levels of NQO1 mRNA were increased by oltipraz treatment 7.6-, 3.5-, 7.4- and 1.7-fold in liver, lung, kidney and forestomach, respectively. In accord with enzymatic activities, basal levels for NQO1 mRNA were repressed in *nrf2*-deficient mice in these different tissues, and inducible levels were completely abolished in the absence of Nrf2. Collectively, these results demonstrate that NQO1 is a highly inducible gene in multiple tissues

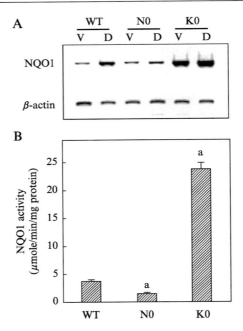

FIG. 2. Levels of NQO1 in wild-type (WT), *nrf2*-deficient (N0) or *keap1*-deficient (K0) MEF cells. (A) NQO1 mRNA levels measured in these cells when using RT-PCR. RT-PCR analysis was performed with total RNA prepared from vehicle (V, DMSO) or D3T (D, 20 μM)-treated cells for 24 h. (B) NQO1 enzyme activity was measured in MEF cells when using a microtiter plate method as described previously[12]; enzyme activity is expressed as μmole of converted menadiol per mg protein/min. [a], $P < 0.05$ compared with vehicle-treated wild-type cells.

following treatment of mice with chemopreventive dithiolethiones and that Nrf2 is an essential transcription factor for the basal and inducible expression of this gene.

Susceptibility to Benzo[a]pyrene-Induced Gastric Tumor Formation in Wild-Type and nrf2-Deficient Mice

Female wild-type and *nrf2*-deficient mice were randomized into groups of 20 mice and treated with vehicle or 500 mg/kg oltipraz by gavage. Benzo(a)pyrene (120 mg/kg in corn oil) was given 48 h after oltipraz treatment, and this sequential administration was repeated once a week for 4 weeks. Animals were killed 30 weeks after the first oltipraz treatment,

TABLE I
EFFECTS OF *nrf2* GENOTYPE ON CHANGES IN mRNA LEVELS FOR NQO1
AFTER OLTIPRAZ TREATMENT

Tissues	WT	*nrf2*-knockout	WT	*nrf2*-knockout
	Relative Constitutive NQO1 mRNA (Vehicle/Vehicle)		Relative Inducible NQO1 mRNA (Oltipraz/Vehicle)	
Forestomach	1	0.08	1.68[a]	0.10
Liver	1	0.80	7.64[a]	1.39
Lung	1	0.21	3.48[a]	0.33
Kidney	1	0.42	7.44[a]	1.17

Levels of NQO1 mRNA were measured by northern hybridization in mice tissues following vehicle or oltipraz treatment (24 h after treatment). Values are the mean of determinations of three to four mice. Levels of NQO1 mRNA of forestomach, liver, lung and kidney were normalized by GAPDH, albumin and β-actin mRNA levels, respectively, and expressed as a ratio over vehicle-treated, wild-type control.
[a] $P < 0.05$ compared with wild-type vehicle-treated control.

and the tumors in the forestomach were counted grossly.[14] Vehicle-treated, wild-type mice had 9.5 ± 1.0 gastric tumors per mouse from benzo(a)-pyrene treatment, while oltipraz pre-treatment decreased tumor multiplicity to 4.6 ± 0.5. However, tumor numbers increased significantly (14.6 ± 1.2) in vehicle-treated, *nrf2*-deficient mice as compared with vehicle-treated wild-type mice, and oltipraz completely lost its protective efficacy in *nrf2*-deficient mice (13.6 ± 1.1). Basal activity of NQO1 in the forestomach was 4 to 10 fold higher than in other tissues (Fig. 3), and *nrf2*-deficient mice had only 10% of NQO1 activity as compared with wild-type mice (Table I). In accordance with this carcinogenesis study, levels of benzo(a)pyrene-DNA adducts in the forestomach were significantly higher in *nrf2*-deficient as compared with wild-type mice.[19] Although the effects of *nrf2* genotype on the overall metabolism of benzo(a)pyrene—and in particular the role of benzo(a)pyrene-quinones in forestomach carcinogenesis— have not been investigated, these results indicate a correlation between Nrf2-dependent expression of NQO1 and susceptibility to chemical carcinogenesis *in vivo*.

[19] M. Ramos-Gomez, P. M. Dolan, K. Itoh, M. Yamamoto, P. Talalay, and T. W. Kensler, *Carcinogenesis* **24**, 461 (2002).

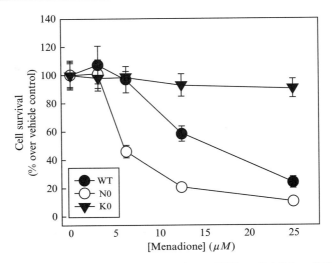

FIG. 3. Menadione-induced cytotoxicity in wild-type (WT), *nrf2*-deficient (N0) or *keap1*-deficient (K0) MEF cells. Cells were incubated with menadione (3.13, 6.25, 12.5 and 25 μM) for 24 h and extent of cytotoxicity was measured by MTT assay. Viable cell numbers are expressed as ratios relative to vehicle-treated (PBS) controls. Values are mean $\pm$ SD from 16 measurements in each group.

Microarray Analysis of Dithiolethione-Inducible Genes in Mouse Liver and Effect of *nrf2* Genotype

Hepatic Gene Expression Patterns Following D3T Treatment in Wild-Type and nrf2-Disrupted Mice

Wild-type and *nrf2*-disrupted mice (male mice, 8 to 10 weeks old) were dosed with vehicle (1% cremophor and 25% glycerol) or D3T (0.5 mmole/kg) by gavage and sacrificed 6 h and 24 h after treatment. Total RNAs were isolated from livers of mice and biotin-labeled cRNAs were prepared by *in vitro* transcription. Fragmented cRNA was hybridized at 45° for 16 h to Murine Genome U74Av2 GeneChip (Affymetrix), and images were scanned and analyzed with Affymetrix Microarray Suite 4.1. D3T treatment increased multiple xenobiotic metabolizing genes in wild-type mice.[20] NQO1 transcript levels were increased rapidly following D3T treatment and showed 5.2- and 6.0-fold increases 6 h and 24 h after treatment with D3T, respectively, in wild-type mice (Table II). However, induction of this gene was not detected in *nrf2*-disrupted mice. Supply of the electron donor NADPH is important for the function of NQO1. Genes involved in NADPH-generation were also increased by D3T treatment. UDP-glucose

TABLE II

Subset of Genes Induced by D3T Treatment in Livers From Wild-Type and *nrf2*-Knockout Mice

Categories of Genes	Description of Genes	Average Fold Change (D3T/Vehicle)		*nrf2*-knockout
		WT		
		6 h	24 h	24 h
Detoxifying Genes	NAD(P)H:menadione oxidoreductase (NQO1)	5.2	6.0	
	Glutathione-S-transferase, mu 1	1.5	3.1	
	Glutathione-S-transferase, mu 2		4.4	
	Glutathione-S-transferase, mu 3	1.7	4.6	
	Glutathione transferase, alpha 4	1.4	3.3	
	Glutathione-S-transferase, alpha 2	2.2	5.9	
	Glutathione S-transferase, theta 2		3.3	
	Microsomal glutathione S-transferase 3	1.5		
	Gamma-glutamylcysteine synthetase, catalytic	3.6	2.8	
	Gamma-glutamylcysteine synthetase, regulatory	1.6	2.4	
NADPH-Generating Enzymes	NADH-glucose dehydrogenase	3.9	7.5	1.7
	Malate oxidoreductase	2.0		
Heat Shock Proteins	Hsp40 homolog, B11		4.1	
	Hsp40 homolog, B9	6.0	3.9	3.2
	Hsp70, protein 8		2.2	1.8
	Hsp70, protein 5		3.3	2.2
	Stress-induced phosphoprotein 1		2.0	

dehydrogenases and malate oxidoreductase were induced 3.8- and 2.0-fold 6 h after treatment and transcript levels of UDP-glucose dehydrogenases remained increased (7.5-fold) 24 h after treatment. These increases were *nrf2*-dependent as the induction of these genes was largely attenuated in *nrf2*-disrupted mice. D3T treatment also increased several heat shock proteins including Hsp 40 homolog and Hsp 70. Recently it has been reported that Hsp 40 and 70 bind to the NQO1 enzyme *in vitro* and *in vivo,* suggesting that stability and function of NQO1 enzyme are enhanced by binding to these chaperones. Facilitation of correct protein folding by these protein chaperones may result in more efficient expression of NQO1 enzyme. Increased expression of heat shock proteins, coupled with elevated intracellular generation of NADPH and NQO1 expression itself, could lead to enhanced resistance to quinone-mediated toxicities.

NQO1 Levels and Cytotoxicity Against Quinone Compound Menadione in Murine Embryonic Fibroblasts (MEF)

Levels of NQO1 in MEF From Wild-Type, nrf2- or Keap1-Disrupted Mice

Murine embryonic fibroblasts were seeded in 96-well plates at a density of 5×10^3 cells/well and cultured for overnight in Iscove's Modified Dulbeco's Medium containing 10% fetal bovine serum and antibiotics. Cells were treated with D3T (10 and 20 μM) for 18 h and harvested for analysis of levels of mRNA of NQO1. For RT-PCR analysis of NQO1 mRNA, cDNA prepared from 100 ng of total RNA was used for PCR amplification with the following primers for mouse NQO1 (ATCCTTCC-GAGTCATCTCTA and CAACGAATCTTGAATGGAGG). D3T increased levels of NQO1 mRNA 4-fold in wild-type cells, while this induction was largely abolished in *nrf2*-deficient cells (Fig. 2A). *Keap1*-deficient cells showed very high basal level of NQO1 mRNA, clearly indicating the inhibitory role of Keap1 on the function of Nrf2. NQO1 activity was also greatly increased (6.4 fold) in *keap1*-deficient cells as compared with wild-type cells, whereas it was reduced to 60% of the activity of wild-type cells in *nrf2*-deficient cells (Fig. 2B).

Cytotoxicity by Menadione in MEF Cells

MEF cells were treated with menadione (3.13, 6.25, 12.5, 25.0 μM) for 24 h and cytotoxicity determined by MTT (3-[4,5-dimethylthiazole-2-yl]-2,5-diphenyl-tetrazolium bromide) assay[21] (Fig. 3). *Nrf2*-deficient cells, having repressed expression of NQO1 were significantly more sensitive to menadione exposure than wild-type cells. It was found that 54% and 79% of cells were dead by 6.25 and 12.5 μM menadione incubation in *nrf2*-deficient cells, while 10% and 46% cells were killed by same concentrations of menadione, respectively, in wild-type cells. However, *keap1*-deficient cells, which express high levels of NQO1, were very resistant to menadione, and only 9% cells were killed by 25 μM menadione. Regulation of NQO1 through the Nrf2-Keap1 pathway is an important determinant of cellular sensitivity to quinones and is an attractive target for chemopreventive interventions.

---

[20] M.-K. Kwak, N. Wakabayashi, K. Itoh, H. Motohashi, M. Yamamoto, and T. W. Kensler, *J. Biol. Chem.* **278,** 8135 (2003).

[21] J. Carmichael, W. G. DeGraff, A. F. Gazdar, J. D. Minna, and J. B. Mitchell, *Cancer Res.* **47,** 936 (1987).

[23]   Chemical Structures of Inducers of Nicotinamide Quinone Oxidoreductase 1 (NQO1)

By ALBENA T. DINKOVA-KOSTOVA, JED W. FAHEY, and PAUL TALALAY

Introduction

More than 40 years ago, Williams-Ashman and Huggins (1961) made the fortuitous observation that the enzyme discovered by Lars Ernster,[1] DT-diaphorase, NAD(P)H:quinone oxidoreductase 1, NQO1 (EC 1.6.99.2), was highly inducible in rat tissues.[2] While developing the now widely used DMBA mammary tumor model in Sprague-Dawley rats, Huggins and his colleagues showed that administration of small doses of aromatic hydrocarbons before challenge with DMBA protected against the lethal effects of this potent carcinogen.[3] This protection was accompanied by elevation of NQO1 activity in the liver and other organs.[4,5] Furthermore, the potencies of structurally related azo dyes in protecting against DMBA carcinogenesis paralleled their effectiveness in inducing NQO1. This finding led Huggins to propose the use of activity of NQO1 as a rapid screening procedure for the identification of protectors against

---

[1] L. Ernster and F. Navazio, *Acta Chem. Scand.* **12,** 595 (1958).

[2] H. G. Williams-Ashman and C. Huggins, *Med. Exper.* **4,** 223 (1961).

[3] C. B. Huggins, *in* "Experimental Leukemia and Mammary Cancer: Induction, Prevention, Cure." The University of Chicago Press, Chicago, 1979.

[4] C. Huggins and R. Fukunishi, *J. Exp. Med.* **119,** 923 (1964).

[5] C. Huggins and J. Pataki, *Proc. Natl. Acad. Sci. USA* **53,** 791 (1965).

carcinogenesis.[3] The ultimate importance of this long-neglected prediction cannot be overestimated.

Two parallel lines of research caused this idea to be revisited. The laboratories of Frankfurt and Wattenberg demonstrated that the phenolic antioxidants 2(3)-*tert*-butyl-4-hydroxyanisole (BHA) and 3,5-di-*tert*-butyl-4-hydroxytoluene (BHT) that are commonly used as food preservatives protect rodents against a wide range of chemical carcinogens.[6,7] Subsequently Bueding and Talalay reported that administration of BHA to mice reduced the formation of mutagenic metabolites from benzo[a]pyrene and raised the activity of NQO1 in many tissues parallel to the increases in the levels of classical phase 2 detoxication enzymes—for example, glutathione *S*-transferases, epoxide hydrolase.[8–11] These observations led to the suggestion that the protective function of phenolic antioxidants against tumor development in animals could be ascribed to the induction of the phase 2 response. In the ensuing years, the evidence supporting this suggestion has become progressively more convincing.[12–16]

Subsequently, structure-activity studies revealed that a wide array of compounds caused coordinate induction of phase 2 enzymes and protected against carcinogenesis. The development of a cultured cell microtiter plate bioassay for NQO1 by Hans Prochaska provided a quick and highly quantitative system for evaluation of the potencies of inducers and for screening pure compounds, as well as complex mixtures such as plant extracts, for their inducer activity.[17–20]

[6] O. S. Frankfurt, L. P. Lipchina, T. V. Bunto, and N. M. Emanuel, *Biull. Eksp. Biol. Med.* **64,** 86 (1967).

[7] L. W. Wattenberg, *Adv. Cancer Res.* **26,** 197 (1979).

[8] R. P. Batzinger, S. Y. Ou, and E. Bueding, *Cancer Res.* **38,** 4478 (1978).

[9] A. M. Benson, R. P. Batzinger, S. Y. Ou, E. Bueding, Y.-N. Cha, and P. Talalay, *Cancer Res.* **38,** 4486 (1978).

[10] A. M. Benson, Y.-N. Cha, E. Bueding, H. S. Heine, and P. Talalay, *Cancer Res.* **39,** 2971 (1979).

[11] A. M. Benson, M. J. Hunkeler, and P. Talalay, *Proc. Natl. Acad. Sci. USA* **77,** 5216 (1980).

[12] P. Talalay, *Adv. Enzyme Regul.* **28,** 237 (1989).

[13] P. Talalay, J. W. Fahey, W. D. Holtzclaw, T. Prestera, and Y. Zhang, *Toxicol. Lett.* **82–83,** 173 (1995).

[14] T. W. Kensler, *Environ. Health Perspect.* **105,** 965 (1997).

[15] P. Talalay, *Biofactors* **12,** 5 (2000).

[16] P. Talalay and A. T. Dinkova-Kostova, *Methods Enzymol.* **382,** 355 (2004).

[17] H. J. Prochaska and A. B. Santamaria, *Anal. Biochem.* **169,** 328 (1988).

[18] H. J. Prochaska, A. B. Santamaria, and P. Talalay, *Proc. Natl. Acad. Sci. USA* **89,** 2394 (1992).

[19] J. W. Fahey, Y. Zhang, and P. Talalay, *Proc. Natl. Acad. Sci. USA* **94,** 10367 (1997).

[20] J. W. Fahey, A. T. Dinkova-Kostova, K. K. Stephenson, and P. Talalay, *Methods Enzymol.* **382,** 243 (2004).

Identification of the Chemical Signals for Monofunctional
Phase 2 Gene Induction

Phenolic Antioxidants and Their Alkyl Ethers as Inducers

The proposal that the protection by BHA and related phenolic antioxidants against chemical carcinogens, mutagens, and other toxic agents resulted from the induction of a family of phase 2 proteins that detoxified the reactive metabolites of these agents, focused attention on the mechanisms of the transcriptional enhancement of these genes. Elucidation of these mechanisms required understanding both of the molecular events involved in regulating these genes, and of the structural specificity of the inducers. This chapter describes the development of our understanding of the nature of the chemical signals involved in the induction of NQO1 and other phase 2 enzymes. In our first efforts to identify the essential structural features of inducers, we compared the inducer potencies of the two isomers of commercial BHA ([2]- and [3]-*tert*-butyl-4-hydroxyanisole), their demethylation product *tert*-butylhydroquinone, and an extended series of synthetic mono- and dialkyl ethers of *tert*-butylhydroquinone $R_1O-[(CH_3)_3C-C_6H_3]-OR_2$ on the levels of NQO1 and glutathione transferases in mouse liver and small intestinal mucosa.[21-23] These studies revealed that there was relatively little structural specificity among these compounds and that the free diphenol, *tert*-butylhydroquinone, was more potent than its alkyl ethers. Moreover, the *tert*-butyl group influenced potency only slightly. We concluded that the proximate inducers were probably the free phenols and that the substituted alkyl phenols required metabolic dealkylation.

Inducer Potency of Phenolic Antioxidants Depends on Oxidative Lability

However, the two phenolic hydroxyl groups of all these compounds were 1,4-*(para)*-oriented. Further experiments with a variety of 1,2-diphenols (catechols), 1,3-diphenols (resorcinols), and 1,4-diphenols (hydroquinones) disclosed the important finding that resorcinols were invariably completely inactive as inducers.[22] Moreover, analogous behavior was observed with 1,2-, 1,3-, and 1,4-phenylenediamines: the 1,3-diamines were not inducers. We concluded from these observations that capacity for oxidation to

[21] H. J. Prochaska, H. S. Bregman, M. J. De Long, and P. Talalay, *Biochem. Pharmacol.* **34,** 3909 (1985).
[22] H. J. Prochaska, M. J. De Long, and P. Talalay, *Proc. Natl. Acad. Sci. USA* **82,** 8232 (1985).
[23] M. J. De Long, H. J. Prochaska, and P. Talalay, *Cancer Res.* **45,** 546 (1985).

quinones or quinoneimines, respectively, was an essential property of inducers. The 1,3-*meta* derivatives cannot undergo such oxidation, whereas both catechols and 1,4-hydroquinones can be easily oxidized to quinones. These experiments did not, however, disclose whether the quinone (or quinoneimine) products were in fact the ultimate inducers or whether the oxidation reactions themselves participated in the induction mechanism.

Monofunctional and Bifunctional Inducers

Related experiments on the mechanism of induction revealed that inducers of phase 2 enzymes are of two types: (1) bifunctional inducers (principally large planar aromatics such as 2,3,7,8-tetrachlorodibenzo-*p*-dioxin [TCDD], polycyclic aromatics, azo dyes, β-naphthoflavone) elevate both phase 2 and phase 1 (certain cytochromes P450) genes; and (2) monofunctional inducers (that differ widely in chemical structure) elevate phase 2 enzymes selectively.[24] In a murine hepatoma cell line (Hepa1c1c7) and its mutants defective in their ability to express certain phase 1 enzymes, and in analogous genetically defined mice, it was shown that monofunctional inducers did not depend on *Ah* (Aryl hydrocarbon) receptor function or aryl hydrocarbon hydroxylase expression, whereas bifunctional inducers did. The current review of inducer chemistry is confined to monofunctional inducers.

Michael Reaction Acceptors as Inducers

The next major advance in understanding of the chemical characteristics of inducers arose from an examination of the components of the structure of coumarin.[25] These experiments led to the realization that indeed many inducers contain a previously unrecognized olefinic or acetylenic linkage conjugated to electron-withdrawing groups, a prime example being quinones that contain olefinic functions conjugated to carbonyl groups (Tables I through V). These functional groups, designated as Michael reaction acceptors, are highly susceptible to attack by nucleophiles (on the β-carbon atom), and form the basis for the Michael reaction that is widely used in organic synthesis. The reactivity of Michael acceptors with nucleophiles depends on the strength of the electron-withdrawing function: NO_2 > COAr > CHO > $COCH_3$ > CN > $CONH_2$ > $CONR_2$. Moreover the potency of inducers of NQO1 was correlated with the power of the electron withdrawing group and consequently with the nucleophilicity of the electrophilic carbon center.

[24] H. J. Prochaska and P. Talalay, *Cancer Res.* **48,** 4776 (1988).
[25] P. Talalay, M. J. De Long, and H. J. Prochaska, *Proc. Natl. Acad. Sci. USA* **85,** 8261 (1988).

TABLE I
INDUCERS OF NQO1: PHENOLIC ANTIOXIDANTS, DIPHENOLS,
AMINOPHENOLS, PHENYLENEDIAMINES

Compound	Tissue or cell line	Reference
3-BHA 2-BHA	Liver, kidney, lung, forestomach, glandular stomach, small intestine, colon, uterus, spleen, bladder, adrenal	Benson et al. [11]; Talalay and Benson [26]; Cha and Heine [27]; De Long et al. [23]; Heine et al. [28]; Siegel et al. [29]; Shertzer and Sainsbury [30]; Munday and Munday [31]; McMahon et al. [32]
BHT 4-Hydroxy-anisole	Liver, kidney, forestomach, glandular stomach, small intestine, colon, Hepa1c1c7	De Long et al. [23, 33]; Prochaska et al. [21, 22]
$R_1 = R_2 = CH_3, C_2H_5, C_3H_7$	Liver, small intestine	Prochaska et al. [21]
(aminophenol/phenylenediamine structures)	Hepa1c1c7	Prochaska et al. [21]; De Long et al. [33]

[26] P. Talalay and A. M. Benson, *Adv. Enzyme Regul.* **20,** 287 (1982).

[27] Y.-N. Cha and H. S. Heine, *Cancer Res.* **42,** 2609 (1982).

[28] H. S. Heine, M. K. Stoskopf, D. C. Thompson, and Y.-N. Cha, *Chem. Biol. Interact.* **59,** 219 (1986).

[29] D. Siegel, A. M. Malkinson, and D. Ross, *Toxicol. Appl. Pharmacol.* **96,** 68 (1988).

[30] H. G. Shertzer and M. Sainsbury, *Food Chem. Toxicol.* **29,** 391 (1991).

[31] R. Munday and C. M. Munday, *Nutr. Cancer* **34,** 42 (1999).

TABLE II
INDUCERS OF NQO1: COUMARINS AND OTHER CYCLIC LACTONES AND ENONES

Compound	Tissue or cell line	Reference
 $R_1 = R_2 = $ H,Coumarin (weak) $R_1 = $ H, $R_2 = $ OH, 7-Hydroxycoumarin $R_1 = $ OH, $R_2 = $ H, 3-Hydroxycoumarin $R_1 = COCH_3$, $R_2 = $ H, 3-Acetylcoumarin	Hepa1c1c7, small intestine	De Long et al. [33]; Talalay et al. [25]; Dinkova-Kostova et al. [34]; McMahon et al. [32]
 α-angelicalactone	Hepa1c1c7, small intestine	De Long et al. [33]; Talalay et al. [25]; McMahon et al. [32]
	Hepa1c1c7	Talalay et al. [25]
	Hepa1c1c7	Prestera et al. [35]

Nine Chemical Classes of Inducers

At first glance, the structural versatility of inducers seemed extraordinary. Today, inducers can be classified into 9 categories.

Class 1: Diphenols, Phenylenediamines, and Quinones

The dietary antioxidants that were first shown to induce NQO1 belong to this class.[9,11,26] Subsequently, we and many other investigators have demonstrated that members of this class induce NQO1 in various rodent

[32] M. McMahon, K. Itoh, M. Yamamoto, S. A. Chanas, C. J. Henderson, L. I. McLellan, C. R. Wolf, C. Cavin, and J. D. Hayes, Cancer Res. 61, 3299 (2001).
[33] M. J. De Long, H. J. Prochaska, and P. Talalay, Proc. Natl. Acad. Sci. USA 83, 787 (1986).
[34] A. T. Dinkova-Kostova, C. Abeygunawardana, and P. Talalay, J. Med. Chem. 41, 5287 (1998).
[35] T. Prestera, Y. Zhang, S. R. Spencer, C. A. Wilczak, and P. Talalay, Adv. Enzyme Regul. 33, 281 (1993).

TABLE III
INDUCERS OF NQO1: MICHAEL REACTION ACCEPTORS: FUMARIC, MALEIC,
ACRYLIC, CROTONIC, FERULIC, AND CAFFEIC ACID DERIVATIVES

Compound	Tissue or cell line	Reference
trans $CH_3OOCCH = CHCOOCH_3$ *trans* $C_2H_5OOCCH = CHCOOC_2H_5$ *cis* $CH_3OOCCH = CHCOOCH_3$ *cis* $C_2H_5OOCCH = CHCOOC_2H_5$ $CH_3OOCC (= CH_2)CH_2COOCH_3$ $CH_3OOCC \equiv CCOOCH_3$ $CH_3CH \equiv CH—R$ $R = COOCH_3, CHO, CHCN$ $CH_3CH \equiv CCOOCH_3$ $CH_2 = CH — R$ $R = COOCH_3, CHO, COCH_3, SO_2CH_3, CN$ $CH \equiv CCOOCH_3$	Forestomach, glandular stomach, intestine, colon, spleen, Hepa1c1c7	Spencer *et al.* [36]; Talalay *et al.* [25]
 $X = CH_3O, NO_2, Br, Cl, F$	Hepa1c1c7	Spencer *et al.* [37]; Dinkova-Kostova *et al.* [34]
 $R_1 = R_2 = R_3 = R_4 = H$ $R_1 = R_4 = H, R_2 = R_3 = OH$ $R_1 = OH, R_2 = R_3 = R_4 = H$ $R_1 = R_3 = R_4 = H, R_2 = OH$ $R_1 = R_2 = R_4 = H, R_3 = OH$ $R_1 = R_4 = H, R_2 = OCH_3, R_3 = OH$ $R_1 = H, R_2 = R_4 = OCH_3, R_3 = OH$	Hepa1c1c7	Talalay *et al.* [25]; Dinkova-Kostova *et al.* [34]
 Caffeic acid phenethyl ester	HepG2	Jaiswal *et al.* [38]
 Ferulic acid	Liver, colon	Kawabata *et al.* [39]

TABLE IV

INDUCERS OF NQO1: CURCUMIN AND OTHER DOUBLE MICHAEL REACTION ACCEPTORS

Compound	Tissue or cell line	Reference
Curcumin	Liver, kidney, Hepa1c1c7	Iqbal et al. [40]; Okada et al. [41]; Singletary et al. [42]; Dinkova-Kostova and Talalay [43]; Gerhäuser et al. [44]
$R_1 = R_2 = R_3 = H$ $R_1 = H, R_2 = R_3 = OH$ $R_1 = OCH_3, R_2 = R_3 = OH$	Hepa1c1c7	Dinkova-Kostova and Talalay [43]
Salicylcurcuminoid	Hepa1c1c7	Dinkova-Kostova and Talalay [43]
Yakuchinone B	Hepa1c1c7	Dinkova-Kostova and Talalay, unpublished
	Hepa1c1c7	Jang et al. [45]
$R_1 = R_2 = H$ $R_1 = H, R_2 = OH$ $R_1 = OH, R_2 = H$	Hepa1c1c7	Dinkova-Kostova et al. [46]
$R_1 = R_2 = H$ $R_1 = R_2 = OH$	Hepa1c1c7	Dinkova-Kostova et al. [46]
$R_1 = R_2 = R_3 = H$ $R_1 = H, R_2 = OCH_3, R_3 = OH$ $R_1 = OH, R_2 = R_3 = H$ $R_1 = OCH_3, R_2 = R_3 = H$	Hepa1c1c7	Dinkova-Kostova et al. [46]

TABLE V
INDUCERS OF NQO1: FLAVONOIDS, WITHANOLIDES, AND NORWITHANOLIDES

Compound	Tissue or cell line	Reference
Flavones	Hepa1c1c7, HepG2, MCF-7, HT-29, LNCaP, HeLa, heart	De Long *et al.* [33]; Talalay *et al.* [25]; Jiang *et al.* [47]; Floreani *et al.* [48]
β-Naphthoflavone	Hepa1c1c7	Dinkova-Kostova [49]
α-Naphthoflavone	Hepa1c1c7	Dinkova-Kostova [49]
γ-Naphthoflavone $R_1 = R_2 = R_3 = R_4 = H$ $R_1 = H, R_2 = R_3 = R_4 = OH$, Quercetin $R_1 = R_3 = H, R_2 = R_4 = OH$, Kaempferol $R_1 = R_2 = R_3 = H, R_4 = OH$, Galangin $R_1 = R_2 = R_3 = R_4 = OH$, Myricetin	Hepa1c1c7, MCF-7	De Long *et al.* [33]; Dinkova-Kostova [49]; Williamson *et al.* [50]; Uda *et al.* [51]; Valerio *et al.* [52]; Fahey and Stephenson [53]
Morin	Liver, tongue, large intestine, Hepa1c1c7	Fahey and Stephenson [53]; Tanaka *et al.* [54]; Kawabata *et al.* [55]
	Hepa1c1c7	Su *et al.* [56]

(continued)

TABLE V *(continued)*

Compound	Tissue or cell line	Reference
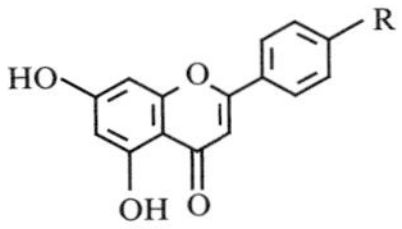 R = H, Chrysin R = OH, Apigenin	Hepa1c1c7	Uda *et al.* [51]; Fahey and Stephenson [53]
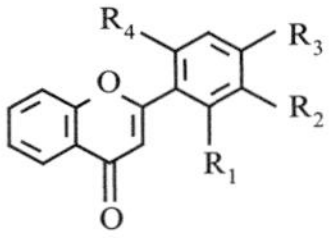 $R_1 = R_2 = R_4 = H, R_3 = Br$ $R_1 = R_2 = R_4 = H, R_3 = Cl$ $R_1 = R_2 = R_4 = H, R_3 = CF_3$ $R_1 = R_3 = R_4 = H, R_2 = Br$ $R_1 = R_3 = R_4 = H, R_2 = F$ $R_1 = R_3 = R_4 = H, R_2 = Cl$ $R_1 = F, R_2 = R_3 = R_4 = H$	Hepa1c1c7	Song *et al.* [57]
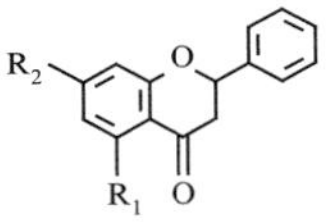 $R_1 = OH, R_2 = OCH_3$, Pinostrobin $R_1 = OCH_3, R_2 = OH$ $R_1 = R_2 = OH$, Pinocembrin $R_1 = R_2 = OCH_3$	Hepa1c1c7	Fahey and Stephenson [53]; Su *et al.* [56]; Gu *et al.* [58]
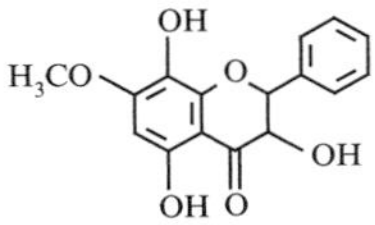	Hepa1c1c7	Su *et al.* [56]
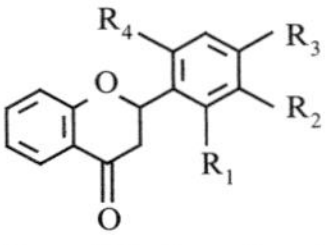 $R_1 = CH_3, R_2 = R_3 = R_4 = H$ $R_1 = OCH_3, R_2 = R_3 = R_4 = H$	Hepa1c1c7	Song *et al.* [57]

(continued)

TABLE V *(continued)*

Compound	Tissue or cell line	Reference
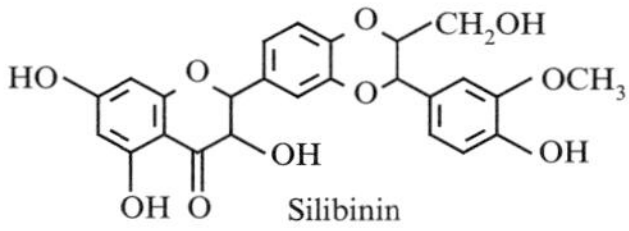 Silibinin	Liver, lung, stomach, small intestine, skin	Zhao and Agarwal [59]

Chalcones

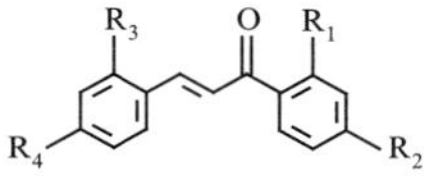

$R_1 = R_2 = R_3 = R_4 = H$
$R_1 = R_2 = R_3 = H, R_4 = OH$
$R_1 = R_2 = R_4 = H, R_3 = OH$
$R_1 = OH, R_2 = R_3 = R_4 = H$
$R_1 = R_2 = R_4 = OH, R_3 = H$
$R_1 = R_3 = OH, R_2 = R_4 = H$
$R_1 = R_2 = R_3 = OH, R_4 = H$

| | Hepa1c1c7 | Dinkova-Kostova *et al.* [34]; Su *et al.* [56] |

| 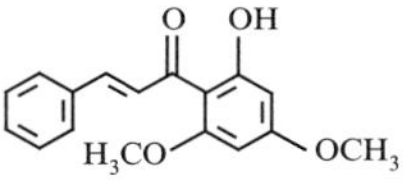 | Hepa1c1c7 | Gu *et al.* [58] |

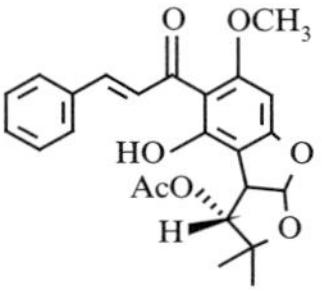

(+)-Tephropurpurin

| | Hepa1c1c7 | Chang *et al.* [60] |

Isoflavones

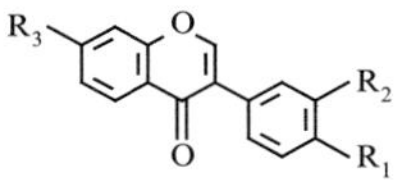

$R_1 = R_2 = H, R_3 = OH$
$R_1 = R_2 = R_3 = OCH_3$

| | Hepa1c1c7 | Su *et al.* [56] |

Withanolides

| 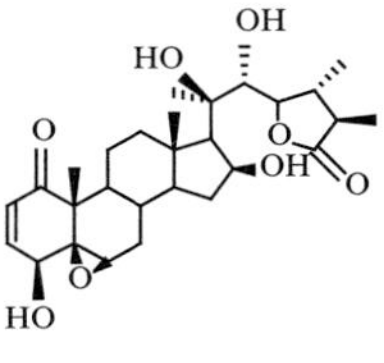 | Hepa1c1c7 | Gu *et al.* [61] |

(continued)

TABLE V (continued)

Compound	Tissue or cell line	Reference
	Hepa1c1c7	Su et al. [62]
	Hepa1c1c7, liver, colon	Misico et al. [63]

[36] S. R. Spencer, C. A. Wilczak, and P. Talalay, Cancer Res. **50,** 7871 (1990).

[37] S. R. Spencer, L. A. Xue, E. M. Klenz, and P. Talalay, Biochem. J. **273,** 711 (1991).

[38] A. K. Jaiswal, R. Venugopal, J. Mucha, A. M. Carothers, and D. Grunberger, Cancer Res. **57,** 440 (1997).

[39] K. Kawabata, T. Yamamoto, A. Hara, M. Shimizu, Y. Yamada, K. Matsunaga, T. Tanaka, and H. Mori, Cancer Lett. **157,** 15 (2000).

[40] M. Iqbal, S. D. Sharma, Y. Okazaki, M. Fujisawa, and S. Okada, Pharmacol. Toxicol. **92,** 33 (2003).

[41] K. Okada, C. Wangpoengtrakul, T. Tanaka, S. Toyokuni, K. Uchida, and T. Osawa, J. Nutr. **131,** 2090 (2001).

[42] K. Singletary, C. MacDonald, M. Iovinelli, C. Fisher, and M. Wallig, Carcinogenesis **19,** 1039 (1998).

[43] A. T. Dinkova-Kostova and P. Talalay, Carcinogenesis **20,** 911 (1999).

[44] C. Gerhäuser, K. Klimo, E. Heiss, I. Neumann, A. Gamal-Eldeen, J. Knauft, G. Y. Liu, S. Sitthimonchai, and N. Frank, Mutat. Res. **523–524,** 163 (2003).

[45] D. S. Jang, E. J. Park, M. E. Hawthorne, J. S. Vigo, J. G. Graham, F. Cabieses, B. D. Santarsiero, A. D. Mesecar, H. H. Fong, R. G. Mehta, J. M. Pezzuto, and A. D. Kinghorn, J. Agric. Food Chem. **50,** 6330 (2002).

[46] A. T. Dinkova-Kostova, M. A. Massiah, R. E. Bozak, R. J. Hicks, and P. Talalay, Proc. Natl. Acad. Sci. USA **98,** 3404 (2001).

[47] Z. Q. Jiang, C. Chen, B. Yang, V. Hebbar, and A. N. Kong, Life Sci. **72,** 2243 (2003).

[48] M. Floreani, E. Napoli, and P. Palatini, Biochem. Pharmacol. **60,** 601 (2000).

[49] A. T. Dinkova-Kostova, Mini Rev. Med. Chem. **2,** 595 (2002).

[50] G. Williamson, G. W. Plumb, Y. Uda, K. R. Price, and M. J. Rhodes, Carcinogenesis **17,** 2385 (1996).

[51] Y. Uda, K. R. Price, G. Williamson, and M. J. Rhodes, Cancer Lett. **120,** 213 (1997).

[52] L. G. Valerio, J. K. Kepa, G. V. Pickwell, and L. C. Quattrochi, Toxicol. Lett. **119,** 49 (2001).

[53] J. W. Fahey and K. K. Stephenson, J. Agric. Food Chem. **50,** 7472 (2002).

[54] T. Tanaka, K. Kawabata, M. Kakumoto, H. Makita, J. Ushida, S. Honjo, A. Hara, H. Tsuda, and H. Mori, Carcinogenesis **20,** 1477 (1999).

tissues.[27–33] Initially, 1,4-diphenols and *tert*-butylhydroquinone were found to raise the levels of NQO1 and GST in mouse liver in a concentration-dependent manner.[22] A series of analogs were synthesized (Table I) and their inducer potencies were compared. It was found that the critical structural feature of these phenolic compounds was their capacity to undergo reversible oxidation-reduction reactions. Thus catechols, hydroquinones, 1,2-, and 1,4-phenylenediamines were inducers, whereas resorcinols or 1,3-phenylenediamines were not. In contrast, the presence or absence of an alkyl side chain had little impact on the inducer potency. It was subsequently established that Michael reaction acceptors are a major class of inducers (see the following). This finding strongly suggested that the electrophilic quinone (or quinoneimine) oxidation products were the ultimate inducers and not the oxidation process *per se*.

Class 2: Michael Reaction Acceptors

Coumarin was one of the first plant compounds that was shown to induce NQO1, although weakly.[33] This led to the evaluation of the inducer potencies of a number of lactones, cinnamates, acrylates, crotonates, and indoles—all bearing olefins (or acetylenes) conjugated to electron-withdrawing groups—that is, Michael reaction acceptors. The potency of these compounds as NQO1 inducers paralleled closely their reactivity in the Michael reaction.[25] The Michael reaction acceptor group is present in the molecules of many plant metabolites and is essential for NQO1 inducer activity of cinnamates, curcuminoids, and flavonoids. Some examples

[55] K. Kawabata, T. Tanaka, S. Honjo, M. Kakumoto, A. Hara, H. Makita, N. Tatematsu, J. Ushido, H. Tsuda, and H. Mori, *Int. J. Cancer* **83**, 381 (1999).

[56] B. N. Su, E. J. Park, J. S. Vigo, J. G. Graham, F. Cabieses, H. H. Fong, J. M. Pezzuto, and A. D. Kinghorn, *Phytochemistry* **63**, 335 (2003).

[57] L. L. Song, J. W. Kosmeder, 2nd, S. K. Lee, C. Gerhäuser, D. Lantvit, R. C. Moon, R. M. Moriarty, and J. M. Pezzuto, *Cancer Res.* **59**, 578 (1999).

[58] J. Q. Gu, E. J. Park, J. S. Vigo, J. G. Graham, H. H. Fong, J. M. Pezzuto, and A. D. Kinghorn, *J. Nat. Prod.* **65**, 1616 (2002).

[59] J. Zhao and R. Agarwal, *Carcinogenesis* **20**, 2101 (1999).

[60] L. C. Chang, C. Gerhäuser, L. Song, N. R. Farnsworth, J. M. Pezzuto, and A. D. Kinghorn, *J. Nat. Prod.* **60**, 869 (1997).

[61] J.-Q. Gu, W. Li, Y.-H. Kang, B.-N. Su, H. H. S. Fong, R. B. van Breemen, J. M. Pezzuto, and A. D. Kinghorn, *Chem. Pharm. Bull. (Tokyo)* **51**, 530 (2003).

[62] B.-N. Su, E. J. Park, D. Nikolic, B. D. Santarsiero, A. D. Mesecar, J. S. Vigo, J. G. Graham, F. Cabieses, R. B. van Breemen, H. H. Fong, N. R. Farnsworth, J. M. Pezzuto, and A. D. Kinghorn, *J. Org. Chem.* **68**, 2350 (2003).

[63] R. I. Misico, L. L. Song, A. S. Veleiro, A. M. Cirigliano, M. C. Tettamanzi, G. Burton, G. M. Bonetto, V. E. Nicotra, G. L. Silva, R. R. Gil, J. C. Oberti, A. D. Kinghorn, and J. M. Pezzuto, *J. Nat. Prod.* **65**, 677 (2002).

are given in Tables II through V. Similarly, the Michael reaction acceptor—that is, the α,β-unsaturated ketone moiety of a number of withanolides isolated from Solanaceae is required for NQO1 inducer activity.[61–63]

It should be noted that some flavanones (e.g., pinostrobin) exhibit inducer activity, implying that such molecules might be precursors rather than the ultimate inducers and that metabolism plays a role in generating the active species. This notion is supported by the fact that pinostrobin is inactive in cells with defective *Ah*-receptor function.[53] Indeed, most flavonoids are bifunctional inducers.[64] In a series of aromatic Michael reaction acceptors the presence of hydroxyl group(s) on the aromatic ring(s) at *ortho*-position(s) to the vinyl carbons enhances the inducer potency enormously, in some cases by more than 200-fold,[34,43,46] and proposed that this structural feature must function to accelerate the rate of addition of the inducer to reactive sulfhydryl(s) on a hypothetical intracellular sensor, which has subsequently been identified as Keap1.[34,46,65]

Class 3: Isothiocyanates, Dithiocarbamates, and Related Sulfur Compounds

The characteristic $-N = C = S$ functionality of all isothiocyanates reacts very readily at its highly electrophilic central carbon atom with sulfhydryl-containing compounds to give dithiocarbamates. Isothiocyanates, thio-, dithiocarbamates, as well as related sulfur compounds, were recognized as NQO1 inducers in the early studies directed towards understanding the mechanism of induction.[25,26,66,67] At about the same time L. Wattenberg showed that, similar to BHA, dietary administration of benzyl isothiocyanate induced glutathione *S*-transferase activity in mouse esophagus, forestomach, and small intestine and this induction was associated with inhibition of benzo[*a*]pyrene-induced neoplasia of the forestomach.[68,69] Isothiocyanates are present in cruciferous vegetables as "inert" glucoside precursors (glucosinolates) and are compartmentally isolated from the enzyme (myrosinase) that catalyzes their conversion to the reactive isothiocyanates—for example, when the plant is injured or chewed. Sulforaphane was isolated as the principal NQO1 inducer from

[64] S. Yannai, A. J. Day, G. Williamson, and M. J. Rhodes, *Food Chem. Toxicol.* **36,** 623 (1998).

[65] A. T. Dinkova-Kostova, W. D. Holtzclaw, R. N. Cole, K. Itoh, N. Wakabayashi, Y. Katoh, M. Yamamoto, and P. Talalay, *Proc. Natl. Acad. Sci. USA* **99,** 11908 (2002).

[66] A. M. Benson, P. B. Barretto, and J. S. Stanley, *J. Natl. Cancer Inst.* **76,** 467 (1986).

[67] P. Talalay, M. J. De Long, and H. J. Prochaska, *in* "Cancer Biology and Therapeutics." Plenum Publishing Corporation, 1987.

[68] V. L. Sparnins and L. W. Wattenberg, *J. Natl. Cancer Inst.* **66,** 769 (1981).

[69] V. L. Sparnins, J. Chuan, and L. W. Wattenberg, *Cancer Res.* **42,** 1205 (1982).

broccoli,[70] and it was subsequently found that 3-day-old broccoli sprouts (grown from selected seeds) contain 20 to 50 times higher levels of its glucosinolate, glucoraphanin than the mature broccoli plant.[19] The demonstration of the protective effect of sulforaphane against cancer in the DMBA mammary tumor model in Sprague-Dawley rats was a proof of the principle that measuring the NQO1 inducer activity is an effective strategy in the search for chemoprotective agents. A variety of isothiocyanates and synthetic analogs have been tested in this laboratory as inducers of NQO1 and their potencies have been compared (Table VI).[71] Sulforaphane remains one of the most potent NQO1 inducers known to date. Importantly, sulforaphane quickly accumulates in cells of various types, reaching intracellular concentrations in the millimolar range, which is probably one of the reasons underlying its extremely high inducer potency.[72] Intracellular accumulation is achieved through conjugation with cellular glutathione,[73] a reaction that is further accelerated by glutathione S-transferases,[74] and the glutathione conjugate is then exported by a transporter-mediated mechanism.[75] Furthermore, NQO1 inducer potency among a series of isothiocyanates correlates with their intracellular accumulation,[76] highlighting that transport mechanisms should also be carefully considered in addition to structure and chemical reactivity in determining inducer potency. We recently isolated 4-(rhamnopyranosyloxy)benzyl isothiocyanate as a very potent NQO1 inducer from *Moringa oleifera*. Depending on the cell type, its inducer potency was comparable to, or even exceeded that of sulforaphane.[20]

Class 4: 1,2-Dithiole-3-thiones, Oxathiolene Oxides, and Other Organosulfur Compounds

In the early 1980s, work from the laboratory of E. Bueding showed that administration of 1,2-dithiole-3-thione derivatives to mice produces biochemical effects similar to those of BHA and protected against toxicity and carcinogenicity.[77] The antischistosomal agent oltipraz and related dithiolethiones (Table VII) were then shown to induce NQO1 in many

[70] Y. Zhang, P. Talalay, C. G. Cho, and G. H. Posner, *Proc. Natl. Acad. Sci. USA* **89**, 2399 (1992).
[71] G. H. Posner, C. G. Cho, J. V. Green, Y. Zhang, and P. Talalay, *J. Med. Chem.* **37**, 170 (1994).
[72] Y. Zhang and P. Talalay, *Cancer Res.* **58**, 4632 (1998).
[73] Y. Zhang, *Carcinogenesis* **21**, 1175 (2000).
[74] Y. Zhang, *Carcinogenesis* **22**, 425 (2001).
[75] Y. Zhang and E. C. Callaway, *Biochem. J.* **364**, 301 (2002).
[76] L. Ye and Y. Zhang, *Carcinogenesis* **22**, 1987 (2001).
[77] S. S. Ansher, P. Dolan, and E. Bueding, *Hepatology* **3**, 932 (1983).

TABLE VI

INDUCERS OF NQO1: ISOTHIOCYANATES, DITHIOCARBAMATES, AND RELATED SULFUR COMPOUNDS

Compound	Tissue or cell line	Reference
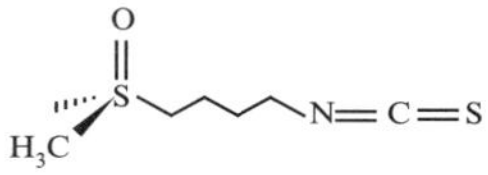 Sulforaphane $H_3C{-}S({=}O){-}(CH_2)\cdots N{=}C{=}S$	Liver, lung, forestomach, glandular stomach, small intestine, colon, Hepa1c1c7, HepG2, MCF-7, HT-29, LNCaP, HeLa	Zhang et al. [70]; McMahon et al. [32]; Jiang et al. [47]
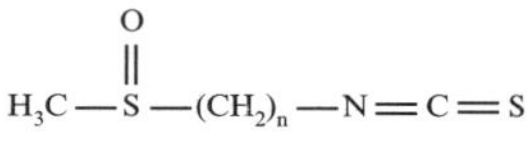 $H_3C{-}S({=}O){-}(CH_2)_n{-}N{=}C{=}S$ n = 3, 4, 5, 6	Liver, lung, forestomach, glandular stomach, small intestine, Hepa1c1c7	Zhang et al. [70]; Rose et al. [78]; Hou et al. [79]; Morimitsu et al. [80]
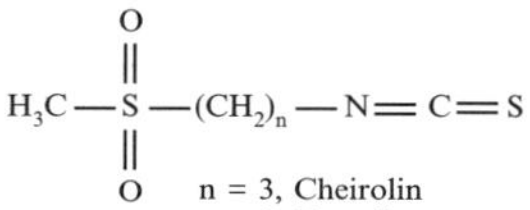 $H_3C{-}S({=}O)_2{-}(CH_2)_n{-}N{=}C{=}S$ n = 3, Cheirolin n = 4, Erysolin n = 5	Liver, forestomach, glandular stomach, small intestine, colon, Hepa1c1c7	Zhang et al. [70]
$H_3C{-}S{-}(CH_2)_n{-}N{=}C{=}S$ n = 3, Iberverin n = 4, Erusin n = 5, Berteroin	Liver, forestomach, glandular stomach, small intestine, colon, Hepa1c1c7	Zhang et al. [70]
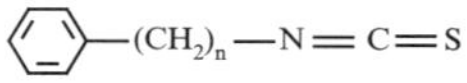 (phenyl)${-}(CH_2)_n{-}N{=}C{=}S$ n = 0, 1, 2, 4, 6	Liver, lung, kidney, forestomach, small intestine, colon, bladder, Hepa1c1c7	Prestera et al. [35]; Benson et al. [66]; Guo et al. [81]
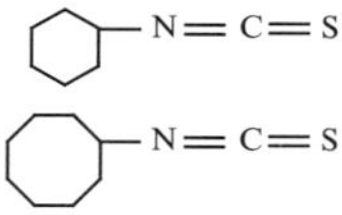 (cyclohexyl)${-}N{=}C{=}S$ (cycloheptyl)${-}N{=}C{=}S$	Hepa1c1c7	Prestera et al. [35]
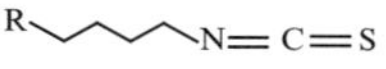 $R{\sim}\!\!\sim\!\!N{=}C{=}S$ R = C_2H_5, N≡C, HOOC, CH_3OOC, CH_3SCO, CH_3CO, n-BuCO, $(CH_3)_2P({=}O)$	Hepa1c1c7	Posner et al. [71]

(continued)

TABLE VI (continued)

Compound	Tissue or cell line	Reference
Sulforaphene	Hepa1c1c7	Posner et al. [71]
R = O_2N, $N\equiv C$, CH_3SO_2, CH_3CO, CH_3OOC, $CH_3CH(OH)$	Hepa1c1c7	Posner et al. [71]
OH	Hepa1c1c7	Posner et al. [71]
R = CH_3SO_2, CH_3CO	Hepa1c1c7	Posner et al. [71]
R = CH_3SO_2, CH_3CO, CH_3OOC	Hepa1c1c7	Posner et al. [71]
CH_2—N=C=S	Liver	Leonard et al. [82, 83]
	Hepa1c1c7 ARPE-19 HepG2 AGS	Fahey et al. [20]
$(C_2H_5)_2N$—C(=S)—SNa	Liver, lung, kidney, forestomach, small intestine, colon, bladder, Hepa1c1c7	Talalay and Benson [26]; Benson et al. [66]; Talalay et al. [67]
R—C(=S)—S—S—C(=S)—R R = $(C_2H_5)_2N$ Disulfiram R = C_2H_5O Bisethylxanthogen	Liver, lung, kidney, forestomach, small intestine, colon, bladder, Hepa1c1c7	Talalay and Benson [26]; Benson et al. [66]; Talalay et al. [67]

TABLE VII
INDUCERS OF NQO1: 1,2-DITHIOLE-3-THIONES, AND OXATHIOLENE OXIDES

Compound	Tissue or cell line	Reference
 1,2-dithiole-3-thione	Liver, lymphocytes	De Long *et al.* [84]; Kensler *et al.* [85]; Gordon *et al.* [86]
 5-(2-pyrazinyl)-4-methyl- 1,2-dithiole-3-thione(oltipraz)	Liver, lung, jejunum	McMahon *et al.* [32]; De Long *et al.* [84]; Ansher *et al.* [87]; Kensler *et al.* [85]
 5-phenyl-1,2-dithiole-3-thione	Liver, lung, jejunum	De Long *et al.* [84]; Ansher *et al.* [87]; Kensler *et al.* [85]
 5-(*p*-methoxyphenyl)- 1,2-dithiole-3-thione	Liver, lung, jejunum	De Long *et al.* [84]; Ansher *et al.* [87]; Kensler *et al.* [85]
 4-phenyl-1,2-dithiole-3-thione	Liver, lung, jejunum	De Long *et al.* [84]; Ansher *et al.* [87]; Kensler *et al.* [85]
 3-(*p*-methoxyphenyl)-4-methyl- 1,2-oxathiol-3-ene-2-oxide	Hepa1c1c7, BNLCL2	Pietsch *et al.* [88]
 3-(*p-t*-butylphenyl)-4-methyl- 1,2-oxathiol-3-ene-2-oxide	Hepa1c1c7, BNLCL2	Pietsch *et al.* [88]

(continued)

TABLE VII *(continued)*

Compound	Tissue or cell line	Reference
 3-cyclohexenyl-4-methyl- 1,2-oxathiol-3-ene-2-oxide	Hepa1c1c7, BNLCL2	Pietsch *et al.* [88]
 4-bromo-5,5-dimethyl- 1,2-oxathiol-3-ene-2-oxide	Hepa1c1c7, BNLCL2	Pietsch *et al.* [88]

organs and tissues,[84,85,87] as well as in human lymphocytes in culture.[86] Oltipraz was subsequently developed as a chemopreventive agent in clinical trials in Qidong, China, an area with a very high incidence of liver cancer attributable in part to consumption of grains contaminated with aflatoxin B_1.[89,90] In such volunteers, administration of oltipraz profoundly affected the excretion of aflatoxin B_1-DNA adducts. At low doses of oltipraz the

[78] P. Rose, P. K. Faulkner, G. Williamson, and R. Mithen R, *Carcinogenesis* **21,** 1983 (2000).

[79] D. X. Hou, M. Fukuda, M. Fujii, and Y. Fuke, *Int. J. Mol. Med.* **6,** 441 (2000).

[80] Y. Morimitsu, Y. Nakagawa, K. Hayashi, H. Fujii, T. Kumagai, Y. Nakamura, T. Osawa, F. Horio, K. Itoh, K. Iida, M. Yamamoto, and K. Uchida, *J. Biol. Chem.* **277,** 3456 (2002).

[81] Z. Guo, T. J. Smith, E. Wang, K. I. Eklind, F. L. Chung, and C. S. Yang, *Carcinogenesis* **14,** 1167 (1993).

[82] T. B. Leonard, J. A. Popp, M. E. Graichen, and J. G. Dent, *Carcinogenesis* **2,** 473 (1981).

[83] T. B. Leonard, J. A. Popp, M. E. Graichen, and J. G. Dent, *Toxicol. Appl. Pharmacol.* **60,** 527 (1981).

[84] M. J. De Long, P. Dolan, A. B. Santamaria, and E. Bueding, *Carcinogenesis* **7,** 977 (1986).

[85] T. W. Kensler, P. A. Egner, P. M. Dolan, J. D. Groopman, and B. D. Roebuck, *Cancer Res.* **47,** 4271 (1987).

[86] G. B. Gordon, H. J. Prochaska, and L. Y. Yang, *Carcinogenesis* **12,** 2393 (1991).

[87] S. S. Ansher, P. Dolan, and E. Bueding, *Food Chem. Toxicol.* **24,** 405 (1986).

[88] E. C. Pietsch, A. L. Hurley, E. E. Scott, B. P. Duckworth, M. E. Welker, S. Leone-Kabler, A. J. Townsend, F. M. Torti, and S. V. Torti, *Biochem. Pharmacol.* **65,** 1261 (2003).

[89] T. W. Kensler, T. J. Curphey, Y. Maxiutenko, and R. D. Roebuck, *Drug Metabol. Drug Interact.* **17,** 3 (2000).

[90] T. W. Kensler, P. A. Egner, J. B. Wang, Y. R. Zhu, B. C. Zhang, G. S. Qian, S. Y. Kuang, S. J. Gange, L. P. Jacobson, A. Munoz, and J. D. Groopman, *Eur. J. Cancer Prev.* **11,** S58 (2002).

spectrum of urinary aflatoxin metabolites changes and is consistent with phase 2 enzyme induction.

Recently, as part of the development of new candidates for chemopreventive agents, the chemically similar oxathiolene oxides (Table VII) were synthesized and shown to be inducers of NQO1.[88] Various mono-, di-, and polysulfides from *Allium* plants (Table VIII), shown in the 1980s to elevate glutathione *S*-transferase activity in the forestomach and to protect against benzo[*a*]pyrene-induced carcinogenesis,[91] also induce NQO1, with diallyl disulfide being the most potent among them.[25,31,92–96] Structure-activity relationship studies revealed that the unsaturated allyl or propenyl functionalities are critical determinants of inducer potency.[25,91–96]

Class 5: Hydroperoxides

Cumene hydroperoxide, hydrogen peroxide, and *tert*-butyl hydroperoxide are, although weak, also inducers of NQO1 (Table IX).[35,37]

Class 6: Trivalent Arsenicals

Trivalent arsenicals—for example, phenylarsine oxide, are extremely potent inducers of NQO1 and this finding gave an important clue to the mechanism involved in the initial sensing of the inducer signal. In their classical studies on the development of antidotes for arsenic poisoning, Stocken and Thompson (1946)[97–99] demonstrated that vicinal dithiols are excellent reagents for trivalent arsenicals and form highly stable five-membered cyclic products, while the reaction with monothiols gives rise to "open chain" much more easily dissociable compounds. These properties of trivalent arsenicals form the basis for the development of arsenical affinity chromatography for proteins with sulfhydryl groups that are positioned close in space.[100] In contrast to trivalent arsenicals, pentavalent arsenicals—for example sodium arsenate, are much weaker NQO1 inducers. This finding correlates with the much higher reactivity of trivalent

[91] V. L. Sparnins, G. Barany, and L. W. Wattenberg, *Carcinogenesis* **9**, 131 (1988).

[92] S. V. Singh, S. S. Pan, S. K. Srivastava, H. Xia, X. Hu, H. A. Zaren, and J. L. Orchard, *Biochem. Biophys. Res. Commun.* **244**, 917 (1998).

[93] D. Guyonnet, M. H. Siess, A. M. Le Bon, and M. Suschetet, *Toxicol. Appl. Pharmacol.* **154**, 50 (1999).

[94] D. Guyonnet, C. Belloir, M. Suschetet, M. H. Siess, and A. M. Le Bon, *Mutat. Res.* **495**, 135 (2001).

[95] R. Munday and C. M. Munday, *Nutr. Cancer* **40** (2001).

[96] R. Munday, J. S. Munday, and C. M. Munday, *Free Radic. Biol. Med.* **34**, 1200 (2003).

[97] L. A. Stocken and R. H. S. Thompson, *Biochem. J.* **40**, 529 (1946).

[98] L. A. Stocken and R. H. S. Thompson, *Biochem. J.* **40**, 535 (1946).

[99] L. A. Stocken and R. H. S. Thompson, *Biochem. J.* **40**, 548 (1946).

[100] R. D. Hoffman and M. D. Lane, *J. Biol. Chem.* **267**, 14005 (1992).

TABLE VIII

INDUCERS OF NQO1: ALKYL AND ALKENYL SULFIDES AND POLYSULFIDES

Compound	Tissue	Reference
Diallyl sulfide	Lung, heart, bladder, forestomach, glandular stomach, duodenum, jejunum, ileum, cecum, colon, brain	Singh et al. [92]; Guyonnet et al. [93, 94]; Munday and Munday [95]; Yang et al. [101]
Diallyl disulfide	Hepa1c1c7, liver, lung, heart, bladder, forestomach, glandular stomach, duodenum, jejunum, spleen, ileum, cecum, colon	Talalay et al. [25]; Singh et al. [92]; Guyonnet et al. [93, 94]; Munday and Munday [31]
Diallyl trisulfide	Liver, lung, heart, kidney, bladder, forestomach, glandular stomach, duodenum, jejunum, spleen, ileum, cecum, colon	Singh et al. [92]; Guyonnet et al. [93, 94]; Munday and Munday [95]; Munday et al. [96]
Diallyl tetrasulfide	Liver, lung, heart, kidney, spleen	Munday et al. [96]
Dipropyl sulfide	Liver, forestomach, glandular stomach, cecum, colon	Singh et al. [92]; Guyonnet et al. [93, 94]
Dipropyl disulfide	Liver, lung, heart, forestomach, duodenum, spleen, cecum, colon	Singh et al. [92]; Guyonnet et al. [93, 94]
Dipropenyl sulfide	Liver, lung, heart, kidney, bladder, forestomach, glandular stomach, duodenum, jejunum, spleen, ileum, cecum, colon	Munday and Munday [95]
Dipropenyl disulfide	Liver, lung, heart, kidney, bladder, forestomach, glandular stomach, duodenum, jejunum, spleen, ileum, cecum, colon	Munday and Munday [95]

arsenicals with sulfhydryl groups that are closely positioned in space (e.g., vicinal). Based on this, Prestera et al. (1993)[35] made the following prediction that has, with modern tools of molecular biology, since been shown to be correct[65]; "The potent induction of NQO1 by such compounds suggests the critical presence of two neighboring sulfhydryl groups on the protein(s) that receive and transmit the inductive signal."

[101] C. S. Yang, S. K. Chhabra, J. Y. Hong, and T. J. Smith, J. Nutr. 131, 1041S (2001).

TABLE IX
INDUCERS OF NQO1: HYDROPEROXIDES, ARSENICALS, MERCURIALS, AND MERCAPTANS

Compound	Tissue or cell line	Reference
H — O — O — H	Hepa1c1c7	Prestera [35, 102]
H_3C + O — O — H (with two CH$_3$ groups)	Hepa1c1c7	Prestera [35, 102]
(phenyl) + O — O — H (with two CH$_3$ groups)	Hepa1c1c7	Prestera [35, 102]; Spencer *et al.* [37]
(phenyl) — As = O	Hepa1c1c7	Prestera [35, 102]
(phenyl) — Hg — Cl	Hepa1c1c7	Prestera [35, 102]
HO — C(=O) — (phenyl) — Hg — Cl	Hepa1c1c7	Prestera [35, 102]
SH, SH, OH	Hepa1c1c7	Prestera [35, 102]
SH, SH	Hepa1c1c7	Prestera [35, 102]
(chain with SH SH and C(=O)OH) Lipoic acid	Astroglial cells	Flier *et al.* [103]

Class 7: Heavy Metals

Divalent metal cations induce NQO1 in an order of potency that correlates with their reactivity with sulfhydryl groups—that is, $Hg^{2+} > Cd^{2+} > Zn^{2+}$ in Hepa1c1c7 and LNCaP cells.[35,104] Hg^{2+} and Cd^{2+} increase NQO1 activity in explants of both first-trimester, as well as full-term

[102] T. Prestera, W. D. Holtzclaw, Y. Zhang, and P. Talalay, *Proc. Natl. Acad. Sci. USA* **90,** 2965 (1993).

[103] J. Flier, F. L. Van Muiswinkel, C. A. Jongenelen, and B. Drukarch, *Free Radic. Res.* **36,** 695 (2002).

[104] J. D. Brooks, M. F. Goldberg, L. A. Nelson, D. Wu, and W. G. Nelson, *Cancer Epidemiol. Biomarkers Prev.* **11,** 868 (2002).

human placentae.[105–107] In addition, the sulfhydryl reagents phenylmercuric chloride and *para*-chloromercuribenzoate are also potent inducers of NQO1.[35] Administration of triethyl lead to rats has been shown to increase the levels of NQO1 in liver and kidney.[108]

Class 8: Vicinal Dimercaptans

The finding that selected nucleophilic mercaptans are also NQO1 inducers was somewhat unexpected since most of the other inducers are electrophilic. The most potent members of this class have two sulfhydryl groups in close proximity—for example, 2,3-dimercapto-1-propanol (BAL) and 1,2-ethanedithiol, containing vicinal dithiols, and dihydrolipoic acid, which has 2 closely spaced thiol groups (Table IX).[35,103] Interestingly, in Hepa1c1c7 cells HgCl$_2$ and BAL act synergistically in inducing NQO1.[109] In contrast, no synergism was observed between HgCl$_2$ and monothiols, suggesting that the ultimate inducer could be a high-affinity chelate complex between Hg^{2+} and the vicinal thiol groups of BAL. Indeed, synthetic equimolar BAL-mercury chelates are much more potent inducers than HgCl$_2$. Such complexes entered and accumulated in cells more efficiently and excess BAL increased the synergism even further.

Class 9: Carotenoids and Related Polyenes

The finding that polyenes (Table X) are NQO1 inducers[110] was intriguing, given their known protective role against oxidants and photooxidative damage. The unsubstituted lycopene is a good NQO1 inducer, indicating that the conjugated polyene chain itself has inducer activity.[110,111] The 7-fold difference in inducer potency between α- and β-carotene is noteworthy since these compounds only differ by the position of a single double bond in the end groups (without any other structural differences)—that is, the more-potent inducer (β-carotene) has 11 conjugated double bonds vs 10 in the molecule of α-carotene. In addition to the extent of conjugation, the presence of hydroxyl as well as α, β-unsaturated ketone

[105] W. Y. Boadi, J. Urbach, E. R. Barnea, J. M. Brandes, and S. Yannai, *Pharmacol. Toxicol.* **68,** 317 (1991).

[106] W. Y. Boadi, J. Urbach, J. M. Brandes, and S. Yannai, *Environ. Res.* **57,** 96 (1992).

[107] W. Y. Boadi, J. Urbach, J. M. Brandes, and S. Yannai, *Toxicol. Lett.* **60,** 155 (1992).

[108] D. A. Daggett, E. F. Nuwaysir, S. A. Nelson, L. S. Wright, S. E. Kornguth, and F. L. Siegel, *Toxicology* **117,** 61 (1997).

[109] R. R. Putzer, Y. Zhang, T. Prestera, W. D. Holtzclaw, K. L. Wade, and P. Talalay, *Chem. Res. Toxicol.* **8,** 103 (1995).

[110] F. Khachik, J. S. Bertram, M.-T. Huang, J. W. Fahey, and P. Talalay, *in* "Antioxidant Food Supplements in Human Health," p. 203. Academic Press, 1999.

[111] V. Breinholt, S. T. Lauridsen, B. Daneshvar, and J. Jakobsen, *Cancer Lett.* **154,** 201 (2000).

TABLE X
INDUCERS OF NQO1: CAROTENOIDS AND RELATED POLYENES

Compound	Tissue or cell line	Reference
α-carotene	Hepa1c1c7	Khachik *et al.* [110]
β-carotene	Hepa1c1c7	Khachik *et al.* [110]
lycopene	Hepa1c1c7, liver	Khachik *et al.* [110]; Breinholt *et al.* [111]
violaxanthin	Hepa1c1c7	Khachik *et al.* [110]
neoxanthin	Hepa1c1c7	Khachik *et al.* [110]
2,6-cyclolycopene-1,5-diol	Hepa1c1c7	Khachik *et al.* [110]
(3R,6′R)-3-hydroxy-β-,ε-carotene-3′-one	Hepa1c1c7	Khachik *et al.* [110]
(6R,6′R)-ε-,ε-carotene-3,3′-dione	Hepa1c1c7	Khachik *et al.* [110]
(3R,3′R,6′R)-lutein	Hepa1c1c7	Khachik *et al.* [110]
(3R,3′R)-zeaxanthin	Hepa1c1c7	Khachik *et al.* [110]

functionalities increases further the inducer potency. The concentrations of the polyenes that induce NQO1 are in the micromolar range and can be achieved through dietary means. Importantly, the xanthophyll carotenoids lutein and zeaxanthin are found in high concentrations in the macula lutea region of the primate retina and are thought to protect this critical region of the eye against degeneration.[112]

Implications of the Chemical Structures of Inducers for their Mechanism of Action

The compounds that were first recognized to induce NQO1—that is, the oxidizable diphenols, were antioxidants. Paradoxically, other inducers—for example, the hydroperoxides, are powerful oxidants. Yet a third class, the isothiocyanates, have no significant redox properties. While nearly all inducers are electrophilic, the dimercaptans represent inducers that are nucleophilic. The only property that is shared by all inducers is their ability to react with sulfhydryl groups. Indeed, in the course of the systematic studies leading to the "Michael reaction acceptor hypothesis" of induction of NQO1,[25] it became striking that among inducers of different chemical classes were the same compounds that had been previously reported to react with glutathione and to serve as substrates for glutathione S-transferases.[113,114] In these classical studies, Chasseaud pointed out that the nature of the metabolic products of xenobiotic electrophiles—for example, α,β-unsaturated aldehydes, ketones, lactones, sulfones, nitro-olefins, as well as isothiocyanates, quinones, mercurials, acrylates, fumarates, allyl and epoxide derivatives—suggests the occurrence of an initial conjugation reaction with glutathione followed by catabolism via the mercapturic acid pathway.[114] Unrelated studies on the cystine and glutamate transport system from the laboratory of S. Bannai showed that while high concentrations of electrophilic agents (e.g., ethacrynic acid, cyclohex-2-en-1-one, diethyl maleate) caused depletion of glutathione and cytotoxicity, at lower concentrations the same compounds increase the intracellular glutathione levels substantially.[115] We now know that these changes in glutathione levels resulted from alkylation and enhanced synthesis, respectively. Indeed, most NQO1 inducers were found to be substrates for glutathione S-transferases.[25,37] Conversely, compounds that are commonly used as substrates for glutathione S-transferases (e.g., CDNB, DCNB, ethacrynic acid) were shown to be NQO1 inducers.[25] Moreover, the potency of

[112] B. R. Hammond, Jr., E. J. Johnson, R. M. Russell, N. I. Krinsky, K. J. Yeum, R. B. Edwards, and D. M. Snodderly, *Invest. Ophthalmol. Vis. Sci.* **38,** 1795 (1997).

[113] L. F. Chasseaud, *Biochem. J.* **131,** 765 (1973).

[114] L. F. Chasseaud, *in* "Glutathione: Metabolism and Function." Raven Press, New York, 1974.

[115] S. Bannai, *J. Biol. Chem.* **259,** 2435 (1984).

inducers to elevate NQO1 paralleled their reactivity in the Michael reaction with nucleophilic donors such as mercaptans.[25,37,46] Further, within a series of structurally closely related compounds—for example, isothiocyanates, or aromatic single and double Michael reaction acceptors—the inducer potency is closely related to the second-order non-enzymatic rate constants of reactions with various sulfhydryl reagents.[46,80]

On the basis of all of these findings, we suggested the existence of a cellular "sensor" protein endowed with highly reactive sulfhydryl groups that recognizes and reacts with inducers in the initial event of the signal transduction leading to induction of phase 2 proteins.[46,65,102] Following the demonstration that the transcription factor Nrf2 is essential for the regulation of phase 2 genes[116] and that under basal conditions it is sequestered in the cytoplasm by the actin-bound cytosolic repressor Keap1,[117] we undertook detailed analysis of these two proteins and their interactions. Perhaps not surprisingly, Keap1 (624 amino acids) is a cysteine-rich protein: it has 25 cysteines that are conserved in its human and rat homologs. After cloning and overexpressing murine Keap1 in *Escherichia coli*, we purified the protein to homogeneity. Under reducing conditions, Keap1 binds to the N-terminal Neh2 domain of Nrf2. The complex can be visualized following non-denaturing polyacrylamide gel electrophoresis. Inducers— for example, sulforaphane, or Michael reaction acceptors—disrupt the complex in a concentration-dependent manner. Kinetic, radiolabeling and UV-spectroscopic studies demonstrated that four highly reactive cysteine residues (C257, C273, C288, and C297) located in the central intervening region (IVR) of the protein are probably the primary cellular sensors that recognize and react with inducers.[65] Subsequent mutagenesis analysis revealed that substitutions of these cysteine residues with alanine render Keap1 unable to sequester Nrf2 in the cytoplasm (our unpublished observations).

These recent experiments provide a satisfying explanation for the broad chemical specificity of phase 2 inducers: all inducers react chemically with specific cysteine thiols of the sensor protein Keap1 (by alkylation, oxidation, or thiolation) and thereby suppress its ability to bind and retain the transcription factor Nrf2 in the cytoplasm. Nrf2 is then able to migrate into the nucleus where it binds to the antioxidant response element and activates (in heterodimeric combinations with members of the small Maf family) the transcription of phase 2 genes.

[116] K. Itoh, T. Chiba, S. Takahashi, T. Ishii, K. Igarashi, Y. Katoh, T. Oyake, N. Hayashi, K. Satoh, I. Hatayama, M. Yamamoto, and Y. Nabeshima, *Biochem. Biophys. Res. Commun.* **236,** 313 (1997).
[117] K. Itoh, N. Wakabayashi, Y. Katoh, T. Ishii, K. Igarashi, J. D. Engel, and M. Yamamoto, *Genes Dev.* **13,** 76 (1999).

[24] Induction of Phase II Enzymes by Aliphatic Sulfides Derived from Garlic and Onions: An Overview

By Rex Munday and Christine M. Munday

Introduction

Plants of the *Allium* family, such as garlic and onions, have been culti-vated for food since earliest times.[1] They have also been employed in folk medicine in many parts of the world.[1] More recently, epidemiological stud-ies have indicated a chemoprotective effect of both onions and garlic, with individuals consuming large amounts of these vegetables showing a lower incidence of cancer, particularly cancer of the stomach and intestine, than those consuming only small amounts.[2] These findings have stimulated much research into the active components of garlic and onions and the mechanism of their chemoprotective action. *Allium* plants contain high concentrations of alk(en)yl cysteine sulfoxides. Garlic and its relatives con-tain predominantly allyl cysteine sulfoxide, while onions, shallots and leeks contain the methyl, propyl and prop-1-enyl derivatives.[3] When the tissue is disrupted by cutting, crushing or chewing, the cysteine sulfoxides are enzymatically degraded to sulfenic acids. The sulfenic acids decompose spontaneously. Allyl, methyl and propyl sulfenic acids yield mainly thiosul-finates, while prop-1-enyl sulfenic acid forms both the corresponding thio-sulfinate and thiopropanal *S*-oxide, the onion lachrymatory factor. Thiosulfinates themselves are unstable, particularly on heating, and break down to a complex mixture of compounds, in which mono-, di-, tri- and tetra-sulfides predominate.[4,5] Both symmetrical and mixed sulfides are formed. The structures of the symmetrical methyl, propyl, allyl and prop-1-enyl sulfides are shown in Fig. 1.

Several of the sulfides formed from the thiosulfinates present in onion and garlic have been shown to protect against a variety of chemical car-cinogens in experimental animals, and it has been suggested that such substances are responsible for the chemoprotective action of these vege-tables.[6–8] Furthermore, it has been proposed that this effect is due, at least

[1] G. R. Fenwick and A. B. Hanley, *CRC Crit. Rev. Food Sci.* **22**, 199 (1985).
[2] E. Ernst, *Phytomedicine* **4**, 79 (1997).
[3] W. Breu and W. Dorsch, *Econ. Med. Plant Res.* **6**, 115 (1994).
[4] E. Block, *Angew. Chem. Int. Ed. Engl.* **31**, 1135 (1992).
[5] L. D. Lawson, *ACS Symposium Ser.* **691**, 176 (1998).
[6] D. Haber-Mignard, N. Suschetet, R. Bergès, P. Astorg, and M.-H. Siess, *Nutr. Cancer* **25**, 61 (1996).

$$CH_3(S)_nSCH_3 \qquad\qquad CH_3CH_2CH_2(S)_nCH_2CH_2CH_3$$

$$CH_2=CHCH_2(S)_nCH_2CH=CH_2 \qquad CH_3CH=CH(S)_nCH=CHCH_3$$

Fig. 1. Structures of dimethyl, dipropyl, diallyl and diprop-1-enyl sulfides. Monosulfides, n = 1; disulfides, n = 2; trisulfides, n = 3; tetrasulfides, n = 4.

in part, to the ability of these sulfides to increase tissue activities of phase II detoxification enzymes.[9,10] These enzymes, which include glutathione S-transferase (GST, EC 2.5.1.18), epoxide hydrolase (EH, EC 3.3.2.3), quinone reductase (QR, DT-diaphorase, NAD[P]H:quinone-acceptor oxidoreductase, EC 1.6.99.2) and UDP-glucuronosyl transferase (UDPGT, EC 2.4.1.17), inactivate many electrophilic substances, including certain carcinogens, and facilitate their elimination from the body.[11] Many experiments have been conducted on the ability of *Allium*-derived sulfides to increase tissue activities of phase II enzymes in animal models, and these experiments, and their relevance to the human situation, are the subject of the present review.

Materials and Assay Methods

Some *Allium*-derived sulfides are commercially available. Pure dimethyl, diallyl and dipropyl monosulfides and dimethyl and dipropyl disulfide may be obtained from several suppliers (e.g., Acros, Lancaster, Aldrich), while diallyl trisulfide is available from LKT Laboratories. Diallyl disulfide is also sold by many suppliers, but it should be noted that the commercial product contains only 75% to 80% of the disulfide; the remainder is largely diallyl sulfide[12] or diallyl trisulfide.[13] Pure diallyl disulfide is easily synthesized.[14] Other compounds are not articles of commerce,

[7] L. W. Wattenberg, V. L. Sparnins, and G. Barany, *Cancer Res.* **49**, 2689 (1989).

[8] H. Sumiyoshi and M. J. Wargovich, *Cancer Res.* **50**, 5084 (1990).

[9] D. Guyonnet, C. Belloir, M. Suschetet, M.-H. Siess, and A.-M. Le Bon, *Mutation Res.* **495**, 135 (2001).

[10] S. V. Singh, S. S. Pan, S. K. Srivastava, H. Xia, X. Hu, H. A. Zaren, and J. L. Orchard, *Biochem. Biophys. Res. Commun.* **244**, 917 (1998).

[11] P. Talalay, *Biofactors* **12**, 5 (2000).

[12] H. S. Marks, J. L. Anderson, and G. S. Stoewsand, *J. Toxicol. Environ. Health* **37**, 1 (1992).

[13] D. Guyonnet, M.-H. Siess, A.-M. Le Bon, and M. Suschetet, *Toxicol. Appl. Pharmacol.* **154**, 50 (1999).

[14] T. A. Hase and H. Peräkylä, *Synth. Comm.* **12**, 947 (1982).

but methods for the synthesis of dipropyl trisulfide,[15] dipropyl and diallyl tetrasulfide,[16] diprop-1-enyl sulfide[17] and diprop-1-enyl disulfide[18] are available. GST has generally been assayed by the method of Habig et al.[19] by using CDNB as substrate. This is a "universal" substrate, reacting with all the major isoenzymes (α, μ, π) of rodent tissues.[20] For assay of QR, the method of Ernster,[21] which uses 2,6-dichlorophenol-indophenol as substrate, has proved very satisfactory for a wide range of tissues. Like the GST assay, this technique can be used with crude tissue homogenates. EH may be assayed in microsomes fluorimetrically[22] or radiometrically[23] and UDPGT fluorimetrically[24] or spectrophotometrically.[25]

Phase II Enzyme Induction by Allium-Derived Sulfides

Twenty-seven experiments on phase II enzyme induction by Allium-derived sulfides have been reported (see references 9, 26–30 and literature cited therein). In all cases, rats or mice have been employed. Many dosing protocols have been used, with a wide range of dose levels, duration of dosing and route of administration. Most studies have employed oral dosing, either by intubation or by dietary feeding, although some have involved injection of the test materials. Most studies have focused on the sulfides that are commercially available, with 22 experiments with diallyl monosulfide, 17 with diallyl disulfide, 11 with dipropyl sulfide, 8 with dipropyl disulfide and 7 with diallyl trisulfide, but only 2 with dipropyl trisulfide and 1 each with diprop-1-enyl monosulfide, diprop-1-enyl disulfide, dipropyl

[15] B. Milligan, B. Saville, and J. M. Swan, *J. Chem. Soc.* 4850 (1961).

[16] C. G. Moore and B. R. Trago, *Tetrahedron* **18**, 205 (1962).

[17] B. A. Trofimov, S. V. Amosova, G. K. Musorin, G. A. Kalabin, V. V. Nosyreva, and M. L. Al'pert, *Sulfur Lett.* **4**, 67 (1986).

[18] H. E. Wijers, H. Boelens, A. van der Gen, and L. Brandsma, *Rec. Trav. Chim. Pays-Bas* **88**, 519 (1969).

[19] W. H. Habig, M. J. Pabst, and W. B. Jakoby, *J. Biol. Chem.* **249**, 7130 (1974).

[20] J. D. Hayes and D. J. Pulford, *Crit. Rev. Biochem. Mol. Biol.* **30**, 445 (1995).

[21] L. Ernster, *Methods Enzymol.* **10**, 309 (1967).

[22] M. E. Herrero and J. V. Castell, *Anal. Biochem.* **230**, 154 (1995).

[23] T. M. Guenthner, P. Bentley, and F. Oesch, *Methods Enzymol.* **77**, 344 (1981).

[24] A. V. Collier, M. D. Tingle, J. A. Keelan, J. W. Paxton, and M. D. Mitchell, *Drug Metab. Disp.* **28**, 1184 (2000).

[25] B. Antoine, J. A. Boutin, and G. Siest, *Biochem. J.* **252**, 930 (1998).

[26] C. Bose, J. Guo, L. Zimniak, S. K. Srivastava, S. P. Singh, P. Zimniak, and S. V. Singh, *Carcinogenesis* **23**, 1661 (2002).

[27] C. C. Wu, L. Y. Sheen, H.-W. Chen, S.-J. Tsai, and C.-K. Lii, *Fd. Chem. Toxicol.* **39**, 563 (2001).

[28] R. Munday and C. M. Munday, *Nutr. Cancer* **34**, 42 (1999).

[29] R. Munday and C. M. Munday, *Nutr. Cancer* **40**, 205 (2001).

[30] R. Munday, J. S. Munday, and C. M. Munday, *Free Rad. Biol. Med.* **34**, 1200 (2003).

tetrasulfide and diallyl tetrasulfide. The methyl derivatives have attracted little attention, with only one study with dimethyl sulfide and dimethyl disulfide being reported.

Several conclusions on the relationship between structure and enzyme inducing ability can be drawn from these studies. Allyl sulfides are more powerful inducers than the methyl or propyl derivatives, and within the allyl series, the disulfide is a more potent inducer than the monosulfide. Because of the very different protocols employed in the early studies, however, quantitative comparison of the various compounds is not feasible. More recently, data on the effects of 10 *Allium*-derived sulfides on phase II enzyme activity in rat tissues became available, and in these experiments,[29,30] the same dose-level of sulfide (500 μmol/kg/day) and the same duration of administration (5 days) were employed in all cases, permitting direct comparison of the efficacy of the different compounds. Results on QR activities in the rat tissues are summarized in Tables I and II. The importance of unsaturation in the alkyl chain was confirmed, with both the allyl and prop-1-enyl derivatives showing a higher level of induction in many tissues than the propyl. The importance of the number of sulfur atoms in the molecule was also confirmed, although the direction of the effect was different in the allyl and prop-1-enyl sulfides. In accord with the previous work, diallyl disulfide was found to be more active than the

TABLE I

INDUCTION OF QUINONE REDUCTASE IN THE LIVER, KIDNEYS, SPLEEN, LUNGS, HEART AND
URINARY BLADDER OF RATS DOSED WITH *ALLIUM*-DERIVED SULFIDES

Sulfide	Quinone Reductase Activity (I. U./g tissue)					
	Liver	Kidney	Spleen	Lungs	Heart	Urinary Bladder
Dipropyl Monosulfide	1.2	0.9	0.9	0.9	1.0	1.0
Dipropyl Disulfide	1.1	0.9	0.9	1.1	0.9	1.0
Dipropyl Trisulfide	1.0	0.9	1.3	1.2	1.2	1.0
Dipropyl Tetrasulfide	1.0	0.9	1.4	1.3	1.2	1.0
Diallyl Monosulfide	1.0	1.4	1.1	1.3	1.3	1.3*
Diallyl Disulfide	2.1*	2.4*	2.6*	5.2*	1.6*	1.6*
Diallyl Trisulfide	2.8*	5.1*	3.0*	6.7*	2.5*	1.4*
Diallyl Tetrasulfide	2.6*	3.3*	2.5*	6.6*	2.0*	2.1*
Diprop-1-Enyl Monosulfide	1.7*	2.4	3.2*	2.5*	1.3	2.9*
Diprop-1-Enyl Disulfide	1.1	1.8	2.1*	1.5	1.2	1.6*

Rats were dosed with the sulfides at 500 μmol/kg/day for 5 days. Figures shown are the QR activities in the specified tissues of the sulfide-dosed rats relative to those in control animals. Figures marked with asterisks indicate that the indicated QR activities were significantly different ($p < 0.05$) than those of control rats.

TABLE II

INDUCTION OF QUINONE REDUCTASE IN THE GASTROINTESTINAL TRACT OF RATS DOSED WITH *ALLIUM*-DERIVED SULFIDES

Sulfide	Quinone reductase activity (I. U./g tissue)								
	Forestomach	Glandular Stomach	Duodenum	Jejunum	Ileum	Cecum	Colon + Rectum
Dipropyl Monosulfide	1.1	1.0	1.0	1.0	1.0	0.8	0.8
Dipropyl Disulfide	1.1	1.5	1.1	1.2	1.0	0.7	0.9
Dipropyl Trisulfide	1.5	0.9	0.6	1.2	0.8	1.1	0.9
Dipropyl Tetrasulfide	0.8	1.0	1.1	1.4	1.1	1.2	0.9
Diallyl Monosulfide	2.1*	1.3	1.6	1.6	1.2	1.2	1.2
Diallyl Disulfide	8.1*	2.4*	7.9*	8.5*	4.5*	4.0*	3.8*
Diallyl Trisulfide	2.7*	2.4*	8.9*	8.3*	4.6*	3.5*	3.7*
Diallyl Tetrasulfide	3.1*	2.1*	8.6*	10.4*	4.0*	3.1*	4.1*
Diprop-1-Enyl Monosulfide	3.6*	2.4*	4.2*	3.9*	3.8*	4.6*	4.4*
Diprop-1-Enyl Disulfide	2.1*	1.5*	3.0*	3.0*	1.7	1.9*	1.6

Rats were dosed with the sulfides at 500 μmol/kg/day for 5 days. Figures shown are the QR activities in the specified tissues of the sulfide-dosed rats relative to those in control animals. Figures marked with asterisks indicate that the indicated QR activities were significantly different ($p < 0.05$) than those of control rats.

monosulfide, but diprop-1-enyl monosulfide was a stronger inducer than the disulfide. In the allyl compounds, activity did not continue to increase with increasing number of sulfur atoms, with diallyl tri- and tetra-sulfides being of similar potency to diallyl disulfide.

In most of the published experiments, only hepatic enzyme activities were determined. While the liver is recognized as a major site of toxin metabolism and contains high levels of phase II enzymes, extrahepatic conjugation reactions may be equally important. In the rat, tissues such as the forestomach, glandular stomach, duodenum, ileum, cecum and lung also contain high levels of phase II enzymes,[31] and it was shown that enzyme activities in these tissues are more sensitive to induction by *Allium*-derived sulfides than those in the liver.[29] As shown in Tables I and II, the effect of such sulfides was most pronounced in the jejunum, duodenum, forestomach, lung, ileum, colon plus rectum, and cecum, with smaller effects being recorded in the kidney, glandular stomach, liver, spleen, heart, and urinary bladder.

The dose levels used in most experiments on enzyme induction by *Allium*-derived sulfides ranged between 500 and 4500 μmol/kg/day, and only a single dose level was generally employed. A detailed dose-response experiment has been reported only for diallyl disulfide. In this study, dose levels between 0.5 and 1500 μmol/kg were employed. In most tissues, significant increases in enzyme activity were seen only at relatively high dose levels (300 μmol/kg/day or above). In contrast, significant increases in QR and/or GST were seen in the forestomach, duodenum, and jejunum at dose levels of only 2 μmol/kg.[28]

Relevance of Animal Studies to the Human Situation

If phase II enzyme induction by sulfides is involved in the chemopreventative action of *Allium* vegetables, it would be expected that effects would be seen in animal models at doses similar to those that could be obtained from dietary intake of these vegetables by humans, and that induction would be seen in those tissues which epidemiological studies show a decrease in cancer incidence.

The average daily intake of garlic in the USA is 1.4 g/day,[32] while individuals with a high intake of this vegetable may consume 20 g/day.[33] The total concentration of thiosulfinates in macerated garlic is of the order of

[31] R. Munday, B. L. Smith, and C. M. Munday, *Chem. Biol. Interact.* **123,** 219 (1999).

[32] E. Block, E. M. Calvey, C. W. Gillies, J. Z. Gillies, and P. Uden, *Rec. Adv. Phytochem.* **31,** 1 (1997).

[33] G. Guhr and P. A. Lachance, *in* "Nutraceuticals: Designer Foods III: Garlic, Soy and Licorice" (P. A. Lachance, ed.), p. 311. Food and Nutrition Press, Trumbull, CT, 1997.

20 μmol/g fresh weight.[34] If these thiosulfinates were converted quantitatively to sulfides, a 70-kilogram individual eating 1.4 g of raw garlic a day would receive approximately 0.4 μmol/kg/day of sulfide, while an intake of 20 g/day would provide around 6 μmol/kg/day. However, much garlic is consumed after cooking, and a large proportion of the sulfides are lost in this process.[35] If one assumes a 75% loss, the human intake of sulfides from the average and high intakes of cooked garlic would be ~0.1 and 1.5 μmol/kg/day, respectively. The average intake of onions in the United States is 13 g/day,[1] while individuals with a high intake of these vegetables may consume 75 g/day.[36] The thiosulfinate content of macerated onions is ~0.2 μmol/g,[34] and again assuming 100% conversion to sulfides, the intake of sulfides by a 70-kg individual with an average intake of raw onions would be 0.04 μmol/kg/day and 0.2 μmol/kg/day for a high intake. The corresponding figures for the cooked vegetable, assuming 75% loss in cooking, are 0.01 μmol/kg/day and 0.05 μmol/kg/day, respectively.

Most of the published experiments have employed dose levels of the sulfides orders of magnitude higher than any conceivable human intake and are thus not directly relevant to the human situation. In contrast, the dose-response study with diallyl disulfide[28] showed significant effects on enzyme activity in rats at dose levels similar to those that would be gained from a high dietary intake of garlic. Furthermore, the sites at which induction was recorded at these low levels were the same as those reported to be protected from cancer by a high intake of the vegetable. In this case, therefore, the evidence is consistent with diallyl disulfide (and the corresponding tri- and tetra-sulfides, which are as potent as the disulfide) exerting a chemopreventative effect via phase II enzyme induction.

The situation with onions is somewhat different. Because of the breakdown of prop-1-enyl cysteine by pathways that do not lead to sulfide formation, the human intake of sulfides from onions is lower than that from garlic. The methyl and propyl sulfides are, without exception, weak inducers of phase II enzymes. Diprop-1-enyl disulfide is a weaker inducer than diallyl disulfide, and while diprop-1-enyl sulfide was shown to be a powerful inducer, a recent dose-response experiment in rats[37] has shown that, while the most sensitive tissues were again the stomach and upper intestine, a dose of at least 10 μmol/kg is needed to increase phase II enzyme activity. At present, therefore, there is no convincing evidence that onion-derived sulfides are involved in the chemoprevention recorded with onions.

[34] E. Block, S. Naganathan, D. Putman, and S.-H. Zhao, *J. Ag. Food Chem.* **40**, 2418 (1992).

[35] L. D. Lawson, *ACS Symposium Ser.* **534**, 306 (1993).

[36] E. Dorant, P. A. van den Brandt, R. A. Goldbohm, and F. Sturmans, *Gastroenterology* **110**, 12 (1996).

[37] R. Munday and C. M. Munday, Manuscript in Preparation.

Toxicity of Sulfides and Relationship to Mechanism of
  Enzyme Induction

Di-, tri- and tetra-sulfides cause hemolytic anemia in laboratory and domestic animals.[30,38] In the presence of reduced glutathione, these substances undergo redox cycling, with concomitant production of "active oxygen" species. The latter species are held responsible for initiating the erythrocyte damage leading to hemolysis.

The induction of phase II enzymes by many substances, of diverse chemical structure, has been shown to involve transcriptional gene activation through covalent alteration of protein thiol groups by alkylation, oxidation or reduction.[39] While the toxicological studies show that the di-, tri- and tetra-sulfides are able to interact with thiol groups and cause thiol-group oxidation, redox reactions cannot account for the inductive activity of the *Allium*-derived sulfides. Redox cycling of prop-1-enyl disulfide is much faster than that of diallyl disulfide, but it is a weaker inducer.[29] Furthermore, both the diallyl and dipropyl tri- and tetra-sulfides undergo rapid redox cycling, yet the propyl derivatives have little or no effect on the phase II enzymes.[30]

Conclusions

Some *Allium*-derived sulfides are potent inducers of phase II enzymes in rodents. It is possible that allyl sulfides are involved in the chemoprotective action of garlic, but compounds of comparable activity have not been isolated from onion. Most of the studies to date have involved short-term administration of *Allium*-derived sulfides. Since humans consume *Allium* vegetables regularly throughout their lives, and since the degree of enzyme induction may increase with prolonged exposure, more work on the long-term effects of sulfides on enzyme activity is needed. Re-investigation of possible inducers in onion is also required. In this regard, the effects of other sulfur-containing materials that are known to be present in processed onion, such as thiophene derivatives and the cepaenes and zweibelanes,[3] would be of considerable interest. Biochemical studies on the mechanism of induction of phase II enzymes by *Allium*-derived sulfides are also needed, with particular reference to the role of the allyl and prop-1-enyl groups in the inductive process.

[38] R. Munday and E. Manns, *J. Ag. Food Chem.* **42,** 959 (1994).
[39] A. T. Dinkova-Kostova, M. A. Massiah, R. E. Bozak, R. J. Hicks, and P. Talalay, *Proc. Natl. Acad. Sci. USA* **98,** 3404 (2001).

[25] Upregulation of Quinone Reductase by Glucosinolate Hydrolysis Products From Dietary Broccoli

By Elizabeth H. Jeffery and Kristin E. Stewart

Introduction

An increasing regard for the health benefits of whole foods has sparked interest in the chemoprotective effects of vegetables. Increased consumption of fruits and vegetables has been repeatedly linked to lower risk for many cancers, including prostate, breast, lung and colon.[1] Of the fruits and vegetables studied, cruciferous vegetables are often found to offer substantially greater chemoprotective benefit than other food groups.[2-8] Inclusion of cruciferous vegetable powders in experimental rodent diets has been associated with a significant decrease in the incidence and size of chemically induced tumors. For example, Stoewsand and colleagues observed a 64% reduction in the incidence of dimethyl benzanthracene-induced mammary tumors in rats fed Brussels sprouts as compared to those fed a semi-purified casein-cornstarch diet during the initiation period of carcinogenesis.[9] One of the proposed mechanisms by which the cruciferous vegetables mediate these chemoprotective effects is by the increased expression and activity of the phase II detoxification enzymes, including glutathione S-transferase and quinone reductase (QR). This article will focus on the upregulation of the detoxification enzyme QR by crucifers, particularly broccoli, as it relates to cancer prevention.

[1] K. A. Steinmetz and J. D. Potter, J. Am. Diet Assoc. 96, 1027 (1996).

[2] E. Benito, A. Obrador, A. Stiggelbout, F. X. Bosch, M. Mulet, N. Munoz, and J. Kaldor, Int. J. Cancer 45, 69 (1990).

[3] D. T. Verhoeven, R. A. Goldbohm, G. van Poppel, H. Verhagen, and P. A. van den Brandt, Cancer Epidemiol. Biomarkers Prev. 5, 733 (1996).

[4] D. S. Michaud, D. Spiegelman, S. K. Clinton, E. B. Rimm, W. C. Willett, and E. L. Giovannucci, J. Natl. Cancer Inst. 91, 605 (1999).

[5] L. N. Kolonel, J. H. Hankin, A. S. Whittemore, A. H. Wu, R. P. Gallagher, L. R. Wilkens, E. M. John, G. R. Howe, D. M. Dreon, D. W. West, and R. S. Paffenbarger, Jr., Cancer Epidemiol. Biomarkers Prev. 9, 795 (2000).

[6] J. H. Cohen, A. R. Kristal, and J. L. Stanford, J. Natl. Cancer Inst. 92, 61 (2000).

[7] S. M. Zhang, D. J. Hunter, B. A. Rosner, E. L. Giovannucci, G. A. Colditz, F. E. Speizer, and W. C. Willett, Cancer Epidemiol. Biomarkers Prev. 9, 477 (2000).

[8] M. P. Zeegers, R. A. Goldbohm, and P. A. van den Brandt, Cancer Epidemiol. Biomarkers Prev. 10, 1121 (2001).

[9] G. S. Stoewsand, J. L. Anderson, and L. Munson, Cancer Lett. 39, 199 (1988).

Crucifers, Cancer Prevention and Quinone Reductase

Numerous animal studies demonstrate an induction in detoxification enzymes by cruciferous vegetables, including broccoli and Brussels sprouts. In one study, the addition of 20% freeze-dried Brussels sprout powder to the diet of Wistar rats resulted in up to 3-fold increases in hepatic and intestinal QR as well as glutathione and glutathione-S-transferase activity.[10] These researchers found a dose relationship between the amount of Brussels sprout powder in the diet (2.5% to 20%) and the induction of QR. Feeding broccoli to rats is also associated with increased colonic and hepatic QR. When 20% freeze-dried broccoli was added to the diet of male Fisher 344 rats, a greater than 4-fold increase was seen in colonic QR activity, although hepatic QR was only increased by 40%.[11]

It has been suggested that the upregulation of QR and other phase II detoxification enzymes may be the means by which cruciferous vegetables provide protection from cancer, either by destroying carcinogens before they have their effects, or more generally, by inhibiting the ability of reactive oxygen species to initiate or support carcinogenesis.[12–16] Compounds that activate a battery of phase II enzymes, without activation of cytochrome P450, have been termed *monofunctional inducers*.[17] Although monofunctional inducers appear to share anticarcinogenic properties, making induction of QR (or other phase II enzyme such as glutathione-S-transferase) a good biomarker of anticarcinogenic action, the mechanistic relationship between increases in QR and decreased cancer risk is not fully elucidated.

There are several advantages to the measurement of QR, rather than other detoxification enzymes, as a bioassay or biomarker of anticarcinogenic activity. Although there are at least two isoenzymes, one predominates during induction, simplifying measurement.[18] In addition, QR activity can be used both as a measure of tissue response to ingestion of a cruciferous

[10] H. M. Wortelboer, C. A. de Kruif, A. A. van Iersel, J. Noordhoek, B. J. Blaauboer, P. J. van Bladeren, and H. E. Falke, *Food Chem. Toxicol.* **30**, 17 (1992).

[11] A. S. Keck, Q. Qiao, and E. H. Jeffery, *J. Agric. Food Chem.* **51**, 3320 (2003).

[12] G. W. Plumb, N. Lambert, S. J. Chambers, S. Wanigatunga, R. K. Heaney, J. A. Plumb, O. I. Aruoma, B. Halliwell, N. J. Miller, and G. Williamson, *Free Radic. Res.* **25**, 75 (1996).

[13] C. Y. Zhu and S. Loft, *Food Chem. Toxicol.* **41**, 455 (2003).

[14] H. J. Prochaska, A. B. Santamaria, and P. Talalay, *Proc. Natl. Acad. Sci. USA* **89**, 2394 (1992).

[15] W. Fahey, Y. Zhang, and P. Talalay, *Proc. Natl. Acad. Sci. USA* **94**, 10367 (1997).

[16] J. W. Fahey and P. Talalay, *Food Chem. Toxicol.* **37**, 973 (1999).

[17] H. J. Prochaska, A. B. Santamaria, and P. Talalay, *Proc. Natl. Acad. Sci. USA* **89**, 2394 (1992).

[18] A. K. Jaiswal, *Arch. Biochem. Biophs.* **375**, 62 (2000).

vegetable and as a bioassay for induction potential of plant extracts.[19–21] While measurement of enzyme induction in animals can be both laborious and expensive, measurement of QR activity in cultured cells functions as a relatively inexpensive, rapid, and reproducible *in vitro* screening tool for anticarcinogenic plant compounds.[14] The *in vitro* system involves direct measurement of NAD(P)H:quinone reductase in murine hepatoma Hepa1c1c7 cells grown in microtiter plates in the presence and absence of the plant extract or other compound under study.[18,19] Use of a screening assay is limited however, since it cannot take into account any degradation in the gut or other issues associated with bioavailability.

Variation in QR levels in response to diet have successfully been measured in human saliva samples.[22] For a number of years, glutathione S-transferase has been estimated in plasma and white blood cells, as a measure of upregulation of detoxification enzymes in response to ingestion of crucifers.[23,24] Future development of QR estimation in human white blood cells or exfoliated cells has the potential to be of great use clinically, by using induction of QR as a biomarker for the cancer-preventive action of crucifers such as broccoli. Interestingly, screening of extracts from a variety of vegetables for QR inducing potency identified cruciferous vegetables, particularly those of the genus *Brassica,* as potent activators of the phase II enzymes.[10] This strongly supports evidence from rodent and epidemiological data that indicates a relationship between consumption of cruciferous vegetables, potency in induction of detoxification enzymes such as QR, and protection from a variety of cancers.

Sulforaphane

Much of the recent research regarding crucifers has focused on the isothiocyanate sulforaphane. Sulforaphane is the hydrolysis product of glucoraphanin, the primary glucosinolate found in broccoli. Sulforaphane has been shown to augment the cells' defense against carcinogenic insult through numerous effects on the cell cycle, as well as triggering upregulation of phase II enzymes, including QR. These increases are seen in both primary and transformed cell lines derived from human and rat prostate,

[19] H. J. Prochaska and A. B. Santamaria, *Anal. Biochem.* **169,** 328 (1988).

[20] H. J. Prochaska and P. Talalay, *Cancer Research* **48,** 4776 (1988).

[21] A. T. Dinkova-Kostova and P. Talalay, *Free Radical Biology and Medicine* **29,** 231 (2000).

[22] L. Sreerama, M. W. Hedge, and N. E. Sladek, *Clin. Cancer Res.* **1,** 1153 (1995).

[23] M. L. Clapper, C. E. Szarka, G. R. Pfeiffer, T. A. Graham, A. M. Balshem, S. Litwin, E. B. Goosenberg, H. Frucht, and P. F. Engstrom, *Clin. Cancer Res.* **3,** 25 (1997).

[24] J. J. Bogaards, H. Verhagen, M. I. Willems, G. van Poppel, and P. J. van Bladeren, *Carcinogenesis* **15,** 1073 (1994).

TABLE I
RELATIVE POTENCIES OF ISOTHIOCYANATES

Sulforaphane[29]	0.4–0.8 μM
Benzyl Isothiocyanate[31]	3.7 μM
Phenethyl Isothiocyanate[30]	5 μM
Allyl Isothiocyanate[28]	5 μM*
Indole-3-Carbinol[40]	50 μM**
Sulforaphane Nitrile[34]	1 mM
Crambene[11]	5 mM

Dose required to induce a 2-fold increase in quinone reductase activity in Hepa1c1c7 (mouse hepatoma) cells.

* Dose required to induce a 1.8-fold increase in quinone reductase activity in PE (mouse skin papilloma) cells.

** Dose required to induce a 2-fold increase in quinone reductase mRNA levels in HepG2 (human hepatoma) cells.

colon, mammary epithelial and liver tissues.[25–28] Sulforaphane appears to be much more potent than many of the other isothiocyanates (ITC; Table I).[29] As little as 0.5 μM sulforaphane caused a doubling in quinone reductase activity, whereas approximately 10-fold this amount of allyl-ITC, benzyl-ITC, and phenethyl-ITC was required to double activity.[26,30,31] Intracellular accumulation may be one mechanism by which sulforaphane gains its potency. When sulforaphane was introduced into the medium of cultured cells, it accumulated rapidly. Thirty minutes' exposure of hepatic cells to 50 to 500 μM sulforaphane resulted in a very high intracellular sulforaphane level, 5 to 9 mM, conjugated to glutathione.[32] Even prostate cells incubated with 5 μM sulforaphane accumulated up to 900 μM in 2 hours. Intracellular sulforaphane levels remained elevated for a longer period and to a greater extent than other ITCs. The degree of accumulation

[25] K. Singletary and C. MacDonald, *Cancer Lett.* **155,** 47 (2000).

[26] C. Bonnesen, I. M. Eggleston, and J. D. Hayes, *Cancer Res.* **61,** 6120 (2001).

[27] J. D. Brooks, V. G. Paton, and G. Vidanes, *Cancer Epidemiol. Biomarkers Prev.* **10,** 949 (2001).

[28] L. Ye and Y. Zhang, *Carcinogenesis* **22,** 1987 (2001).

[29] Y. Zhang, P. Talalay, C. G. Cho, and G. H. Posner, *Proc. Natl. Acad. Sci. USA* **89,** 2399 (1992).

[30] P. Rose, K. Faulkner, G. Williamson, and R. Mithen, *Carcinogenesis* **21,** 1983 (2000).

[31] T. Prestera, W. D. Holtzclaw, Y. Zhang, and P. Talalay, *Proc. Natl. Acad. Sci. USA* **90,** 2965 (1993).

[32] Y. Zhang, *Carcinogenesis* **21,** 1175 (2000).

correlated with the observed increase in QR activity.[28] These high rates of intracellular accumulation may account for sulforaphane's high potency compared to other glucosinolate hydrolysis products.

Sulforaphane Nitrile

In intact plant tissue, the glucosinolates are sequestered in the vacuoles of the plant cell. Upon chewing or chopping, the crucifer tissue is disrupted, breaking the vacuole and releasing the glucosinolate, allowing interaction with the plant enzyme myrosinase (thioglucoside hydrolase, EC 3.2.3.1). Myrosinase hydrolyzes the β-thioglucoside linkage of the glucosinolate molecule, producing glucose, sulfate, and an unstable aglucon intermediate. The aglucon intermediate undergoes spontaneous rearrangement to form the isothiocyanate. However, this is not the sole product. For example, in broccoli, the primary glucosinolate glucoraphanin is converted to two principle aglucon products, sulforaphane and sulforaphane nitrile (Fig. 1). If glucoraphanin is extracted from broccoli and hydrolyzed by using a purified myrosinase from raphanus, sulforaphane is the sole product.

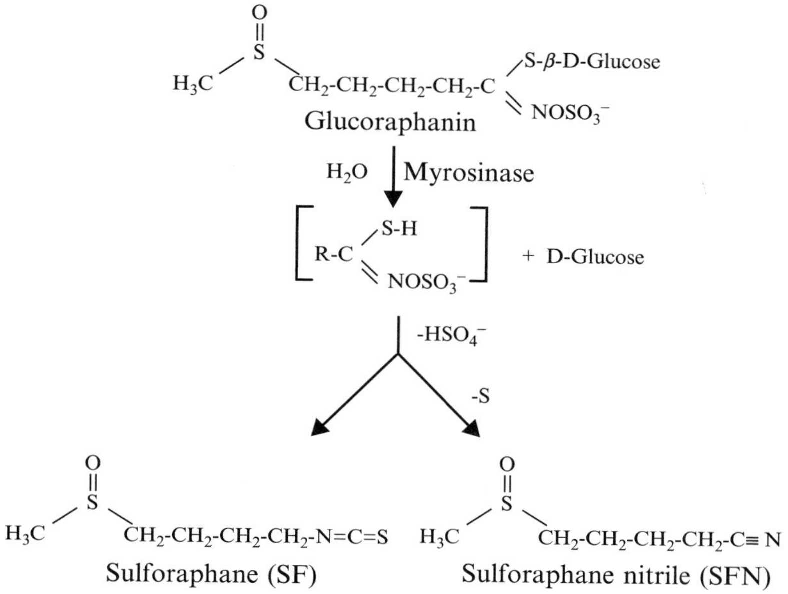

FIG. 1. Enzymatic conversion of glucoraphanin to sulforaphane and sulforaphane nitrile. (Adapted from Matusheski and Jeffery, 2001.[34])

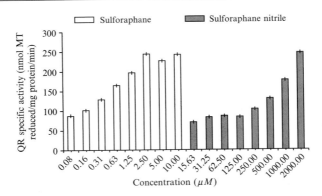

Fig. 2. Relative bioactivity of sulforaphane and sulforaphane nitrile. Quinone reductase activity (measured as nmol of MTT reduced/min/mg of protein) in Hepa1c1c7 cells treated for 24 hours. (Adapted from Matusheski and Jeffery, 2001.[34])

This 100% conversion has been used successfully to estimate glucoraphanin.[33] However, a recent publication shows that hydrolysis of the glucosinolates by endogenous myrosinase, brought about by homogenizing fresh broccoli in water, results in significant accumulation of sulforaphane nitrile, which lacks the bioactivity of the isothiocyanate compounds.[34] Broccoli tissue homogenized with only endogenous myrosinase exhibited 25% or lower conversion of glucoraphanin to sulforaphane, depending on variety, with nitrile production accounting for the remainder. The addition of exogenous myrosinase to a broccoli preparation led to an increased ratio of sulforaphane to sulforaphane nitrile, indicating the exogenous myrosinase could shift conversion of glucoraphanin toward sulforaphane. In Brigadier broccoli, the addition of white mustard myrosinase increased SF from 1.9 μmol/g to 3.4 μmol/g, while SFN decreased from 8.7 μmol/g to 3.9 μmol/g. Sulforaphane nitrile was found to be several orders of magnitude less potent than sulforaphane, both in cell culture and in orally treated rats (Fig. 2).[22] Some varieties of crucifer are associated with nitrile formation (crambe, rape, white cabbage), while others are not (horseradish, white mustard, dikon).[35–37] Therefore nitrile formation may be associated with an additional gene product, present only in some varieties and

[33] T. Prestera, J. W. Fahey, W. D. Holtzclaw, C. Abeygunawardana, J. L. Kachinski, P. Talalay, *Anal. Biochem.* **239,** 168 (1996).
[34] N. V. Matusheski and E. H. Jeffery, *J. Agric. Food Chem.* **49,** 5743 (2001).
[35] R. J. Petroski and H. L. Tookey, *Phytochemistry* **21,** 1903 (1982).
[36] A. J. MacLeod and J. T. Rossiter, *Phytochemistry* **24,** 1895 (1985).
[37] R. A. Cole, *Phytochemistry* **15,** 759 (1976).

absent from others. These data can be interpreted to mean that nitrile formation is not essential for plant survival and that hybridization technology may be used to produce a variety that has high glucoraphanin levels and little or no nitrile-forming characteristics.

Synergistic Effects of Glucosinolate Metabolites

In addition to sulforaphane, broccoli contains the glucosinolates glucobrassicin and progoitrin, which form the bioactive hydrolysis products indole-3-carbinol and crambene (1-cyano-2-hydroxy-3-butene), respectively. Crambene, like sulforaphane, behaves as a monofunctional inducer, increasing phase II enzyme activity, with no change in CYP 1A activity.[38] Both compounds transcriptionally upregulate phase II enzymes through triggering an antioxidant response element (ARE)-mediated mechanism.[39,40] The compounds apparently disrupt the interaction between Nrf2 and the sequestering protein Keap1, allowing Nrf2 to translocate to the nucleus and interact with the ARE, increasing transcription of target genes.[41]

In the acidic environment of the stomach, indole-3-carbinol can undergo self-condensation reactions to form a group of compounds termed acid condensates, which include 3,3'-diindoylmethane (DIM) and indolo[3,2-b]carbazole (ICZ).[42] Indole-3-carbinol has been classified as a bifunctional inducer, inducing both phase I and phase II enzymes.[15,43,44] Induction has been shown to depend on the presence of a xenobiotic response element (XRE) in the 5' regulatory region of these genes.[45,46] Transcriptional activation through the XRE is achieved by ligand-dependent aryl hydrocarbon receptor (AhR) binding. In confirmation of earlier work, we find that unlike acid condensation products, indole-3-carbinol does not activate an XRE driven reporter construct, even though it is reported to be a weak ligand of the AhR.[47,48] The acid condensates of indole-3-carbinol are

[38] R. Staack, S. Kingston, M. A. Wallig, and E. H. Jeffery, *Toxicol. Appl. Pharmacol.* **149**, 17 (1998).

[39] T. Prestera and P. Talalay, *Proc. Natl. Acad. Sci. USA* **92**, 8965 (1995).

[40] C. W. Nho, Ph.D. Thesis, University of Illinois, 2001.

[41] K. Itoh, N. Wakabayashi, Y. Katoh, T. Ishii, K. Igarashi, J. D. Engel, and M. Yamamoto, *Genes Dev.* **13**, 76 (1999).

[42] C. A. Bradfield and L. F. Bjeldanes, *J. Toxicol. Environ. Health* **21**, 311 (1987).

[43] W. D. Loub, L. W. Wattenberg, and D. W. Davis, *J. Natl. Cancer. Inst.* **54**, 985 (1975).

[44] L. W. Wattenberg, *Cancer Res.* **35**, 3326 (1975).

[45] I. Chen, S. Safe, and L. Bjeldanes, *Biochem. Pharmacol.* **51**, 1069 (1996).

[46] P. H. Jellinck, P. G. Forkert, D. S. Riddick, A. B. Okey, J. J. Michnovicz, and H. L. Bradlow, *Biochem. Pharmacol.* **45**, 1129 (1993).

[47] L. F. Bjeldanes, J. Y. Kim, K. R. Grose, J. C. Bartholomew, and C. A. Bradfield, *Proc. Natl. Acad. Sci. USA* **88**, 9543 (1991).

[48] I. Chen, S. Safe, and L. Bjeldanes, *Biochem. Pharmacol.* **51**, 1069 (1996).

capable of activation of the XRE reporter gene. The acid condensates DIM and ICZ have been shown specifically to activate *cyp1A1*, a gene that shows XRE dependent activation.[26] In addition, DIM and ICZ were shown to modestly increase QR activity in human colon cancer cell lines.

While much of the published research focuses on the identification of individual compounds in the vegetable that may provide protective benefits, the chemical composition of the whole vegetable is far more complex, and the presence of multiple bioactive components undoubtedly provides additional health benefit. Research in our laboratory has focused on the synergistic interactions of the multiple isothiocyanates found in broccoli. When administered to rats, the mixture of crambene and indole-3-carbinol caused an increase in QR activity as well as an increase in hepatic glutathione S-transferase activity and mRNA levels, all of which were significantly greater than the induction seen with either treatment alone. Individually, administration of crambene or indole-3-carbinol to rats led to a 1.4- and 1.5-fold increase in hepatic glutathione-S-transferase, respectively, while the mixture resulted in a 2.1-fold increase. Hepatic QR activity was induced 4-fold by the mixture of crambene and indole-3-carbinol, while the individual compounds only resulted in 1.8- and 2.1-fold increases.[38,49] These data indicate a potential synergistic interaction between crambene and indole-3-carbinol in the induction of phase II enzymes.

In several genes, including that for quinone reductase, the 5′ regulatory region possesses both ARE and XRE consensus sequences, exposure to multiple isothiocyanates may result in a synergistic upregulation of gene transcription.[50–53] We hypothesize that exposure to multiple isothiocyanates may result in a synergistic upregulation of gene transcription, since transcriptional increases in these genes may occur via combined activation of the ARE and XRE. When human hepatoma (HepG2) cells were transfected with a QR reporter construct and treated with crambene and acidified indole-3-carbinol, a greater-than-additive increase in activity was observed.[40] Whereas studying the effects on QR of the individual bioactive components found in broccoli may provide useful insight into key mechanisms of enzyme induction, these data indicate that combinations of bioactive compounds found in the whole vegetable may afford additional protection via the synergistic induction of chemoprotective enzymes. This suggests the possibility that the health benefits of whole vegetables may be greater than those of the isolated compounds. Such synergistic

[49] C. W. Nho and E. Jeffery, *Toxicol. Appl. Pharmacol.* **174,** 146 (2001).

[50] R. M. Bayney, M. R. Morton, L. V. Favreau, and C. B. Pickett, *J. Biol. Chem.* **264,** 21793 (1989).

[51] L. V. Favreau and C. B. Pickett, *J. Biol. Chem.* **266,** 4556 (1991).

[52] T. H. Rushmore and C. B. Pickett, *Methods Enzymol.* **206,** 409 (1991).

[53] T. H. Rushmore and C. B. Pickett, *J. Biol. Chem.* **256,** 14648 (1990).

activation has not been evaluated for the combination of sulforaphane and indole-3-carbinol.

Variability in the Effect of Dietary Cruciferous Vegetables on Induction of QR

As described previously, a majority of the chemoprotective benefits of crucifers have been attributed to the presence of isothiocyanates. These are hydrolysis products of a family of compounds called glucosinolates that are relatively unique to cruciferous vegetables.[54] The glucosinolates are not biologically active but are hydrolyzed to active products on mastication or mechanical disruption of the plant tissue, which brings the hydrolyzing enzyme myrosinase into contact with the glucosinolate. This lack of biological activity of the glucosinolates has been confirmed by incubation of purified glucosinolates in cell culture: no change in quinone reductase was seen until a myrosinase enzyme was introduced into the medium.[55] The potent isothiocyanate sulforaphane is produced during myrosinase-dependent hydrolysis of glucoraphanin, the predominant glucosinolate in broccoli. Thus the amount of an individual bioactive component such as sulforaphane that arrives at the site of action depends on a number of variables, including the glucosinolate content of the broccoli (regulated by both genotype and environment), the inherent myrosinase activity, the storage and processing of the broccoli and the microflora of the gut, which play a key role in hydrolysis of glucosinolates in the absence of active myrosinase (i.e., boiled broccoli).

Genotype Variation in Glucosinolate Content of Cruciferous Vegetables

Glucosinolate profiles of cruciferous vegetables vary widely between species and, within the species, between varieties. The most abundant glucosinolates in dietary crucifers include the indolyl glucosinolate glucobrassicin, the phenyl glucosinolate gluconastutiin, and the aliphatic glucosinolates—sinigrin, progoitrin and glucoraphanin. A comprehensive analysis of 50 broccoli, 4 Brussels sprouts, 6 cabbage, 3 cauliflower and 2 kale varieties shows the relative distribution of key glucosinolates among the species and varieties.[56] Broccoli contains glucoraphanin as the primary

[54] G. R. Fenwick, R. K. Heaney, and W. J. Mullin, *Crit. Rev. Food Sci. Nutr.* **18,** 123 (1983).

[55] C. Nastruzzi, R. Cortesi, E. Esposito, F. Menegatti, O. Leoni, R. Iori, and S. Palmieri, *J. Agric. Food Chem.* **44,** 1014 (1996).

[56] M. M. Kushad, A. F. Brown, A. C. Kurilich, J. A. Juvik, B. P. Klein, M. A. Wallig, and E. H. Jeffery, *J. Agric. Food Chem.* **47,** 1541 (1999).

glucosinolate, whereas Brussels sprouts, cabbage, cauliflower and kale express higher levels of sinigrin and progoitrin and little or no glucoraphanin. Upon hydrolysis, sinigrin releases allyl isothiocyanate, responsible for the spicy bite of raw cabbage. Within the species, the glucosinolate profiles of the vegetables vary widely among varieties, indicating genetic differences that allow for selection of varieties particularly high in an individual glucosinolate. The range of glucoraphanin content among the 50 varieties of broccoli evaluated was 0.8 to 21.7 μmol/g DW. A study of 10 varieties over 4 seasons found that 61% of the variation in aliphatic glucosinolates was due to genotype, whereas only 12% of variation in indolyl glucosinolates—including glucobrassicin, neoglucobrassicin and methoxyglucobrassicin—was due to genotype.[57] This ability to manipulate the glucosinolate profile of broccoli through hybridization has been harnessed to produce a broccoli particularly high in glucoraphanin.[58] This work has recently resulted in the patenting of a high-glucoraphanin broccoli by one of the major broccoli seed companies.

Environmental Effects on Glucosinolate Content of Crucifers

In addition to differences in glucosinolate profile due to species and variety, cultivation conditions and life stage have been found to impact glucosinolate concentrations. The seed has very high glucosinolate content, and this is reflected in the content remaining high for the first few days of growth. However, the seedling does not appear to synthesize new glucosinolate, so the content per unit mass falls to normal adult levels by approximately the tenth day of seedling growth. This higher level of glucosinolate has been capitalized upon in the production of broccoli sprouts as a dietary supplement. Some sprouts may have 10- to 100-fold more glucosinolate than the mature broccoli.[15] Furthermore, seedlings cultivated at higher temperatures than normal (30°/15°, day/night) possess significantly higher levels of glucosinolates. Additionally, the seed was found to possess the greatest concentration of glucosinolates and consequently had the greatest QR induction potential. Glucosinolate levels decrease as the sprouts germinate and develop, with the younger stages of the plant consistently expressing higher glucosinolate levels, regardless of propagation temperature.[59] Glucosinolate content also varies substantially with growing

---

[57] A. F. Brown, G. G. Yousef, E. H. Jeffery, B. P. Klein, M. M. Kushad, M. A. Wallig, and J. A. Juvik, *J. Am. Soc. Hort. Sci.* **127,** 807 (2002).

[58] R. Mithen, K. Faulkner, R. Magrath, P. Rose, G. Williamson, and J. Marquez, *Theor. Appl. Genet.* **106,** 727 (2003).

[59] F. M. Pereira, E. Rosa, J. W. Fahey, K. K. Stephenson, R. Carvalho, and A. Aires, *J. Agric. Food Chem.* **50,** 6239 (2002).

conditions and environmental stresses, such as drought. However, glucora-phanin and other aliphatic glucosinolates vary less with environment than glucobrassicin.[57] These data are consistent with the known role of GB in protecting the plant from browsing: the stress of browsing is associated with a rapid increase in GB synthesis.

Microbial Conversion

Cooked broccoli contains no myrosinase activity, yet sulforaphane mer-capturate is found in the urine after ingestion of cooked broccoli, and QR is upregulated. In the absence of myrosinase, glucosinolates can be hydro-lyzed to isothiocyanates by the bacteria that constitute the gut microflora. Human fecal material cultured *in vitro* supports hydrolysis of progoitrin This myrosinase-like activity appeared to be widely distributed among bac-terial genera, including paracolobactrum, *Proteus vulgaris, Bacillus subtilis,* and *Escherichia coli.*[60] Gnotobiotic rats harboring a whole human fecal flora were capable of hydrolysis of progoitrin from a rape seed meal. In addition, introduction of an *E. coli* or *B. vulgaris* strain to the digestive tract showed similar conversion.[61]

Recent studies have looked at the effects of human gut microflora on glucosinolate hydrolysis of crucifers in which cooking had destroyed the plant's endogenous myrosinase. Incubation of cooked watercress juice with human fecal cultures results in approximately 20% conversion to isothio-cyanates. Additionally, thiol conjugates of the ITCs (the primary form ex-creted in urine) were present in the urine of human subjects after consumption of cooked watercress, indicating that hydrolytic release of iso-thiocyanates had occurred after ingestion of the vegetable.[62] Rats fed anti-biotics do not respond to cooked broccoli, whereas when germ-free rats were inoculated with a human digestive strain of *B. thetaiotaomicron,* they became capable of converting sinigrin to allyl-ITC. Similarly, pooled human gut microflora cultures have been shown to carry out hydrolytic conversion.[63,64] While the previous results point to the possibility of con-version of glucoraphanin to sulforaphane by gut microflora, specific data on microbial conversion of the glucosinolates in broccoli does not exist.

[60] E. L. Oginsky, A. Stein, and M. Greer, *Proc. Soc. Exp. Biol. Med.* **119,** 360 (1965).
[61] S. Rabot, L. Nugon-Baudon, P. Raibaud, and O. Szylit, *Br. J. Nutr.* **70,** 323 (1993).
[62] S. M. Getahun and F. L. Chung, *Cancer Epidemiol. Biomarkers Prev.* **8,** 447 (1999).
[63] L. Elfoul, S. Rabot, N. Khelifa, A. Quinsac, A. Duguay, and A. Rimbault, *FEMS Microbiol. Lett.* **197,** 99 (2001).
[64] C. Krul, C. Humblot, C. Philippe, M. Vermeulen, M. van Nuenen, R. Havenaar, and S. Rabot, *Carcinogenesis* **23,** 1009 (2002).

Further research in the area of microbial isothiocyanate production from cruciferous vegetables is needed.

Effects of Processing on Quinone Reductase-Inducing Activity

The content of bioactive components, and therefore the efficacy of a serving of it, is greatly affected by storage and cooking of broccoli and other crucifers. An active area of research involves the study of how one might optimize or conserve the bioactive components during the food preparation process. Individual research groups have focused on the impact of post-harvest storage conditions and cooking on glucosinolate content in broccoli. Open-air box storage of broccoli, a common method of storage during transport to the supermarket and the retail sale period, was found to result in a 55% loss of glucosinolate content after 3 days.[65] A similar loss (56%) was seen after 7 days, when broccoli was stored in plastic bags at 22°. The study found that glucosinolate levels were most stable when the broccoli was stored in low-density polyethylene bags without perforations and refrigerated at 4°.[65] Glucosinolate content was conserved for at least 10 days. However, a second study found that when film-wrapped broccoli was stored for 7 days at 1° or at 15° for 3 days, glucosinolate loss was 70% to 80%.[66] Neither study evaluated isothiocyanate levels; therefore one cannot determine whether the loss of glucosinolate was due to conversion to isothiocyanate or whether no isothiocyanates were formed, attributing the loss to another form of degradation. One study found decreased sulforaphane after storage of broccoli in perforated polyethylene vegetable bags at 4° for 21 days post-harvest.[67] However, this study did not measure glucosinolates. Unfortunately, there are not studies estimating the effects of storage on both glucosinolates and isothiocyanates.

Cooking conditions also can have a significant impact on glucosinolate concentrations and conversion of glucosinolates to the bioactive isothiocyanates. Recent research indicates that 5 minutes of microwave cooking at full power results in a 74% loss of glucosinolate content from broccoli florets, while high-pressure cooking and boiling led to 33% and 55% losses, respectively. Steaming appeared to only minimally impact glucosinolate concentrations of the florets.[68] Further studies are needed on the effect of

[65] N. Rangkadilok, B. Tomkins, M. E. Nicolas, R. R. Premier, R. N. Bennett, D. R. Eagling, and P. W. Taylor, *J. Agric. Food Chem.* **50**, 7386 (2002).

[66] F. Vallejo, F. Tomas-Barberan, and C. Garcia-Viguera, *J. Agric. Food Chem.* **51**, 3029 (2003).

[67] L. A. Howard, E. H. Jeffery, M. A. Wallig, and B. P. Klein, *Journal of Food Science* **62**, 1098 (1998).

[68] F. Vallejo, F. Tomas-Barberan, and C. Garcia-Viguera, *Eur. Food Res. Technol.* **215**, 310 (2002).

processing and cooking conditions on content of isothiocyanates and their glucosinolate precursors, if we are to optimize the induction of quinone reductase and other health benefits from inclusion of broccoli in the diet.

Effects of Vegetable Tissue Matrix on Bioavailability of Sulforaphane

A recent study compared the effects of feeding rats for 5 days, either purified sulforaphane, or fresh, freeze-dried broccoli containing intact glucosinolates, free to be hydrolyzed during digestion by both the plant myrosinase and the intestinal microflora. Induction of QR was similar between these groups, in both liver (1.4 fold over control) and in colon (4.5 fold over control).[11] Yet the sulforaphane content of the diet in those rats receiving the purified sulforaphane (5 mmol/kg diet) was at least 2-fold greater than the glucoraphanin content of the broccoli diet (2.2 mmol/kg diet). This similarity in bioactivity was reflected in a similarity in urinary sulforaphane mercapturate excretion, suggesting that purified sulforaphane was very much less bioavailable than sulforaphane within the matrix of broccoli. However, when rats were fed a mixture of purified sulforaphane together with fresh freeze-dried broccoli, neither bioactivity—measured as quinone reductase—nor bioavailability—measured as urinary excretion of sulforaphane mercapturate—appeared to have been improved for the purified sulforaphane. The idea supported by these experiments, that purified sulforaphane is less bioavailable/bioactive than sulforaphane within broccoli, is supported by the sparse literature on potency of sulforaphane in the whole animal. It appears that $\sim$100 μmol/rat/day is necessary to see quinone reductase upregulation in rat tissue, an amount that would be present in 50 g to 1 kg broccoli, depending on the variety. Yet normal dietary intake of broccoli is associated with decreased cancer risk and a 20% freeze-dried broccoli diet (equivalent to $\sim$20 g fresh broccoli/rat/day) causes induction of quinone reductase in rats. It remains to be determined whether these bioactive agents can be successfully used in purified form or whether they are more suited to improving our health through diet.

Section IV

Quinones and Age-Related Diseases

[26]   Therapeutic Effects of Coenzyme Q_{10} in
Neurodegenerative Diseases

By M. FLINT BEAL

Introduction

There is increasing interest in the potential usefulness of coenzyme Q_{10} (CoQ_{10}) to treat both mitochondrial disorders as well as neurodegenerative diseases such as Parkinson's disease (PD), Huntington's disease (HD), and amyotrophic lateral sclerosis (ALS). CoQ_{10} may also be useful in treating Freidriech's ataxia, which has been shown to be caused by a mutation in the protein frataxin, which is localized to mitochondria. CoQ_{10} is composed of a quinone ring and a 10-isoprene unit tail and is distributed in all membranes throughout the cell. CoQ_{10} serves as an important cofactor of the electron transport chain, where it accepts electrons from complexes I and II. It is initially reduced to the semi-ubiquinone radical and then transfers electrons one at a time to complex III of the electron transport chain.[1,2] CoQ_{10}, which is also known as ubiquinone, serves as an important antioxidant in both mitochondria and lipid membranes. It mediates some of its antioxidant effects through interactions with alpha-tocopherol.[1,3]

Effects in the Central Nervous System

The importance of CoQ_{10} for central nervous system function is corroborated by neuromuscular disease, which occurs in patients who have a CoQ_{10} deficiency. A report of two sisters included symptoms of encephalopathy, proximal weakness, myoglobinuria, and lactic acidosis.[4] Another patient report was that of a 35-year-old woman who developed proximal weakness, premature exertional fatigue, complex partial seizures, and myoglobinuria.[5] On muscle biopsy she was found to have reductions in complex I-III and II-III activities of the electron transport chain. There was also a marked reduction in muscle CoQ_{10} content. A 4-year-old boy presented with progressive muscle weakness, seizures, cerebellar ataxia, and elevated

[1] R. E. Bayer, *Biochem. Cell. Biol.* **70**, 390 (1992).

[2] G. Dallner and P. J. Sindelar, *Free Radic. Biol. Med.* **29**, 285 (2000).

[3] H. Noack, U. Kube, and W. Agugustin, *Free Radic. Res.* **20**, 375 (1994).

[4] S. Ogasahara, A. G. Engel, D. Frens, and D. Mack, *Proc. Natl. Acad. Sci. USA* **86,** 2379 (1989).

[5] S. C. Sobereira, M. Hirano, S. Shanske, R. K. Keller, R. G. Haller, E. Davidson, F. M. Santorelli, A. F. Miranda, E. Bonilla, D. S. Mojon, A. A. Barreira, M. P. King, and S. Dimauro, *Neurology* **48,** 1238 (1997).

METHODS IN ENZYMOLOGY, VOL. 382

CSF lactate concentrations.[6] His muscle biopsy demonstrated reductions in complex I-II and II-III activities as well as reduced mitochondrial Q_{10} content.

A recent report documented two siblings in whom nystagmus, visual loss, neural deafness, progressive ataxia, dystonia with amyotrophy and lower extremity spasticity occurred.[7] Both patients developed nephrotic syndrome followed by renal failure. They were found to have widespread deficiencies in CoQ_{10} content as well as respiratory chain activities, and there was clinical improvement after oral CoQ_{10} supplementation. Musumeci et al. reported that treatment of CoQ_{10} at doses of 300 to 3000 mg daily benefited patients with muscle CoQ_{10} deficiency and cerebellar ataxia.[8] DiGiovanni et al. reported on two brothers who presented with myopathy and one with seizures.[9] Analyses of the muscles revealed ragged red fibers as well as reductions in electron transport enzyme activities. The patients were treated with 200 mg or 300 mg daily of oral CoQ_{10} with resolution of symptoms and change in the muscle morphology to a more normal pattern. These studies indicate that CoQ_{10} deficiency can result in both neurologic as well as neuromuscular dysfunction and that oral supplementation with CoQ_{10} can provide substantial benefits in these patients.

Coenzyme Q_{10} at 300 mg/d improved Leigh's encephalopathy in 2 sisters.[10] One patient was a 31-year-old woman with growth retardation, ataxia, deafness, lactic acidosis and increased signal in the basal ganglia on magnetic resonance imaging. After CoQ_{10} supplementation she resumed walking, gained weight, underwent puberty and grew 20 cm. She and her sister were thought to have primary CoQ_{10} deficiency.

Mitochondrial Myopathies

It has been suggested that mitochondrial abnormalities occur in myotonic dystrophy, which is an autosomal dominant disease caused by a CTG expansion on chromosome 19. These patients have significant reductions in plasma CoQ_{10} levels which inversely correlate with CTG

[6] E. Boieter, F. Degoul, I. Desguerre, C. Charpentier, D. Francois, G. Ponsot, M. Diry, P. Rustin, and C. Marsac, J. Neurol. Sci. 156, 41 (1998).

[7] A. Rotig, E. L. Appelkvist, V. Geromel, D. Chretien, N. Kadhom, P. Edery, M. Lebideau, G. Dallner, A. Munnich, L. Ernster, and P. Rustin, Lancet 356, 391 (2000).

[8] O. Musumeci, A. Naini, A. E. Slonim, N. Skavin, G. L. Hadjigeorgiou, N. Krawiecki, B. M. Weissman, C. Y. Tsao, J. R. Mendell, S. Shanske, D. C. De Vivo, M. Hirano, and S. DiMauro, Neurology 56, 849 (2001).

[9] S. Di Giovanni, M. Mirabella, A. Spinazzola, P. Crociani, G. Silvestri, A. Broccolini, P. Tonali, S. Di Mauro, and S. Servidiei, Neurology 57, 515 (2001).

[10] L. Van Maldergem, F. Trijbels, S. DiMauro, P. J. Sindelar, O. Musumeci, A. Jannsen, X. Delberghe, J. J. Martin, and Y. Gillerot, Ann. Neurol. 52, 750 (2002).

expansions and lactate levels after exercise.[11] Investigators have studied the effectiveness of CoQ as a treatment for known mitochondrial disorders. These have met with variable success. Patients have been studied who had Kearns-Sayre syndrome, mitochondrial encephalopathy, lactic acid, stroke-like episodes (MELAS), myoclonic epilepsy with ragged red fibers (MERRF), and Leber's hereditary optic neuropathy. The most success appears to have been with patients with MELAS. The results in patients with other mitochondrial disorders have been much less consistent. In five patients with Kearns-Sayre syndrome, there were low plasma levels of CoQ_{10}, and daily administration enhanced abnormal metabolism of pyruvate in NADH oxidation and in skeletal muscle.[12] However, there was no significant clinical improvement. Treatment of 44 patients with heterogeneous mitochondrial disorders with 2 mg/kg daily resulted in increases in CoQ_{10} in plasma and platelets.[13] Sixteen of the patients had at least a 25% decrease in post-exercise lactate levels between 0 and 6 months of therapy, and they were then entered into a blinded placebo-controlled trial in which there was no significant benefit.

Peterson reported 16 patients with a variety of mitochondrial disorders that shared impaired complex-I activity in fresh isolated skeletal muscle mitochondria.[14] The patients were treated with CoQ_{10} at doses of 30 to 120 mg daily as well as other antioxidants and methylpredisolone for up to 15 years. It was the author's impression that the patients appeared to survive longer with less functional disability and medical complications than typically seen in his clinical practice. Barbiroli treated 6 patients with mitochondrial cytopathies including 4 with chronic external opthalmoplegia and 2 with Leber's hereditary optic neuropathy with CoQ_{10} at 150 mg daily.[15] Phosphorous nuclear magnetic resonance spectroscopy in the occipital lobe showed significant improvement in measures of mitochondrial function. Two patients with MELAS improved based on oxygen consumption as assessed by non-invasive tissue oximetry.[16] A 32-year-old man who had bilateral visual loss developed choreiform movements associated with

[11] G. Siciliano, M. Mancuso, D. Tedeschi, M. L. Manca, M. R. Renna, V. Lombardi, A. Rocchi, F. Martelli, and L. Murri, *Brain Res. Bull.* **56,** 405 (2001).

[12] S. Ogasahara, S. Yorifuji, Y. Nishikawa, M. Takahashi, K. Wada, T. Hazama, Y. Nakamura, S. Hashimoto, N. Kono, and S. Tarui, *Neurology* **35,** 372 (1985).

[13] N. Bresolin, L. Bet, A. Binda, M. Moggio, G. Comi, F. Nador, C. Ferrante, A. Carenzi, and G. Scarlato, *Neurology* **38,** 892 (1988).

[14] P. L. Peterson, *Biochim. Biophys. Acta* **1271,** 275 (1995).

[15] A. Babiroli, C. Frassineti, P. Martinelli, S. Lotti, R. Lodi, P. Cortelli, and P. Montagna, *Cell Mol. Biol.* **43,** 741 (1997).

[16] K. Abe, H. Fujimura, Y. Nishikawa, S. Yorifuki, T. Mezaki, N. Hirono, N. Nishitani, and M. Kameyama, *Acta Neurol. Scand.* **83,** 356 (1991).

hypointense lesions in the subthalamic nucleus on MRI.[17] Treatment with CoQ$_{10}$ at 250 mg per day resulted in recovery of the movement disorder, normalization of the lactate/pyruvate ratio, and disappearance of the MRI lesions 3 years later. Treatment of another patient with MELAS with CoQ$_{10}$ at 120 to 300 mg per day caused a gradual improvement in neurological deficits and a reduction in CSF lactate and pyruvate.[16] In a recent study of 5 patients with MELAS, CoQ$_{10}$ at 3 mg/kg and nicotinamide 50 mg/kg decreased blood lactate and pyruvate concentration, but there was little clinical improvement.[18] In this study, 2 patients died suddenly and unexpectedly, but this was not beyond the authors' experience in similarly severely affected patients.

A number of recent studies also showed improvement in patients with mitochondrial disorders after treatment with a CoQ$_{10}$ analog idebenone. Idebenone protects against excitotoxic cell death in cultured cortical neurons.[19] Treatment of a 36-year-old man with MELAS with idebenone produced improved oxygen extraction without increasing cerebral blood flow as assessed by PET.[20] Another patient with Leber's optic atrophy developed spastic paraparesis with white matter lesions on MRI.[21] After administration of idebenone, the patient showed reversal of his paraparesis, which was correlated with normalization of serum lactate and improvement on brain and muscle phosphorous MRS.

Pharmacokinetics of Orally Administered CoQ$_{10}$

A number of studies have evaluated the pharmacokinetics of CoQ$_{10}$ in humans.[22–25] A single oral dose of CoQ$_{10}$ is followed by two peaks in serum levels. The first peak occurs at approximately 5 to 6 hours and the second, a much smaller peak, occurs approximately 24 hours after the oral dose. The

[17] P. Chariot, P. Brugieres, M. C. Eliezer-Vanerot, C. Geny, M. Binaghi, and P. Cesaro, *Mov. Disorders* **14,** 855 (1999).
[18] B. M. Remes, E. V. Limatta, S. Winqvist, U. Tolonen, J. A. Ranua, K. Reinikainen, I. E. Hassinen, and K. Majaama, *Neurology* **59,** 1275 (2002).
[19] V. Bruno, G. Battaglia, A. Copani, M. A. Sortino, P. L. Canonico, and F. Nicoletti, *Neurosci. Lett.* **178,** 193 (1994).
[20] Y. Ikejiri, E. Mori, K. Ishii, K. Nishimoto, M. Yasuda, and M. Sasaki, *Neurology* **47,** 583 (1996).
[21] P. Cortelli, P. Montagna, G. Pierangeli, R. Lodi, P. Barboni, R. Liguori, V. Carelli, S. Lotti, P. Zaniol, E. Lugaresi, and B. Barbiroli, *J. Neurol. Sci.* **148,** 25 (1997).
[22] Y. Tomono, J. Hasegawa, T. Seki, K. Motegi, and N. Morishita, *Int. J. Clin. Pharmacol. Ther. Toxicol.* **24,** 536 (1986).
[23] T. Okamoto, T. Matsuya, Y. Fukunaga, T. Kishi, and T. Yamagami, *Int. J. Vitam. Nutr. Res.* **59,** 288 (1989).
[24] V. Mohr, W. Bowry, and R. Stocker, *Biochim. Biophys. Acta* **1126,** 247 (1992).
[25] M. Weiss, S. A. Mortensen, M. R. Rassing, J. Moller-Sonnergaard, G. Poulsen, and S. N. Rasmussen, *Mol. Aspects Med.* **15** (Suppl.), s273 (1994).

explanation for the second peak has been proposed to be uptake by the liver and subsequent resecretion. Absorption of CoQ_{10} is improved by inclusion of lipid in the formulation and by taking CoQ_{10} with food. The elimination half-life has been estimated to be between 30 and 50 hours. Studies showed that a dose of 30 mg has negligible effects on plasma levels.[26] In this study, coadministration of vitamin E at 700 mg daily impaired CoQ_{10} absorption.

The Antioxidant Properties and Effects of CoQ_{10} Supplementation in Animals

Several studies have shown that oral administration of CoQ_{10} can produce protection in experimental models of cerebral ischemia or against mitochondrial toxins. These studies, however, have been controversial, since there are reports that CoQ_{10} administration does not increase levels in either muscle or brain. In young rats, alpha-tocopherol supplementation produced increases in tissue levels of alpha-tocopherol in plasma, liver, kidney, muscle, and brain, however, CoQ_{10} supplementation increased CoQ_{10} levels only in plasma and liver.[27] Oral administration of CoQ_{10} and alpha-tocopherol alone or together increased plasma levels in another study, and CoQ_{10} increased mitochondrial levels of both CoQ_{10} and alpha-tocopherol, consistent with a sparing effect of alpha-tocopherol.[28] It was suggested that it is only possible to increase CoQ_{10} levels in muscle and brain when there is a deficiency.[29]

There is evidence for decreasing CoQ_{10} levels with aging in both human and rat tissues. This may be a consequence of either decreased synthesis or increased oxidative damage. In rats there is a significant decrease in brain CoQ_{10} levels as early as 5 months of age.[30,31] Decreases with aging in man have also been documented.[32] We examined whether feeding with CoQ_{10} at 200 mg/kg/d for two months could increase brain CoQ_{10} levels in 12-month-old rats.[33] Oral administration of CoQ_{10} produced a significant

[26] J. Kaikkonen, T. P. Tuomainen, K. Nyyssonen, and J. T. Salonen, *Free Radic. Res.* **36,** 389 (2002).

[27] Y. Zhang, M. Turunen, and E.-L. Appelkvist, *J. Nutr.* **126,** 2089 (1996).

[28] M. Lass, J. Forster, and R. S. Sohal, *Free Radic. Biol. Med.* **26,** 1375 (1999).

[29] W. H. Ibrahim, H. N. Bhagavan, R. K. Chopra, and C. K. Chow, *J. Nutr.* **130,** 2343 (2000).

[30] M. Battino, A. Gorini, R. F. Villa, M. L. Genova, C. Bovina, S. Sassi, G. P. Littaru, and G. Lenaz, *Mech. Ageing Dev.* **78,** 173 (1995).

[31] R. E. Beyer, B.-A. Burnett, K. J. Cartwright, D. E. Edington, M. J. Falzon, K. R. Kreitman, T. W. Kuhn, B. J. Ramp, S. Y. S. Rhee, M. J. Rosenwasser, M. Stein, and L. C. I. An, *Mech. Ageing Dev.* **32,** 267 (1985).

[32] A. Kalen, E.-L. Appelkvist, and G. Daliner, *Lipids* **24,** 579 (1989).

[33] R. T. Matthews, S. Yang, S. Browne, M. Baik, and M. F. Beal, *Proc. Natl. Acad. Sci. USA* **95,** 8892 (1998).

35% to 40% increase in both oxidized and reduced forms of CoQ_9 and CoQ_{10}, restoring concentrations to those seen in young animals. We also documented an increase in mitochondrial levels in the cerebral cortex of 12-month-old rats. There was a non-significant trend toward an increase in alpha-tocopherol levels consistent with the alpha-tocopherol sparing effect reported in other studies.

A recent study examined the effects of CoQ_{10} supplementation at rats at 150 mg/kg/d in their diets for 4 or 13 weeks.[34] Thirty 15-month-old Sprague-Dawley rats were treated with either CoQ_{10}-supplemented or CoQ_{10}-unsupplemented diets. CoQ_{10} levels were measured is plasma, tissue homogenates and mitochondria. CoQ_{10} plasma levels increased 6-fold at 4 weeks and 9-fold at 13 weeks. In heart and skeletal muscle homogenates, CoQ_{10} content increased by 23% and 45%, respectively, and there were significant increases in heart and skeletal muscle mitochondria of 20% to 30%. There were increases in both homogenates and mitochondria of liver and kidney. Both homogenates and mitochondria from brain showed significant increases in CoQ_{10} levels after 13 weeks of treatment. Interestingly, similar to our findings, CoQ levels also increased in tissue homogenates and mitochondria. There was a 40% decrease in protein carbonyls in skeletal muscle mitochondria and a 30% decrease in liver mitochondria. The plasma oxidized glutathione content decreased by 60%, consistent with an antioxidant effect. Surprisingly, however, there was no effect on H_2O_2 production from heart, skeletal muscle and brain mitochondria.

Another recent study examined uptake of [^{3}H] CoQ_{10} after a single i.p. administration to rats.[35] There was efficient uptake into plasma, spleen, liver and white blood cells; lower concentrations in adrenals, ovaries, thymus and heart, and practically no uptake in kidney, muscle and brain. The majority of the metabolites were found in urine, which were phosphorylated, with small amounts in feces. This study would appear to contradict our results and those discussed previously. There are, however, significant differences in the studies. We and Kwong *et al.* (2002) studied 12–15 month-old rats treated with 1 to 3 months of CoQ_{10} administration, whereas Bentinger *et al.* (2003) studied a single injection in 45-day-old rats. We and others previously showed that CoQ_{10} levels build up slowly and progressively in tissue and that the increase is very difficult to detect in younger animals.

[34] L. K. Kwong, S. Kamzalov, I. Rebrin, A. C. Bayne, C. K. Jana, P. Morris, M. J. Forster, and R. S. Sohal, *Free Radic. Biol. Med.* **33,** 627 (2002).
[35] M. Bentinger, G. Dallner, T. Chojnacki, and E. Swiezewska, *Free Radic. Biol. Med.* **34,** 563 (2003).

CoQ_{10} is recognized as being an important antioxidant in the inner mitochondrial membrane, where it can scavenge radicals directly.[36] A direct action of ubiquinol with nitric oxide has also been documented.[37] There is substantial evidence that ubiquinol may act in concert with alpha-tocopherol reducing alpha-tocopherol radical to alpha-tocopherol.[38–40] In rat liver mitochondrial preparations, CoQ_9 levels were oxidized prior to the onset of massive lipid peroxidation, with subsequent depletion of alpha-tocopherol.[3] Succinate results in a reduction in CoQ to ubiquinol ($CoQH_2$), and in the absence of succinate, CoQ is oxidized.[41] When mitochondria are depleted of alpha-tocopherol there is also oxidation of CoQ. This suggests that alpha-tocopherol is the direct radical scavenger and ubiquinol acts to regenerate alpha-tocopherol. CoQ_{10} also interacts with dihydrolipoic acid. Dihydrolipoic acid reduces ubiquinone to ubiquinol by the transfer of a pair of electrons, thereby increasing antioxidant capacity of CoQ_{10} in biomembranes.[42] Lipoic acid was shown to maintain a normal ratio of reduced to oxidized CoQ_{10} after MPTP administration *in vivo*.[43]

Oral supplementation with CoQ_{10} or alpha-tocopherol results in increased total CoQ_{10} content, and alpha-tocopherol increased by 5-fold in mitochondria when administered at 200 mg/kg daily.[28] After administration in mice, the rate of superoxide generation from submitochondrial particles was inversely related to alpha-tocopherol content but unrelated to CoQ_{10} content. These findings therefore provide *in vivo* evidence that part of the antioxidant effects of CoQ_{10} are due to its ability to reduce the alpha-tocopherol radical. CoQ_{10} with alpha-tocopherol inhibits atherosclerosis in apolipoprotein-deficient mice and significantly decreases tissue lipid hydroperoxides.[44] CoQ_{10} supplementation in human lymphocytes *in vitro* decreases oxidative DNA damage.[45] A potentially very interesting mechanism of CoQ_{10} is by interactions with uncoupling proteins. CoQ_{10}

[36] V. Kagan, E. Serbinova, and L. Packer, *Biochem. Biophys. Res. Commun.* **169,** 851 (1990).
[37] J. J. Poderoso, M. C. Carreras, F. Schopfer, C. L. Lisdero, N. A. Riobo, C. Giulivi, A. D. Boveris, A. Boveris, and E. Cadenas, *Free Radic. Biol. Med.* **26,** 925 (1999).
[38] V. E. Kagan, E. A. Serbinova, G. M. Koynova, S. M. Kitanova, V. A. Tyurin, T. S. Stoytchev, P. J. Quinn, and L. Packer, *Free Radic. Biol. Med.* **9,** 117 (1990).
[39] J. J. Maguire, V. Kagan, B. A. Ackrell, E. Serbinova, and L. Packer, *Arch. Biochem.* **292,** 47 (1992).
[40] K. Mukai, H. Morimoto, S. Kikuchi, and S. Nagaoka, *Biochim. Biophys. Acta* **1157,** 313 (1993).
[41] H. Shi, N. Noguchi, and E. Niki, *Free Radic. Biol. Med.* **27,** 334 (1999).
[42] V. Kozlov, L. Gille, K. Staniek, and H. Nohl, *Arch. Biochem. Biophys.* **363,** 148 (1999).
[43] M. E. Gotz, A. Dirr, R. Burger, B. Janetzky, M. Weinmuller, W. W. Chan, S. C. Chen, H. Reichmann, W. D. Rausch, and P. Reiderer, *Eur. J. Pharmacol.* **266,** 291 (1994).
[44] S. R. Thomas, S. B. Leichtweis, K. Pettersson, K. D. Croft, T. A. Mori, A. J. Brown, and R. Stocker, *Arterioscler. Thromb. Vasc. Biol.* **21,** 585 (2001).
[45] M. Tomasetti, G. P. Littarru, R. Stocker, and R. Alleva, *Free Radic. Bio. Med.* **27,** 1027 (1999).

has been shown to be an obligatory cofactor for uncoupling protein function.[46,47] Uncoupling proteins have been demonstrated to exert neuroprotective effects.[48,49] They regulate ATP levels, NAD to NADH ratio and the amount of superoxide production in mitochondria. Increased uncoupling proteins reduces free radical damage.[49] It is therefore possible that CoQ_{10} supplementation may be acting through this mechanism. CoQ_{10} may also have effects on the mitochondrial permeability transition, which is linked to cell death.[50,51]

For CoQ_{10} to function effectively as an antioxidant it must be maintained in a reduced form ($CoQ_{10}H_2$). A number of enzymes as well as selenium may play a role. Selenium is an essential cofactor of glutathione peroxidase and thioredoxin reductase. An initial study showed that dietary deficiency of selenium reduced CoQ_9 and CoQ_{10} levels by about 50% in liver and 15% in heart.[52] In a follow-up study, 18 months of selenium deficiency reduced liver CoQ_9 content by 40% and CoQ_{10} content by 67%.[53]

Recently, thioredoxin reductase was shown to reduce CoQ_{10} to $CoQ_{10}H_2$ with either NADPH or NADH as cofactors.[54] The reduction was selenium dependent and occurred at physiologic conditions and was the most efficient yet described. In human kidney, cell lines overexpressing thioredoxin reductase mirrored their ability to reduce CoQ_{10}. Both lipoamide dehydrogenase and glutathione reductase also reduce CoQ_{10}, but at lower rates as compared with thioredoxin reductase at physiologic conditions. The pH optimum for these enzymes, however, is 6.0, and they are stimulated by zinc.[55,56] Another enzyme that can reduce CoQ_{10} is DT-diaphorase, but it is unknown whether this occurs under physiologic conditions.[57]

[46] K. S. Echtay, E. Winkler, and M. Klingenberg, *Nature* **408,** 609 (2000).

[47] K. Echtay, E. Winkler, K. Frischmuth, and M. Klingenberg, *Proc. Natl. Acad. Sci. USA* **98,** 1416 (2001).

[48] D. Ricquier and F. Bouillaud, *Biochem. J.* **345,** 161 (2000).

[49] P. G. Sullivan, C. Dube, K. Dorenbos, O. Steward, and T. Z. Baram, *Ann. Neurol.,* published online (2003).

[50] L. Walter, V. Nogueira, X. Leverve, M. P. Heitz, P. Bernardi, and E. Fontaine, *J. Biol. Chem.* **275,** 29521 (2000).

[51] E. Fontaine, F. Ichas, and P. Bernardi, *J. Biol. Chem.* **273,** 25734 (1998).

[52] S. Vadhanavikit and H. E. Ganther, *Biochem. Biophys. Res. Comm.* **190,** 921 (1993).

[53] S. Vadhanavikit and H. E. Ganther, *Molec. Aspects Med.* **15,** s103 (1994).

[54] L. Xia, T. Nordman, J. M. Olsson, A. Damdimopoulos, L. Bjorkhem-Bergman, I. Nalvarte, L. C. Eriksson, E. S. Arner, G. Spurou, and M. Bjornstedt, *J. Biol. Chem.* **278,** 2141 (2003).

[55] J. M. Olsson, L. Xia, L. C. Eriksson, and M. Bjornstedt, *FEBS Lett.* **448,** 190 (1999).

[56] L. Xia, M. Bjornstedt, T. Nordman, L. C. Eriksson, and J. M. Olsson, *Eur. J. Biochem.* **268,** 1486 (2001).

[57] R. E. Beyer, J. Segura-Aguilar, S. Di Bernardo, M. Cavazzoni, R. Fato, D. Fiorentini, M. C. Galli, M. Setti, L. Landi, and G. Lenaz, *Proc. Natl. Acad. Sci. USA* **93,** 2528 (1996).

Neuroprotective Effects in Animal Models of Neurodegeneration

CoQ_{10} administration protects myocardium from ischemia-reperfusion injury and preserves mitochondrial function.[58] CoQ_{10} has been demonstrated to exert neuroprotective effects in animal models of neuronal injury in the central nervous system. Experimental ischemia can be produced by intracerebroventricular administration of the potent vasoconstrictor endothelin. Administration of CoQ_{10} at a dose of 10 mg/kg i.p. resulted in a significant attenuation of ATP and glutathione depletion and diminished neuronal injury in the hippocampus.[59] CoQ_{10} protects cultured cerebellar neurons against excitotoxin-induced degeneration.[60]

We studied the effects of administration of CoQ_{10} on lesions produced by mitochondrial toxins. Oral administration of CoQ_{10} produced dose-dependent neuroprotective effects against striatal lesions produced by the mitochondrial toxin malonate. It attenuated malonate-induced lesions as well as depletions of ATP and increases in lactate concentrations.[61] CoQ_{10} exerted additive neuroprotective effects against malonate lesions when administered with MK801, an N-methyl-D-aspartate receptor antagonist.[62] We also found that CoQ_{10} administration attenuated striatal lesions produced by aminooxyacetic acid.[63] This compound produced mitochondrial defects by blocking the malate-aspartate shunt. Administration of CoQ_{10} produced significant protection against dopamine depletions induced by MPTP administration to 24-month-old mice.[62] It also significantly protected against depletion of tryosine hydroxylase immunostained neurons.

We examined whether CoQ_{10} can exert neuroprotective effects against systemic administration of 3-nitropropionic acid.[33] This is a compound that is an irreversible succinate dehydrogenase. It is known to produce selective striatal lesions in grazing animals in the Western hemisphere. Outbreaks of encephalopathy have occurred in Chinese children who have ingested mildewed sugar cane. Administration of 3-nitropropionic acid produces selective striatal lesions in both rats and primates that closely resemble those found in HD.[64,65] The striatal lesions are characterized by sparing of

[58] J. A. Crestanello, N. M. Doliba, N. M. Doliba, A. M. Babsky, K. Niborii, M. D. Osbakken, and G. J. Whitman, *J. Surg. Res.* **102,** 221 (2002).

[59] R. P. Ostrowski, *Brain Res. Bull.* **53,** 399 (2000).

[60] A. Favit, F. Nicoletti, U. Scapagnini, and P. L. Canonico, *J. Cereb. Blood Flow Metab.* **12,** 638 (1992).

[61] M. F. Beal, R. Henshaw, B. G. Jenkins, B. R. Rosen, and J. B. Schulz, *Ann. Neurol.* **36,** 882 (1994).

[62] M. F. Beal and R. T. Matthews, *Mol. Asp. Med.* **18,** s169 (1997).

[63] E. Brouillet, D. R. Henshaw, J. B. Schulz, and M. F. Beal, *Neurosci. Lett.* **177,** 58 (1994).

[64] M. F. Beal, E. Brouillet, B. G. Jenkins, R. J. Ferrante, N. W. Kowall, J. M. Miller, E. Storey, R. Srivastava, B. R. Rosen, and B. T. Hyman, *J. Neurosci.* **13,** 4181 (1993).

NADPH-diaphorase neurons, which are also spared in HD postmortem tissue. In primates, the toxin produces both cognitive deficits as well as a choreiform movement disorder that resembles those found in HD patients. The lesions are accompanied by focal increases in lactate confined to the basal ganglia, and they are attenuated by antioxidants. We found that oral administration of CoQ_{10} for one week prior to coadministration of 3-nitropropionic acid resulted in a significant 90% neuroprotection against the 3-nitropropionic acid induced lesions.[33] CoQ_{10} administration also significantly attenuated reductions of CoQ_9H_2 and $CoQ_{10}H_2$ after 3-nitropropionic administration in the same animals.

We recently examined the effects of CoQ_{10} in transgenic mouse models of neurodegenerative diseases. We examined its effects in a transgenic mouse model of familial ALS which is produced by a point mutation in Cu/Zn superoxide dismutase. These point mutations have been associated with dominant inherited ALS. Overexpression of the mutant enzyme in transgenic mice leads to motor neuron degeneration and reduced survival. An early pathologic feature in these mice is mitochondrial swelling and vacuolization. This correlated well with the subsequent onset of rapid loss of motor neurons. We found that oral administration of CoQ_{10} starting at 50 days of age in these mice significantly increased their life span.[33]

Similarly in HD a major advance has been the development of transgenic mouse models. Transgenic mice overexpressing exon 1 of the human HD gene with an expanded CAG repeat develop a progressive neurological disorder.[66] At approximately 6 weeks of age, the R6/2 mice develop loss of brain and body weight, and at 9 to 11 weeks they develop an irregular gait, stuttering stereotypic movements, resting tremors and epileptic seizures. The brains of the R6/2 mice show progressive striatal atrophy as well as neuronal intranuclear inclusions that are immunopositive for both huntingtin and ubiquitin.

We examined the effects of CoQ_{10}, either alone or in combination with the NMDA receptor antagonist remacemide.[67] The compounds were administered in the diet starting at 21 days of age. We administered CoQ_{10} at a dose of 0.2% in the diet. Remacemide was administered at a dose of 0.007% in the diet. The calculated dose for CoQ_{10} was 400 mg/kg per day, and for remacemide, it was 14 mg/kg per day.

[65] E. Brouillet, P. Hantraye, R. J. Ferrante, R. Dolan, A. Leroy-Willig, N. W. Kowall, and M. F. Beal, *Proc. Natl. Acad. Sci. USA* **92**, 7105 (1995).
[66] L. Mangiarini, K. Sathasivam, M. Seller, B. Cozens, A. Harper, C. Hetherington, M. Lawton, Y. Trottier, H. Lehrach, S. W. Davies, and G. P. Bates, *Cell* **87**, 493 (1996).
[67] R. J. Ferrante, O. A. Andreassen, A. Dedeoglu, K. L. Ferrante, B. G. Jenkins, S. M. Hersch, and M. F. Beal, *J. Neurosci.* **22**, 1592 (2002).

We found that the mean survival in the CoQ_{10}-treated R6/2 mice increased by 14.5%, the percentage increase in survival when using remacemide in the diet was 15.5% and the combined treatment when using CoQ_{10} with remacemide increased survival by 32%. This is consistent with additive neuroprotective effects of the 2 agents, which are presumed to act on different disease mechanisms. We also found that administration of either CoQ_{10} or remacemide significantly delayed the development of motor deficits, weight loss, cerebral atrophy and neuronal inclusions. The combination of the 2 agents together, CoQ_{10} and remacemide, was more efficacious than either compound alone in attenuating the neuropathologic deficits. We also studied the effects of these compounds by using magnetic resonance spectroscopy. The administration of CoQ_{10} with remacemide significantly attenuated the progressive decrease in N-acetylaspartate concentrations, and development of striatal atrophy which occur in these mice.

The Effects of CoQ_{10} Supplementation in Patients With Neurodegenerative Diseases

We and others have previously examined a number of aspects of CoQ_{10} in patients with neurodegenerative diseases. We measured CoQ_{10} levels in mitochondria isolated from platelets of PD patients.[68] We found a significant reduction in CoQ_{10} levels that directly correlated with decreases in complex I activity. Oral administration of CoQ_{10} to the PD patients was well tolerated and resulted in dose-dependent significant increases in plasma CoQ_{10} levels.

We studied the effects of CoQ_{10} on elevated striatal and occipital cortex lactate concentrations in HD patients.[69] There was a significant increase in lactate concentrations in these patients in both basal ganglia as well as numerous areas of cerebral cortex. We administered CoQ_{10} at a dose of 360 mg per day to patients for 1 to 2 months. We obtained a baseline lactate concentration both before, during and after discontinuation of CoQ_{10} therapy. CoQ_{10} therapy led to a significant 37% reduction in occipital cortex lactate concentrations that reversed after discontinuation of therapy, indicating a therapeutic effect of CoQ_{10}. A tolerability study of CoQ_{10} in HD patients showed that there were minimal adverse effects at doses of 600 to 1200 mg daily.[70]

[68] C. W. Shults, R. H. Haas, D. Passov, and M. F. Beal, *Annals Neurol.* **42,** 261 (1997).

[69] R. J. Ferrante, O. A. Andreassen, A. Dedeoglu, K. L. Ferrante, B. G. Jenkins, S. M. Hersch, and M. F. Beal, *J. Neurosci.* **22,** 1592 (2002).

[70] A. Feigin, K. Kieburtz, P. Como, C. Hickey, K. Claude, D. Abwendere, C. Zimmerman, K. Steinberg, and I. Shoulson, *Mov. Disorders* **11,** 321 (1996).

These findings led to a clinical trial in HD patients. This was the CARE-HD trial carried out by the Huntington's Study Group.[71] The trial encompassed 360 patients treated for 30 months. They were randomized to CoQ_{10} at 600 mg per day, remacemide at 600 mg per day or the combination in a 2×2 factorial design. The primary outcome variable was change in the Unified Huntington's Disease Rating Scale. In this trial, CoQ_{10} slowed the progression on the total functional capacity measure scale by approximately 14% over the 30 months. Remacemide had no effect. CoQ_{10} was very well tolerated whereas remacemide was associated with significant side effects. CoQ_{10} produced a slowing on HD functional assessment and HD independence scales. A number of neuropsychological tests were examined as secondary end points. CoQ_{10} had a trend towards beneficial impact on two of the cognitive tasks, stroop (color-naming p = 0.01, and word-reading p = 0.09) and brief tests of attention, p = 0.02. There was also a trend towards a beneficial impact of CoQ_{10} on behavior as assessed by the behavior frequency and severity scale, p = 0.08.

These findings are of interest, however, the results did not reach significance. The trial was powered to be able to determine a 40% attenuation of total functional capacity decline with 80% power. Since the magnitude of the change detected did not reach this, the results did not reach significance. We recently measured CoQ_{10} levels in the patients before and during the trial. These demonstrated that CoQ_{10} plasma levels significantly increased to values of approximately 2.4 μg per ml. This is consistent with levels we observed in Parkinson's disease patients at 600 mg/day.

Friedriech's ataxia is the most common cause of early-onset inherited ataxias and occurs in 1 in 50,000 cases of hereditary ataxia among Caucasians. Friedriech's ataxia is transmitted as an autosomal-recessive trait.[72] Disease symptoms typically appear between ages 8 and 15, but some patients develop symptoms earlier in childhood. In rare instances, patients develop symptoms as late as the third or fourth decade. Gait ataxia is the most common presenting symptom. Patients develop dysarthria, areflexia, pyramidal weakness of the legs, extensor plantar responses, and distal loss of joint position and vibration sense. There is degeneration both in the cerebellar nuclei as well as the dorsal root ganglia. Patients suffer from a cardiomyopathy which is frequently the cause of premature death. The gene responsible for Friedriech's ataxia was identified and encodes a protein of 210 amino acids, with the most common mutation being an expansion of a GAA trinucleotide repeat within the first intron. Affected individuals have more than 100 repeats, and there is reduced synthesis of

[71] The Huntington Study Group, *Neurology* **57**, 397 (2001).
[72] J. Kaplan, *Neurochem. Int.* **40**, 553 (2002).

the protein frataxin. The expanded repeat induces a triple helical structure that lowers the rate of transcription. There is compelling evidence that Friedriech's ataxia results from a mitochondrial defect.[72] Frataxin has been localized to the mitochondrial matrix in all species examined, and deficiency of the yeast homolog of frataxin results in a respiratory-deficient phenotype. Studies in patients with Friedriech's ataxia showed evidence of mitochondrial deficits when using phosphorous NMR spectroscopy.[73] There were reduced levels of phosphocreatine as well as reduced recovery of ATP after exercise.

The mutant protein leads to an accumulation of mitochondrial iron. It has been hypothesized that this will lead to aberrant Fenton chemistry, which will generate free radicals. There is evidence for mitochondrial enzyme deficiencies in muscle biopsy samples of patients. In particular, aconitase is markedly diminished and aconitase is very susceptible to free radical damage. Decreased expression of the yeast gene that is homologous to frataxin decreases mitochondrial iron export. It also decreases the synthesis of iron sulfur clusters, which may occur prior to the accumulation of mitochondrial iron.[74] Other evidence suggests that the yeast frataxin is an iron storage protein.[75] Evidence of oxidative damage in Friedriech's ataxia patients has been demonstrated by findings of reduced plasma glutathione and increases in concentration of 8-hydroxy-2-deoxyguanosine, a marker of oxidative DNA damage in urine, and a significant increase in antioxidant enzymes in the blood of patients with Friedriech's ataxia.[76-78]

It was suggested that the CoQ analog idebenone may be a useful treatment. We found that oral administration of idebenone significantly decreased urinary 8-hydroxy-2-deoxyguanosine concentrations in Friedriech's ataxia patients.[77] Idebenone reduces cardiac hypertrophy in Friedriech's ataxia patients.[79] After six months, cardiac ultrasound shows a reduction in left ventricular mass of more than 20% in about half the patients.[80]

[73] R. Lodi, J. M. Cooper, J. L. Bradley, D. Manners, P. Styles, D. J. Taylor, and A. H. Schapira, *Proc. Natl. Acad. Sci. USA* **96**, 11492 (1999).

[74] O. S. Chen, S. Hemenway, and J. Kaplan, *Proc. Natl. Acad. Sci. USA* **99**, 12321 (2002).

[75] O. Gakh, J. Adamec, A. M. Gacy, R. D. Twesten, W. G. Owen, and G. Isaya, *Biochemistry* **41**, 6798 (2002).

[76] G. Tozzi, M. Nuccetelli, M. Lo Bello, S. Bernardini, L. Bellincampi, S. Ballerini, L. M. Gaeta, C. Casali, A. Pastore, C. Federici, E. Bertini, and F. Piemonte, *Arch. Dis. Child* **86**, 376 (2002).

[77] J. B. Schulz, T. Dehmer, L. Schols, H. Mende, C. Hardt, M. Vorgerd, K. Burk, W. Matson, J. Dichgans, M. F. Beal, and M. B. Bogdanov, *Neurology* **55**, 1719 (2000).

[78] F. Piemonte, A. Pastore, G. Tozzi, D. Tagliacozzi, F. M. Santorelli, R. Carrozzo, C. Casali, M. Damiano, G. Federici, and E. Bertini, *Eur. J. Clin. Invest.* **31**, 1007 (2001).

[79] P. Rustin, J. C. von Kleist-Retzow, K. Chantrel-Groussard, D. Sidi, A. Munnich, and A. Rotig, *Lancet* **354**, 477 (1999).

Idebenone was administered at a dose of 5 mg per kg daily to 38 patients. A study of the effects of CoQ_{10} in Friedriech's ataxia patients showed improvement of cardiac and skeletal muscle bioenergetics.[81] Coenzyme Q_{10} was administered at 400 mg daily with vitamin E, 2000 units daily. After 3 months of treatment, the cardiac phosphocreatine to ATP ratios showed a mean relative increase to 178% of initial values. The maximum rate of skeletal muscle mitochondrial ATP production increased 139%. The improvement was sustained after 6 months of therapy. There were no consistent improvements in neurological evaluations, but further follow-up has shown no deterioration in neurologic function for up to 3 years.

A phase-II clinical trial in Parkinson's disease patients was recently performed.[82] The clinical trial was carried out by the Parkinson's Study Group. The trial enrolled 80 patients who were randomly assigned to placebo or CoQ_{10} at doses of 300, 600 or 1200 mg per day. The primary outcome measure was the Unified Parkinson's Disease Rating Scale (UPDRS), which was administered at screening, baseline, and 1, 4, 8, 12 and 16 months. The subjects were patients with early PD who did not require treatment (levodopa) for their disability. They were followed up for 16 months or until disability requiring treatment with levodopa had developed. Only one patient was lost to follow-up. The subjects were very well matched with regard to their ages, gender and clinical disability at baseline. No significant side effects were encountered. The findings were an increase in the adjusted mean UPDRS of 12 points for the placebo group, 8.8 for the 300 mg per day, 10.8 for the 600 mg per day and 6.7 for the 1200 mg per day group. The primary analysis was a test of a linear trend that showed a significant effect by the prespecified criteria. Secondary analysis was a comparison of each treatment group with placebo. The difference between the 1200 mg and placebo groups was significant with a p = 0.04. The overall slowing of disability in the 1200 mg per day group was 44% at 16 months.

A number of other biochemical end points were examined. Plasma CoQ_{10} levels showed dose-dependent increases. Interestingly, the 300 and 600 mg doses produced increases in plasma concentrations that were roughly comparable, at approximately 2 μg per ml. The 1200 mg dose,

[80] B. O. Hausse, Y. Aggoun, D. Bonnet, D. Sidi, A. Munnich, A. Rotig, and P. Rustin, *Heart* **87,** 346 (2002).

[81] B. Lodi, P. E. Hart, B. Rajagopalan, D. J. Taylor, J. G. Crilley, J. L. Bradley, A. M. Blamire, D. Manners, P. Styles, A. H. Schapira, and J. M. Cooper, *Ann. Neurol.* **49,** 590 (2001).

[82] C. W. Shults, D. Oakes, K. Kieburtz, M. F. Beal, R. Haas, S. Plumb, J. L. Juncos, J. Nutt, I. Shoulson, J. Carter, K. Kompoliti, J. S. Perlmutter, S. Reich, M. Stern, R. L. Watts, R. Kurlan, E. Molho, M. Harrison, and M. Lew, Parkinson Study Group, *Arch. Neurol.* **59,** 1541 (2002).

however, produced a two-fold increase above the 300 and 600 mg doses, to approximately 4 μg per ml. An assay of mitochondrial complex I activity, which depends on the endogenous levels of CoQ$_{10}$ in the mitochondria, showed increases in all of the treated groups.

Interestingly, all 3 subscores of the UPDRS showed improvement. This included the mental score, the activities of daily living score, as well as the motor score. The initial changes in the total UPDRS score were comparable in varying groups at the first month, and only at the 4-month and later time points were there increasing separations consistent with a possible neuroprotective effect. The results of this trial appear extremely promising. The studies showed dose-dependent slowing of disease progression. There was improvement in complex I activities, suggesting that the beneficial effects may be mediated by improvements in mitochondrial function. It remains possible, however, that the effects are mediated by a reduction in oxidative damage. A much larger phase III trial is planned to confirm these results.

Conclusions

CoQ$_{10}$ administration can increase brain concentrations in mature and older animals. It can also increase brain mitochondrial concentrations. There is substantial evidence that CoQ$_{10}$ can act in concert with alpha-tocopherol as an antioxidant within mitochondria. CoQ$_{10}$ administration has been demonstrated to be efficacious in experimental models of neurodegenerative diseases. It is neuroprotective against lesions produced by mitochondrial toxins including malonate, 3-nitropropionic acid and MPTP. CoQ$_{10}$ extends survival in a transgenic mouse model of ALS, and it extends survival and exerts significant neuroprotective effects in a transgenic mouse model of HD. CoQ$_{10}$ and idebenone appear to be promising for treatment of Friedriech's ataxia. Initial clinical trials in both HD and PD have shown beneficial effects. Clinical trials in ALS are being planned. The initial results of utilizing CoQ$_{10}$ administration for treatment of neurodegenerative disease appear very promising as a treatment to slow the inexorable progression of these disorders.

[27] Neuroprotective Actions of Coenzyme Q_{10} in Parkinson's Disease

By S. Sharma, M. Kheradpezhou, S. Shavali, H. El Refaey,
J. Eken, C. Hagen, and M. Ebadi

Introduction

Although the exact etiopathogenesis of Parkinson's disease (PD) remains unknown, it has been hypothesized that the demise of nigrostriatal dopaminergic (DA-ergic) neurons could occur because of the production of endogenous neurotoxins such as tetrahydroisoquinolines (THIQs) or by exposure to environmental neurotoxins, such as rotenone.[1] These neurotoxins produce down-regulation of mitochondrial complex-1 (ubiquinone NADH-oxidoreductase),[2,3] reduce glutathione in the substantia nigra (SN) and enhance the risk of free radicals (mainly $^\bullet OH$ and NO) overproduction in PD. Overproduction of NO may induce deleterious consequences on the complex-1 activity.[4,5] In addition to NO, accumulation of iron in the substantia nigra (SN) has been implicated in the DA-ergic cell neurotoxicity.[6,7] Peroxynitrite ($ONOO^-$) ions, generated in the mitochondria by Ca^{2+}-dependent NOS activation, readily react with lipids, aromatic amino acids, or metalloproteins, and inhibit complex-1 activity during oxidative- and nitrative stresses.[8,9]

We have reported that oxidative and nitrative stresses are involved in neurodegeneration, whereas metallothioneins (MTs) provide coenzyme Q_{10}-mediated neuroprotection in DA-ergic neurons.[10] Pretreatment with coenzyme Q_{10} attenuated MPTP-induced intra-mitochondrial accumulation of metal ions (Cu^{2+}, Fe^{3+}, Zn^{2+} and Ca^{2+}) in the DA-ergic neurons,

[1] M. Ebadi, P. Govitropong, S. Sharma, D. Muralikrishnan, S. Shavali, L. Pellet, R. Schaffer, C. Albano, and J. Ekens, *Biological Signals and Receptors* **10**, 224 (2001).

[2] E. Fosslien, *Ann. Clin. Lab. Sci.* **31**, 25 (2001).

[3] R. N. Rosenberg, *Arch. Neurol.* **59**, 1523 (2002).

[4] U. Bringold, P. Ghafourifar, and C. Richter, *Free Radic. Biol. Med.* **29**, 343 (2000).

[5] P. Ghafourifar, U. Bringold, and S. D. Klein, *Biol. Signal. Recept.* **10**, 57 (2001).

[6] M. B. Youdim, D. Ben-Shachar, G. Eshel, J. P. Finberg, and P. Riederer, *Adv. Neurol.* **60**, 259 (1993).

[7] M. B. Youdim, L. Lavie, and P. Riederer, *Ann. N.Y. Acad. Sci.* **738**, 64 (1994).

[8] M. B. Youdim and L. Lavie, *Life Sci.* **55**, 2077 (1994).

[9] F. Torreilles, S. Salman-Tabcheh, M. Guerin, and J. Torreilles, *Brain. Res. Rev.* **30**, 153 (1999).

[10] M. Ebadi and S. Sharma, *Antioxidants and Redox Signaling* **5**, 319 (2003).

suggesting its neuroprotective action in PD.[11] Coenzyme Q_{10} attenuated MPTP-induced loss of dopamine and dopaminergic axons in aged mice; hence it may be employed as an effective treatment of PD.[12,13]

We have reported that selegiline, a monoamine oxidase B inhibitor, provided neuroprotection via MT gene over-expression.[14] Furthermore, we discovered that MT gene over-expression in metallothionein transgenic (MT_{trans}) mouse brain inhibits MPTP-induced nitration of α-synuclein (α-Syn) and preserves mitochondrial coenzyme Q_{10} levels, hence providing neuroprotection.[15]

MTs attenuated 3-morpholinosydnonimine (SIN-1)-induced oxidative and nitrative stresses[16] and suppressed 6-hydroxy dopamine-induced $^{\bullet}$OH radical generation in DA-ergic neurons.[17] Since the involvement of iron-induced oxidative- and nitrative stresses are now advocated in the etio-pathogenesis of PD,[17–21] a detailed study was needed to explore the exact molecular mechanism of coenzyme Q_{10}-mediated neuroprotection in PD. SIN-1, a potent NO and peroxynitrite ion generator and soluble guanyl cyclase stimulator, produces not only oxidative stress but also nitrative stress in DA-ergic neurons, portending to play an important role in understanding the exact molecular mechanism of PD. In view of the previous, we have determined brain regional coenzyme Q_9 and Q_{10} levels from control$_{wt}$, MT_{trans}, MT_{dko} and α-Syn$_{ko}$ mice and in human dopaminergic (SK-N-SH) neurons transfected with sense and antisense oligonucleotides to MT-1. We have also explored the neuroprotective potential of coenzyme Q_{10} in aging mitochondrial genome knockout (RhO_{mgko}) neurons by transfecting with the complex-1 gene and have established that coenzyme Q_{10} nullifies the MPP^{+-} or SIN-1–induced free radical generation providing neuroprotection.

[11] M. Ebadi, S. Sharma, D. Muralikrishnan, S. Shavali, J. Ekens, P. Sangchot, B. Chetsawang, and L. Brekke, *Proc. West Pharmacol. Soc.* **45,** 1 (2002).

[12] M. F. Beal, *Biofactors* **9,** 261 (1999).

[13] M. F. Beal, *Free Radic. Res.* **36,** 455 (2002).

[14] M. Ebadi, M. Hiramatsu, W. J. Burke, D. G. Folks, and M. A. El-Sayed, *Proc. West Pharmacol. Soc.* **41,** 155 (1998).

[15] S. K. Sharma and M. Ebadi, *Antioxidants and Redox Signaling* **5,** 251 (2003).

[16] M. Ebadi, S. Sharma, S. Shavali, and H. El ReFaey, *J. Neurosci. Res.* **27,** 379 (2002).

[17] P. Sangchot, S. Sharma, B. Chetsawang, P. Govitrapong, and M. Ebadi, *Dev. Neurosci.* **22,** 143 (2002).

[18] V. L. Dawson and T. M. Dawson, *Prog. Brain Res.* **118,** 215 (1998).

[19] D. S. Bredt, *Free Radic. Res.* **31,** 577 (1999).

[20] M. Gerlach, D. Blum-Degen, J. Lan, and P. Riederer, *Adv. Neurol.* **80,** 239 (1999).

[21] E. C. Hirsch and S. Hunot, *Trends Pharmacol. Sci.* **21,** 163 (2002).

Materials and Methods

Chemicals

Cell culture materials including powdered Dulbecco's Modified Eagle's Medium (DMEM), fetal bovine serum (FBS), strepto-penicillin, Ham's F12 medium, trypsin, and specific primers for cell transfection were purchased from GIBCO/BRL Life Technologies (Rockville, MD). Vector (pEGFP-N1) was purchased from BD Bioscience Clon-Tech (Palo Alto, CA). Liposome-based Effectine Transfection reagent with DNA enhancer was purchased from Qiagen Inc. (Stanford, CA). Protein assay dye was purchased from Bio-Rad Laboratories (Hercules, CA). Mitochondrial $\Delta\Psi$ fluorochrome; 5,5',6,6'-tetrachloro-1,1',3.3'-tetraethylbenzimidazolo-carbocyanide iodide (JC-1); nucleocytoplasmic fluorochromes; 4',6-dia-midino-2-phenylindole dihydrochloride (DAPI); ethidium bromide; acridine orange; and FITC-conjugated antimouse IgG were purchased from Molecular Probes (Eugene, Oregon). All other chemicals were of reagent-grade quality and were purchased from Sigma Chemical Co. (St. Louis, MO).

Experimental Animals

Experimental animals were housed in temperature- and humidity-controlled rooms with a 12-hr day and night cycle and were provided with commercially prepared chow and water ad libitum. The animals were acclimated to laboratory conditions for at least 4 days prior to experimentation. Care was also taken to avoid any distress to animals during the period of experiment. Breeder pairs of control$_{wt}$ C57BJ6, MT$_{dko}$, MT$_{trans}$, and α-Syn$_{ko}$ mice, weighing 25 to 30 g, were purchased from Jackson's Laboratories (Minneapolis, MN). The animals were maintained in an air-conditioned animal house facility in hepta-filtered cages with free access to water and lab chow. The zinc, copper, and iron contents in the lab chow were monitored by atomic absorption spectrophotometer to maintain their adequate supply.

Neuronal Culture

Human neuroblastoma (SK-N-SH) cell lines were cultured in Dulbecco's modified Eagle's medium (DMEM) supplemented with high glucose, glutamine, 3.7 g/L sodium bicarbonate, and 10% fetal bovine serum (pH 7.4). The cells were incubated in a Forma Scientific CO_2 incubator set at $37°$ with 5% CO_2 and 95% oxygen supply and a humidified environment under aseptic conditions. The cells were grown in small T_{25} flasks to avoid contamination and were grown for 48 hr in the above medium to a subconfluent stage. The cellular monolayer was detached by using a 0.25% trypsin-EDTA solution for 2 min at 37°. Trypsin from detached cells was neutralized with 10% fetal bovine serum, and the cells were spun at 1200 rpm for 5 min. The cell pellet was washed thrice with Dulbecco's phosphate buffered saline solution (d-PBS, pH 7.4) and stored at −80° before analysis.

Preparation of Mitochondrial Genome Knock-Out (RhO$_{mgko}$) Neurons

Mitochondrial genome knock-out DA neurons were prepared by supplementing Dulbecco's modified Eagle's complete medium with 5 μg/L of ethidium bromide. This treatment for 4 to 6 weeks selectively suppressed the mitochondrial genome (without affecting the nuclear genome), as confirmed by its absence when using RT-PCR with specific primer sets encoding complex-1 activity. RhO$_{mgko}$ neurons mimicked a cellular model of aging and exhibited typical characteristics of genetic susceptibility and neurodegeneration in response to Parkinsonian neurotoxins. We characterized these cells by estimating mitochondrial superoxide dismutase (SOD), catalase, and glutathione peroxidase and then determined their proliferative potential. Under phase-contrast microscopy, these cells exhibit typical granular appearance that is due to mitochondrial membrane aggregation. By using fluorescence microscopy, we observed a reduction in mitochondrial membrane potential ($\Delta\Psi$), peri-nuclear accumulation of MT, and α-Syn. The procedure to prepare aging RhO$_{mgko}$ cells is routinely performed and is well established in our lab. Although these cells can divide and afford neuroprotection in response to neuroprotective antioxidants, their bioenergetics are typically compromised, and they exhibit genetic susceptibility to excitoneurotoxins and partial neuronal recovery in response to antioxidants. Transfection with mitochondrial genome, however, significantly improved their metabolic as well as growth potential.

High-Performance Liquid Chromatography (HPLC-UV)

High-performance liquid chromatography (HPLC) was performed by using an ISCO pump and UV detector set at 275 nm spectral wavelength. Coenzyme Q_9 and Q_{10} were estimated before and after overnight exposure to MPP$^+$ (100 μM) and/or MT and/or selegiline. Coenzyme Q_{10} levels were estimated quantitatively by using the computer software and by preparing the overlay chromatograms before and after antioxidant treatment. To confirm MPP$^+$-induced coenzyme Q_{10} depletion, SK-N-SH neurons were incubated with MPP$^+$ for 72 hr. The cell pellet was washed with Dulbecco's phosphate-buffered saline solution and suspended in 0.5 ml of 10% ethanol. The cell suspension or brain tissue (25 mg) was sonicated for

30 sec at low wattage, and 1 ml of hexane was added. The samples were shaken vigorously in the dark chamber and centrifuged at 14,000 rpm for 20 min at 4°. The supernatant was dried in a nitrogen environment, and the residue was reconstituted in Millipore-filtered mobile phase (25% hexane:75% methanol). Ten microliters of filtered sample was injected in HPLC to detect coenzyme Q_9 and Q_{10} levels in the mouse brain regions and coenzyme Q_{10} levels from the SK-N-SH neurons. The data were quantitated by employing JCL-6000 computer software.[22]

Cell Transfection

The SKN-SH neurons were transiently transfected by using liposome-based Effectine reagent with DNA enhancer, with transfection buffer using sense and antisense oligonucleotides to MT-1, and mitochondrial genome encoding complex-1. For stable transfection, pEGFP-N1 vector was used as per manufacturer's recommendations. The reporter gene (green fluorescent protein) analysis was done by using real-time digital fluorescence imaging following selection with G-418 (250 μg/ml) for 4 weeks in enrichment medium. The cells were selected by using limiting dilution technique and grown further to study the expression of mitochondrial genome by routine radioimmunoprecipitation, immunoblotting, and RT-PCR with specific primers. The control and transfected cells were used to study SIN-1-induced apoptosis and coenzyme Q_{10}-induced neuroprotection by using multiple fluorochrome analysis as described later.

Digital Fluorescence Microscopy and the Neuroprotective Potential of Coenzyme Q_{10}

Control$_{wt}$, MT-1 sense and antisense transfected, and aging RhO$_{mgko}$ SK-N-SH were grown on glass cover slips or multi-chambered microscopic slides and maintained in Dulbecco's modified Eagle's Medium (DMEM) supplemented with 10% fetal bovine serum. The cells were grown for 48 hr to the subconfluent stage and were exposed to SIN-1 (10 μM) and/ or coenzyme Q_{10} (10 μM) overnight. After incubation, they were washed thrice in D-PBS and incubated at 37° for 30 min to stain with the mitochondrial marker JC-1 (5 nM) and then counter-stained for 30 seconds in 10 μg/ml of acridine orange, DAPI and/or ethidium bromide, washed thrice with Dulbecco's PBS, mounted on microscopic slides with aquamount supplemented with a photo-bleach inhibitor, and observed under digital fluorescence microscope (Leeds Instruments, Minneapolis, MN) set at three (blue for DAPI, green for acridine orange or fluorescein

[22] C. B. Albanao, D. Muralikrishnan, and M. Ebadi, *Neurochem. Res.* **27**, 359 (2002).

isothiocyanate [FITC] and red for ethidium bromide or JC-1) spectral wavelengths. (DAPI preferentially stain structurally intact nuclear DNA, while ethidium bromide stains fragmented DNA). The fluorescence images were captured by using a SpotLite digital camera and Image-Pro computer software. Target accentuation and background inhibition computer software were utilized to improve the quality of images. The images were merged to estimate mitochondrial versus nuclear apoptosis simultaneously.

Statistical Analysis

Repeated measures analysis of variance (repeated ANOVA) was employed for the statistical evaluation of the experimental data by using Sigma-Stat (version 2.03). The number of observations made in each experimental group are presented in the figure legends. Values of $p < 0.05$ were considered statistically significant.

Observations and Results

General

MT_{trans} mice were lean, agile, with shiny smooth black coat and increased vigilance status, whereas MT_{dko} mice were obese, lethargic, with dull brown coat and reduced vigilance status. MT_{dko} mice were genetically susceptible, whereas MT_{trans} mice exhibited genetic resistance to 1-methyl, 4-phenyl, 1,2,3,6-tetrahydropyridine (MPTP) parkinsonism. This was represented by severe body tremors, followed by immobilization in the MT_{dko} mice. Twenty-five percent of aging MT_{dko} mice exhibited skin lesions represented by facial and abdominal de-pigmentation (leukoderma), whereas aging MT_{trans} mice were still lean, agile, with shiny smooth black coat and increased vigilance status. On average, control$_{wt}$ mice lived 2.3 ± 0.25 years. There was no significant difference in age between control$_{wt}$ and MT_{dko} mice. However, MT_{trans} lived on an average 2.8 ± 0.5 years. Thus MT_{trans} mice lived long as compared with control$_{wt}$ and MT_{dko} mice ($p < 0.05$). A very old MT_{trans} female mouse developed cataracts and lived 3.2 years. Control$_{wt}$ and MT_{dko} mice exhibited severe skin lesions, loss of body hair, morbidity and mortality at this age. These symptoms were ameliorated by coenzyme Q_{10} (10 mg/kg, i.p.) treatment for 7 days.

Coenzyme Q_9 and Coenzyme Q_{10}

We estimated brain regional microdistribution of coenzyme Q_9 and Q_{10} from genetically engineered mice and coenzyme Q_{10} levels from the human dopaminergic (SK-N-SH) neurons because of their relative abundance and

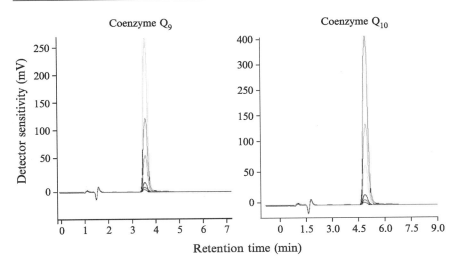

FIG. 1. Overlay chromatograms of coenzyme Q_9 (left panel) and coenzyme Q_{10} (right panel) at 6.25 ng, 12.5 ng, 25 ng, 50 ng, 100 ng and 200 ng concentrations. Standards were prepared in the mobile phase, preserved in the dark microcentrifuge tubes, and analyzed with high-performance liquid chromatography (HPLC) with UV detection at 275 nm spectral wavelength at room temperature. An ISCO HPLC pump and JC1-6000 computer software were used for the quantitative analysis of data as described in the text. (See color insert.)

physiological significance in the etiopathogenesis of PD. Overlay chromatograms were prepared to demonstrate the accuracy and reproducibility of HPLC-UV detection. Overlay chromatograms of six different concentrations (6.25 ng, 12.5 ng, 25 ng. 50 ng, 100 ng, and 200 ng) of coenzyme Q_9 and coenzyme Q_{10}, respectively, are presented in Fig. 1. Minimum detection limits of coenzyme Q_9 and Q_{10} were 3.2 ± 0.3 ng and 3.5 ± 0.45 ng, respectively. Although we estimated Q_2, Q_4 and Q_6 in addition to Q_9 and Q_{10}, their natural abundance was low as compared with coenzyme Q_9 and Q_{10}. A representative chromatogram of coenzymes Q_2, Q_4, Q_6, Q_9 and Q_{10} is presented in Fig. 2A (upper left panel).

Chromatograms demonstrating serial dilutions of coenzyme Q_9 (arrows) are presented in Fig. 2B (middle left panel), and its standard curve is shown in Fig. 2C (lower left panel). Chromatograms demonstrating serial dilutions of coenzyme Q_{10} (circles) are presented in Fig. 2D (upper right panel), and its standard curve is shown in Fig. 2E (middle right panel). Chromatograms of control$_{wt}$, MT$_{dko}$, MT$_{trans}$ and α-Syn$_{ko}$ mice striatum are presented in Fig. 2F (lower right panel).

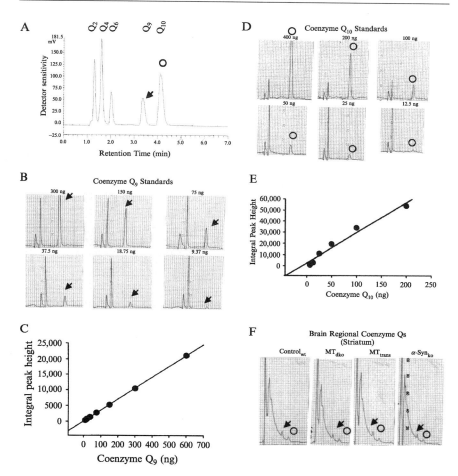

FIG. 2. (A) Upper left panel: Standardization of coenzyme Q (Q_2, Q_4, Q_6, Q_9, and Q_{10}) employing HPLC with UV detection at 275 nm spectral wavelength. (B) Middle left panel: Chromatograms of coenzyme Q_9 at different concentrations. (C) Lower left panel: Standard curve of coenzyme Q_9 at different concentrations, demonstrating straight line with a correlation coefficient of 0.98. (D) Chromatograms of coenzyme Q_{10} at different concentrations. (E) Middle right panel: Standard curve of coenzyme Q_{10} at different concentrations, demonstrating straight line with a correlation coefficient of 0.95. (F) Lower right panel: Chromatograms illustrating coenzyme Q_9 (arrows) and Q_{10} (circles) concentrations of control$_{wt}$, MT_{dko}, MT_{trans} and α-Syn$_{ko}$ mouse striatum.

Quantitative Estimation of Coenzyme Q_9 and Q_{10}

Brain regional coenzyme Q_9 and Q_{10} concentrations were estimated from cerebral cortex, hippocampus, striatum, and cerebellar cortex from contol$_{wt}$, MT$_{dko}$, MT$_{trans}$ and α-Syn$_{ko}$ mice, with a primary objective of establishing their functional significance in the etiopathogenesis of PD. Human DA-ergic (SK-N-SH) neurons possessed predominantly coenzyme Q_{10}, whereas control$_{wt}$ and genetically engineered mice possessed both coenzyme Q_9 and Q_{10}. It is important to point out that metabolic activity is higher in rodents than in humans; hence they possess higher concentrations of coenzyme Q_9 as compared with coenzyme Q_{10}. Hence estimation of both coenzyme Q_9 and Q_{10} would provide an overall estimate of coenzyme Q metabolism in these animals. The rodents exhibit higher mobility, which requires more energy as compared with humans, hence higher concentrations of coenzyme Q_9 and Q_{10} in the cerebellum compensate for the increased demands of ATP in rodents. However, on genetic manipulation of MTs and/or α-Syn, coenzyme Q_9 and Q_{10} contents are significantly altered, which could also modulate the genetic predisposition to MPTP-Parkinsonism.

Coenzyme Q_9 and coenzyme Q_{10} followed a similar pattern of brain regional microdistribution. When coenzyme Q_9 levels were high in certain brain regions, coenzyme Q_{10} levels were also high, and vice versa. Coenzyme Q_9 and Q_{10} contents were very high in the cerebral cortex and in the cerebellum. The highest concentrations of coenzymes Q_9 and Q_{10} were observed in the cerebellum. Physiologically susceptible brain regions, including the hippocampus and the striatum, possessed significantly reduced coenzyme Q_9 and Q_{10} contents. Striatal and hippocampal concentrations of coenzyme Q_9 and Q_{10} were significantly reduced in MT$_{dko}$ and α-Syn$_{ko}$ mice as compared with MT$_{trans}$ mice, suggesting that brain regional MT induction protects the DA-ergic neurons from oxidative and nitrative stress of aging by enhancing coenzyme Q synthesis. Coenzyme Q_9 and Q_{10} contents from cerebral cortex, striatum, hippocampus, and cerebellar cortex of different genotypes are presented in Table I. Coenzyme Q_{10} concentrations were significantly ($p < 0.05$) higher in MT$_{trans}$ mice striatum as compared with control$_{wild-type}$ and MT$_{dko}$ mice (Fig. 3; left panel). Their concentrations were significantly ($p < 0.01$) reduced in aging RhO$_{mgko}$ neurons (Fig. 3; right panel). Transfection of aging RhO$_{mgko}$ neurons with complex-1 significantly ($p < 0.01$) enhanced coenzyme Q_{10} synthesis (Fig. 3; right panel). Overnight treatment with selegiline (10 μM) also enhanced coenzyme Q_{10} synthesis, while MPP$^+$ (100 μM) significantly ($p < 0.05$) reduced coenzyme Q_{10} synthesis. Selegiline pretreatment attenuated MPP$^+$-induced reductions in coenzyme Q_{10} synthesis (Fig. 4, upper left

TABLE I
BRAIN REGIONAL CONCENTRATIONS OF COENZYME Q_9 AND Q_{10}

S. No.	Control$_{wt}$	MT$_{dko}$	MT$_{trans}$	α-Syn$_{ko}$
		Coenzyme Q_9		
Cerebral Cortex	10.4 ± 0.4	8.0 ± 0.5	12.0 ± 0.6	9.5 ± 0.3
Hippocampus	8.4 ± 0.5	7.0 ± 0.6[*]	9.0 ± 0.7	7.5 ± 0.5[*]
Striatum	7.0 ± 0.3	6.5 ± 0.4[*]	7.8 ± 0.7	6.8 ± 0.4[*]
Cerebellum	12.7 ± 0.5	10.0 ± 0.7[*]	13.7 ± 0.8	11.2 ± 0.7[*]
		Coenzyme Q_{10}		
Cerebral Cortex	2.6 ± 0.3	2.0 ± 0.2	3.0 ± 0.3	2.3 ± 0.2
Hippocampus	2.1 ± 0.2	1.8 ± 0.3[*]	2.5 ± 0.4	1.5 ± 0.3[*]
Striatum	1.7 ± 0.4	1.3 ± 0.5[*]	1.9 ± 0.2	1.2 ± 0.2[*]
Cerebellum	4.0 ± 0.3	3.0 ± 0.4[*]	5.0 ± 0.5	3.5 ± 0.3[*]

Data are mean ± SD of 5 determinations (ng/mg tissue) in each experimental group.
([*]p value < 0.05, repeated measures ANOVA).

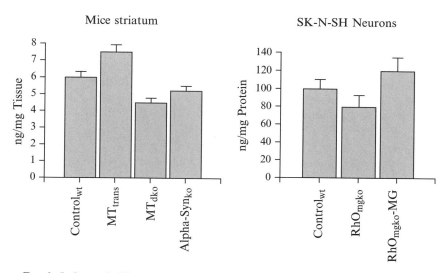

FIG. 3. Left panel: Histograms representing striatal coenzyme Q_{10} levels in contol$_{wt}$, MT$_{trans}$, MT$_{dko}$ and α-Syn$_{ko}$ mice. Striatal coenzyme Q_{10} levels were significantly reduced in MT$_{dko}$ as compared with control$_{wt}$ and MT$_{trans}$ mice. Right panel: Histogram demonstrating coenzyme Q_{10} levels in control$_{wt}$, aging RhO$_{mgko}$ and RhO$_{mgko}$ neurons transfected with the mitochondrial genome encoding complex-1. Coenzyme Q_{10} levels were significantly reduced in aging RhO$_{mgko}$ neurons as compared with control$_{wt}$ neurons. Transfection with complex-1 gene increased coenzyme Q_{10} levels in RhO$_{mgko}$ neurons. Data are mean ± SD of 8 determinations in each experimental group (p < 0.05, repeated measures ANOVA).

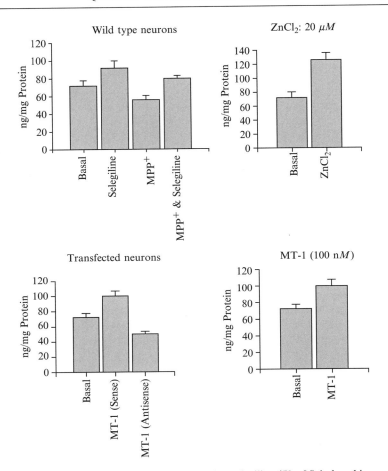

FIG. 4. Upper left panel: A histogram representing selegiline (50 μM)-induced increase in coenzyme Q_{10} and its reduction after overnight exposure to MPP^+ (100 μM). Selegiline pretreatment attenuated MPP^+-induced reductions in coenzyme Q_{10} levels. Upper right panel: A histogram demonstrating zinc chloride (20 μM)-induced increase in coenzyme Q_{10} synthesis in SK-N-SH neurons. Lower left panel: A histogram representing increase in coenzyme Q_{10} on transfection of SK-N-SH with MT-1$_{sense}$ and reduced synthesis in MT-1$_{antisense}$ oligonucleotide-transfected neurons. Lower right panel: A histogram demonstrating increased coenzyme Q_{10} synthesis after 48 hr exposure to MT-1 (100 nM) in SK-N-SH neurons. Data are mean ± SD of 6 determinations in each experimental group ($p < 0.05$, repeated measures ANOVA).

panel). Transfection with sense oligonucleotides to MT-1 increased, while with antisense oligonucleotides suppressed coenzyme Q_{10} synthesis (Fig. 4; lower left panel). Exogenous administration of zinc chloride (Fig. 4, upper right panel) or MT-1 (Fig. 4, lower right panel) also increased coenzyme Q_{10} synthesis within 48 hr in SK-N-SH neurons.

Digital Fluorescence Imaging Microscopy

We prepared mitochondrial genome knock out (RhO_{mgko}) dopaminergic (SK-N-SH) neurons with a primary intention to mimic the cellular model of aging and PD. The RhO_{mgko} neurons exhibit granular appearance caused by mitochondrial aggregation and down-regulation of machinery involved in oxidative phosphorylation. They exhibit a spherical or rounded shape with reduced neuritogenesis, and their coenzyme Q_{10} as well as MT-1 levels are significantly reduced. Upon transfection with mitochondrial genome encoding complex-1 or MT-1 sense oligonucleotides, the RhO_{mgko} neurons exhibit neuritogenesis and mitochondrial synthesis, and their coenzyme Q_{10} levels are also increased, suggesting the physiological significance of mitochondrial complex-1 (ubiquinone-NADH-oxido-reductase) and the neuroprotective action of coenzyme Q_{10} in PD. Digital fluorescence images of (A) control$_{wt}$, (B) aging RhO_{mgko}, and (C) RhO_{mgko} neurons transfected with mitochondrial complex-1 (ND-1) gene are presented in Fig. 5.

Cell Transfection

To authenticate further the neuroprotective potential of coenzyme Q_{10} in control$_{wt}$ and RhO_{mgko} dopaminergic neurons, we transfected them with MT-1$_{sense}$, MT-1$_{antisense}$, and mitochondrial genome encoding complex-1 (ND-1 loop; ubiquinone-NADH oxidoreductase) and have evaluated the neuroprotective potential of coenzyme Q_{10} on SIN-1-induced apoptosis. Aging RhO_{mgko} as well as MT-1$_{antisense}$ oligonucleotide-transfected SK-N-SH neurons were highly susceptible to SIN-1-induced apoptosis, which was characterized by spherical appearance, suppressed neuritogenesis, phosphatidyl serine externalization, plasma membrane perforations, $\Delta\Psi$ collapse, mitochondrial aggregation, and nuclear DNA fragmentation and condensations. Overnight exposure of SIN-1 induced degeneration of mitochondrial membranes and their peri-nuclear accumulation, particularly in MT-1$_{antisense}$ oligonucleotide-transfected and aging RhO_{mgko} neurons. These neurons exhibited a necrotic appearance with plasma membrane ballooning and perforations. Pretreatment with coenzyme Q_{10} provided neuroprotection from SIN-1-induced apoptosis. Coenzyme Q_{10} provided efficient neuroprotection in MT-1$_{sense}$ and complex-1 gene-transfected SK-N-SH neurons. Coenzyme Q_{10}-mediated neuroprotection was compromised in MT-1$_{antisense}$ and aging RhO_{mgko} neurons (Fig. 6), suggesting that MT induction provides coenzyme Q_{10}-mediated

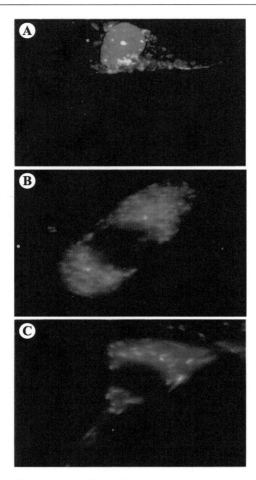

Fig. 5. Multiple fluorochrome digital fluorescence imaging microscopic pictures of (A) control$_{wt}$, (B) aging RhO$_{mgko}$ and (C) RhO$_{mgko}$ neurons transfected with complex-1 gene. The RhO$_{mgko}$ neurons exhibit rounded or spherical appearance, increased granularity, reduced neuritogenesis and mitochondrial aggregation in the peri-nuclear region. Neurito-genesis reappeared in complex-1 transfected RhO$_{mgko}$ neurons. Digital fluorescence images were captured with a SpotLite digital camera on an Olympus microscope and were analyzed with ImagePro computer software. Target accentuation and background inhibition software was employed to improve the quality of images. Fluorochromes: Green, acridine orange (nucleo-cytoplasmic marker especially for RNA and proteins); red, JC-1 (mitochondrial membrane potential $\Delta\Psi$ marker); blue, DAPI (nuclear DNA marker). (See color insert.)

neuroprotection, while its down-regulation in MT_{dko} mice as well as in MT-$1_{antisense}$ oligonucleotide-transfected dopaminergic neurons renders them highly susceptible to SIN-1-induced apoptosis.

Discussion

The present study was conducted with a primary objective of exploring the molecular mechanism of neuroprotection provided by coenzyme Q_{10} in PD. We hypothesized that in MT_{trans} mice, coenzyme Q_{10} which is involved in mitochondrial stability, remains preserved in the presence of excess MT. Indeed, MT_{trans} mice were lean and agile, with shiny smooth black coat and increased vigilance status, whereas MT_{dko} mice were obese and lethargic, with dull brown coat and reduced vigilance status. MT_{dko} mice were more susceptible, whereas MT_{trans} mice exhibited genetic resistance to MPTP parkinsonism. To examine the molecular mechanism(s) of genetic suscepti-bility of MT_{dko} and α-Syn_{ko} mice and resistance of MT_{trans} mice to MPTP parkinsonism, we estimated brain regional coenzyme Q_9 and Q_{10} levels. Coenzyme Q_{10} levels were increased in MT_{trans} mice striatum as compared with $control_{wt}$ and MT_{dko} animals. Coenzyme Q_{10} levels were high in cere-bral cortex and cerebellum and low in hippocampus and striatum. Hippo-campal and striatal coenzyme Q_{10} levels remained elevated in MT_{trans} mice as compared with MT_{dko} and α-Syn_{ko} mice. Exogenous administra-tion of MT-1 also increased coenzyme Q_{10} in the SK-N-SH neurons. Trans-fection with sense oligonucleotides to MT-1 increased, while that with antisense oligonucleotides suppressed coenzyme Q_{10} synthesis. Pretreat-ment with selegiline also increased coenzyme Q_{10} and MT-1 synthesis. Aging RhO_{mgko} neurons possessed reduced coenzyme Q_{10} and MT-1 ex-pression. Transfection of complex-1 gene in RhO_{mgko} neurons ameliorated mitochondrial functions by augmenting coenzyme Q_{10} synthesis, and they became resistant to SIN-1-induced apoptosis. These data are interpreted to suggest that oxidative and nitrative stresses are involved in the etio-pathogenesis of PD, whereas exogenous and/or endogenous induction of MT preserves coenzyme Q_{10} to provide neuroprotection.

Brain regional coenzyme Q concentrations may fluctuate depending on species, age, sex, disease, metabolic demand, and circadian metabolic rhythm. We have now demonstrated that biodistribution of coenzyme Q would depend on the metabolic demand and physiological activity of an or-ganism. Rodents are nocturnal animals and metabolically more active at night as compared with humans. They possess higher levels of coenzyme Q_9 and Q_{10} in cerebellum and reduced levels in the hippocampus and striatum. Rodents utilize coenzyme Q_9, while humans require a higher form of coenzyme Q (i.e., Q_{10}) for mitochondrial function. Based on these

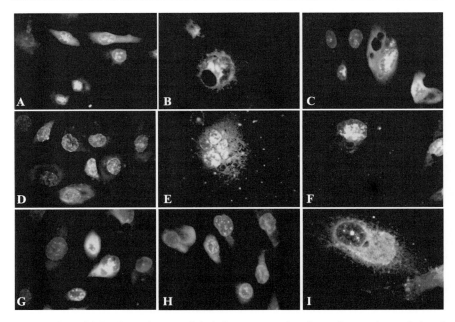

FIG. 6. Multiple fluorochrome digital fluorescence imaging microscopic analysis of (A) control$_{wt}$; (B) SIN-1 (100 μM) exposed; (C) SIN-1 with coenzyme Q$_{10}$ treatment; (D) aging RhO$_{mgko}$, $\Delta\Psi$ RhO$_{mgko}$ neurons exposed overnight to SIN-1; (F) RhO$_{mgko}$ neurons transfected with MT-1$_{sense}$ oligonucleotide; (G) RhO$_{mgko}$ neurons transfected with complex-1; (H) control$_{wt}$ neurons transfected with complex-1 and exposed overnight with SIN-1; RhO$_{mgko}$ neuron transfected with MT-1$_{antisense}$ oligonucleotide and exposed to SIN-1 overnight; and coenzyme Q$_{10}$ (10 μM), illustrating genetic resistance of MT-1$_{sense}$ transfected and complex-1 transfected control$_{wt}$ neurons and susceptibility of aging RhO$_{mgko}$ and MT-1$_{antisense}$ oligonucleotide-transfected aging RhO$_{mgko}$ neurons to SIN-1-induced oxidative stress and apoptosis. SIN-1 induces phosphatidyl serine externalization, plasma membrane perforations, mitochondrial aggregation, and nuclear DNA fragmentation and condensations. The apoptotic changes are attenuated on treatment of the neurons with coenzyme Q$_{10}$ and by transfecting with either complex-1 or MT-1$_{sense}$ oligonucleotides, indicating that complex-1 gene is down-regulated during oxidative and nitrative stress, while up-regulation of antioxidant genes (complex-1 and/or MT-1) may provide neuroprotection in aging and PD. Fluorochromes: Green, fluorescein isothiocyanate-conjugated-annexin-V (ApoAlert; for demonstrating phosphatidyl serine externalizatation); red, JC-1 (for estimating $\Delta\Psi$); and blue, DAPI (nuclear DNA stain to demonstrate structurally intact vs. fragmented DNA). Digital fluorescence images were captured and processed as described above. Fluorescence images were first digitized individually and merged together to obtain overall information about plasma membrane, $\Delta\Psi$ and nuclear DNA during SIN-1-induced apoptosis and coenzyme Q$_{10}$-mediated neuroprotection. (See color insert.)

observations, it is hypothesized that coenzyme Q and its natural abundance may have functional significance. Previous studies have shown that coenzyme Q_{10} levels correlated with the complex-1 activity.[23] Recent clinical trials have shown that coenzyme Q_{10} enhanced mitochondrial activity, was well-tolerated even at higher doses (1200 mg) and prevented functional decline in PD patients[24] and that its beneficial effects were dose-dependent.[25,26] These reports, including our own, provide further support regarding the therapeutic potential of coenzyme Q_{10} in PD.

By employing digital fluorescence imaging microscopy, we have now established that SIN-1 down-regulated complex-1 activity, while pretreatment with coenzyme Q_{10} or transfecting with MT-1_{sense} oligonucleotides suppressed this neurotoxicity in DA-ergic neurons. Recently we have reported that SIN-1 induced apoptosis was attenuated in MT_{trans}, MT-1, and complex-1 gene-transfected DA-ergic neurons. MPTP-induced neurodegenerative changes were also suppressed by exogenous administration of selegiline, MT-1, or coenzyme Q_{10}. MPTP enhanced α-Syn expression and induced genes (*c-fos, c-jun,* and caspase-3) involved in apoptosis. Furthermore, pretreatment with coenzyme Q_{10} or MT-1 suppressed intra-mitochondrial and intra-nuclear biosynthesis of 8-OH-2dG and translocation of caspase-3, affording neuroprotection.[15]

As described earlier, SIN-1 is a potent peroxynitrite ($ONOO^-$) generator, produces oxidative and nitrative stresses and causes apoptosis in DA-ergic neurons. Coenzyme Q_{10} inhibited SIN-1-induced apoptosis, suggesting its neuroprotective potential in PD. Generation of $ONOO^-$ and metal nitrosyl adducts might have pathophysiological consequences in PD.[27] Peroxynitrite ions have been implicated in progressive neurodegeneration in stroke, ischemia, AD, PD, multiple sclerosis, motor neuron disease and inflammatory diseases.[28,29] Nitration of protein tyrosine residues and neurofilaments is a molecular marker of $ONOO^-$ production and is enhanced by SOD over-production.[30]

[23] C. W. Shults, R. H. Haas, D. Passov, and M. F. Beal, *Ann. Neurol.* **42,** 261 (1997).

[24] C. W. Shults, M. F. Beal, D. Fontaine, K. Nakano, and R. H. Haas, *Neurology* **50,** 793 (1998).

[25] C. W. Shults, R. H. Maas, and M. F. Beal, *Arch. Neurol.* **59,** 1523 (2002).

[26] C. W. Shults, D. Oakes, K. Kieburtz, M. F. Beal, R. Haas, S. Plumb, J. L. Juncos, J. Nutt, I. Shoulson, J. Carter, K. Kompoliti, J. S. Perlmutter, S. Reich, M. Stern, R. L. Watts, R. Kurlan, E. Molho, M. Harrison, and M. Lew, *Arch. Neurol.* **59,** 1541 (2002).

[27] J. S. Stamler, D. L. Singel, and J. Loscalzo, *Science* **258,** 1898 (1998).

[28] J. S. Beckman, M. Carson, C. D. Smith, and W. H. Koppenol, *Nature* **364,** 584 (1993).

[29] J. S. Beckman, *Ann. N.Y. Acad. Sci.* **738,** 69 (1994).

[30] J. S. Beckman and W. H. Koppenol, *Am. J. Physiol.* **271,** C1424 (1996).

Superoxide ions (O_2^-), produced by the reduction of O_2 have one unpaired electron, which can rapidly combine with the unpaired electron of NO to form $ONOO^-$. O_2^- levels are kept low by superoxide dismutase (SOD), which dismutases O_2^- to H_2O_2 and O_2^-. SOD catalyzes the dismutation reaction of O_2^-.

In the presence of H_2O_2, SOD can produce potentially damaging $^{\bullet}OH$ radicals, and high levels of SOD can be damaging to the cell.[31] $ONOO^-$ can also inhibit the mitochondrial electron transport chain.[32] Therefore, the over-expression of complex-1 or MT-1 gene in aging RhO_{mgko} neurons attenuates the MPTP- or SIN-1-induced lipid peroxidation and caspase-3 activation, suggesting the neuroprotective potential of MT-induced coenzyme Q_{10} synthesis. Exogenous administration of MPTP enhanced NOS activity as well as NO production in the substantia nigra, while NOS knock-out animals exhibit neuroprotection against MPTP neurotoxicity,[33] indicating the involvement of oxidative and nitrative stresses in the etiopathogenesis of PD.

Studies from our lab have established the neuroprotective and antioxidant role of MT in PD.[34,35] MT-1 mRNA expression was enhanced in response to acute 6-OH-DA exposure,[36] while chronic treatment with MPP^+ suppressed MT-1 mRNA expression in the rat striatum.[37,38] Recently we have reported that MPTP-enhanced α-Syn expression as well as intra-mitochondrial accumulation of metal ions are attenuated by pretreatment with MT-induced coenzyme Q_{10} synthesis or selegiline.[15] Selegiline protects human DA-ergic neurons from apoptosis via inhibition of $ONOO^-$ and NO.[39,40] Peroxynitrite-induced DNA fragmentation was also attenuated by MT-1.[41] We have now demonstrated that selegiline-induced

[31] R. H. Brown, *Cell* **80,** 687 (1995).

[32] R. Radi, M. Rodriguez, L. Castro, and L. Telleri, *Arch. Biochem. Biophys.* **308,** 89 (1994).

[33] H. Y. Yun, V. L. Dawson, and T. M. Dawson, *Mol. Psychiatry* **2,** 300 (1997).

[34] M. Ebadi, R. F. Pfeiffer, L. C. Murrin, and H. Shiraga, *Proc. West Pharmacol. Soc.* **34,** 285 (1991).

[35] M. Ebadi, M. P. Leuschen, H. E. I. Refaey, F. M. Hamada, and P. Rojas, *Neurochem. Int.* **29,** 159 (1996).

[36] P. Rojas, D. R. Cerutis, H. K. Happe, L. C. Murrin, R. Hao, R. F. Pfeiffer, and M. Ebadi, *Neurotoxicology* **17,** 323 (1996).

[37] P. Rojas, J. Hidalgo, M. Ebadi, and C. Rios, *Prog. Neuropsychopharmacol. Biol. Psychiatry* **24,** 143 (2001a).

[38] P. Rojas, J. Rojas-Castaneda, R. M. Vigueras, S. S. Habeebu, C. Rojas, C. Rios, and M. Ebadi, *Neurochem. Res.* **25,** 503 (2001b).

[39] W. Maruyama, T. Takahashi, and M. Naoi, *J. Neurochem.* **70,** 2510 (1998).

[40] M. Naoi, W. Maruyama, K. Yagi, and M. Youdim, *Neurobiology (BP)* **8,** 69 (2000).

[41] M. B. Mattammal, J. H. Haring, H. D. Chung, G. Raghu, and R. Strong, *Neurodegeneration* **4,** 271 (1995).

neuroprotection of DA-ergic neurons is mediated through MT-induced co-enzyme Q$_{10}$ synthesis. Selegiline as well as MT can also afford neuroprotection by acting as a modulator of cytokine expression. Both selegiline and MT trigger the DNA cell cycle and enhance cell proliferation and differentiation, because these agents are also involved in growth and maintenance of nigrostriatal DA-ergic neurotransmission. Indeed, by employing microdialysis, we observed enhanced release of striatal dopamine from the MT$_{trans}$ as compared with MT$_{dko}$ mice.

DOPAL, a dopamine oxidation product, has also been shown to suppress striatal DA-ergic neurotransmission and occurs in progressive neuro-degeneration.[42–44] SIN-1-induced DOPAL synthesis was suppressed by MT gene induction. Hence, treatment with coenzyme Q$_{10}$ or over-expression with MT may have therapeutic potential in PD.

Molecular mechanism of NO-mediated neurodegeneration in PD remains unknown. NO interacts with 6-OHDA to induce neurodegeneration via enhanced •OH radical production.[45] NO also enhance MPP$^+$-induced •OH radical generation via activation of NOS in the rat striatum.[46] Similarly, MPTP induced DA-ergic neurotoxicity through iNOS induction.[47,48] Peroxynitrite- and nitrite-induced oxidation of DA causes loss of DA cells in PD,[49] while inhibition of nNOS prevents MPTP-induced parkinsonism in baboons.[50]

Furthermore, iNOS knock-out mice exhibit genetic resistance to MPTP neurotoxicity,[51] while iNOS-induced inflammation causes DA-ergic neuro-degeneration.[52] Peroxynitrite ions also inhibit DOPA synthesis in PC-12 cells,[53] and a selective nNOS inhibitor, 7-nitroindazole, protects mice

[42] B. S. Kristal, A. D. Conway, A. M. Brown, J. C. Jain, P. A. Ulluci, S. W. Li, and W. J. Burke, *Free Radic. Biol. Med.* **30,** 924 (2001).

[43] I. Lamensdorf, G. Eisenhofer, J. Harvey-White, Y. Hayakawa, K. Kirk, and I. J. Kopin, *J. Neurosci. Res.* **60,** 552 (2001).

[44] I. Lamnesdorf, G. Eisenhofer, J. Harvey-White, A. Neuchustan, K. Kirk, and I. J. Kopin, *Brain Res.* **868,** 191 (2001).

[45] N. A. Riobo, F. J. Schopfer, A. D. Boveris, E. Cadenas, and J. J. Poderoso, *Free Radic. Biol. Med.* **32,** 115 (2002).

[46] T. Obata and Y. Yamanaka, *Brain. Res.* **902,** 223 (2001).

[47] S. Przedborski, V. Jackson-Lewis, R. Yokoyama, T. Shibata, V. L. Dawson, and T. M. Dawson, *Proc. Natl. Acad. Sci. USA* **93,** 4565 (1996).

[48] J. P. Kiss, *Brain Res. Bull.* **52,** 459 (2002).

[49] M. J. LaVoie and T. G. Hastings, *J. Neurochem.* **73,** 2546 (1999).

[50] P. Hantraye, E. Brouillet, R. Ferrante, S. Palfi, R. Dolan, R. T. Matthews, and M. F. Beal, *Nat. Med.* **2,** 1017 (1996).

[51] T. Grunewald and M. F. Beal, *Nat. Med.* **5,** 1354 (1999).

[52] M. M. Iravani, K. Kashefi, P. Mander, S. Rose, and P. Jenner, *Neuroscience* **110,** 49 (2002).

[53] H. Ischiropoulos, D. Duran, and J. Horwitz, *J. Neurochem.* **65,** 2366 (1995).

against methamphetamine-[54] and MPTP-induced neurotoxicity.[55] In culture, MPP^+ increases the vulnerability to oxidative stress in SHS-Y-5Y cells.[56] NO-induced neurotoxicity is mediated through iNOS-mediated neurodegeneration of DA-ergic neurons in the MPTP model of PD.[57] Significantly elevated levels of nitrites in the CSF would indicate the involvement of NOS activation and $ONOO^-$ generation in PD.[58] MPTP-induced NO production could induce mitochondrial dysfunction in the DA-ergic neurons as observed in PD.[59] MPTP-induced neurotoxicity is associated with microglial iNOS induction; hence, inhibition of iNOS may be a promising target for the treatment of PD.[60]

Indeed, a blockade of microglial cell activation induces neuroprotection against MPTP mouse model of PD.[61] After administration of MPTP to mice, there occurs a robust gliosis in the SNc, associated with significant up-regulation of iNOS. These findings tend to suggest the involvement of microglial cell iNOS activation in the etiopathogenesis of PD and the therapeutic potential of MT-induced coenzyme Q_{10} synthesis in PD.

Repairing the Brain in Parkinson's Disease and Providing Neuroprotection

The question of where Parkinson's disease pathology begins in the brain suggests that the substantia nigra is not the induction site in the brain of the neurodegenerative process underlying Parkinson's disease.[62] In spite of the fact that Parkinson's disease (PD) is the most widespread neurodegenerative movement disorder of the aging human nervous system, the clinical diagnosis—particularly the differential diagnosis with respect to

[54] Y. Itzhak and S. F. Ali, *J. Neurochem.* **67,** 1770 (1996).

[55] J. B. Schulz, R. T. Matthews, M. M. Muqit, S. E. Browne, and M. F. Beal, *J. Neurochem.* **64,** 936 (1995).

[56] H. S. Lee, C. W. Park, and Y. S. Kim, *Exp. Neurol.* **165,** 164 (2002).

[57] G. T. Liberatore, V. Jackson-Lewis, S. Vukosavic, A. S. Mandir, M. Vila, W. G. McAuliffe, V. L. Dawson, T. M. Dawson, and S. Przedborski, *Nat. Med.* **5,** 1403 (1999).

[58] G. A. Qureshi, S. Baig, I. Bednar, P. Sodersten, G. Forsberg, and A. Siden, *Neuroreport* **6,** 1642 (1995).

[59] J. B. Schulz, R. T. Matthews, T. Klockgether, J. Dichgans, and M. F. Beal, *Mol. Cell. Biochem.* **174,** 193 (1997).

[60] D. C. Wu, V. Jackson-Lewis, M. Vila, K. Tieu, P. Teismann, C. Vadseth, D. K. Choi, H. Ischiropoulos, and S. Przedborski, *J. Neurosci.* **22,** 1763 (2002).

[61] C. H. Hawkes, B. C. Shephard, and S. E. Daniel, *J. Neurol. Neurosurg. Psychiatry* **62,** 436 (1997).

[62] T. Dehmer, J. Lindenau, S. Haid, J. Dichgans, and J. B. Schulz, *J. Neurochem.* **74,** 2213 (2001).

related disorders—is not unproblematic and requires postmortem verification. In the course of PD, susceptible regions and vulnerable nerve cell populations become progressively impaired owing to the extensive presence of Lewy neurites (LNs) and Lewy bodies (LBs). The abnormal inclusions—consisting of aggregates within cellular processes and cell somata of involved nerve cells—contain α-synuclein (α-SN), proteolytic stress proteins (such as ubiquitin and sometimes αB-crystallin) and phosphorylated neurofilaments.

Although it is probable that the deterioration of the somatomotor system is attributable not only to nigrostriatal damage but also to detrimental events within other nerve cell populations and systems of the brain, no definitive explanation exists for the apparent selective vulnerability (pathoklisis) on the part of some classes of neurons in the brains to PD.[63]

Increasing awareness of PD as a multiple system neurodegenerative disorder has grown out of research demonstrating the existence of considerable extranigral pathology. The question, of course, arises whether any of these extranigral sites are affected before the substantia nigra. Some of the literature calls attention to the presence of inclusion bodies in the dorsal motor nucleus of the glossopharyngeal and vagal nerves. Clinical experience, on the other hand, points to hyposmia as an early phenomenon in PD.[61,64,65]

An important issue with respect to the pathogenesis of PD in the brain is not only how the pathological process begins but also where. Granted, α-SN-immunoreactive inclusion bodies are known to occur in amyotrophic lateral sclerosis, Hallervorden-Spatz disease and multiple system atrophy. Nonetheless, PD is the only illness that always presents postmortem with LBs and LNs as the sole pathological markers in specific types of nerve cells at predisposed sites. In the other neurodegenerative disorders, the inclusion bodies are restricted to either glial oligodendrocytes in multiple-system atrophy, astrocytes and Schwann cells in amyotrophic lateral sclerosis and/or other neuronal populations at other predilection sites within the nervous system (e.g., nerve cells of the pons, inferior olives).

Although the pathological changes and motor dysfunction characterizing this disease are well documented, the mechanism(s) responsible for the death of these neurons is not established. Extensive postmortem studies have provided evidence to support the notion that oxidative stress is involved in the pathogenesis of PD—in particular, alterations in the antioxidant protective system such as superoxide dismutase (SOD) and

[63] K. D. Tredici, U. Rüb, R. A. I. DeVos, J. R. E. Bohl, and H. Braak, *J. Neuropath. Exp. Neurol.* **61,** 413 (2002).

[64] R. L. Mesholam, P. J. Moberg, R. N. Mahr, and R. L. Doty, *Arch. Neurol.* **55,** 84 (1998).

[65] C. D. Ward, W. A. Hess, and D. B. Calne, *Neurology* **33,** 943 (1983).

reduced levels of glutathione (GSH) and evidence of oxidative damage to lipids and proteins. The SNc is a dopamine (DA)-rich brain region that contains neuromelanin (NM) and exhibits a significantly elevated tissue iron content. Because DA, iron, and to some extent NM can induce oxidative stress, one or the interaction of two or more of these factors may be involved in the pathological mechanism that underlies the relatively specific neurodegeneration seen in PD.[11,66,67]

PD is the only neurodegenerative disorder in which pharmacological intervention has resulted in a marked decrease in morbidity and a significant delay in mortality. However, the medium- to long-term efficacy of this pharmacotherapy, mainly consisting of dopaminomimetics like L-dopa and dopamine receptor agonists, suffers greatly from the unrelenting progression of the disease process underlying PD—that is, the degeneration of neuromelanin-containing dopaminergic neurons in the substantia nigra. Efforts concentrated on understanding the mechanisms of dopaminergic cell death in PD have led to identification of a large variety of pathogenetic factors, including excessive release of oxygen free radicals during enzymatic dopamine breakdown, impairment of mitochondrial function, production of inflammatory mediators, loss of trophic support, and apoptosis. Therapeutic approaches aimed at correcting these abnormalities, including the efficacy of coenzyme Q_{10} and metallothionein, are currently being evaluated on their efficacy as neuroprotectants for Parkinson's disease.

Stem cells have recently received a lot of attention, both from the scientific community and the lay press, which has meant that many claims are being made, only some of which have been substantiated scientifically. Stem cells are defined as cells that are capable of self-renewal, and in the case of neural stem cells, these are found not only in the developing brain (embryonic neural stem cells) but also at certain sites within the adult brain. Repairing the parkinsonian brain is entering an exciting new era, as proof of the concept that it can be done has been shown with several studies using human fetal ventral mesencephalic tissue.

Summary and Conclusions

Parkinson's disease is the second most common neurodegenerative disorder after Alzheimer's disease, affecting approximately 1% of the population older than 50 years. There is a worldwide increase in disease prevalence that is due to the increasing age of human populations. A definitive

[66] M. Ebadi, S. K. Srinivasan, and M. Baxi, *Prog. Neurobiol.* **48,** 1 (1996).
[67] A. Barzilai, E. Melamed, and A. Shirvan, *Cellular and Molecular Neurobiol.* **21,** 215 (2001).

neuropathological diagnosis of Parkinson's disease requires loss of dopaminergic neurons in the substantia nigra and related brain stem nuclei and the presence of Lewy bodies in remaining nerve cells. The contribution of genetic factors to the pathogenesis of Parkinson's disease is increasingly being recognized. A point mutation that is sufficient to cause a rare autosomal-dominant form of the disorder has been recently identified in the α-synuclein gene on chromosome 4 in the much more common sporadic, or "idiopathic," form of Parkinson's disease, and a defect of complex I of the mitochondrial respiratory chain was confirmed at the biochemical level. Disease specificity of this defect has been demonstrated for the parkinsonian substantia nigra. These findings, and the observation that the neurotoxin 1-methyl-4-phenyl-1,2,3,6-tetrahydropyridine (MPTP)—which causes a Parkinson-like syndrome in humans—acts via inhibition of complex I have triggered research interest in the mitochondrial genetics of Parkinson's disease.

1-Methyl-4-phenyl-1,2,3,6 tetrahydropyridine (MPTP) becomes oxidized to 1-methyl-4-phenyl-2,3-dihydropyridinium ion ($MPDP^+$) and finally to methyl-phenyl-tetrahydro-pyridinium ion (MPP^+), which generates free radicals and causes parkinsonism in human beings. A deficiency of NADH:ubiquinone oxidoreductase (EC 1.6. 5.3; complex I) also causes striatal cell death. A deficiency of complex I may signify that an MPTP-like neurotoxin is generated endogenously, enhancing the vulnerability of striatum to oxidative stress reactions. The neurotoxic effects of MPP^+ are blocked by metallothionein, a zinc-binding protein that scavenges hydroxyl radicals; by deferoxamine, an iron-binding compound that inhibits fenton reaction; and by α-phenyl-ter-butyl-nitrone, which traps free radicals. Neurons containing high concentrations of calbindin D28K are relatively resistant to MPP^+. Excitatory amino acid receptor blockers such as dizocilpine attenuate MPP^+-induced neurotoxicity. Amfonelic acid and mazindol, preventing the uptake of MPTP into dopaminergic neurons, and selegiline, preventing the formation of MPP^+, prevent the neurotoxic effects of MPTP.[1,16]

Acknowledgments

The studies reported here have been supported in part by a grant from USPHS NIA 17059-04 (M.E.) and NINDS 2R01 NS 34566-09 (ME). The authors express their heartfelt appreciation to Dani Stramer for her excellent secretarial skills in typing this manuscript.

Author Index

Kohno, T., 134(151), 138
Kohnoe, S., 176
Koike, H., 275
Kolesar, J. M., 128(103; 106), 135(154), 136, 139
Kolonel, L. N., 132(135), 138, 457
Komai, T., 144, 359
Komatsuda, A., 214
Kompoliti, K., 486, 503
Kon, H., 224
Kong, A. N., 256, 397, 431(47), 434, 438(47)
Kong, L.-Y., 302(51), 303
Kono, N., 475
Konoshima, T., 413
Kontush, A., 81, 82, 82(3)
Kopelovich, L., 387, 397(51)
Kopin, I. J., 505
Koppenol, W. H., 181, 503
Korach, K. S., 302(44), 303, 307, 307(44)
Kornguth, S. E., 445
Korpijaakko, T. A., 176
Kosmeder, J. W., 381, 384, 388
Kosmeder, J. W. II, 406, 432(57), 435
Koster, A. S., 183, 186, 187(61), 187(76)
Kostic, S., 129(114), 137
Kotlyar, A. B., 9
Koutalos, Y., 119, 284, 323
Kowall, N. W., 481, 481(65), 482
Koynova, G. M., 94, 479
Kozlov, A. V., 68, 83, 93
Kozlov, V., 479
Kozuka, T., 106
Krafft, W., 175
Krailo, M., 176
Krajinovic, M., 133(91; 96), 134(143; 148), 136, 138, 141(91; 96), 363
Kramer, B. S., 383, 403(39)
Kramer, D. M., 21, 24, 26, 27, 27(28), 28, 28(19), 29(19), 30(19), 33(19; 28), 34, 36(59), 40(59)
Kramer, R. A., 380, 383(3)
Kranis, K. T., 22
Kraulis, P., 147
Krawiecki, N., 474
Krediet, R. T., 330, 339(79)
Kreitman, K. R., 477
Krinsky, N. I., 447
Krishcher, J. P., 176
Krishnamoorthy, R., 132(136), 138
Krishnaswamy, M., 364, 380(6)

Kristal, A. R., 457
Kristal, B. S., 505
Kristensen, V. N., 135(155), 139
Kristiansen, O. P., 130(119), 137
Krivit, W., 176
Kroemer, G., 280, 288(23)
Krohn, R. I., 416
Krokan, H. E., 232
Kromhout, D., 406
Kruk, J., 90
Krul, C., 467
Krummeck, A., 123, 129(73), 324, 362
Ku, H. H., 31
Kuang, S. Y., 328, 441
Kubbutat, M. H., 278
Kube, U., 105, 473, 479(3)
Kubota, Y., 131(93), 136, 141(93)
Kudo, S., 222
Kuehl, B. L., 128(107; 108), 136
Kuhn, J. G., 128(103; 106), 136
Kuhn, T. W., 477
Kuippala, T. A., 176
Kulys, J. J., 264, 266, 271
Kumagai, T., 438(80), 441
Kumar, G. S., 222, 222(13), 223
Kumar, S., 302(43), 303, 307(43)
Kummer, D., 175
Kun, E., 7
Kunchandy, M. A., 399
Kunitoh, H., 134(151), 138
Kunze, B., 22
Kuppusamy, P., 30
Kuras, R., 34
Kurilich, A. C., 465
Kurisu, K., 196
Kuriyama, K., 133(90), 136, 141(90)
Kurlan, R., 486, 503
Kuroda, S., 106
Kuschel, B., 134(147), 138
Kushad, M. M., 465, 466, 467(57)
Kusumoto, T., 187
Kuwabara, K., 95, 102(31)
Kwak, M.-K., 402, 414, 415, 423
Kwong, L. K., 109, 478

L

Labuda, D., 133(91; 96), 134(143; 148), 136, 138, 141(91; 96), 363
Labuda, M., 133(91), 136, 141(91), 363

Subject Index

A

Allium-derived sulfides
  DT-diaphorase induction
    diallyl disulfide, 399
    diallyl trisulfide, 399
  phase II enzyme induction
    animal studies, 451–454
    human intake considerations, 454–456
    overview, 450
    sulfide preparation, 450–451
  toxicity and relationship to enzyme
    induction, 456
  types and structures, 449–450
ALS, *see* Amyotrophic lateral
  sclerosis
Amyotrophic lateral sclerosis,
  coenzyme Q_{10} therapy, 474, 482, 487
Apoptosis, *see also* p53
  p53-dependent apoptosis assays
    buffers, 291
    cell culture, 290–291
    DNA fragmentation assay, 292
    fluorescence-activated cell sorting,
      292–293
    irradiation induction of apoptosis, 291
    light microscopy of morphology,
      291–292
    overview, 289–290
    Western blot, 293
  SIN-1-induced apoptosis inhibition by
    coenzyme Q_{10}, 503
5-(Aziridin-1-yl)-2,4-nitrobenzamide
  clinical trials, 211
  DNA interactions of cytotoxic
    compound, 200, 202, 220–221
  DT-diaphorase activation
    human enzyme inefficiency, 204–205
    rat enzyme comparison with human
      enzyme, 205–207
    reaction, 199–200
    reduction product identification and
      cytotoxicity analysis, 202–204

NAD(P)H:quinone oxidoreductase
  type 2 activation, 211–212,
    215–218, 221
  targeted prodrug therapy using
    Escherichia coli B nitroreductase,
    207–210, 220
  tumor chemotherapy potential, 197–199

B

BHA, *see* Butylated hydroxyanisole
Brassinin, DT-diaphorase induction and
  derivatives, 397–398, 404
Butylated hydroxyanisole, DT-diaphorase
  induction, 356–359, 424–425

C

Carcinogenesis, chemical
  cancer chemoprevention approach, 383,
    414–415
  carcinogen structure and DNA adduction
    specificity, 296
  catechol estrogen quinones, *see* Estrogen
  catechol quinones in cancer initiation,
    316–318
  depurinating adducts and tumorigenic
    mutations, 298, 300
  error-prone repair in mutagenesis,
    300–302, 313–315
  history of study, 293–294
  nuclear cytochrome P450 catalysis, 295
  phase II enzyme protection, 381–383
  polycyclic aromatic hydrocarbon tumor
    initiation mechanism, 294
  sequence specificity of mutagenesis,
    296–298
  stable versus depurinating DNA
    adducts, 298
Catechol quinones, *see* Carcinogenesis,
  chemical; Estrogen
CB 1954, *see* 5-(Aziridin-1-yl)-2,
  4-nitrobenzamide

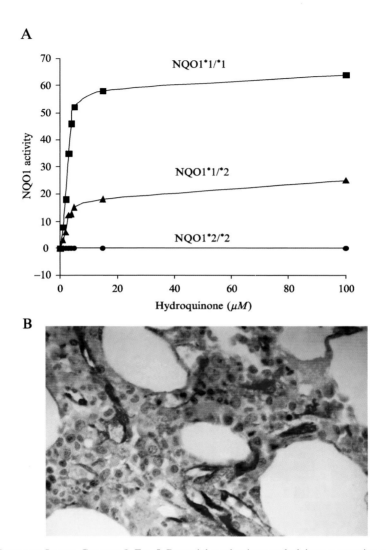

Ross and Siegel, Chapter 8, Fig. 5. Potential mechanisms underlying a protective role of NQO1 against benzene toxicity. (A) Induction of NQO1 in human bone marrow mononuclear cells after exposure to hydroquinone depends on NQO1 genotype. Catechol induces a similar induction of NQO1, and induction of NQO1 by hydroquinone was also detected in bone marrow CD34[+] enriched cell populations. The inability of polyphenolic metabolites of benzene to increase NQO1 activity in NQO1*2/*2 cells was presumably related to the instability of the mutant NQO1*2 protein rather than any effect on induction. (B) Presence of NQO1 in bone marrow endothelial cells. Although NQO1 could not be detected in aspirated human bone marrow cells, it could be detected by immunohistochemistry in bone marrow endothelial cells.

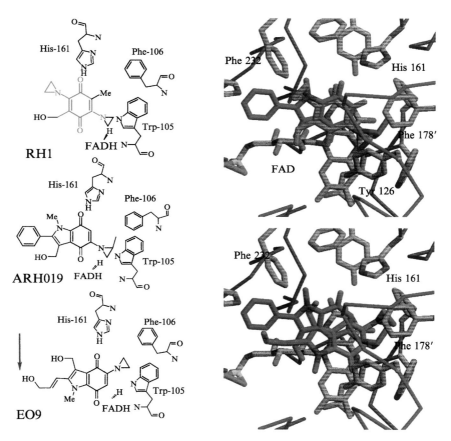

BIANCHET *ET AL.*, CHAPTER 9, FIG. 15. Comparison of the binding of ARH019 with RH1 and EO9. Left panel: Schematic representation of the residues involved in binding of these prodrugs and their relative positions. Note that ARH019 and EO9, despite being highly similar, bind in opposite orientations. Right panel: Binding site of NQO1 showing the overlap of the structures of RH1 and ARH019 (top) and of EO9 and ARH019 (bottom).

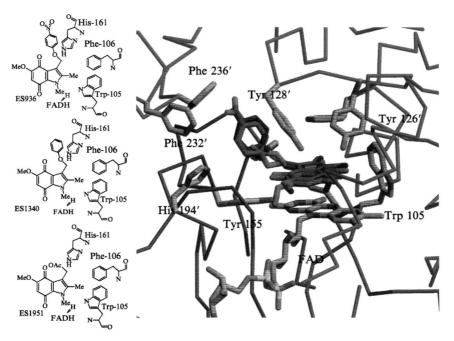

BIANCHET *ET AL.*, CHAPTER 9, FIG. 16. *Binding of inhibitors and poor substrates to hNQO1.* Left: Schematic representation of the residues involved in binding ES936 (top), ES1340 (middle), and ES1951 (bottom). Right: Overlap of the positions of the three drugs in the binding site of hNQO1. The three compounds bind in highly similar positions. What determines whether they are a poor substrate or an inhibitor is the nature of the group at the 3-position: good leaving groups make compounds inactivate the enzyme.

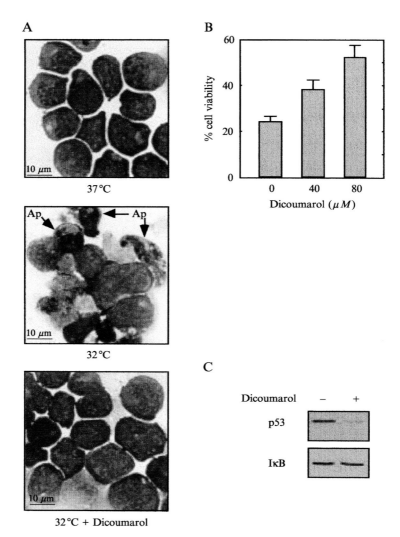

ASHER *ET AL.*, CHAPTER 16, FIG. 2. Dicoumarol decreases p53 level and p53-dependent apoptosis in M1-t-p53 cells. (A) M1-t-p53 myeloid leukemic cells were cultured at $37°$ or $32°$ for 23 h without or with 100 μM dicoumarol and stained with May-Grünwald Giemsa. Ap, Apoptotic cells. (B) Analysis of the percent of viable M1-t-p53 cells at $32°$ treated with different concentrations of dicoumarol for 23 h. Incubation of M1-t-p53 cells for 23 h with concentrations above 125 μM dicoumarol are cytotoxic. (C) Immunoblot analysis of p53 and IkB proteins in M1-t-p53 cells cultured at $32°$ for 5 h without ($-$) or with ($+$) 200 μM dicoumarol. Similar results were obtained by culture of M1-t-p53 cells for 16 h with 100 μM dicoumarol.

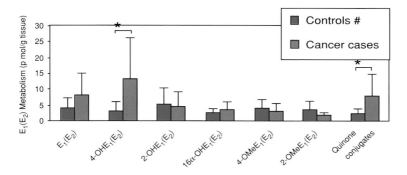

Cavalieri *ET AL.*, Chapter 17, Fig. 5. Estrogen metabolites and estrogen conjugates in human breast tissue from women with and without breast carcinoma. Quinone conjugates are 4-OHE$_1$(E$_2$)-2-Cys, 4-OHE$_1$(E$_2$)-2-NAcCys, 2-OHE$_1$(E$_2$)-(1 + 4)-Cys and 2-OHE$_1$(E$_2$)-(1 + 4)-NAcCys. [#]Controls are benign fatty breast tissue and benign fibrocystic changes. [*]Statistically significant differences were determined by using the Wilcoxon rank sum test, $p < 0.01$.

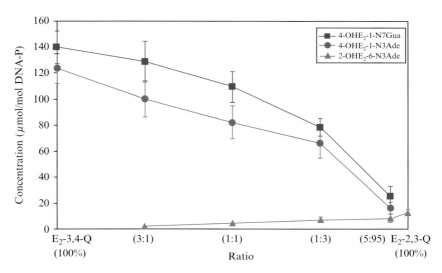

Cavalieri *ET AL.*, Chapter 17, Fig. 8. Depurinating adducts formed after 10 h by mixtures of E$_1$(E$_2$)-3,4-Q and E$_1$(E$_2$)-2,3-Q reacted with DNA. The levels of stable adducts formed in the mixtures ranged from 0.1% to 1% of the total adducts.

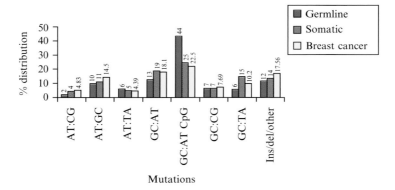

Cavalieri *ET AL.*, Chapter 17, Fig. 9. Distribution of p53 gene mutations. The data are taken from the IARC database, which contains 17,914 p53 mutations in all cancers, consisting of 225 germline and 17,689 somatic mutations. The latter include 1523 breast cancer mutations.

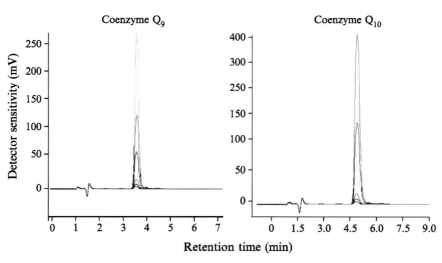

Kheradpezhou, *ET AL.*, Chapter 27, Fig. 1. Overlay chromatograms of coenzyme Q_9 (left panel) and coenzyme Q_{10} (right panel) at 6.25 ng, 12.5 ng, 25 ng, 50 ng, 100 ng and 200 ng concentrations. Standards were prepared in the mobile phase, preserved in the dark microcentrifuge tubes, and analyzed with high-performance liquid chromatography (HPLC) with UV detection at 275 nm spectral wavelength at room temperature. An ISCO HPLC pump and JC1-6000 computer software were used for the quantitative analysis of data as described in the text.

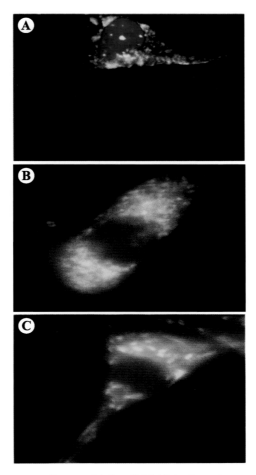

KHERADPEZHOU, *ET AL.*, CHAPTER 27, FIG. 5. Multiple fluorochrome digital fluorescence imaging microscopic pictures of (A) control$_{wt}$, (B) aging RhO$_{mgko}$ and (C) RhO$_{mgko}$ neurons transfected with complex-1 gene. The RhO$_{mgko}$ neurons exhibit rounded or spherical appearance, increased granularity, reduced neuritogenesis and mitochondrial aggregation in the peri-nuclear region. Neuritogenesis reappeared in complex-1 transfected RhO$_{mgko}$ neurons. Digital fluorescence images were captured with a SpotLite digital camera on an Olympus microscope and were analyzed with ImagePro computer software. Target accentuation and background inhibition software was employed to improve the quality of images. Fluorochromes: Green, acridine orange (nucleo-cytoplasmic marker especially for RNA and proteins); red, JC-1 (mitochondrial membrane potential $\Delta\Psi$ marker); blue, DAPI (nuclear DNA marker).

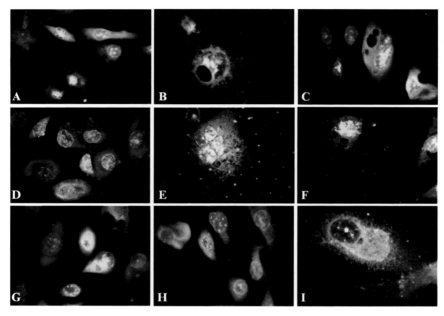

KHERADPEZHOU, *ET AL.*, CHAPTER 27, FIG. 6. Multiple fluorochrome digital fluorescence imaging microscopic analysis of (A) control$_{wt}$; (B) SIN-1 (100 μM) exposed, (C) SIN-1 with coenzyme Q_{10} treatment, (D) aging RhO$_{mgko}$, $\Delta\Psi$ RhO$_{mgko}$ neurons exposed overnight to SIN-1; (F) RhO$_{mgko}$ neurons transfected with MT-1$_{sense}$ oligonucleotide; (G) RhO$_{mgko}$ neurons transfected with complex-1; (H) control$_{wt}$ neurons transfected with complex-1 and exposed overnight with SIN-1; RhO$_{mgko}$ neuron transfected with MT-1$_{antisense}$ oligonucleotide and exposed to SIN-1 overnight; and coenzyme Q_{10} (10 μM), illustrating genetic resistance of MT-1$_{sense}$ transfected and complex-1 transfected control$_{wt}$ neurons and susceptibility of aging RhO$_{mgko}$ and MT-1$_{antisense}$ oligonucleotide-transfected aging RhO$_{mgko}$ neurons to SIN-1-induced oxidative stress and apoptosis. SIN-1 induces phosphatidyl serine externalization, plasma membrane perforations, mitochondrial aggregation, and nuclear DNA fragmentation and condensations. The apoptotic changes are attenuated on treatment of the neurons with coenzyme Q_{10} and by transfecting with either complex-1 or MT-1$_{sense}$ oligonucleotides, indicating that complex-1 gene is down-regulated during oxidative and nitrative stress, while up-regulation of antioxidant genes (complex-1 and/or MT-1) may provide neuroprotection in aging and PD. Fluorochromes: Green, Fluorescein isothiocyanate-conjugated-annexin-V (ApoAlert; for demonstrating phosphatidyl serine externalizatation); red, JC-1 (for estimating $\Delta\Psi$); and blue, DAPI (nuclear DNA stain to demonstrate structurally intact vs. fragmented DNA). Digital fluorescence images were captured and processed as described above. Fluorescence images were first digitized individually and merged together to obtain overall information about plasma membrane, $\Delta\Psi$ and nuclear DNA during SIN-1-induced apoptosis and coenzyme Q_{10}-mediated neuroprotection.